Business Data Communications & IT Infrastructures

2nd Edition

Manish Agrawal
University of South Florida, Tampa, Florida

Rekha Sharma
HP Enterprises, Tampa, Florida

Founded in 2014, Prospect Press serves the academic discipline of Information Systems by publishing innovative textbooks across the curriculum including introductory, emerging, and upper level courses. Prospect Press offers reasonable prices by selling directly to students. Prospect Press provides tight relationships between authors, publisher, and adopters that many larger publishers are unable to offer. Based in Burlington, Vermont, Prospect Press distributes titles worldwide. We welcome new authors to send proposals or inquiries to Beth.golub@prospectpressvt.com.

Cover Design: Maddy Lesure
Cover Photo: © istock.com/4X-image
Production Management: Kathy Bond Borie
Editor: Beth Lang Golub

eTextbook ISBN: 978-1-943153-11-4 • Available from Redshelf.com and VitalSource.com
Printed Paperback ISBN: 978-1-943153-12-1 • Available from Redshelf.com

Contents

List of Figures *x*
Preface *xi*

Chapter 1 Introduction **1**
Overview 1
Introduction 1
Utility of Computer Networking 3
Technology Milestones 6
Packetization 14
Layering 18
TCP/IP Stack 23
OSI Model 24
Principles of Internet Protocols 27
Typical Computer Network 29
Summary 30
Example Case—Domino's Pizza 31
References 33
Review Questions 34
Example Case Questions 34
Hands-on Exercise—Traceroute 35
Critical-thinking Exercise—The Power of Exchange Formats 37
IT Infrastructure Design Exercise—Identifying Uses 37

Chapter 2 Physical Layer **41**
Overview 41
Functions of the Physical Layer 41
Physical Media and their Properties 44
Data vs. Signals 56
Signals and their Properties 56
Impact of Noise and the Importance of Binary Signals 63
Transmission and Reception of Data Using Signals 67
Multiplexing 70
Summary 75
Example Case—Smart Grids 76
References 79
Review Questions 79
Example Case Questions 80

Hands-on Exercise—Amplitude Shift Keying 80
Critical-thinking Exercise—The Value of Carriers 82
IT infrastructure Design Exercise—Media Selection 82

Chapter 3 Data-link Layer 83
Overview 83
Functions of the Data-link Layer 83
Ethernet 85
CSMA/CD 91
Error Detection 95
Ethernet Frame Structure 101
Switched Ethernet—State of the Market 106
Example Case—Networks Helping Big Data Applications 109
References 110
Summary 110
Review Questions 111
Example Case Questions 112
Hands-on Exercise—OUI Lookup 112
Critical-thinking Exercise—Broadcast and Search 114
IT Infrastructure Design Exercise—Ethernet Diagram 115

Chapter 4 Network Layer 117
Overview 117
Functions of the Network Layer 117
Overview of the Internet Protocol (IP) 120
IP Addresses 126
Classless Inter-domain Routing (CIDR) 141
Obtaining IP Addresses 143
IPv6 144
Example Case—Retailing 148
References 152
Summary 152
Review Questions 153
Example Case Questions 154
Hands-on Exercise—ipconfig and ping 154
Critical-thinking Exercise—Genetic Code 159
IT Infrastructure Design Exercise—Estimating CIDR Requirements 159

Chapter 5 Transport Layer 161
Overview 161
The Need for a Transport Layer 161
Transmission Control Protocol (TCP) 163
TCP Functions 163
TCP Header 180
UDP 182

	Example Case—Financial Industry	183
	References	185
	Summary	185
	Review Questions	186
	Example Case Questions	187
	Hands-on Exercise—netstat	188
	Critical-thinking Exercise—Flow Control of Distractions	190
	IT Infrastructure Design Exercise—Estimating Data Requirements	190
Chapter 6	**Application Layer**	**193**
	Overview	193
	Application Layer Overview	193
	HTTP	195
	E-mail	207
	FTP (File Transfer Protocol)	219
	IM	221
	Example Case—Google Ads	222
	References	224
	Summary	224
	Additional Notes	225
	Review Questions	226
	Example Case Questions	227
	Hands-on Exercise—Wireshark	227
	Critical-thinking Exercise—Protesting SOPA	233
	IT Infrastructure Design Exercise—Identifying Market Leaders	234
Chapter 7	**Support Services**	**235**
	Overview	235
	DHCP	235
	Non-routable (RFC 1918) Addresses	242
	Network Address Port Translation (NAPT)	244
	Address Resolution Protocol (ARP)	247
	DNS	249
	Home Networking	259
	Example Case—DNS and Virtual Hosts	261
	References	264
	Summary	264
	Review Questions	265
	Example Case Questions	266
	Hands-on Exercise—nslookup	266
	Critical-thinking Exercise—Nissan Computer Corp.	267
	IT Infrastructure Design Exercise—Start Infrastructure Diagram	268

Chapter 8 Routing 269
Overview 269
Introduction 270
Autonomous Systems 273
Routing Tables 275
Viewing Routes 277
Simplifying Routing Tables—Route Aggregation 282
Multiprotocol Label Switching (MPLS) 285
OpenFlow—Software-defined Networking 288
Routing as a Metaphor for the Aerotropolis 289
Example Case—Disasters, Katrina, and 9/11 290
References 294
Summary 294
Review Questions 294
Example Case Questions 296
Hands-on Exercise—bgplay 296
Critical-thinking Exercise—Internet Censorship 297
IT Infrastructure Design Exercise—Failover 298

Chapter 9 Subnetting 299
Overview 299
Why Subnetting 299
Three-part IP Addresses with Subnetting 303
Subnetting the Network Address Block 305
Subnet Masks 310
Benefits of Subnetting within Subnets 314
Representative Subnetting Computations 316
Subnetting in IPv6 318
Example Case—An ISP in Texas 318
Acknowledgments 320
Summary 320
Review Questions 321
Example Case Questions 322
Hands-on Exercise—Subnet Mask 322
Critical-thinking Exercise—Subnet Design 323
IT Infrastructure Design Exercise—Subnet Design 324

Chapter 10 Wide-area Networks 325
Overview 325
Introduction 325
Point-to-point WANs 328
Statistically Multiplexed WANs 331
TDM WANs 337
FDM WANs 337
WANs and the TCP/IP Stack 338

Example Case—Unmanned Aerial Vehicles 339
References 341
Summary 342
Review Questions 342
Example Case Questions 343
Hands-on Exercise—Web Page Debugging 344
Critical-thinking Exercise—Professional WAN 345
IT Infrastructure Design Exercise—WAN Design 345

Chapter 11 Network Security 347
Overview 347
Introduction 347
Network Security 351
Network Security Controls for Incoming Information 352
Network Security Controls for Outgoing Information 360
Example Case—T.J. Maxx 372
References 374
Summary 374
Review Questions 375
Example Case Questions 376
Hands-on Exercise—https 376
Critical-thinking Exercise—Identifying Threats 377
IT Infrastructure Design Exercise—Adding Security 378

Chapter 12 Computing Infrastructures 379
Overview 379
Origins—von Neumann Architecture 379
The Operating System 381
Scaling Up—VMs and WSCs 394
Cloud Computing 401
Acknowledgment 404
Summary 405
Example Case—40 servers to 5,000 in 3 days 405
References 405
Review Questions 406
Example Case Questions 407
Hands-on Exercise—perfmon 407
Hands-on Activity Questions 410
Critical-thinking Exercise—Nano Scale Computers 410
IT Infrastructure Design Exercise—Using the Cloud 411

Chapter 13 Services Delivery 413
Introduction 413
IT Services Management 413
Service Delivery Disciplines 415

High-availability Concepts 417
High-availability Architectures 420
Business Continuity and Disaster Recovery 427
Summary 430
Example Case—Chaos Monkey at Netflix 430
Review Questions 430
Hands-on Exercise—Device Uptime 431
Critical-thinking Exercise—Personal High Availability 432
IT Infrastructure Design Exercise—Including Active Replication 432
Example Case Exercise 432

Chapter 14 Managerial Issues 433
Overview 433
Introduction 433
Network Design 434
Maintenance 440
Standards 444
Government Involvement, Legal Issues 447
Example Case—Telework, Telemedicine 451
References 453
Summary 454
Review Questions 454
Example Case Questions 455
Hands-on Exercise—Standards Development Review 456
Critical-thinking Exercise—Patents 456
IT Infrastructure Design Exercise—Hire Yourself 456

Note: The following two supplementary chapters are included in the eTextbook only. Users of the printed textbook can freely access and download Supplementary Chapters 15 and 16 at the title's website: Visit http://www.ProspectPressVT.com/titles/business-data-communications-2nd-edition/, then click on Student Resources in the gray box on the right.

Supplementary Chapter 15: Wireless 457
Overview 457
Introduction 457
ISM Frequency Bands 459
Wireless Local-area Networks (the 802.11 series) 462
Personal-area Networks (the 802.15 series) 470
802.16—Wireless Metropolitan-area Networks 476
Example Case—The Oil Industry 479
References 481
Summary 481
Review Questions 482
Example Case Questions 483
Hands-on Exercise—AirPCap Wireshark Captures 483

Critical-thinking Exercise—Ubiquitous Wi-Fi 485
IT Infrastructure Design Exercise—Add Wi-Fi 485

Supplementary Chapter 16: Phone Networks 487
Overview 487
Introduction 487
Phone Network Components 488
Phone Signals 490
Legal Developments 491
Digital Subscriber Line (DSL) 494
Cell Phones 496
Example Case—Cell Phones and Global Development 505
References 506
Summary 506
Review Questions 507
Example Case Questions 509
Hands-on Exercise—CDMA 510
Critical-thinking Exercise—Other Three Billionaires 512
IT Infrastructure Design Exercise—Switch to VoIP 512

Appendix: Networking Careers A1
Glossary G1
Index I1

List of Figures

Note: A List of Figures is available at the title's website: Visit http://www.ProspectPressVT.com/titles/business-data-communications-2nd-edition/, then click on Student Resources in the gray box on the right.

Preface

The mind is not a vessel that needs filling, but wood that needs igniting.

Plutarch

This text is designed for a one-semester course in Business Data Communications and IT Infrastructures. IT infrastructure is one of the growth areas of the economy because information exchange is becoming an increasingly important part of people's lives. This book focuses on providing working knowledge of data communication and IT infrastructure concepts that most students are likely to encounter within the first five years after graduation.

This book tries to most effectively utilize the time frame of a semester-long class with about 40 hours of instruction time. Unlike many other information systems classes, which typically have a unifying theme—for example, SQL in Database Development, or the Waterfall model in Systems Analysis and Design—the Data Communications class easily transforms into developing familiarity with an alphabet soup of technologies—QAM, ASK, 802.3, IP, TCP, UDP, HTTP, SMTP, IMAP, DHCP, DNS, NAT, ARP, RFC 1918, subnetting, BGP, and 802.11, to name a few. The challenge in this class is to develop a unifying theme to help a student who is new to these technologies, so that the student is able to see how all these technologies are components of a unified system that enables data communications. All the design choices made while developing this text are based on this goal of communicating a unifying theme underlying contemporary IT infrastructure. The theme in the book is "efficiency of resource utilization."

In pursuit of this goal, at every possible occasion, the book addresses why networking technologies and IT infrastructures are designed to work in their current form. This brings out the central role of resource utilization efficiency in IT infrastructures. Hopefully, this focus on why will also help students recognize opportunities for profitable innovation in their careers, even in contexts that are unrelated to IT infrastructures.

New in this Edition

Several updates have been made to this edition of the text:

1. There are two all-new chapters—Chapter 12, Computing Infrastructures, and Chapter 13, IT Services Delivery. These chapters add the computing component to the book so students can get complete coverage of the IT infrastructure in the course. The Computing Infrastructures chapter is inspired by the "operating systems concepts" knowledge unit, which is one of the core knowledge units for four-year programs aspiring to the NSA CAE designation. Together with the discussion on networks, this chapter covers this knowledge unit, helping schools with their NSA CAE efforts.

With the addition of these chapters, the book also meets the requirements for the IS 2010.4 IT Infrastructure core course.

2. Two supplementary chapters have been added: Chapter 15, Wireless, and Chapter 16, Phone Networks. Instructors who wish to include these chapters in their course content have access to this material at the book's website: http://www.prospectpressvt.com/titles/business-data-communications-2nd-edition/. (See Instructor Resources or Student Resources.)
3. Facts have been updated throughout the book.
4. New concepts that were deemed necessary from the use of the first edition and instructor feedback have been added. This includes Shannon's theorem (Chapter 2), Spanning tree protocol (Chapter 3), IPv6 addresses (Chapter 4), multi-path TCP (Chapter 5), software-defined networking (Chapter 8), IPv6 Unicast field sizes (Chapter 9), standards-essential patents (Chapter 14), and 802.11ac (supplement).
5. Critical-thinking questions have been added to every chapter in the text.
6. Example cases have been updated as appropriate to introduce more contemporary topics, such as mobile payments.
7. Numerous examples and contemporary references have been added throughout the book to improve student comprehension and maintain student interest.

Key Features

The book incorporates a number of features to improve student comprehension. These include:

Use of research-based learning principles to the extent possible. The Eberley Center at Carnegie Mellon University (CMU) has identified seven principles that impact student learning.[1] These include: (1) Students' prior knowledge can help or hinder learning; (2) How students organize knowledge influences how they learn and apply what they know; (3) Students' motivation determines, directs, and sustains what they do to learn; (4) To develop mastery, students must acquire component skills, practice integrating them, and know when to apply what they have learned; and (5) Goal-directed practice coupled with targeted feedback enhances the quality of students' learning. These recommendations also implicitly assume that students need to eventually become self-directed learners. These principles guide everything we do in this text. Connections to the larger business context are used to help explain topics. Hands-on exercises give students immediate feedback and motivate students to learn more. The example cases and critical-thinking exercises help students organize what they learn.

Reinforcement of a unifying theme throughout the book: "efficiency of network and computing resource utilization." Knowledge organization, one of these learning principles, is likely to improve if the concepts are arranged around a unifying theme, to build and reinforce these connections across concepts. The theme adopted in this text is that all modern IT is motivated by the need to most efficiently utilize the extremely expensive IT

1 https://www.cmu.edu/teaching/principles/learning.html (accessed Oct. 2015).

infrastructure—cables, bandwidth, routers, exchange points, CPU cycles, even IP addresses. If at first a technology makes no sense to a student, they should be encouraged to step back for a minute and try to assess how this technology improves IT resource utilization. Hopefully, by the end of the course, students will recognize that organizations would not invest in using a technology unless the technology was absolutely necessary to improve resource utilization in some meaningful way. This idea of improving resource utilization is reinforced in almost every chapter. Examples include multiplexing in Chapter 2, broadcast in Chapter 3, IP addressing in Chapter 4, flow control in Chapter 5, address reuse in Chapter 7, and point-to-point communication in Chapter 10.

A focus on describing why technologies have been designed to work the way they do. When students understand why a technology works, it helps improve comprehension, long-term recall, and, potentially, helps deployment of the idea in other contexts. Resource utilization efficiency (as described above) is the primary explanation for IT infrastructure design. Supporting factors that facilitate this goal are covered in the appropriate sequence. Examples in the book include layered architectures in Chapter 1, signaling in Chapter 2, multi-part addressing in Chapter 4, three-way handshake in Chapter 5, hierarchical naming in Chapter 7, and modulation in the supplement.

A focus on covering a core set of data communication and infrastructure technologies. Trying to cover every possible technology infrastructure concept within one class would be overwhelming and actually hinder student comprehension. Accordingly, the book takes the minimal set of technologies that are absolutely necessary to enable computer networking and IT infrastructures in organizations—Ethernet, TCP/IP, ARP, NAT, DNS, DHCP, routing, subnetting, security, virtualization, and availability—and focuses on showing what each of these technologies does, why each of these technologies is necessary, and how each technology works. Focusing on the most essential topics enables more detailed coverage of important topics such as packetization, IP addresses, subnetting, route aggregation, DNS, security, and service management. Every topic that has been covered is discussed in reasonable detail so students feel confident in their abilities to apply them at work and to discuss them with professional experts in the industry and in job interviews.

Hands-on exercises with every chapter. Students repeatedly state that they understand more from hands-on exercises than from lectures. Every chapter in this book has hands-on exercises that help students use and understand the IT infrastructure concepts covered in the chapter. All personal computers come with easy-to-use networking utilities that students can use to learn more about the capabilities of their computers. Many exercises are based on these utilities. Examples include tracert in Chapter 1; ipconfig in Chapter 3, Chapter 4, and Chapter 9; ping in Chapter 4; netstat in Chapter 5; and nslookup in Chapter 7. There are also spreadsheet exercises that help students understand amplitude modulation (Chapter 2) and CDMA (supplement). Finally, some end-of-chapter hands-on exercises use other interesting software. These include Wireshark (Chapter 6 and supplement), and BGPlay (Chapter 8).

Some of these hands-on exercises are unique to the book. For example, the Wireshark exercise in the supplement uses wireless packets captured using AirPCap, so that students can see the fields in the radio header to see the transmission frequency channel selected.

The spreadsheet exercises demonstrate technical concepts such as modulation and multiplexing, using simple spreadsheet exercises.

Business example case in each chapter. To show students the business use for the technologies covered in the book, each chapter ends with an example case study that shows the business or social impact of the technology covered in the chapter.

Threaded network and IT infrastructure design case integrated throughout the text. To help students see how all the technologies covered in the book integrate with each other in an enterprise IT infrastructure, there is a threaded IT infrastructure design case that runs throughout the book. In each chapter, students make design choices to meet user requirements for the technology covered in the chapter. The finished exercise can also be included in students' portfolios for job interviews.

Critical-thinking exercise in each chapter. It is useful and interesting to give students the opportunity to reflect on the bigger picture and how principles covered in the chapter may have bigger implications in society. A critical-thinking exercise accompanies each chapter, to accomplish this goal.

Book Outline

The book may be seen to have four main sections: introduction, communications, computations, and managerial issues.

The introduction (Chapter 1) provides a high-level overview of the book, the need for computer networks, and their evolution to their current form. It describes why layering and packetization are done to deliver information. Chapters 2–11 cover the communication technologies. Chapters 12–13 cover the computing infrastructure technologies. Finally, Chapter 14 covers managerial issues in IT infrastructure, such as standardization, legal issues, and network design.

Supplements

The following supplements are available to help instructors and students:

- PowerPoint slides are available for all the chapters. These slides include all the figures used in the book and can be used to highlight the key points in the chapters.
- There is a companion website for the book, which is referenced at various places in the text. The site includes technology standards, particularly the easy-to-read RFCs such as IP, TCP, HTTP, SMTP, and NAT. The goal of the site is to get students to devote a meaningful amount of time outside class hours going over the assigned readings and to get a broader understanding of data communications. A side benefit of this approach is that by the end of the class, students can become comfortable reading these kinds of technical articles and reports. This companion website can be accessed at http://www.prospectpressvt.com/titles/businessdata-communications-2nd-edition/, in the Student Resources area.

- There is an instructor's manual with answers to the end-of-chapter questions. There is also a test bank with 25 or more multiple choice questions per chapter for use in tests. Instructors may also e-mail the author and ask for any of the Visio drawings for adaptation to their context. For example, I have used the USF IP address block 131.247.0.0/16 in most places. Instructors may want to replace these addresses with their own preferred address blocks.

Acknowledgments

I would like to acknowledge three people who are especially important in planting the seeds for this book and for giving me the confidence to write it. My PhD advisor, Professor H.R. Rao at the University at Buffalo, encouraged me to teach Business Data Communications when it was time for me to teach a class while doing my PhD. Without that start, I probably would never have taught this topic in the first place. Over the years at the University of South Florida (USF), Joe Rogers, the university's network expert, has been very patient in sharing his expertise with me and responding to my questions about computer networking. Joe also went through the first edition of the book to help remove technical errors. Finally, a friend from college and former McKinsey consultant, Bhasker Natarajan, gave me the idea to write a book when I asked him for productive ideas to keep busy in my spare time. He assured me it would not take longer than two weeks to complete the first draft. That encouragement was enough to get me started, though it certainly took longer than two weeks to complete the first draft of the book.

This text is the result of my own adaptations, over the years, of material covered in existing data communications texts. I thank these authors for developing a structure for this core class in the MIS curriculum. The result here seems to work well for my students, and hopefully it will work for everyone who uses the book. Professor Clinton Daniel at USF has been very helpful with his ongoing suggestions, based on his extensive industry experience.

For the videos that are listed in the Chapter 1 design case, I would like to thank Christine Brown, Diana Trueman, and Ian Crenshaw with USF Innovative Education for their enthusiasm and creativity. For quick reference, please check out the videos at the YouTube channel of the Muma College of Business at: https://www.youtube.com/playlist?list=PLGd4OzfxSKJbSdUWi3GM_VS9FqKreNnUW.

The first edition of this book was published by Wiley. I am grateful to Beth Lang Golub, the Information Systems editor at Wiley at the time, for having faith in me, a first-time textbook author, when I approached her with the idea of a new text in Business Data Communications. For this new second edition, I am pleased to be working once again with Beth, now at her new company, Prospect Press.

Huge thanks also to Kathy Bond Borie, who managed the production process with patience and skill to produce an awesome final product.

Special thanks are due to my department colleagues at the Muma College of Business at the University of South Florida for their constant encouragement and support. Inputs from Clinton Daniel have been especially helpful. Finally, I would like to thank the reviewers of this and the previous edition of this book, who gave very constructive feedback and pointed out errors in earlier drafts. The book is better because of their efforts.

List of Reviewers

I also wish to thank the following who provided input for the development of this second edition:

Clinton Daniel, *University of South Florida*
Sven Hahues, *Florida Gulf Coast University*
Harvey Hyman, *Florida Polytechnic University*
George M. Kasper, *Virginia Commonwealth University*
Jean-Pierre Kuilboer, *University of Massachusetts Boston*
Diane Lending, *James Madison University*
Brandon Phillips, *Texas A&M University, Central Texas*
Martin Weiss, *University of Pittsburgh*

About the Colophon[2]

It is unlikely that a single book can teach everything worth learning in any subject. An approach that tries to fill a "bucket of knowledge" is therefore necessarily futile. A more productive approach is to recognize the boundless potential and spirit of inquiry of the human brain. Students can therefore adapt the ideas in this book to many diverse contexts. Recognizing this, at every opportunity this book tries to show why Internet technologies work the way they do, what the challenges have been, and how the adopted solutions solve these challenges. It is hoped that this approach will ignite students' curiosity, motivate them to look for common principles underlying computer networking, and maybe even improve the ways we currently operate and use computer networks. Maybe some ideas from the technologies discussed in the book could even be usable in entrepreneurial ventures.

Hopefully, the colophons in each chapter will serve to motivate you to reflect on the concepts covered in the chapter.

> Remarks are shaded and boxed.

Notations

Definitions are in italics.

> *Lighter remarks (attempts at humor) are shaded, italicized, and boxed. These are generally sourced from* Boys' Life *magazine, the magazine of the Boy Scouts of America, http://www.boyslife.org. They are mostly directed at younger boys, but I (first author) find them very interesting.*

2 A colophon is a brief comment, usually located at the end of the text, providing finishing touches to the text.

CHAPTER 1

Introduction

Any sufficiently advanced technology is indistinguishable from magic.

Arthur C. Clarke

Overview

This chapter describes why IT infrastructures and computer networking are important and why you should care to know about computer networking. It introduces the TCP/IP technology stack that underlies most computer networks. Transmission Control Protocol (TCP) is the popular transport layer protocol, and Internet Protocol (IP) is the standard protocol used by all networks at the network layer. At the end of the chapter you should know:

- why computer networks are important
- why data is sent on computer networks as packets of information
- the important tasks that must be performed to deliver each packet of information to its destination without error
- the standard technologies that accomplish these tasks (the TCP/IP stack)

After the TCP/IP stack is introduced in this chapter, the next few chapters of the book will describe how each of the technologies in the TCP/IP stack works. We will then discuss the computing component of IT infrastructures.

Introduction

Business Data Communications is the movement of information from one computer application on one computer to another application on another computer by means of electrical or optical transmission systems. In everyday language, business data communications is also called computer networking. The two terms will be used interchangeably in the text. *A transmission system that enables computer networking is a data network.* For example, the computers in your home, together with the wireless router, are a data network. All the cables, routers, and other data-carrying equipment of your Internet Service Provider (ISP) constitute another data network. The computers in your university are also organized as a network. Figure 1.1 shows the three networks. Networking and IT infrastructure spending accounts for a large part of the IT

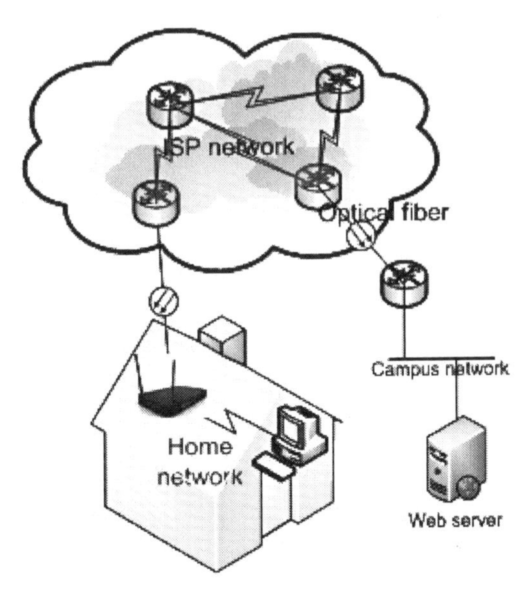

Figure 1.1: Data networks

industry, which in the 2010s is estimated to account for more than $3.5 trillion in annual spending.[1] Of this, approximately $1.2 trillion is for telecommunication services, and another 1 trillion is for IT services.

Practically speaking, a focus on data networks is very useful because it helps us understand how the popular computer networking applications such as web browsing and e-mail work. But an even bigger benefit of focusing on computer networks comes from a learning perspective. It turns out that a common set of technologies supports all computer networking applications, not only web and e-mail, but also all distributed applications such as client-server database applications. The basic principles behind data network technologies have not changed in almost three decades, and are not likely to change in the immediate future. Therefore, understanding the fundamental principles behind these technologies will not only help us understand how modern data networks work, but will also be useful to understand any developments that may occur in computer networks in the near future.

As you will hopefully see throughout the book, computer networking involves some very interesting challenges and equally interesting solutions. The result is that one device—a humble desktop computer—can perform most conceivable communication tasks: web, e-mail, video and music downloads, chat, instant messaging, VoIP telephony, and queries on remote databases. In all likelihood the desktop will also be capable of running new communication applications that entrepreneurs like you will create in the future. This makes computer networking an exciting technology to become familiar with. This book teaches you where the different components of an information system are located in an organization, and how they are integrated to work together.

1 http://www.gartner.com/technology/research/it-spending-forecast/.

Most homes and organizations have another network—the phone network. Phone network technologies are different from data network technologies in many ways, and therefore, it can be confusing to be introduced to phone networks alongside data networks. However, phone networks are extremely important because a lot of business information is exchanged in phone conversations. For this reason an overview of phone networks is provided in the supplement.

Utility of Computer Networking

Let's start with the basics: Why should you care about computer networking? After all, when people talk about IT and careers in information systems, the picture that usually comes to mind is of large databases and rich user interfaces, not computer networks. So, where do data networks fit in from a professional point of view? Are they important enough for you to spend an entire college course learning about them?

It turns out that computer networking is one of the most essential components, not only of IT, but also of the modern economy. Most corporate computer applications use computer networks for storing and retrieving data from databases. Almost every organization depends on computer networks for communication using e-mail or chat. Almost every home and office depends upon computer networks to share resources such as printers and files. Almost every large retail chain depends upon computer networks to get real-time data on the inventory in its stores for just-in-time fulfillment. Almost everybody with a computer depends upon the Internet to search for information.

In fact, computer networks are so pervasive that they play a vital role even in places where their role is not very obvious. For instance, the example case in Chapter 2 shows how the plans for improving energy efficiency in the US depend upon the use of sensors and computer networks to exchange real-time information between power consumers and generators.

Personal expenses comparison

Cell phone expenses have been crowding out spending on food, apparel, and entertainment. In the US between 2007 and 2011, annual telephony expenses went up by $116, while expenses on food away from home fell by $48, apparel fell by $141, and entertainment fell by $126.[2]

Use this exercise to quickly compare your expenses in the following categories to see how critical IT infrastructure expenses are in your life.

Rent = ?

Transportation = ?

Food = ?

Networking and computing = ?

---

2 Anton Troianovski, "Cellphones are eating the family budget," *Wall Street Journal*, Sept. 28, 2012.

Another measure of the importance of networking in general (including voice and video) is the fact that the information-exchange industry is one of the world's largest industries by revenue. Firms in this category include all the large carriers such as AT&T, Verizon, and Bell South. It is easy to see why this industry may command such large revenues. If you look at your own monthly expenses and add up all the bills you pay to stay connected (broadband Internet connection, phone, cable, and cell phone), you are likely to find that your information-exchange expenses are one of your largest monthly expenses. Add up the expenses of most families and businesses in the country to stay connected and you get a sense of the size of the revenues collected by the information-exchange industry. This course introduces you to the technologies and business issues underlying one of the largest industries in the world.

The importance of networking has also been measured from another angle. Evidence suggests that telecommunications infrastructure is a driver of economic activity. Economic progress requires telecommunication infrastructure to support it. This has been observed in both developing and developed countries, even though only the developed countries have well-developed sectors of the economy that depend heavily on telecommunications. Though this evidence is based on data from 1970–1993, and the telecommunications infrastructure studied was telephone networks and not computer networks, it points to the role of information exchange in economic development.[3] The US Department of State has estimated that by 2016, the digital economy will have contributed over $4.2 trillion to the global economy.[4] The Internet is not a toy anymore; rather it has become an extremely significant component of the US economy.

Not all effects of computer networks are so benign, though. Computer networks expose organizations to security attacks from around the globe. More than 75 million credit card records were stolen from the databases of J.P. Morgan Chase in a cyberattack. Computers at national security targets such as the Pentagon and NASA are routinely attacked in attempts to steal information or simply to demonstrate technical competence. In recent years, this has led to significant hiring of information security experts.

With so many users online, the total volume of traffic on the Internet is huge and continues to grow rapidly. An estimate made by Cisco, one of the largest companies in this sector, suggests that the total volume of Internet traffic in 2014 was 59,848 petabytes per month.[5] Written numerically, this number is 59,848,000,000,000,000,000 bytes of data on the Internet per month. And this number is growing at the rate of 23% per year. There are few other business statistics that are already so large and yet are experiencing such high growth rates. Internet traffic growth is being driven by both consumers and businesses, with consumers driving a larger share of the growth recently. This is because of the emergence of video as a major Internet application. Video files are huge, and a single video file downloaded by a home user can be equivalent to more than a month of e-mail traffic volume generated by a business user. Traffic volume is now the largest in Asia-Pacific,

---

3 Dutta, A., "Telecommunications and Economic Activity: An Analysis of Granger Causality," *Journal of Management Information Systems*, 17(4) (2001): 71–95.

4 http://www.state.gov/e/oce/rls/2014/233848.htm (accessed Dec. 3, 2015).

5 Cisco Systems, *Cisco Visual Networking Index—Forecast and Methodology, 2014–2019.*

and all regions of the world are experiencing high growth rates. The highest growth rates are in Africa, driven by the entry of new Internet users. Table 1.1 shows the growth rates in Internet traffic by geography and user category.[6]

Table 1.1: Global IP traffic forecast by Cisco (petabytes/month)

	2014	2019	CAGR 2014–2019
By type			
Fixed Internet	39,909	111,899	23%
Mobile data	2,514	24,221	57%
By segment			
Consumer	47,740	138,415	24%
Business	12,108	29,563	20%
By geography			
Asia-Pacific	20,729	54,434	21%
North America	19,628	49,720	20%
Western Europe	9,601	24,680	21%
Latin America	4,297	12,870	25%
Middle East and Africa	1,505	9,412	44%

Careers

As impressive as these statistics are, as a student at a professional school your primary concern should be: What's in it for me? Why should I study computer networking? Will it help me in my job search? Is computer networking a rewarding career path? The numbers are again very encouraging. All indicators suggest that computer networking is indeed a very rewarding career path. According to estimates prepared by the Bureau of Labor Statistics of the US government, the median pay in 2012 for "network and computer system administrators" was $72,560 per year. There were 366,400 people employed in this category in 2012, a number expected to grow by 12% (42,900 additional jobs) by 2022.[7] Another comparison is provided in Table 1.2, which summarizes statistics on three familiar companies, two in the IT infrastructure space, and one in retail. It can be seen that the revenues/employee are approximately one order of magnitude higher in the IT space compared to retail. This is reflected in salaries.

---

6 A good resource for such Internet traffic projections is the Cisco Visual networking Index.

7 Bureau of Labor Statistics, U.S. Department of Labor, *Occupational Outlook Handbook*, 2014–2015 edition, Network and Computer Systems Administrators, on the Internet at http://www.bls.gov/ooh/computer-and-information-technology/network-and-computer-systems-administrators.htm (accessed Nov. 7, 2015).

Table 1.2: Comparison of financial statistics on three familiar companies

	AT&T	Verizon	McDonalds
Revenue	$139.1 Bn	$130.6 Bn	$25.64 Bn
Employees	281,240	177,900	420,000
Revenue/employee	$494,700	$733,900	$61,060

Students reading this text should aspire to be the CIO at an organization. CIOs are increasingly reporting directly to CEOs[8] and a working knowledge of the IT infrastructure is essential to that function. Therefore, if you choose to build a career in computer networking, the prospects are very favorable. Not only is computer networking vital to the economy, it can also be great for your career. Sound interesting?

Technology Milestones

Hopefully, your curiosity to learn more about computer networks has been stirred. We now begin a discussion of the technology behind computer networks. Modern computer networks are extremely capable. To attain these capabilities, networks have also become quite complex. To better understand modern, complex networks, it helps to start with the simplest early networks. We will look at the important developments to see how each development improved the capabilities of computer networks. While innumerable developments have contributed to the present state of networking, four developments stand out—the telegraph, multiplexing, switching, and packetizing. Each development has improved the efficiency of computer networks in terms of increasing the data-carrying capacity of computer networks, while using the same network resources. Remember, this is the theme of this book—all networking technologies aim to improve the efficiency of utilizing network resources.

The Telegraph

One of the earliest instances of using networks to send data was the telegraph. Samuel Morse patented the telegraph in the US in 1840. This was almost 40 years before the invention of the light bulb, which was patented by Thomas Edison in 1880. The telegraph is important because, for the first time, we were able to use wires to send information to far-off places. The system was simple and looked as shown in Figure 1.2.

To send a message, the sender would connect and release the switch as required. When the switch was connected by the sender, the electromagnet at the receiver's station would be energized and pull the marker to one side. This scratched a line on the paper at the receiver. To send meaningful messages using this system, Samuel Morse also developed a code, called Morse code, which coded letters and numbers as dots and dashes. For example, the letter *a* is

8 Rachel Feintzeig, "Add COOs to the endangered species list," *Wall Street Journal*, June 13, 2014.

Figure 1.2: Telegraph

coded as a dot and a dash.[9] To send *a*, the sender would connect the switch for a short time, release it, and then connect it again for a slightly longer time. This would create a short line and a long line on the paper at the receiving end. An operator skilled in Morse code could interpret this set of short and long lines as the letter *a*.

> The first telegraph message was sent by Samuel Morse on May 24, 1844, over a government-funded experimental line between Washington, DC, and Baltimore, MD. The message was "what hath god wrought?"
>
> The telegraph has great sentimental value to the first author of this book. He got his first job offer by telegram.
>
> Use of the telegraph is now winding down around the world. On July 14, 2013, the telegraph service was closed down in India, one of the largest markets for the service at the time, after almost 160 years in service. At its peak in the 1980s, more than 600,000 telegrams were delivered in India every day.[10]

The simple telegraph system has two major ideas that are relevant for modern networks. The first is that information is carried as energy on the wire. The sender powers the line on or off depending upon the message to be transmitted. The receiver interprets the varying patterns of electrical power on the line as the message. While this appears to be a strange way to communicate, we will see in Chapter 2 that information transmission requires some detectable change at the receiver in response to changes created by the sender. The first telegraph accomplished this using pencil and paper.

Just like the telegraph, modern networks also transmit information as varying patterns of energy on a wire. This is the only known way of causing some detectable change by the receiver in response to some action by the sender. This is called signaling. Signaling is covered in Chapter 2.

---

9 The Morse code was developed by Samuel Morse and Alfred Vail, an apprentice with Samuel Morse. They realized that they could save strokes by using shorter sequences for the most common letters. To find out which letters were the most common, Vail visited his local newspaper office in New Jersey and found that the typesetter had 12,000 E's, 9,000 T's and only 200 Z's. Morse and Vail created the code combinations for the different letters accordingly. Experts have estimated that this rough method is within 15% of an optimal arrangement for English text. Source: James Gleick; *The Information: A History, a Theory, a Flood* (Vintage, 2012).

10 "India sends its last telegram. Stop," *The Telegraph*, July 15, 2013.

Morse code exercise

Morse code was created by Samuel Morse in 1836 and its use was phased out on December 31, 1997. The *Titanic* used Morse code to communicate its distress signals.[11] Wikipedia has a very good page on Morse Code at https://en.wikipedia.org/wiki/Morse_code.

It may be a fun exercise to create some text in Morse code. So, here are some suggestions.[12]

1. Spell your Valentine's name in Morse code: ____________________
2. Express your affection to your Valentine in Morse code:

____

The telegraph also shows that systems reflect technological capabilities of the time. Later generations of the telegraph replaced the electromagnet and paper strip with speakers, which vastly improved the speed of the telegraph. However, speakers did not exist at the time of the Morse patent. When you see a technology solution in this text or elsewhere, it often helps to ask why things are done the way they are. Often, they are work-arounds just to make things work. NAT (Chapter 7) is one such work-around. Or, they reflect the limitations of technology at the time of its creation. The relatively small packet size (1,500 bytes) in Ethernet (Chapter 3) is one such example. Removing these work-arounds and technological limitations is an excellent opportunity for students to apply their knowledge towards innovation and entrepreneurship.

The second idea demonstrated by the telegraph is the need for coding. If information can only be sent as energy transmission being turned on or off, we obviously need to express all letters and numbers in terms of energy transmission going on or off. This expression is called coding. The telegraph used the Morse code, which expressed characters as dots and dashes.[13] Converting back from dots and dashes into human-readable messages is called decoding. Though Morse code is no longer used, modern networks do express letters, numbers, characters, pictures, sounds, etc., as codes. Instead of dots and dashes, we now use 0s and 1s to build coding schemes such as ASCII and UNICODE.

11 See http://www.hf.ro/#trd for a transcript (accessed Nov. 7, 2015).

12 A good way to get Morse is www.telehack.com, type morse at the command line. There are some fun commands at telehack, including jokes and starwars. The site is great fun anyway; you can learn about it at http://telehack.com/telehack.html.

13 The Morse code is not strictly a binary code because the space between the dot and the dash is also an important character. In fact, the space between characters is different from the space between words, which means that the Morse code actually has four signs. For an interesting read on this and related topics, check out *The Information: A History, a Theory, a Flood*, by James Gleick (Vintage, 2012).

> **ASCII exercise**
>
> You can repeat the Morse code exercise, but this time using ASCII. You can see the ASCII codes on the Wikipedia page, https://en.wikipedia.org/wiki/ASCII. The alphabets are part of the printable characters.
>
> 1. Spell your Valentine's name in ASCII: ______________________________
> 2. Express your affection to your Valentine in ASCII:
>
> ____

Multiplexing

The second major telecommunication innovation was multiplexing. While it is useful to be able to send a message over a wire, it is even more useful to be able to send more than one message over a wire at the same time. *The ability to combine multiple channels of information on a common transmission medium is called multiplexing.* Thomas Edison invented a multiplexing telegraph in 1874. This device carried four messages at the same time (two in each direction), allowing one cable to send and receive two messages simultaneously. Again, recall the central theme of this course, that networking technologies try to improve the efficiency of utilizing network resources.

Why is multiplexing so important? The answer lies in the economics of communication networks. The biggest costs in networking are often the costs of laying cables. Multiplexing allows the fixed costs of installing and maintaining one long-distance cable to be amortized over multiple messages. This is quite similar to the idea of dividing a single interstate expressway into multiple lanes. It is very expensive to build highways. Highways become affordable only because lanes enable multiplexing by allowing more than one vehicle to use the interstate at the same time.

In fact, although technologically feasible, data communication would be prohibitively expensive without multiplexing. Consider an example of what would happen without multiplexing. If multiplexing were not possible, your cable company would need to bring 200 cables into your neighborhood if they offered a 200-channel package (one cable per channel) and wanted to let users choose any of these 200 channels at any time of the day. Without multiplexing, the cable company would also need to bring one cable into your home from the neighborhood access point for each additional TV you installed at home. Without multiplexing, signals from the different radio stations would interfere with each other and there could only be one radio station in one location. Multiplexing allows multiple radio stations to beam their signals over the same area where listeners can tune into the station of their choice. Multiplexing is used in modern networks to aggregate the data being sent by all the users in one location and send the aggregated data to the next destination over a single link. For example, a single fiber-optic cable carries all the traffic leaving your university to the ISP.

> For a humorous take on Morse code, you may like to watch the Morse code vs. text messaging contest from the Jay Leno show:
>
> https://www.facebook.com/atomicdc/videos/206[illegible]2868164/.

Switching

The third major development in communication was switching. Multiplexed networks can become even more efficient if the same wire can be used to connect different locations on an as-needed basis. *Transmitting data between selected points in a circuit is called switching.* Consider the example in Figure 1.3, where we have four locations to be networked together. Without switching, we would need six cables to connect these four locations. In the example, all data from A to B would go over the cable A-B and data from A to C would go over cable A-C. By adding a switch to the network, we only need four cables—one cable from each destination to the switch. Clearly, adding a switch can reduce the cost of networking in this case if the cost of laying two cables exceeds the cost of adding a switch.

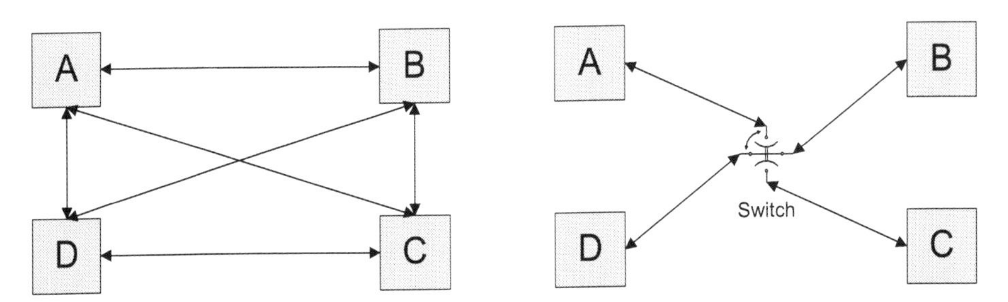

Figure 1.3: Switching

As the number of networked locations increases, the advantages of switching increase further. Adding a new location to a non-switched network requires connections from the new location to all the other locations. For example, adding a fifth location to the non-switched network would require four new cable links (one each to A, B, C, and D). In the switched network, the fifth location would only need one more link—from this new location to the switch. The earliest switches were human-operated; you may have seen operators working on switches in war-time movies.[14, 15] Later switches automated the switching process.[16]

---

14 A good example is the 1937 audiovisual clip *Spot News*, available at http://www.archive.org/details/spotnews1937. The use of switching begins around the 2:15 mark.

15 The telephone industry has played a role in influencing the composition of the modern workforce. James Gleick writes in *The Information*: "The first switchboard operators were teenage boys, the cheapest source of labor at the time. But exchanges everywhere discovered that boys were wild, given to clowning and practical jokes and more likely to be found wrestling on the floor than sitting on stools to perform the exacting, repetitive work of a switchboard operator. A new source of reliable and efficient labor was identified and by 1881, virtually every telephone operator was a woman."

16 Innovation often proceeds from the need to scratch an itch. Almon Strowger, who patented the first automatic telephone exchange in 1891, was an undertaker (funeral director) by profession. However, he was convinced that his business was being hurt by the wife of a competitor, who used her privileged position as a manual telephone exchange operator to direct calls for Strowger's business to her husband. In modern parlance, this would be called a "man-in-the-middle" attack. The automatic exchange was developed to eliminate this interception, and eventually incorporated as the Automatic Electric Company (https://en.wikipedia.org/wiki/Strowger_switch, accessed Nov. 5, 2015).

You can see that the network is beginning to get complex. The switched network adds a device to the network—a switch whose sole function is to facilitate communication. The network is also becoming more difficult for lay users to understand because they don't see all components of the network, in this case the switch, even though all communication passes through the switch.

Packetization and Packet Switching

The previous three developments in networking (telegraph, multiplexing, and switching) were completed by the end of the 19th century and led to the rapid expansion of phone networks. Compared to the next development—packet switching—these three developments are also quite intuitive to understand.

The next major development in networking was packet switching. *Packet switching is the process of routing data using addressed packets so that a channel is occupied only during the transmission of the packet. Upon completion of the transmission, the channel is made available for the transfer of other traffic.* The first packet transmission occurred on October 29, 1969, in California.[17] This was almost 80 years after the most popular circuit switch—the Strowger switch, was patented. The long gap between the two developments indicates the conceptual and practical challenges involved in packet switching. Packet switching underlies all modern communication networks, and is the focus of this text.

Packetization squeezes out further efficiencies in the switched network. A traditional switch establishes connections between end nodes based on network needs. For example, when user A dials user C, the switch connects user A to user C. Now A and C can send data to each other, but if A wants to send data to B or D, he must wait for the switch to close the connection between A and C and to create a new connection from A to B or D. By contrast, in packet switching, all users are always connected to all other users. When A sends packets of information to the packet switch for delivery to C, the switch receives these packets and forwards them on to C. If A simultaneously has some data to send to B, it can pass the data for B to the packet switch, and the switch will instantaneously forward the data to B.

The difference between traditional and packet switching may be seen by comparing the phone and mail networks. On the phone, you dial the number, wait for the connection to be established, and then talk. In sending mail, you just dump everything into the mailbox and trust that the mail system will get each item to its destination.

Why is packet switching more efficient than traditional switching? If A and C are conventional phone users, the advantages of packetization are not very significant. After all, if you are A, and are talking to C, there is generally no reason for you to also be connected to D at the same time. However, if A, B, C, and D are Internet users, the advantages of packet switching are immediately obvious. User A may be working on a document stored on file server B, sending a page in the document for printing to printer D, while listening to music stored on media server C. Packet switching can allow all these communications

17 Chris Sutton, "Internet Began 35 Years Ago at UCLA with First Message Ever Sent Between Two Computers," http://web.archive.org/web/20080308120314/http://www.engineer.ucla.edu/stories/2004/Internet35.htm (accessed Nov. 11, 2015). The Kleinrock Center for Internet Studies at UCLA preserves the first Internet node as a museum. You can read more and consider contributing at its website http://internetstudies.ucla.edu.

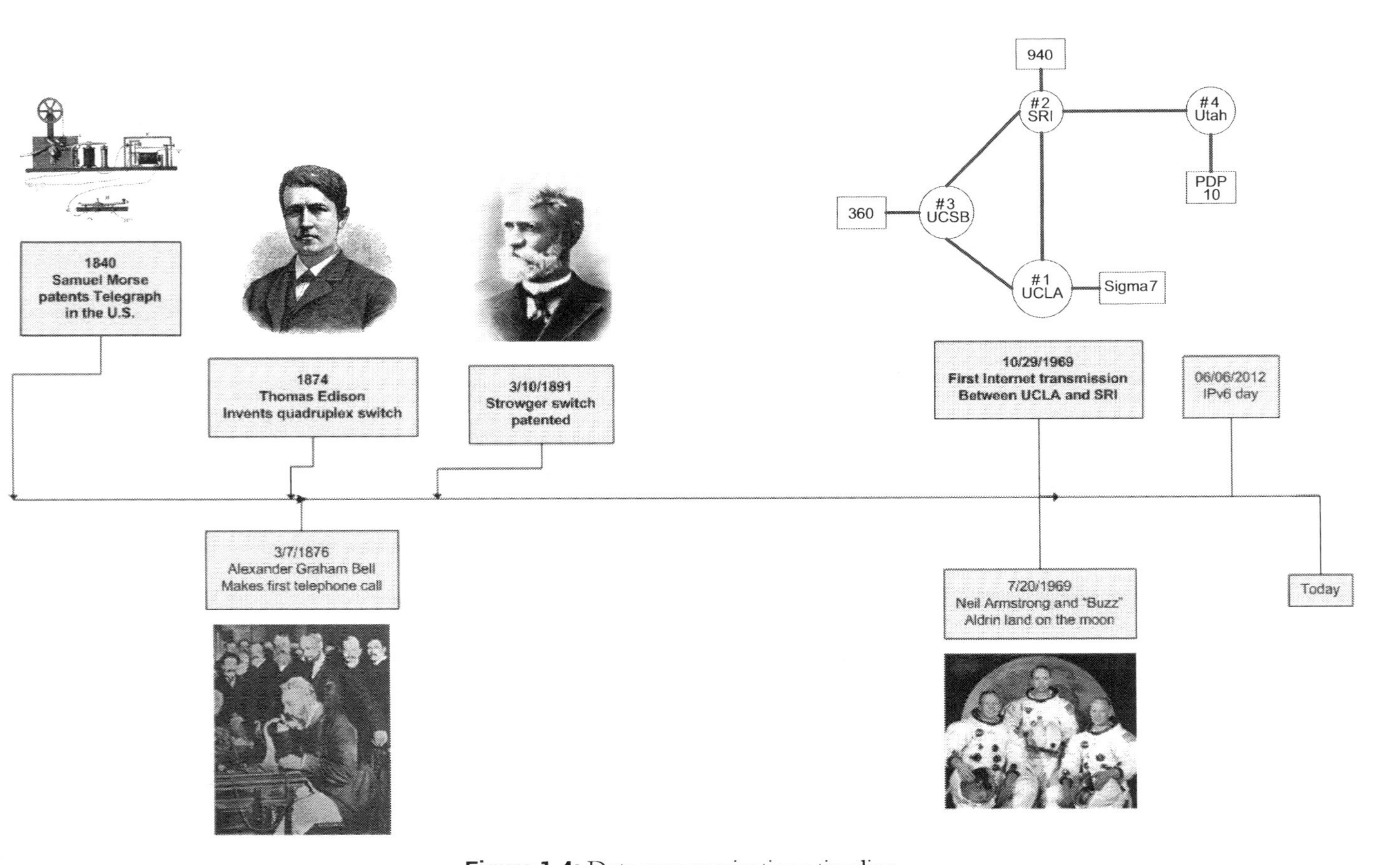

Figure 1.4: Data communications timeline
(Source: Telegraph and Edison photos: © istock.com/ilbusca; astronaut photo: Wikimedia Commons)

to occur simultaneously. In large networks with multiple users, the advantages of packet switching are even more compelling. If A, B, C, and D represent large networks, John in network A could be chatting with Joe in network C, while Jane, also in network A, could be chatting with Alice in network D. A traditional switched network would have to alternate between connecting the pairs John and Joe, similarly Jane and Alice. Quite likely, users will experience slow performance as the switch tears down the previous connection and establishes the new one. In a packet switched network, John, Joe, Jane, and Alice would simply send packets of information to the switch, and the switch would forward the packets to networks A, C, and D, depending upon the destination of the packet. Users are likely to see instantaneous response with such a setup.

Packet switching is quite complex. At the sending end, the data needs to be broken down into small segments, and each of these segments has to be labeled and addressed correctly to create packets of data. Each individual packet has to then reach the destination across the network. At the receiving end, the data in the packets has to be rearranged back into the correct order with the data in other packets. This is a very complex operation. Complexity is almost always undesirable and is tolerated only when it is absolutely necessary. Modern networks introduce the complexity of packet switching only because it offers considerable efficiency over circuit switching. Experiments suggest that packet switching[18] can be up to 100 times more efficient than circuit switching.[19]

Figure 1.4 shows the four developments described in this section on a timeline. For reference, two landmark developments—the first phone call and man's landing on the moon—are shown below the timeline. It can be seen from Figure 1.4 that circuit-switching technologies served global communication needs for almost a century. Packet switching has only become widespread in the last decade or two.

Circuit Switching vs. Packet Switching

Traditional switching (used in phone networks) is called circuit switching because it establishes circuits between senders and receivers. *A circuit is an electronic closed-loop path among two or more points for signal transfer.* Figure 1.2 is an example of a circuit. Packet switches between networks are called routers. *Routers are devices used to interconnect two or more networks.* Routers are very sophisticated devices and can route packets between senders and receivers even if they (the routers) are not directly connected to either the sender or the receiver. They do this by locating other routers on the network that are closer to the destination and by transferring packets to these routers for further routing.

An easy way to identify whether your network is circuit switched or packet switched is to see whether the network is "always on." If the network needs to dial in to its destination, it is a circuit-switched network. If the network is "always on," it is a packet-switched network.

The difference between circuit switching and packet switching is analogous to the difference between the landline phone and messaging. On the phone, you dial a number and wait for the call to be connected. In messaging, you simply click on a friend and send a message. There is no connection set-up or tear-down in packet switching.

18 The term "packet switching" is attributed to Donald W. Davies of the British National Physical Laboratory, who used the term in 1965 to describe a network that chopped data into packets of 1024 bits each.

19 L.G. Roberts, "The evolution of packet switching," *Proceedings of the IEEE*, 66(11) (1978): 1307–1313.

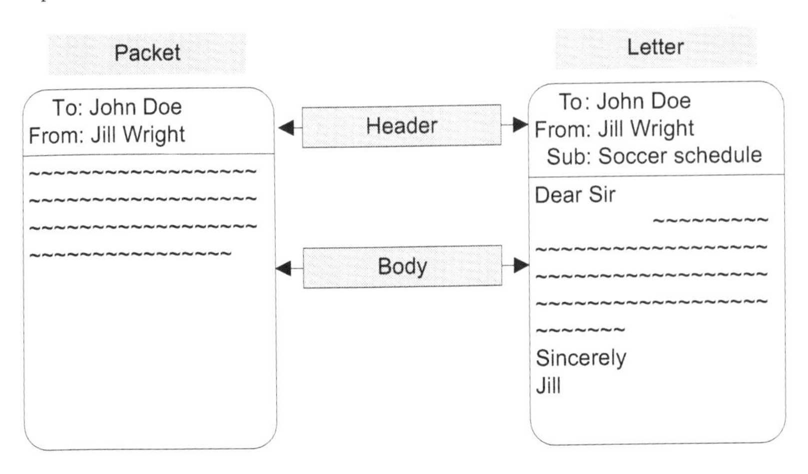

Figure 1.5: Comparing the structure of data packets and letters

Because routers can locate other routers closer to the destination and can do so in real time, packet switching also improves the robustness of networks. A router will stay connected to the network as long as it can find at least one other router to connect to. This robustness of packet switching was the primary motivation for the US Department of Defense to fund the development of packet switching. By contrast, if a link fails in a circuit-switched network while a call is on, the result is a dropped call.

The communication chapters of this entire text focus on how packet transmission works in modern networks.

Packetization

Packetization is the process of breaking down user data into small segments and packaging these segments appropriately so that they can be delivered and reassembled across the network. Data segments produced by packetization are generally called packets, although as you will see in Chapter 3 and Chapter 5, the terms frames and datagrams are also used to denote some specific kinds of packets. Packetization is therefore the idea of sending information as small blocks of information. By contrast, circuit switching sends data as a continuous stream of information.

Each packet has two parts. The body of the packet is the segment from the original data carried by the packet. The header is the information added during packetization to aid the delivery of the packet to the correct destination and for the body of the packet to be correctly reassembled with the bodies delivered by other packets.

Packets are analogous to letters sent in the mail inside envelopes (Figure 1.5). The useful information in the mail is the body of the letter. However, to ensure that the letter reaches the correct mailbox and is read by the intended recipient at the destination, we add additional information such as an addressee name and mailing address to the letter. When the recipient reads the letter, he generally does not care about most of this additional information. He is only interested in the body of the letter. Overhead, such as a mailing address, is added only to ensure that the letter body reaches the intended recipient.

Figure 1.6: Packetization in business knock-down kits
(Source: Top and center photos: author; bottom left: © istock.com/vitpho; bottom right: © istock.com/Pixelci)

Similarly, once packets are processed by the receiver, the additional information, such as addressee, destination address, etc., is discarded by the receiver. The header information added during packetization is also called packet overhead because it is the unavoidable information cost for packet delivery. The receiver does not care about most of the information in the header, but the body cannot be delivered to the receiver without the header. Figure 1.5 shows a comparison between data packets and letters.

There is a very good analogy for packetization from business: knock-down kits. Traditional furniture stores have to deliver every item of furniture in specialized trucks, often by specially trained personnel, which raises costs. In knock-down kits, large and unwieldy items such as furniture and home gyms are broken down into kits for reassembly at the place of use. Each box in the kit can be lifted by an average-bodied person. Knock-down kits demonstrate another advantage of packetization: ease of transport. Once knocked down into small-sized packets, even large furniture items can be transported by end users in domestic vehicles (Figure 1.6). The trade-off in using knock-down kits is the effort involved in reassembling the furniture at home using instructions that come with the kit. Just as knock-down kits can carry objects of any size using a common transport infrastructure, modern packet networks serve as a common infrastructure for all forms of information, including phone, video, and data. By contrast, traditional circuit-switched networks are generally task-specific, with separate networks for voice and video.

Factors Favoring Packet Switching over Circuit Switching

The natural advantage of packet switching over circuit switching in improving network utilization is further enhanced by two factors. The first factor is that the typical user of Internet applications demonstrates very "bursty" behavior. *Burstiness refers to short periods in which large volumes of data are uploaded or downloaded by the user followed by long periods of minimal activity.* Consider a typical web browser. Users typically click on a link to download a web page, and then spend considerable time reading the page. While they are reading the web page, there is no data being sent or received by their computer. After a few minutes, they may find an interesting link on the page and follow it to reach another site, generating another burst of data.

Packetization at IKEA

Reportedly, IKEA introduced the idea of knock-down kits in furniture retailing. In 1952, Gillis Lundgren, an employee at the fledgling retailer, took the legs off a table to fit the table into a customer's car. Beginning the following year, the company began to mass-produce furniture using flat-pack design with a focus on making it easy for customers to transport the furniture in personal vehicles from the store. The reassembly effort at home was worth it because lower transport costs sharply reduced furniture prices and increased sales.[20] The rest is history. IKEA's flat-pack methodology rocketed them past the competition. Company founder Ingvar Kamprad has written, "After that [table] followed a whole series of other self-assembled furniture, and by 1956 the concept was more or less systematized. The more 'knockdown' we could produce, the less damage occurred during transport and the lower freight costs were."[21]

Circuit switching is very inefficient at handling bursts of data. Imagine how inefficient it would be if each time a user followed a web link, the computer had to set up a connection to the target server, download data, wait for some time in case the user wanted to revisit the same target server, and finally disconnect from the target. Not only would the wait time to set up the connection test the patience of the end user, the connection would be idle most of the time when the user was reading the page. Packet switching is very efficient at handling data bursts because no new connections need to be established on a per-packet basis. All packets are forwarded as they arrive at the routers. Routers are always connected to other neighboring routers through connections that can be shared by all the users of the router.

---

20 http://www.ikea.com/ms/en_US/about_ikea/the_ikea_way/history/1940_1950.html.

21 Jason Jennings, *Less Is More: How Great Companies Use Productivity* (Portfolio, 2002).

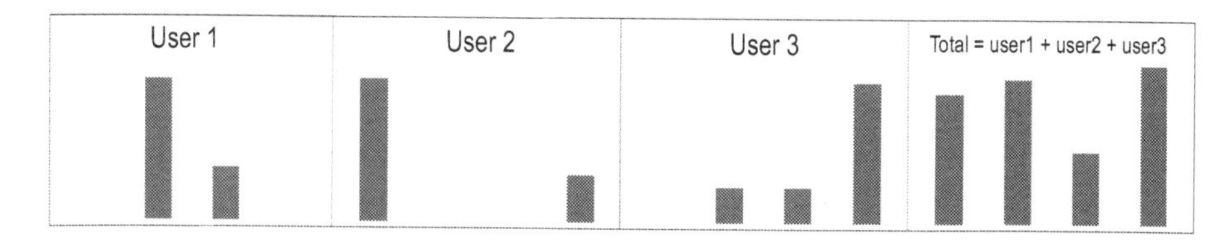

Figure 1.7: Aggregating traffic from multiple users

By handling packets from multiple senders addressed to different receivers, packet switching aggregates network data traffic. Aggregation is a simple technique to "smooth" traffic because it is generally observed that different users have different traffic patterns and when all these patterns are aggregated together, the average traffic volume is much less bursty than the individual traffic volumes. Figure 1.7 shows an example of traffic being generated by three users browsing the Internet. The heights of the bars indicate the traffic volume generated by each user in each time period. In the figure, each individual traffic pattern is bursty. However, the sum of the three traffic patterns, shown with the title "Total" is less bursty. Network designers can then design data links such that the relatively steady, aggregate traffic volume represents, say, 70% of the overall traffic capacity of the link, ensuring suitable link utilization. Figure 1.8 shows aggregate network traffic at two locations, USF, the authors' institution, and the New York International Internet Exchange in 2015. USF's traffic is more bursty than the aggregated traffic at NYIIX.

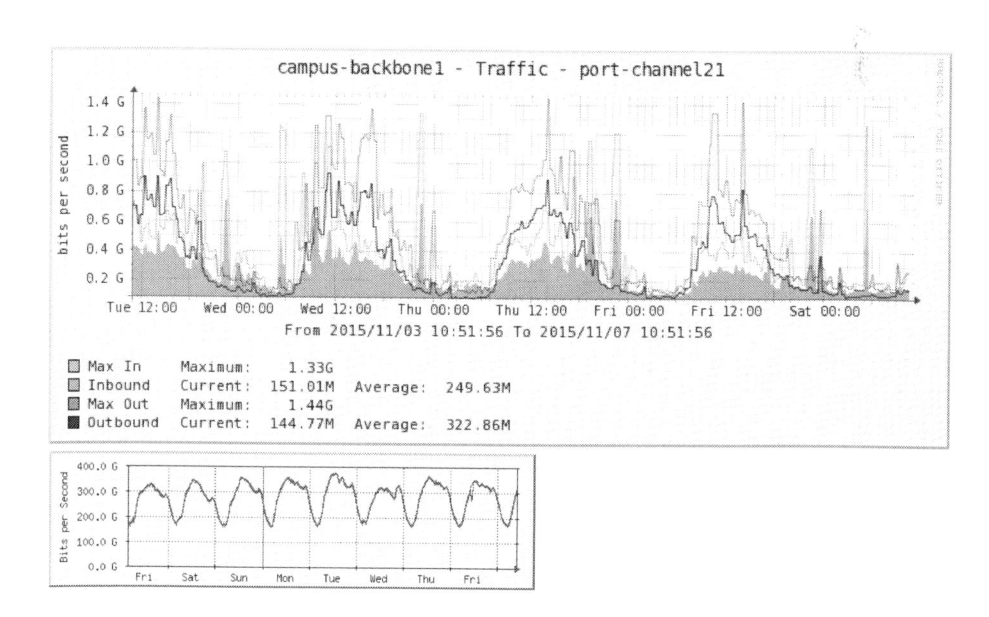

Figure 1.8: Network statistics—USF vs. NYIIX
(Source: [*top*] https://mhb-mon.net.usf.edu/graph_view.php; [*bottom*] http://www.nyiix.net/mrtg/sum.html)

The second factor that has led to the popularity of packet switching is Moore's law. As computers become more capable while getting increasingly cheaper, communication link costs have become the highest-cost components of networks. Therefore, ensuring high utilization of communication links is the top priority for modern network carriers. Since computers are now inexpensive, we can afford to dedicate routers, which are essentially specialized computers, solely to the task of routing packets. Routers aggregate packets from multiple users and improve link utilization in bursty networks. This can make packet switching 3–100 times more efficient than circuit switching.

As a bonus feature, packet switching also improves the reliability of the network over circuit switching. This is described in more detail in Chapter 8 on routing. Routers can automatically discover alternate routes, skirting around dead or damaged routers. End users obtain network connectivity as long as at least one path between the sender and the receiver can be found. Though carriers are now interested in packet switching primarily for its efficiency of link utilization, reliability was the primary reason the federal government was interested in packet switching during Cold War times and funded its early development. Packet switching promised information networks that would survive even if some intermediate communication nodes were destroyed by the enemy in war.

Layering

Now that we know we are going to packetize data, we can begin looking at how packetization is actually done. The hardware and software that performs packetization is designed using a layered architecture. *Layering is the practice of arranging functionality of components in a system in a hierarchical manner such that lower layers provide functions and services that support the functions and services of higher layers.* The end user interacts with the topmost layer. The lowest layer interacts with the wires carrying the signals.

Layering is common when extensive coordination and monitoring is needed to accomplish goals. For example, most organizations are layered, with each layer given well-defined goals. At the top level, CEOs meet stakeholder goals. CEOs are helped by VPs, who may be helped by area managers, who may in turn be helped by fresh college graduates acting as foot soldiers in the organization (Figure 1.9).

Layering is also common in computer software. At the topmost layer are the computer applications and at the lowest layer is the computer hardware. End users interface with computer applications such as web browsers. Applications depend upon the operating system for tasks such as accessing hardware resources for printing, saving files, etc. The operating system, in turn, depends upon device drivers to interact with hardware.

For example, when you click on the Print button on your browser, your browser requests the operating system to print the document. Your operating system sends the document to the driver of your default printer, which sends the correct sequence of commands and information to the printer to print the page (Figure 1.10).

There is a strong tradition of using layering to add functionality to computer software. Hence, networking software is also organized in layers. Layering facilitates modularity in software, enabling easy upgrades and substitution of components. For example, you may have upgraded a version of your web browser or switched to a different web browser.

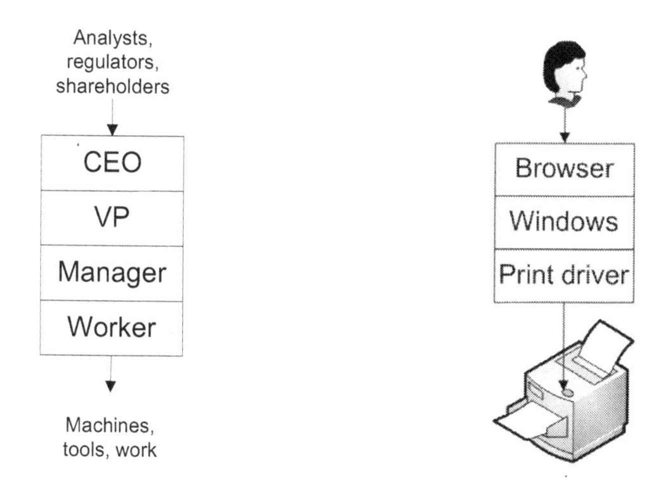

Figure 1.9: Layering in organizations

Figure 1.10: Layering in software

You could do this without changing any other network capabilities on your computer because the browser application is a layer of software that is independent of other layers that enable networking. It is also trivial to move between wired and wireless computer networks without changing browsers or e-mail clients. Again, this is possible because the network access functionality is defined in a layer independent of the applications layer.

Arranging software functionality in layers offers many advantages. An obvious advantage during development is specialization. Experts can focus on delivering the functionality of each layer, interoperating with neighboring layers through well-defined specifications. Interface designers working on the topmost layer can focus on human-computer interaction issues. Programmers developing device drivers can focus on transaction speed and printing features. However, the greatest advantages of software layering are during operation. Individual layers of software can be upgraded without changing other layers. In the software world, we all know how most applications even offer to upgrade themselves automatically. This is possible because the applications are an independent software layer and application upgrades do not require operating system changes. Similarly, upgrading graphics capabilities or adding sound capabilities to computers does not require modifications to existing applications. It is comparably easy to introduce new technology at any layer if it does not affect other layers.

Figure 1.11 shows the overall context of the layered network system. The end user uses applications such as web browsers to access network resources. The application interacts with the topmost layer of the layered network system. The layers process the request layer by layer until the lowest layer transmits the user request as signals over the physical medium. The browser and the physical medium are considered to be outside the layered network software.

Typical Packet Structure

At this point we know that computer networks send information as packets. If you visit a web page, your browser gets that page as packets from the web server. These packets are

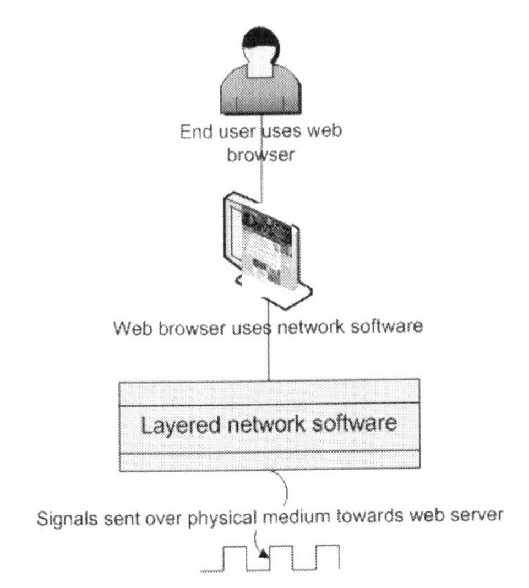

Figure 1.11: Network layers

reassembled by your computer before the page is displayed on your browser. Each packet contains a part of the web page, and some additional information, called header.

We also know that networking software is organized in layers. So, how does layering affect packetization?

Layering affects packetization by organizing the information in the packet header in a layered manner. The big picture is provided here, and the following chapters will look at each layer in detail.

For packetization to work reliably, five essential tasks need to be performed. Together, these five tasks ensure that packets will reach the correct destination, have no data errors, and serve user needs. The five tasks are:

1. specifying user commands
2. segmentation and reassembly of packets
3. identifying and locating the destination on the network
4. error control to remove errors during signaling
5. signaling

Networks perform these five tasks in a well-defined five-layered sequence as follows. When a user needs network service, he invokes the appropriate user command and passes the appropriate data to the command (for example, the URL of a web page). This happens at the topmost layer and completes task 1—specifying user commands. The header at this layer carries the user command.

If the data is large, it needs to be broken up. This is done in the next layer, completing task 2—segmentation and reassembly of packets. The header at this layer keeps track of the different segments and helps the receiver reassemble the packets. Each of these packets has to be delivered to the correct destination. To do this, the next layer takes each segment and adds the appropriate destination address to the segment. This completes task 3—locating the destination. Routers along the way look at these destination addresses and send the packet to the next router. *Each link from one router to the next router is called a hop.* We have to ensure that no errors are introduced while the packet is being transferred over a hop. Therefore, the next layer transfers packets carefully toward the destination, network by network, router by router, hop by hop, from your home network to your ISP's network, and to the target web server's network, ensuring that the packets are not damaged as they cross a network. This completes task 4—error control. To perform error-checks, this layer adds error-checking information at each hop. Finally, within each hop, the last layer sends data as signals necessary for information transfer. This completes task 5—signaling.

You may note that each of the five tasks generally requires some header information. For example, the destination address must be included in the header for packets to be delivered to the correct destination. Each of the five layers in data communications performs one of the five tasks. Each of the next five chapters covers each of these layers and tasks in detail. At this time, try to get the overall picture: users pass commands and data, the data is packetized if necessary, and the packets are transferred to the destination without error. For reference, Table 1.3 shows the five networking layers, the tasks performed at each layer, and the important header information added at each layer.

Table 1.3: Network layer tasks and important header information

Networking task	Supporting information in header
User needs	User commands
Segmentation and reassembly of packets	Sequence numbers
Identifying and locating destination	Address
Error control	Error check
Signaling	None

Please note that while the tasks listed in Table 1.3 are the most important tasks at each layer, most layers also do other tasks to improve network performance. These additional tasks will also be introduced when we look at each layer in detail in the next five chapters.

Figure 1.12 shows how information from a higher layer is carried as part of the data structure delivered by the next lower layer. For example, the user command to get the web page is carried as part of the transport layer information.

After all the layers have added the required headers, at a high level, the packet structure looks as shown in Figure 1.13. Note how the headers are arranged in the same sequence as the layers in Table 1.3.

Data-Link layer
Error check: 4395
Network layer
Destination: www.yahoo.com
Transport layer
Seq. no.: 1839
Application layer
User command: Get index.html

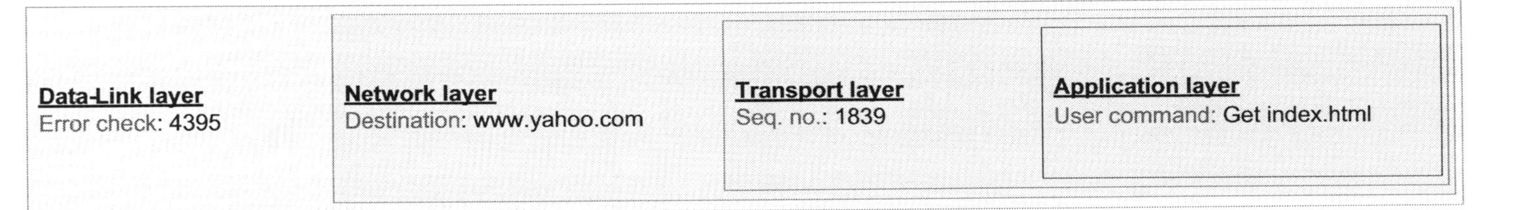

Figure 1.12: Typical packet structure

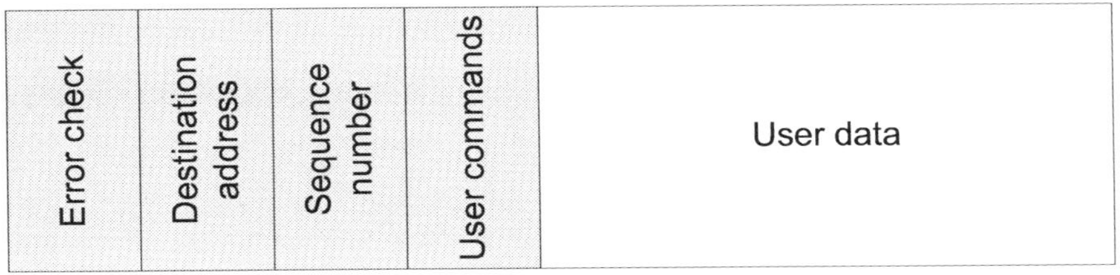

Error check
Destination address
Sequence number
User commands
User data

Figure 1.13: Typical packet header information at a high level

TCP/IP Stack

Since the five-layered network architecture of Table 1.3 is so widely used, each layer has been given a name that closely reflects the function of that layer. Often in industry, these layers are simply addressed by their position in the stack, starting from the lowest layer. Thus, the network layer is widely called layer 3. These names and numbers are shown in Table 1.4 below.

> *Jackie: Which vegetable is great with social media?*
> *Baine: Tell me?*
> *Jackie: A Tweet potato!*
>
> Source: *Boys' Life* magazine, June 2013

Whereas various technologies are available for each layer, the five-layered networking software shown in Table 1.4 is called the TCP/IP stack in honor of the two core technologies used universally. TCP operates at layer 4 and performs segmentation and reassembly of packets. IP operates at layer 3 and identifies and locates the destination. After years of planning, all networks connected to the Internet moved to TCP/IP on New Year's day, 1983, greatly simplifying interconnections. While it is difficult to visualize the significance of this accomplishment, consider that today, most Internet users are members of multiple corporate-sponsored networks that do not talk to each other—Facebook, Google, iCloud, Microsoft. This forces users to maintain their membership on multiple networks. The unification of TCP/IP is comparable in significance to a hypothetical unification of these contemporary walled gardens.

Table 1.4: Network layer names and tasks in TCP/IP stack

Layer number	Layer name	Networking task	Supporting information in header
	Application	User needs	User commands
4	Transport	Segmentation and reassembly of packets	Sequence numbers
3	Network	Identifying and locating destination	Address
2	Data-link	Error control	Error check
1	Physical	Signaling	None

Common technologies used at the different layers are shown in Figure 1.14. In the most common implementation, TCP and IP are combined with another technology called Ethernet, which performs error control and signaling (layers 2 and 1 respectively). This is the set of technologies you have used most of the time when connecting to the Internet.

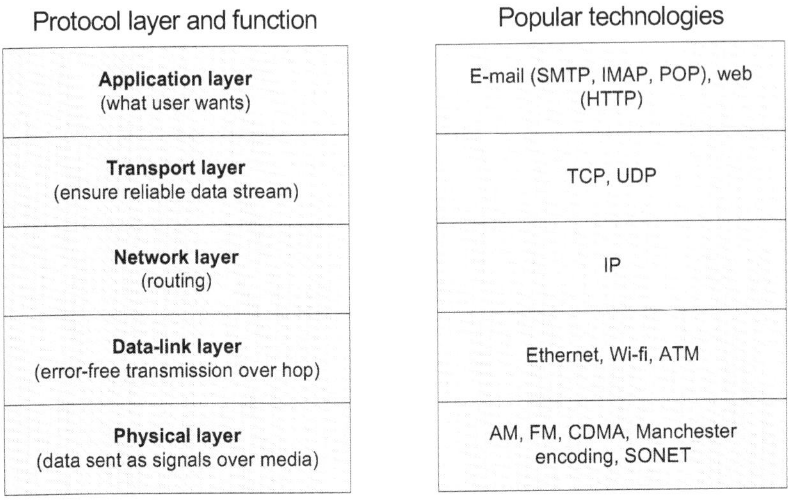

Protocol layer and function	Popular technologies
Application layer (what user wants)	E-mail (SMTP, IMAP, POP), web (HTTP)
Transport layer (ensure reliable data stream)	TCP, UDP
Network layer (routing)	IP
Data-link layer (error-free transmission over hop)	Ethernet, Wi-fi, ATM
Physical layer (data sent as signals over media)	AM, FM, CDMA, Manchester encoding, SONET

Figure 1.14: TCP/IP layers and technologies

Since the Ethernet and TCP/IP combination is so widely used and because both these technologies are publicly documented, this text focuses on Ethernet for the data-link layer (Chapter 3), IP for the network layer (Chapter 4), and TCP for the transport layer (Chapter 5).

OSI Model

What is missing from the neat five-layered network architecture introduced in the previous section is the tortuous path that led to the creation of the architecture. In the early days, many firms developed their own proprietary technologies for computer networking. Technologies such as SNA, DECnet, Appletalk, and IPX/SPX were vying for popularity and commercial success. Each of these technologies was a complete networking solution, and was usually not properly layered. Unfortunately, all these technologies were proprietary and every vendor had an interest in locking in customers into their technology. As a result, IT managers had a hard time getting the different networking technologies to interoperate with each other. If a firm adopted the networking technology of one company, it was forced to buy all computer equipment from the same company. The result was that the choice of a networking technology often made organizations captive customers of the technology vendor, which could charge steep prices for subsequent sales.

Difficulties in interoperation between networking technologies created communication islands within organizations. Groups that adopted, say, SNA could not exchange information over the network with groups in the same organization that adopted, say, AppleTalk. The challenge, therefore, was to ensure interoperability between these competing technologies.

To overcome this challenge, the International Standards Organization (ISO) came up with the Open Standards Interconnect (OSI) model. *The OSI model is a logical structure for communications networks standardized by the International Organization for Standardization (ISO).* As the name suggests, this was a model, not a real technology solution for networking. However, for the first time in the domain of computer networking, the OSI model specified the concept of layers and defined the services to be offered at each layer. Compliance with the OSI model enables any system to communicate with any other OSI-compliant system. The OSI model defined seven layers. These layers and their functions are shown in Figure 1.15. Note the correspondence between the functions of the layers with the same names in Figure 1.14 (TCP/IP) and Figure 1.15 (OSI). The OSI model helped the development of computer networking because technology developers were able to define their technologies in terms of the layers served by their technology. In the end, all proprietary technologies were seen as operating at the data-link layer. IP and TCP were used for network and transport layer functionality and to enable any end-user application to use any of these proprietary technologies.

OSI mnemonic

A popular class exercise is to ask students to develop a mnemonic to memorize the names of the seven layers of the OSI model. My favorite in-class effort has been "All Pizza Served Today Needs Double Pepperoni." Try coming up with your own.

OSI model layer	Layer function
Application layer	Request-reply mechanism for remote operations across a network
Presentation layer	Syntax conversion from host-specific syntax to syntax for network transfer
Session layer	Create and terminate connection; establish synchronization points for recovery in case of failure
Transport layer	Segmentation, reassembly of packets in one connection, multiplexing connections on one machine
Network layer	Routing and network addressing
Data-link layer	Error-free data transmission over a single link
Physical layer	Convert data to signals for transmission over physical media

Figure 1.15: Network layer names and tasks in OSI model

OSI model today

These days, the OSI model is used as an introduction to computer networking even though real-world technologies follow the TCP/IP architecture. TCP/IP may be seen as a simplified version of the OSI model. The correspondence between the layers of the OSI model and TCP/IP is shown in Figure 1.16.

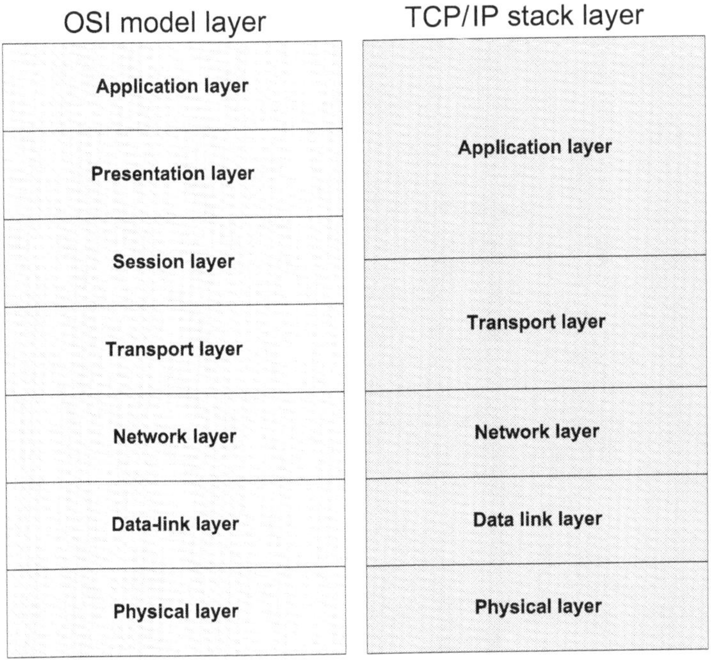

OSI model layer	TCP/IP stack layer
Application layer	Application layer
Presentation layer	
Session layer	
Transport layer	Transport layer
Network layer	Network layer
Data-link layer	Data link layer
Physical layer	Physical layer

Figure 1.16: OSI and TCP/IP

Layering gotcha

When asked to describe the functions of layers in the OSI model or TCP/IP stack, students sometimes write that the function of a layer, say the presentation layer, is to pass data from the session layer to the application layer. This is an unprofessional response. Please remember, no layer exists just to pass data from one layer to another. If that were the case, the layer in question could have been eliminated and the lower layer could have been modified to pass data directly to the upper layer.

Every layer has a well-defined task. These tasks have been summarized in Figure 1.14 (TCP stack) and Figure 1.15 (OSI model).

Figure 1.16 shows that the lowest three layers in both stacks perform similar functionality. However, the TCP transport layer performs some of the functionality of the session layer in addition to the functionality of the OSI transport layer. The TCP/IP application layer performs all the functions of the OSI application layer. In addition, it also performs the functions of the OSI presentation layer and some functions of the OSI session layer.

Principles of Internet Protocols

If you look at the names of the different technologies in Figure 1.14, you will note that most technologies have names that end in "p," e.g. HTTP, TCP, SMTP, UDP, IP, etc. So, what does this "p" represent? Is it some special feature of networking technologies? The "p" in these technology names stands for "protocol."

What are protocols? *Protocols are a set of rules that permit information systems (ISs) to exchange information with each other.* Each protocol is designed to facilitate interoperation within a communication layer. For communication to occur, the communicators must have a common language with a shared interpretation of the words in the language. The protocol is the shared language. We have all heard cryptic conversations by taxi drivers, aircraft pilots, and police officers over the radio. You may have noticed that these radio conversations are very different from face-to-face conversations and mean little to the lay person. Yet they obey well-defined rules, and words in the conversations have well-defined meanings to the operators. Speakers announce their names, deliver the messages, confirm reception, etc. The rules followed in these channels are an example of protocols. Since the communication channel is shared by many speakers, the messages are short and coded to quickly make way for other speakers.[22] Similarly, each layer in the network stack has a well-defined set of tasks, and the protocol software in each layer is capable of performing these tasks very efficiently.

The success of the Internet in general and TCP/IP in particular is based on some key design principles adopted by the designers of these protocols. Two of these stand out. The first is that at the very outset, the designers of TCP/IP visualized a federation of networks, where each network could adopt any arbitrary design or technology. This principle allows wireless Ethernet networks in homes to interoperate with fiber-optic networks in carriers. Interoperation can be expected even with networking technologies that do not exist yet. The designers of the Internet were forced to adopt this principle since they had no control over the underlying data-transmission technologies such as the phone line. This was in stark contrast to the traditional phone network, where AT&T engineers had complete control over every aspect of the specifications of the phone system. The second principle is that multiple applications, even those that have not yet been conceived, can use the Internet for network connectivity.[23]

Together, these two principles led to the development of a common packet format that can be transferred by any network and that serves all applications. This common packet

22 This example is from L. Pouzin and H. Zimmermann, "A tutorial on protocols," *Proceedings of the IEEE*, 66(11) (1978): 1346–1370.

23 These principles have led to numerous recommendations including "protocol implementations should be conservative in their sending behavior and liberal in their receiving behavior" (RFC 706).

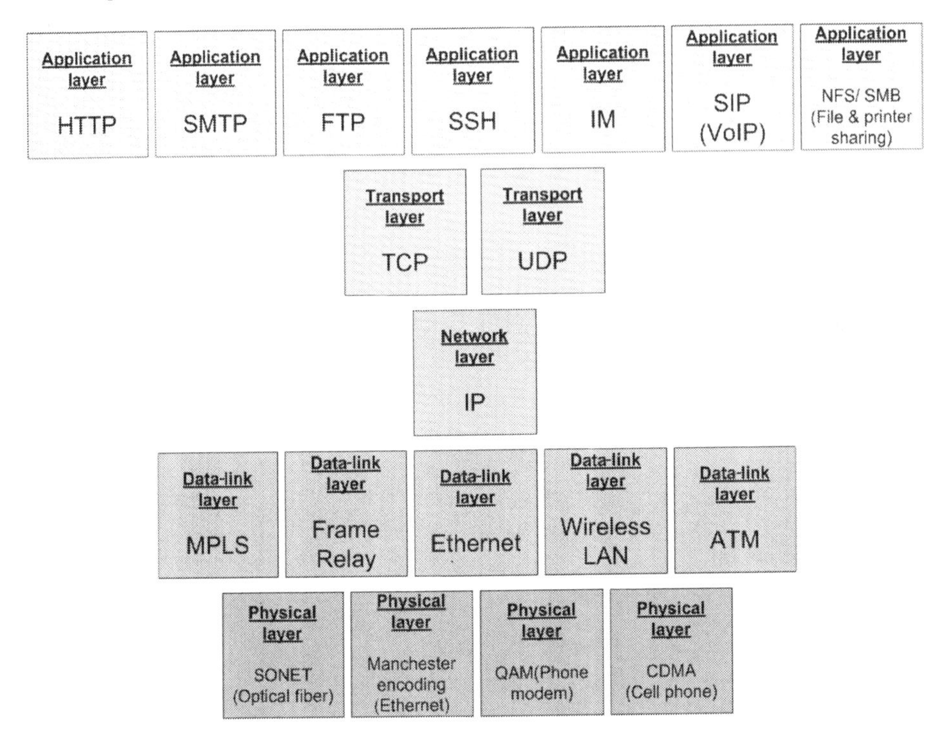

Figure 1.17: Data-delivery technologies at TCP/IP architecture layers

format is the IP packet format. Data from all applications is broken down into IP packets and all network technologies know how to transfer IP packets. This is shown in Figure 1.17, which illustrates that information from different applications such as the web and e-mail is transferred as IP packets. (Readers may note the similarity between this mechanism and the postal service's flat rate packages. The box can be used to ship anything at a fixed price as long as it fits into the box.)

These IP packets may be carried by networks running on any underlying technology. A visual inspection of Figure 1.17 also shows why IP is sometimes called the "thin waist" of the network stack. This model of Internet operation is quite analogous to package shipping by the postal service, as shown in Figure 1.18. There, items of different types are fitted into a common package format (box), which in turn can be shipped using any available delivery method (e.g. road or air).

We experience the advantages of these two design principles of TCP/IP on a daily basis when we use the Internet for network connectivity. One computer, a laptop or PC, is able to run any network communication application because one packet format, IP, can serve all these applications and can be transferred across all networks. This contrasts with dedicated networks such as the phone and TV networks which can only serve one specific application. These two design principles have also led to the longevity of TCP and IP. When TCP and IP were created, the Internet had a few thousand users who connected to each other using

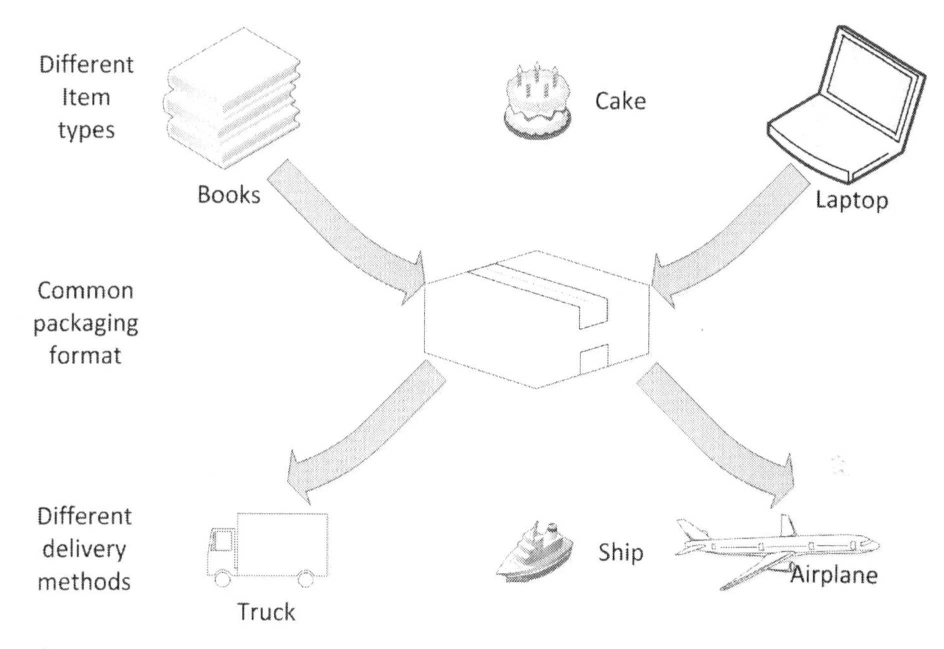

Figure 1.18: Package delivery model in the postal service

phone lines, which offered data rates of about 50 Kbps. Today, the Internet has more than 500 million users connected using high-speed lines with data rates over 10 Gbps. While IT has evolved at a very rapid pace, all modern networks continue to run on TCP and IP, technologies that are largely unchanged since they were first defined almost 40 years ago.

Typical Computer Network

Putting all the ideas in the chapter together and to set the stage for the remaining chapters, Figure 1.19 shows a typical computer network. Please refer to this figure anytime you are not sure where a specific technology fits in the overall network.

The figure considers a subscriber served by Verizon, an ISP (Internet Service Provider), connecting to a website at a university, which is connected to the Internet through Cogent, an ISP. A PC in the home is connected to the Internet through the home router. This router acts as the edge between the home network and the ISP network. A dedicated connection links the home router to the ISP network. Within the ISP network are a large number of routers that move the data packets successively closer to their destination by passing them on to the best neighboring router. Along the way, packets may be handed off to other networks at peering points. Finally, the packets are delivered to the destination network. For example, if you visit your university home page, the destination network is your university network. The destination network locates the web server hosting the web page and delivers these packets to the web server. The response from the web server takes the reverse path back to the home PC. The transaction sequence is also shown in more detail in the inside front cover of the book.

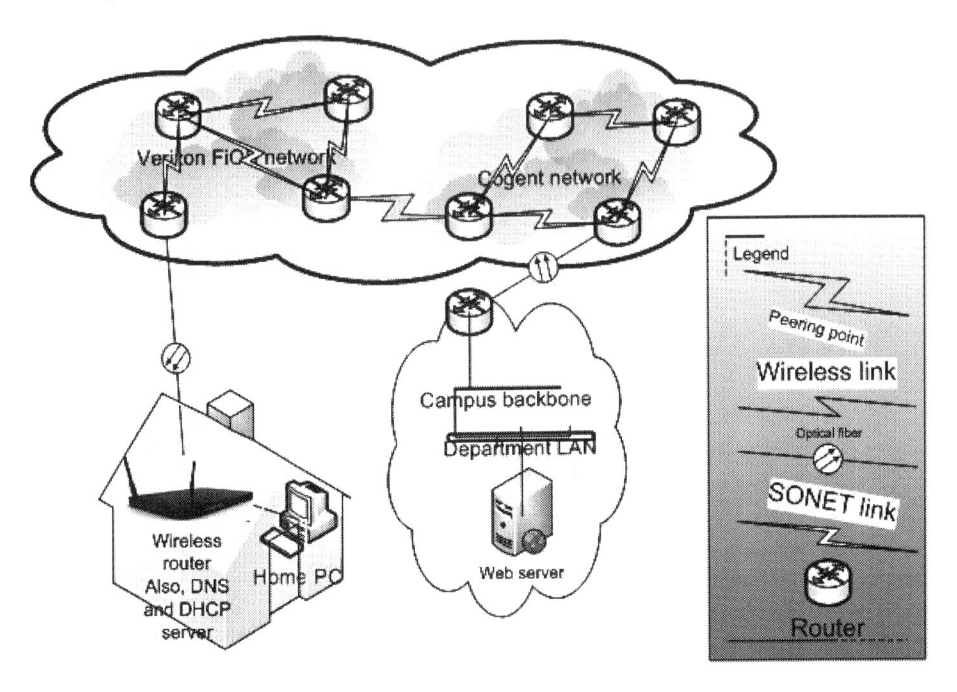

Figure 1.19: A typical computer network

Summary

This chapter and the book began with a discussion of the utility of computer networks to demonstrate how computer networks are useful to individuals and businesses. We saw why it is to your professional advantage to understand the basics of how computer networks operate. We looked at the major milestones in the evolution of computer networks leading to the deployment of circuit-switched networks around the globe. We saw how modern networks are moving to packet switching because packet switching is very well suited to handle the traffic patterns on the Internet. To set the stage for the later chapters, we introduced the idea of layering, using examples to show how layering is a common method to organize complicated functionality in computer software and hardware. The TCP/IP stack, which is the standard implementation of layering for packet switching, was also introduced. Finally, to place these ideas in context, we saw how multiple networks are connected together to form the Internet.

The next few chapters of the book are organized by the layers of the TCP/IP stack. Since lower layers offer services to upper layers, we go from bottom up on the stack. Chapter 2 covers the physical layer; Chapter 3 covers the data-link layer; Chapter 4 the network layer; the transport layer is discussed in Chapter 5, and the application layer in Chapter 6.

The following set of chapters looks at the technologies that support the TCP/IP stack. Chapter 7 covers important support services such as DHCP and DNS that enhance the capabilities of computer networks or make networks easier to use. Chapter 8 and

Chapter 9 cover routing and subnetting, respectively. Two kinds of data-link layer technologies—WANs and wireless LANs—are covered in Chapter 10 and the supplement. Telephony is also covered in the supplement. Finally, issues surrounding networks—security and legal issues—are also covered in Chapter 14 and the supplement.

About the Colophon

Sir Arthur C. Clarke was a British science fiction author and futurist. He is most famous for the novel *2001: A Space Odyssey*. As a futurist, he attracted great fame for his proposal in 1945 for geostationary satellite communication systems. While it was considered a fantasy at the time, it became a reality within 20 years when Intelsat was launched in 1965, marking the beginning of satellite TV in the country.[24] Sir Clarke formulated three laws of predicting the future, the third of which is best known: any sufficiently advanced technology is indistinguishable from magic.

The Internet seems to fit this definition of an advanced technology very well. Pause to think about it. Isn't it magical that every time we use the Internet, our computers routinely break data down into packets; each of these packets independently locates its destination and finds its way there; and the receiver is able to reassemble all these packets of data back into a coherent whole as a web page, e-mail, instant message, speech, video, bank statement, or other artifact? Even better, isn't it wonderful that all these processes happen reliably day in and day out, and within milliseconds? We take all of this for granted, but isn't that part of the magic?

The rest of this book is dedicated to describing how this magic works.

EXAMPLE CASE—Domino's Pizza

When we think of ways in which computer networks help businesses, the first thought that often comes to mind is e-commerce. Companies such as Amazon and Netflix have made it commonplace for consumers to buy various categories of goods and services online. Companies have had less success at selling food over the Internet. But people are becoming more comfortable with the Internet and more people are online. Internet adoption has increased significantly in the United States, reaching almost 84% in 2014. This may finally be increasing our willingness to buy food online, beginning with a food category that people are already comfortable ordering for delivery—pizza.

Domino's Pizza was founded in 1960 by brothers Thomas and James Monaghan, who borrowed $900 to purchase a small pizza store in Ypsilanti, Michigan. Today, Domino's Pizza is the number one pizza-delivery company in the United States, with a 17.5% share of the pizza-delivery market based on reported consumer spending. On average, more than one million pizzas are sold each day throughout the system, with deliveries covering approximately 10 million miles per week. Domino's Pizza pioneered the pizza-delivery business and has built the Domino's Pizza brand into one of the most widely recognized consumer brands in the world.

---

24 Arthur C. Clarke, "Extra-terrestrial relays," *Wireless World*, 1945, pp. 305–308, http://lakdiva.org/clarke/1945ww/ (accessed Nov. 2009).

As of June 2015, the company had 11,900 stores worldwide in more than 80 countries. Of these, 5,098 stores were in the United States. At the end of 2014, the company had approximately 11,000 employees. Including those in franchises, an estimated 240,000 people worked in the Domino's Pizza system.

The company has three reportable business segments: (1) domestic stores (2014 revenues: $578 million), (2) supply chain (2014 revenues: $1,260 million), and (3) international franchise (2014 revenues: $152.6 million).

Domino's Pizza operates in the quick-service restaurant (QSR) industry. With sales of $32.9 billion in the 12 months ended November 2014, the US QSR pizza category is the third-largest category within the $261.9 billion US QSR industry. The US QSR pizza category is large and fragmented. The primary domestic competitors are regional and local companies as well as national chains: Domino's, Pizza Hut, and Papa John's. These three national chains accounted for approximately 55% of US pizza delivery in 2014, based on reported consumer spending, with the remaining 45% attributable to regional chains and individual establishments. The share of the largest chains has grown from 47% in 2008 to 55% in 2014.

Over the years Domino's has developed a simple business model focused on its core strength of delivering quality pizza in a timely manner. This business model includes a delivery-oriented store design with low capital requirements, and a focused menu of pizza and complementary side items. The majority of its domestic stores are located in populated areas in or adjacent to large or mid-sized cities, or on or near college campuses. The company believes that its pizza-delivery model provides a significant competitive advantage because most of its stores do not offer dine-in areas. As a result, they typically do not require expensive real estate, are relatively small, and are relatively inexpensive to build and equip. They also benefit from lower maintenance costs, as store assets have long lives and updates are not frequently required. The typical Domino's pizza store is relatively small, occupying approximately 1,000 to 1,300 square feet, and is designed with a focus on efficient and timely production of consistently high-quality pizza for delivery. The store layout has been refined over time to provide an efficient flow from order-taking to delivery. The entire order-taking and pizza-production process is designed for completion in approximately 12–15 minutes.

Competition in the pizza-delivery market is intense and is generally based on product quality, location, image, service, price, convenience, and concept. The industry is often affected by changes in consumer tastes, economic conditions, demographic trends, and consumers' disposable income.

Pizza Tracker

A large fraction of Domino's customer base is young. This customer base is increasingly connected to the network using cell phones and laptops and has a strong preference for using online search and Internet ordering for products and services. Internet ordering better fits the work flow of this audience than phone ordering. Therefore, strength in online order-taking can help quick-service restaurants reach this customer base.

Domino's introduced Pizza Tracker in 2007 to improve its online-ordering capabilities. Pizza Tracker has been integrated into PULSE, the company's proprietary point-of-sale system. Customers access Pizza Tracker through the build-your-own-pizza application on the company's home page. The application photographically simulates the pizza as customers select a size, choose a sauce, and add toppings. Customers can add coupon codes and are shown the final price before they place their order. Pizza Tracker also allows customers to track the progress of the pizza as it enters the oven, or when it leaves the store. At the time of its introduction, Domino's had the most feature-rich, online pizza-ordering application among the three national pizza chains.

Installing the online ordering system costs approximately $20,000 per store. Many franchisees were initially hesitant to incur these costs. Resistance increased further when many of the bells and whistles in the online-ordering application did not work or were incompatible with the ways the franchisees operated their stores. In response, in developing the online-ordering system, Domino's CIO decided to focus on reliably executing core tasks such as taking orders, scheduling workers, and mapping delivery routes. This has helped overcome franchisee resistance.

Online orders are directly displayed to the right spot on the pizza-assembly line in the store. To provide status updates to customers, employees hit a button on their computer screens to update the status of the order. Another update can be generated when the delivery driver leaves the store with the pizza.

The online application makes it easy for connected customers to locate a nearby Domino's pizza and order their preferred pizza in a single visit to the company's website. The online-ordering system stores contact information and menu preferences for customers who have ordered online. This information can be used to simplify ordering for repeat customers and also to encourage adding items to orders.

The online-ordering system has proved quite popular. Almost 40% of all orders to the large pizza chains (Domino's, Papa John's and Pizza Hut) in 2014 originated online. Anecdotal evidence from adopting franchisees suggests that online ordering has boosted overall orders and increased the average amount of each transaction. CEO of Domino's, Patrick Doyle, says that "in a lot of ways, we're really a technology company." Most employees at headquarters are computer programmers and technicians monitoring production and feedback in real time, and finding ways to improve the process to reduce delivery time.

By comparison, at smaller mom-and-pop shops that can't afford to invest in online ordering, sales have fallen by as much as 20%, especially from younger customers. Since 2001, the number of independent pizza shops has fallen by 6.6%, although the overall number of restaurants has only fallen by 2.9%. One successful response from smaller shops has been gourmet and wood-fired pizzas.

References

1. Jargon, Julie. "Domino's IT Staff Delivers Slick Site, Ordering System." *Wall Street Journal*, Nov. 24, 2009.
2. Jargon, Julie. "Apps are wrecking mom-and-pop pizza shops." *Wall Street Journal*, Feb. 6, 2014.
3. McNamara, Paul. "What percentage of Domino's Pizza is ordered online." *Network World*, May 8, 2014.
4. Moore, Stephen. "How Pizza became a growth stock." *Wall Street Journal*, Mar. 13, 2015.

REVIEW QUESTIONS

1. What is business data communications?
2. What are some of the ways in which computer networks are used in large businesses?
3. What are some of the ways in which small businesses can benefit from computer networks?
4. Look at the websites of some departments of your county government. What three services offered at these sites do you find most interesting?
5. Describe how the three online government services you chose could be helpful to you.
6. Describe the major trends in the growth of Internet traffic.
7. What is packetization?
8. Why is packetization useful in business data communication?
9. What are some of the factors that make packet switching more complex technologically than traditional phone circuits (circuit switching)?
10. Provide a high-level overview of the structure of a typical data packet. What are the kinds of information you are likely to find in the header of a typical packet?
11. Consider a typical office memo as a data packet. What information in the memo would be characterized as header information? What information would be characterized as the body of the packet?
12. Why is Internet traffic considered "bursty" compared to voice traffic?
13. What is layering?
14. Why is layering useful in organizations?
15. Why is layering useful in computer networking?
16. What are the five layers of the TCP/IP model?
17. What are the primary functions of each layer in the TCP/IP model?
18. TCP is often considered the most important layer of the TCP/IP model. What are the primary responsibilities of TCP?
19. What devices does a packet typically encounter in its journey from source to destination?
20. What is the OSI model? What was the motivation for the development of the OSI model?
21. How has the OSI model been useful in the development of computer networks?
22. What are the seven layers of the OSI model?
23. What are the primary functions of each of the seven layers of the OSI model?
24. The packets used to transmit voice on the Internet are similar to the packets that are used to send e-mail. What are some of the advantages of this approach?
25. What are the two design principles behind Internet protocols?

EXAMPLE CASE QUESTIONS

1. Visit the websites of Domino's pizza and one other quick-service pizza restaurant. How long does it take for you to order a pizza of your choice from each of these restaurants? (Of course, you don't have to actually place the order for this exercise. Just get to the point where you are ready to place the order.)
2. What features of the websites do you find most useful as a customer?
3. Based on your experience with ordering pizza online, or from visiting the site, what is one feature you would add to the site to improve your experience?

HANDS-ON EXERCISE—Traceroute

Figure 1.19 showed a typical arrangement of networks that form part of the Internet. In this lab exercise, we will use a command-line utility called traceroute to view the networks that a packet crosses as it travels the network from your computer to some popular websites.

To use traceroute, open up the command prompt on your computer. On Windows, you can do Start → Run and type `cmd` in the dialog box. On the Mac or on Linux desktops, you can open the terminal application to get a command prompt (on the Mac, the terminal application is at Applications → Utilities → Terminal).

At the prompt, enter the command `tracert` followed by the URL of a website you want to trace. In the example of Figure 1.20, we use tracert to trace a route from our computer to the website of the Illinois Institute of Technology (www.iit.edu). Each row in the output has five columns. The first column is the serial count of the routers encountered in the path to the destination. This gives us a count of the number of hops required to reach the destination. For example, from Figure 1.20, we can see that there are 12 hops from the source to the destination. The next three columns provide the round-trip times for three probe packets sent to the router at the location. For example, in Figure 1.20, we see that the round-trip time for each of the three probe packets to reach the router on the 8th hop is 97 milliseconds. The last column gives us the name or numerical address of the router reached on the hop. For example, the router on the 8th hop in Figure 1.20 is chic-hous-67.1ayer3.nlr.net. Many ISPs give routers meaningful names that help network administrators quickly identify the location of the router on a map. For example, the tpa in

```
C:\Windows\system32\cmd.exe

C:\Temp>tracert www.iit.edu

Tracing route to www.iit.edu [216.47.150.245]
over a maximum of 30 hops:

  1    51 ms    50 ms    50 ms  edu-vpn1-172-public.net.usf.edu [131.247.250.35]

  2    58 ms    50 ms    50 ms  131.247.250.62
  3    51 ms    50 ms    50 ms  vlan254.campus-backbone2.net.usf.edu [131.247.25
4.46]
  4    50 ms    50 ms    50 ms  131.247.254.242
  5    52 ms    50 ms    51 ms  tpa-flrcore-7609-1-te31-v1601-1.net.flrnet.org [
198.32.166.93]
  6    55 ms    55 ms    55 ms  tlh-flrcore-7609-1-te21-1.net.flrnet.org [198.32
.155.14]
  7    71 ms    74 ms    71 ms  hous-te0501-v513.layer3.nlr.net [198.32.155.254]

  8    97 ms    97 ms    97 ms  chic-hous-67.layer3.nlr.net [216.24.186.25]
  9   104 ms   107 ms   107 ms  74.114.96.13
 10    97 ms    97 ms    97 ms  216.47.141.25
 11    97 ms    97 ms    97 ms  216.47.159.5
 12    97 ms    97 ms    97 ms  www.iit.edu [216.47.150.245]

Trace complete.

C:\Temp>_
```

Figure 1.20: Use of tracert command to find route to www.iit.edu

```
Windows PowerShell
PS C:\Temp> tracert www.u-tokyo.ac.jp

Tracing route to www.u-tokyo.ac.jp [133.11.114.194]
over a maximum of 30 hops:

  1    50 ms    49 ms    49 ms  edu-vpn1-172-public.net.usf.edu [131.247.250.35]
  2    50 ms    50 ms    50 ms  131.247.250.62
  3    50 ms    50 ms    50 ms  vlan254.campus-backbone1.net.usf.edu [131.247.254.45]
  4    50 ms    50 ms    50 ms  vlan256.wan-msfc.net.usf.edu [131.247.254.81]
  5    50 ms    50 ms    50 ms  atm900-128.enb-msfc.net.usf.edu [131.247.254.230]
  6    50 ms    50 ms    54 ms  tpa-flrcore-7609-1-te31-v1601-1.net.flrnet.org [198.32.166.93]
  7    55 ms    55 ms    55 ms  tlh-flrcore-7609-1-te21-1.net.flrnet.org [198.32.155.14]
  8    72 ms    71 ms    71 ms  hous-te0501-v513.layer3.nlr.net [198.32.155.254]
  9   102 ms   102 ms   102 ms  losa-hous-87.layer3.nlr.net [216.24.186.30]
 10   101 ms   101 ms   101 ms  transpac-1-lo-jmb-702.lsanca.pacificwave.net [207.231.240.136]
 11   216 ms   217 ms   216 ms  tokyo-losa-tp2.transpac2.net [192.203.116.146]
 12   217 ms   216 ms   217 ms  vlan53-cisco2.notemachi.wide.ad.jp [203.178.133.142]
 13   217 ms   216 ms   217 ms  ve-51.foundry6.otemachi.wide.ad.jp [203.178.141.141]
 14   217 ms   217 ms   217 ms  ve-42.foundry4.nezu.wide.ad.jp [203.178.136.66]
 15   218 ms   217 ms   217 ms  ra37-vlan566.nc.u-tokyo.ac.jp [133.11.125.237]
 16   217 ms   218 ms   217 ms  ra3a-gi1-0-0.nc.u-tokyo.ac.jp [133.11.206.146]
 17   217 ms   217 ms   217 ms  ra39-vlan336.nc.u-tokyo.ac.jp [133.11.206.153]
 18   217 ms   217 ms   217 ms  ra35.nc.u-tokyo.ac.jp [133.11.127.41]
 19   217 ms   217 ms   217 ms  www.u-tokyo.ac.jp [133.11.114.194]

Trace complete.
PS C:\Temp>
```

Figure 1.21: Tracing the route to the University of Tokyo

the fifth router indicates that it is located in Tampa and the tlh in the sixth router indicates that it is located in Tallahassee, Florida.

The router names also indicate the networks the routers belong to, so that the names of the organizations operating the routers can be easily checked on the web. For example, routers 1 and 3 are located in USF, as seen from their usf.edu name. Routers 5 and 6 belong to the Florida Research and Educational Network, as can be checked by visiting the website www.flrnet.org. Routers 7 and 8 are operated by the National Lambda Rail, as can be checked by visiting the website www.nlr.net.

Figure 1.21 shows another route trace, this time to the University of Tokyo. We see that the packets take the path from USF → flrnet → nlr → pacificwave → transpac2 → Wide project (wide.ad.jp) → U. Tokyo.

The geographic names of the routers in Figure 1.21 also indicate the geographic path taken by the packets. Starting from the East Coast of the United States in Tampa, Florida, the packets travel via Tallahassee and Houston to reach Losa, California, in hop 9. Hops 10 and 11 take the packets to Tokyo, Japan, from Losa, California, over a transpacific route. In Japan, the Wide project routes packets to the University of Tokyo. Hops 15–19 are within the University of Tokyo as the packets take many small hops within the university to reach its web server.

When packets cross submarine routes (e.g. hops 10, 11 in Figure 1.21), they typically encounter relatively larger delays because of the longer distances that need to be covered. Delays may also be caused when the network on a hop is very busy handling other traffic.

Answer the following questions:

1. Briefly describe what traceroute does and how it is useful (you may find articles on traceroute in sources such as Wikipedia useful).

2. Use traceroute to trace the route from your home or work computer to your university's website. Show the traceroute output. What networks were encountered along the way? What information about the geographic locations of the routers can you infer from the trace? What was the longest mean delay on any one hop along the way?
3. Use traceroute to trace the route from your home or work computer to the website of a university or company on another continent. Show the traceroute output. What networks were encountered along the way? What information about the geographic locations of the routers can you infer from the trace? What was the longest mean delay on any one hop along the way? What factors do you think caused this delay?
4. Visit the websites of the ISPs on one route. Show the network coverage map of any one ISP on the route that provides this information. (Note: This information is useful in marketing, therefore most ISPs provide some graphical information about their coverage areas.)

CRITICAL-THINKING EXERCISE—The Power of Universal Exchange Formats

The idea of using a universally agreed-upon format is fairly common for exchange and greatly facilitates exchange. Ideas gain global traction when they are expressed in English because direct translations exist between almost every language and English and so ideas expressed in English can be easily conveyed to speakers of other languages. Similarly, most international commercial transactions are settled in US dollars for various reasons, including the fact that almost every country has an interest in trading with the US, as well as the perception of the dollar being a well-managed currency. This has given the US dollar the status of the most widely used reserve currency.

Being the universal exchange format can also be influential. Though IP, the common exchange format on the Internet, has no special "powers," the owner of a reserve currency has considerable political influence since it can unilaterally impose penalties on entities in other countries for violating conditions imposed by the owner of the reserve currency.

1. What is a reserve currency?
2. Does being a reserve currency offer any benefits to the owner of the currency?
3. Does being a reserve currency have any trade-offs to the owner of the currency?

IT INFRASTRUCTURE DESIGN EXERCISE—Identifying Uses

We saw in this chapter that computer networks are useful to organizations in various ways. Subsequent chapters deal with the technology components used in these networks. To help you understand how the information in the book is used in the real world, the IT infrastructure design exercise will build the IT infrastructure for a fictional company. Using an example of a fictitious multinational manufacturing firm headquartered in your town, in every chapter you will choose and configure the right technology from the available options in a manner that best meets the business needs of the firm.

Overall information about the company is presented here. Where required, the exercises in the chapters will provide more details specific to the chapter. You will be asked to draw infrastructure diagrams in many chapters. Visio is a popular software for drawing networks. If you do not have access to Visio, the readings for this chapter at the companion website have a document that includes a world map and network icons that you can use to create the diagrams using word processors.

TrendyWidgets Inc., is a successful manufacturing company in your town. The company manufactures and sells widgets to other businesses around the world. As a result of its success, it has offices around the world. Since the company operates from multiple locations, it relies on a communications infrastructure to support business processes that are dispersed across these offices. As a consultant to TrendyWidgets's CIO, you are charged with designing the entire network infrastructure for the company.

Firm Details

TrendyWidgets maintains design, manufacturing, fulfillment, repair, and customer service facilities around the world. The locations are shown in Table 1.5 and Figure 1.22.

The corporate offices have an assortment of mobile and stationary users. Mobile users are assigned laptops to conduct business from locations in the office and at customer sites or home offices. These users are usually connected to the network for 8–10 hours each day. The applications they access include e-mail and a wide host of resources, including marketing, human resources, R&D, product literature, and customer and vendor information. The call center and service center have stationary users. They typically access e-mail, product literature, and customer history.

To help with the design exercise that you will see throughout this text, the multimedia and IT experts at USF collaborated to create a video that you may find useful:

Part 1: https://www.youtube.com/watch?v=QXN1uhGmOx4.
Part 2: https://www.youtube.com/watch?v=HLLd3qT_v88.
Part 3: https://www.youtube.com/watch?v=cgXUj_hLFOo.
Complete: https://www.youtube.com/playlist?list=PLGd40zfxSKJbSdUWi3GM_VS9FqKreNnUW.

Answer the following question. (To complete this exercise, follow the questions at the end of each chapter. You do not have to draw anything for this chapter.)

1. What are the ways in which computer networks can help TrendyWidgets in its business? Think of all the different ways in which the company can use computer networks at all of its different offices.

Table 1.5: TrendyWidgets office locations and staffing

<table>
<tr>
<td><Your town>, United States
• Building #1 with 4 floors, 300 employees, with 100 employees per floor on the first three floors and a lights-out data center on the fourth floor.
• Each floor has a dimension of 100 feet × 150 feet, with a telecommunications closet located near the center of each floor.
• The two lower floors of this building are used for producing widgets.
• The third floor has various corporate offices for HR, finance, IT, and marketing departments.
• Data center hosts data from payroll, accounting, production, and other corporate functions.
• The corporate website is also hosted here.</td>
<td>Amsterdam, The Netherlands
• Building #1 with 2 floors, 100 employees on the first floor.
• The first floor of this building houses a service center for the EMEA (Europe, Middle-East, and Africa) region.
• The firm's backup database is located on the second floor of the building.

Mumbai, India
• Building #1 with two floors, 200 employees, 100 employees on each floor.
• The building houses the company's primary call center, operating 24 hours a day.</td>
</tr>
<tr>
<td colspan="2">Singapore, Singapore
• Building #1 with 3 floors, 200 employees, with 100 employees on the first floor and 50 employees each on the other two floors.
• The first floor is used by the marketing group for the AP (Asia-Pacific) region. The other two floors house a service center for the AP region.</td>
</tr>
</table>

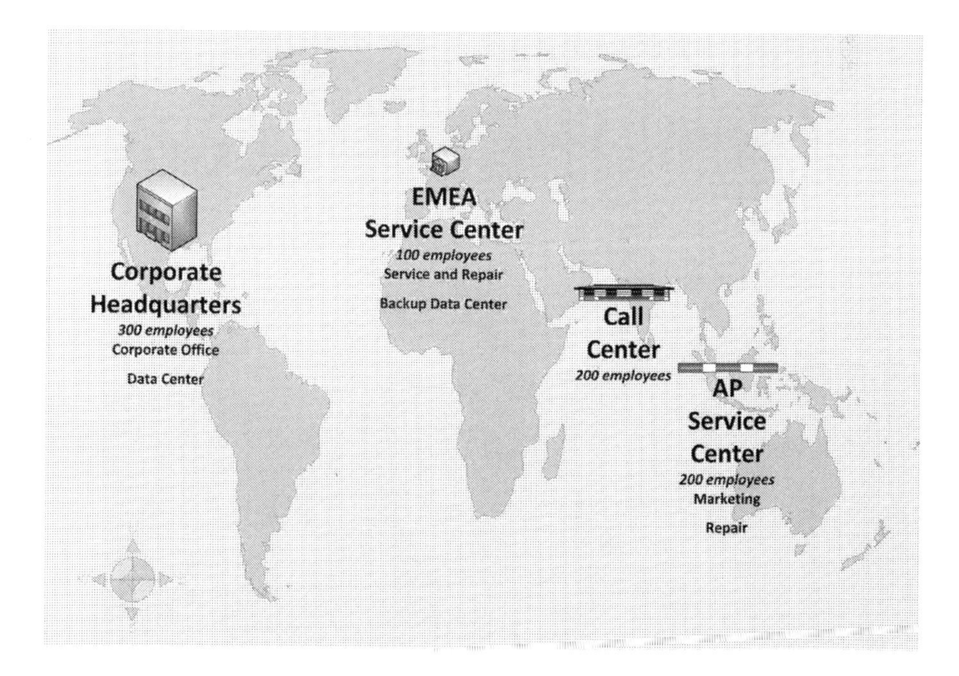

Figure 1.22: TrendyWidgets office locations and staffing

CHAPTER 2

Physical Layer

To accomplish almost anything worthwhile, it is necessary to compromise between the ideal and the practical.

Franklin D. Roosevelt

Overview

This chapter starts at the bottom of the TCP/IP technology stack that was introduced in Chapter 1 and describes the functions of the physical layer. The physical layer sends data as signals over a physical medium. At the end of this chapter you should know:

- the functions of the physical layer
- what a physical medium is and why copper and optical fiber are important physical media
- what signals are and why they are necessary
- an example of using signals to transfer data
- an example of multiplexing

Functions of the Physical Layer

Chapter 1 described the function of the physical layer as *signaling* (Table 1.4). More formally, the function of the physical layer is defined as *providing transparent transmission of a bit stream over a circuit built from some physical communications medium*. Both definitions convey the same idea about the functions of the physical layer—the physical layer gets data from the data-link layer in the form of a stream of data bits, and the physical layer is responsible for transmitting these bits as signals over a wire, optical fiber, wireless, or other medium. *Converting data to signals for transmission over physical media* is called signaling.

This is a good point to introduce the telecom glossary.[1] The telecom glossary is an American National Standard that is developed and maintained by the Alliance for Telecommunications Industry Solutions (ATIS). ATIS incorporates the expertise of the nation's largest communications companies. If you find yourself looking for definitions of

1 Institute for Telecommunication Sciences. *ATIS Telecom Glossary*. http://www.atis.org/glossary/.

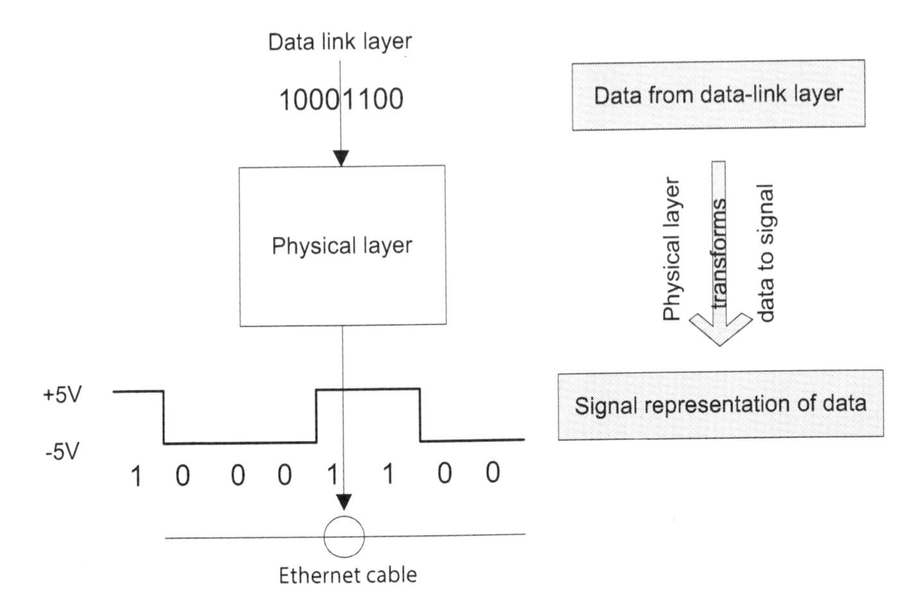

Figure 2.1: Physical layer function

technical terms related to networking for class or professional reports, the ATIS glossary would be an excellent starting point. Wherever possible, definitions in this text will be drawn from the ATIS telecom glossary.

Figure 2.1 shows the physical layer in operation. Recall from Table 1.4 that the physical layer is situated directly below the data-link layer in the TCP/IP stack. This position in the layered architecture means that when the data-link layer is ready to send data, it will pass the data on to the physical layer so that the physical layer can perform its service. Let's say that the data-link layer would like to send 8 bits of data (1 byte)—10001100. What does the physical layer do with this data? What service does the physical layer perform? How does it add value?

The problem the physical layer solves is that data cannot be sent over a wire. The only known method of transferring data over a medium is to convert data into a signal and transmit the signal. We addressed this issue while describing the telegraph and noting why it was such a critical development for data communications. We cannot take a bit of data, pass it into one end of a wire and hope for the bit to somehow emerge at the other end as a data bit. The only known method of sending information over a wire is to convert the information into a form that can be transmitted over the wire. This converted form of information is called a signal. In simple terms, *a signal is a change that can be detected at the receiving end.* The telecom glossary defines a signal as *detectable transmitted energy that can be used to carry information.* In the telegraph example, the signal was created by the closing and opening of the switch, which caused the voltage on the line to rise or fall. This change in voltage was detected at the receiving end by the movement of the marker to one side.

By interpreting the marks on the paper, the receiver could reconstruct the data sent by the sender.

Think of a more familiar example of the need for signals. Can you directly transfer thoughts and ideas that arise in your mind, even to your nearest and dearest ones? No. The only known method of transferring thoughts is to first convert thoughts and ideas into speech, signs (thumbs up, winks, etc.), or written text, and send the speech, signs, or text to the receiver. In this context, the thoughts in the mind are data, and the text or speech are signals. The receivers interpret the sound or text as thoughts in their own minds. Right now when you are reading this text, your eyes are receiving signals in the form of black and white patterns on this page. All your education allows your brain to interpret these black and white patterns (signals) as letters of the alphabet, words, sentences, and paragraphs (data).

Learning to read and reading to learn

In elementary school, the focus of language arts in grades K–3 is learning to read, which is learning to decode alphabet symbols and memorizing basic words. Though students simultaneously use this knowledge to understand what they are reading, the focus is on learning to code and decode the alphabet. Later, beginning in grade 4, the focus in language arts education is on using this knowledge to comprehend new information, i.e. reading to learn.

The learn-to-read and read-to-learn cycle persists throughout life. In later grades and thereafter, we learn new word roots, figures-of-speech, and other details, which are then helpful in interpreting new information.

Figure 2.1 provides a visual description of the functions of the physical layer. The physical layer receives data in the form of numbers from the data-link layer and converts the data into signals that can be transmitted over a wire.

Special Feature of the Physical Layer

So, the physical layer adds value to data communications by converting data into signals for transmission over media. Before we begin discussing media and signals, it is useful to note that one important characteristic distinguishes the physical layer from the other four layers of the TCP/IP stack. The physical layer is the only layer in the stack that interacts with nature. The physical layer is the layer where the data finally hits the wire after being processed by the other layers. The physical layer, therefore, has to confront the limitations and capabilities of nature in a way that other layers do not. All other layers deal with data inside the computer. The physical layer deals with signals outside the computer.

In data communications, the properties of nature relevant to the physical layer are the signal-transmitting properties of the physical medium used. The physical layer has to generate signals that comply with the constraints specified by the medium. For example, if copper wire requires that data for transmission be passed to it in the form of electrical signals, the physical layer will send data as electrical signals over copper wire. If optical fiber requires that data be passed in the form of light signals, the physical layer will convert data to light signals for transmission over optical fiber. If copper has a bandwidth of about 100 MHz,[2] the physical layer has to satisfy itself with whatever data rates it can attain from this available bandwidth.

All the other layers add suitable header content to data packets to do their job. The physical layer does not add any header content. Rather, the physical layer converts data to suitable signals for transmission to the destination over copper wires, optical fiber, wireless, or another medium.

We are, in fact, quite fortunate that it is possible to send information over a wire, even if it requires the transformation of data into a signal at the sending end and conversion of the signal back to data at the receiving end. Seen this way, the activities of the physical layer are a successful exploitation of the signal-carrying properties of transmission media.

This point of dealing with nature has been brought up because you may find the discussion of signals later in this chapter more mathematical than expected. Sine waves come up repeatedly in the section on signals because sine waves are the most elementary signals. We will keep the math to a minimum, only to the extent absolutely necessary to understand signals. Later in this chapter, if you ever find yourself wondering why you are dealing with sine waves in a class on computer networks, please remind yourself that if signals were not used, data could not pass through a wire.

To summarize this section: The physical layer takes data from the data-link layer and converts it to signals for transmission over transmission media. The next two sections look at transmission media and signals in more detail.

The physical layer and the search for Osama Bin Laden

You may recall how Osama Bid Laden was eventually traced by US intelligence organizations by identifying his messenger and following him to his home. If Bin Laden was to communicate with his troops, he had to make that physical layer connection with the world. And that connection was eventually traced.

The physical layer connection to the wider world also helped in the initial capture of Mexican drug kingpin Joaquin "El Chapo" Guzman, when his satellite phone was tracked down.[3]

Physical Media and their Properties

A physical medium in data communications is the transmission path over which a signal propagates. Common media include wire pairs, coaxial cable, optical fiber, and wireless radio paths. Media used in data communications are often called physical media because media are generally physical objects such as copper or glass. A physical object is something you can touch and feel, and it has physical properties such as weight and color. You can touch and feel copper and glass, and for a long time these were the only media used in computer

2 Hz is an abbreviation for hertz, the unit of frequency. This is discussed a little later in the chapter. In general, the greater the bandwidth, the higher the data rates possible.

3 Jose de Cordoba and Santiago Perez, "How Mexico nabbed a drug kingpin," *Wall Street Journal*, Feb. 23, 2014.

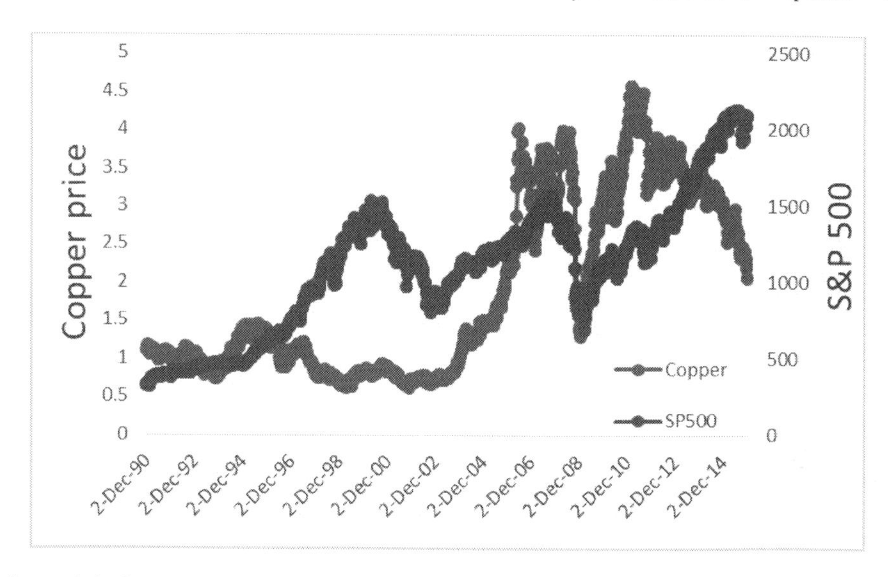

Figure 2.2: Comparing copper and S&P 500 (1990–2015)

networking. Hence, transmission media came to be known in the industry as physical media. A more recent communication medium in computer networking is wireless, and it has no physical properties. It is therefore not strictly appropriate anymore to use the term physical media to refer to all transmission media. However, once terms become popular in the industry it is very difficult to change them, and the term physical media is commonly used in networking to refer to all transmission media, including wireless. Therefore, the terms transmission media and physical media will be used interchangeably in this text.

Not all materials are useful as media. From the previous section it is clear that media must be good at carrying signals. Most plastics are not useful as media because there is no known way of sending signals through these plastics. Media must also be economical to use. Silver is the best known conductor of electricity. But it is not useful as a communication medium because it is prohibitively expensive. In any case, if you had network cables made of silver, you would spend an unacceptable amount of money guarding the cables against theft, rendering these cables useless for business purposes.

Based on industry experience with various materials, copper and glass are common physical media used in computer networking. Both are economical and very efficient at carrying signals. Copper is commonly used in homes and offices to connect desktops to the network. This makes copper extremely important to the economy, and copper prices are often used as a signal of the strength of the economy. For this reason, copper is sometimes called Dr. Copper in financial circles. Figure 2.2 compares the price of copper and the value of the S&P 500 index over a 15-year period (1990–2015). This period included two well-recognized financial bubbles (the dot-com bubble and the housing bubble). Though the two series are volatile with different trends, we see that they catch up periodically, with (importantly) drops in copper prices often predicting drops in stock prices. Optical fiber is commonly used by ISPs in the core of the network to carry signals between networks.

Optical fiber is also beginning to be used in office networks. In some parts of the country, ISPs are connecting homes directly with optical fiber, replacing the copper phone lines. Let's now look at these two transmission media.

Copper Wire as a Physical Medium

Copper wire is one of the most common physical media used in networking. The entire phone network was built using copper wires. Almost every desktop is connected to the network using copper wire.

Copper has many properties that make it useful as a physical medium for data communications. Copper is the second-best known conductor of electricity, next only to silver. This allows copper wires to carry signals to great distances using relatively low amounts of power. In fact, copper is the industry standard benchmark for electrical conductivity. Further, copper is relatively abundant on earth. Of the known worldwide resources of copper, only about 12% have been mined throughout history.[4] Thus, copper provides an excellent combination of signal-transmission capability and economy for use as a transmission medium.

Copper industry resources point out some interesting trivia about copper. Copper is the oldest metal known to man. Its discovery dates back more than 10,000 years. Wires are the only major use of copper that use newly refined metal. More than three-fourths of all other applications of copper use recycled metal. Chile is the world's largest mine producer of copper. The Chilean mining incident of 2010 where 33 miners were trapped underground for 69 days happened at a copper-gold mine in Chile (Figure 2.3).

Figure 2.3: Chilean President Sebastian Piñera with rescued miners
(Source: Wikimedia Commons)

4 http://www.copper.org.sg and http://www.copper.org.

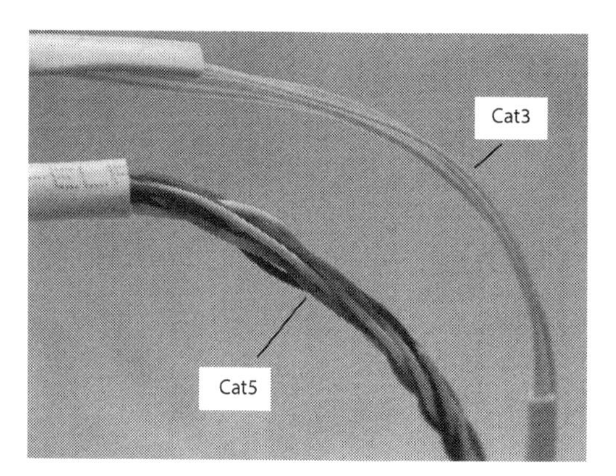

Figure 2.4: Cat5 and Cat3 cables
(Source: Author photo)

The most common cable used in computer networking has eight strands of copper wire. The wires are organized in a manner called unshielded twisted pair, or UTP. As the name suggests, the eight wires are organized as four pairs. Generally, two of these pairs are used for information exchange. One pair of wires is used for the forward data path and another pair is used for the reverse data path. The other two pairs have traditionally been unused, though newer technologies are beginning to use all four wire pairs.

The term *twisted* in UTP refers to the fact that the two wires in each pair of wires are twisted around each other. This is seen in Figure 2.4. Twisting cancels electrical noise from external sources such as electronic appliances and neighboring wire pairs. The tighter the twisting, the greater the noise resistance of the cable, and the higher the data rates the cable can support. In computer networking, the two wires in each pair carry signals that are the opposites of each other. This is called a balanced line and in a balanced line the sender ensures that the voltage on one wire of the pair is the exact opposite of the voltage on the other wire of the pair. A balanced line further improves the noise resistance of twisted copper wires.

The term *unshielded* in UTP refers to the fact that the wire pairs do not have a metallic shield as in cable TV wires. Figure 2.5 shows a picture of a shielded cable. Metallic shields block external interference and improve the signal-carrying capabilities of wires. However, shielding adds to the cost and weight of cables, and twisted balanced pairs provide sufficient noise resistance to eliminate the need for shielding in current networks. Therefore, shielded cables are declining in use for computer networks.

Figure 2.5: Shielded cable
(Source: Wikimedia Commons)

Categories of UTP Cables

Since tighter twisting improves the data rates of copper cable, as computer networks have become faster, cables with tighter twisting have become available. The traditional phone system used cabling with three to four twists *per foot* of cable. This was adequate for slower computer networks as well, supporting data rates of up to 10 million bits per second (10 Mbps). But with the development of faster networks, cables with three to four twists *per inch* (more than 10 times tightly wound than the previous generation) became common. These cables could support data rates of up to 100 Mbps.

Copper cables are specified by category. The older cable, with three to four twists *per foot*, is called category 3 cable. The newer cable, with three to four twists *per inch* is called category 5 cable. Category 5 has been further enhanced and is known as category 5e cable. Category is abbreviated as "Cat" for marketing purposes, so that category 5e cable is called Cat5e. Phone cable is called Cat3. Figure 2.4 shows Cat3 and Cat5 cables. Please note that cabling standards[5] do not specify the number of twists per inch on a cable. Standards only define the signal-transmission properties of the wire. Cable manufacturers use the appropriate twist density to achieve the required signal-carrying specifications of the standard.

Each of the eight wires in a Cat5e cable is color coded to facilitate use. Four colors—blue, orange, green, and brown—are used to identify each of the four twisted pairs. Within each twisted pair, one of the wires uses the pure color, the other uses a colored band on a white background (Figure 2.6). This is commonly written as shown in Table 2.1.

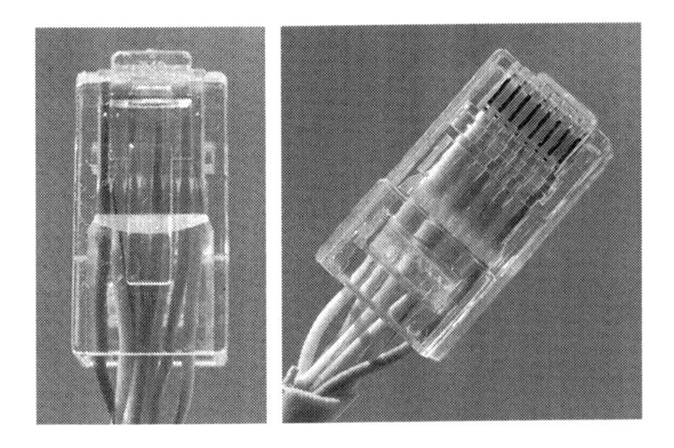

Figure 2.6: Cat5e cable terminating in RJ45 connector
(Source: Author photo)

The example in Figure 2.6 is not the correct way to terminate a cable. The cable has been loosened up to show the wires in the cable. This can create unacceptable interference noise and significantly reduce the maximum data rate of the cable. In a properly terminated cable, the jacket goes all the way into the connector so that the wires are twisted till the very end to minimize interference.

5 TIA/EIA-568-B: *Commercial Building Telecommunications Cabling Standard*, 2001.

Table 2.1: Color codes for Cat5e wires

Conductor identification	Color code	Abbreviation
Pair 1	White-Blue Blue	(W-BL) (BL)
Pair 2	White-Orange Orange	(W-O) (O)
Pair 3	White-Green Green	(W-G) (G)
Pair 4	White-Brown Brown	(W-BR) (BR)

Cat5e cables support gigabit speeds (1,000 Mbps). Higher categories than Cat5e are now available. Category 6 supports data rates of up to 10,000 Mbps (10 Gbps). A relatively newer standard called Cat6a and category 7 also supports up to 10 Gbps. Since Cat5e is already very tightly twisted, Cat6 and Cat7 do not increase the twisting rate over Cat5e. Category 6 achieves higher data rates by using thicker wire and superior connectors. Category 7 achieves higher data rates by shielding the wires.

Copper cables for communication cost in the order of $1/foot. For normal installations, they are usually used in 5-foot, 7-foot, 15-foot, and 25-foot sections. Rolls of 1,000 feet of Cat6 cables cost about $0.25 per foot.

Cable Connectors

UTP cables are connected to computers and other communication devices through plugs called RJ45 jacks. The term RJ stands for *registered jack* and RJ45 is one of the many available cable connectors. Another common such connector, used in phone networks, is called the RJ11 jack. Figure 2.6 shows the top and bottom views of a Cat5e cable ending in a RJ45 jack.

To ensure that any cable may be used to connect any computer to a network, the placement of each of the eight wires on a RJ45 jack has been specified as a standard called the EIA/TIA 568b standard color code. Figure 2.7 shows the positions at which each of the eight wires in a standard Cat5e cable terminates in a standard RJ45 jack. When both ends of the cable are terminated as shown in Figure 2.7, we get a "straight-through" cable, which is appropriate for Ethernet use. The wire positions in Figure 2.7 may be compared to those in the bottom view of Figure 2.6. Table 2.2 shows the pin positions and functions of each wire.

In commercial use, UTP cable with RJ45 jacks has many advantages for desktop use. The cable is flexible and can be easily wired into any corner of any home or office. The cable is quite rugged and is not likely to break when stepped on, even in high-traffic areas. In short segments, UTP cable is relatively light and can be carried by hand, without any heavy equipment.

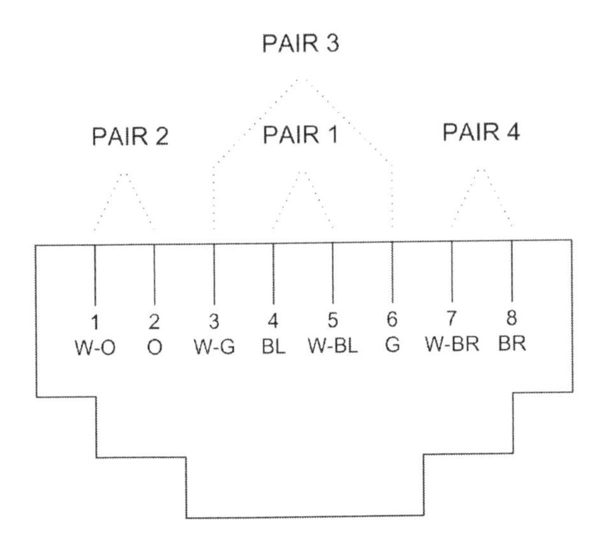

Figure 2.7: Cat5e wire positions on RJ45 plug

Table 2.2: Data transmission on Cat5e

Conductor ID	Pin position	Color	Use (100 Mbps Ethernet)	Use (1 Gbps Ethernet)	Use (Power over Ethernet)
Pair 1	5	White-Blue/	None	Send/receive	+ 48 volts, 25.5W*
	4	Blue			
Pair 2	1	White-Orange/	Transmit/ receive data	Send/receive	Receive data
	2	Orange			
Pair 3	3	White-Green/	Transmit/ receive data	Send/receive	Transmit data
	6	Green			
Pair 4	7	White-Brown/	None	Send/receive	- 48 volts, 25.5W
	8	Brown			

* Defined by the 802.3at standard.

The RJ45 plug uses an extremely simple press-and-click mechanism to connect to communication equipment. Since this design is extremely user friendly, even techno-phobic users can connect printers, home routers, and computers using off-the-shelf cables with RJ45 jacks. This lowers network maintenance costs.

Straight-through vs. crossover cables

Cat5 cables can be wired in two ways—straight through and crossover. The connection scheme shown above is for crossover cables. Straight-through cables connect pin 1 to pin 1, pin 2 to pin 2, and so on. Traditionally, straight-through cables were used to connect different types of devices (e.g. laptops to switches) and crossover cables were used to connect the same type of devices (e.g. laptop to laptop). However, devices are smart today and can use either kind of wiring.

For straight-through cables, the 568b color scheme is used on both sides of the cable. For crossover cables, one end uses the 568a color scheme.

Computer networks have traditionally used only two of the available four pairs of wires. These networks operate at speeds of up to 100 Mbps and two pairs are adequate to provide these speeds. The use of the wires in a Cat5 cable has evolved over the years. Table 2.2 shows how the wires in a Cat5e cable are typically used. As the table shows, only two pairs are in use in most Ethernets (100 Mbps Ethernet). However, all four pairs become necessary to achieve even higher data rates of 1 Gbps and greater. The spare wires can also be used to carry DC power and can power devices that consume up to 13 watts of power. Such devices include surveillance cameras and telephones. Though this appears trivial, in business contexts the savings from eliminating the need for power cables can be substantial.

Optical Fiber as Physical Medium

Whereas copper is wildly popular on the desktop, it has serious disadvantages when used in the core or backbone of the network. The highest data rates that can be attained on a copper cable are limited to about 10 Gbps. Copper also experiences serious signal-reception problems at lengths beyond a few hundred meters. Therefore, long-distance copper cables require repeaters at frequent intervals, which add to the costs of the communication system. Finally, long copper cables can become very heavy, easily weighing a few tons. Optical fiber overcomes these obstacles and has emerged as the most commonly used transmission medium for long-distance communications.

Optical fiber is a thin strand of glass that guides light along its length. Most large offices use optical fiber to bring network connectivity to a central location in buildings, and from there network connectivity is distributed to desktops over copper. Almost every long-distance phone call is carried over optical fiber between the central offices of telecom carriers. In many cities, optical fiber is beginning to replace the traditional copper link to the home. In fact, other than the end-user desktop, optical fiber is becoming the most common transmission medium for every other network link.

Four major factors favor optical fiber over copper—data rates, distance, installation, and costs. It is well known that optical fiber can carry huge amounts of data compared to copper. Less well known is the fact that, whereas signals need periodic amplification on copper links, optical fiber can be run hundreds of miles without the need for signal repeaters. This reduces maintenance costs and improves the reliability of the communication system because repeaters are a common source of failures in networks. Glass is lighter than copper and a single optical fiber cable weighing a few hundred pounds can carry data equivalent to hundreds of copper lines. Therefore, there is less need for specialized heavy-lifting equipment when installing long-distance optical fiber. Finally, optical fiber for typical indoor applications costs approximately a dollar a foot, the same as copper.

Optical fiber may also be more earth-friendly. Whereas copper has to be mined from the earth, optical fiber is made out of glass, which is made out of sand, plentiful in most parts of the world.

Copper continues to be favored on the desktop because it can better handle end-user abuse than fiber. Also, adapters to convert electrical signals inside the computer to light signals on the fiber cable cost in the range of $50. Given that most corporate PCs only cost in the range of $500–$1,000, fiber to the desktop adds 5%–10% to the cost of each desktop. Copper is therefore more economical for the desktop right now. However, it is economical to carry aggregated traffic from multiple desktops over fiber as shown in Figure 2.8.

Modern global communication depends upon vast lengths of optical fiber that have been laid out on the ocean floor to handle global information traffic. The map of submarine cables is shown in Figure 2.9.

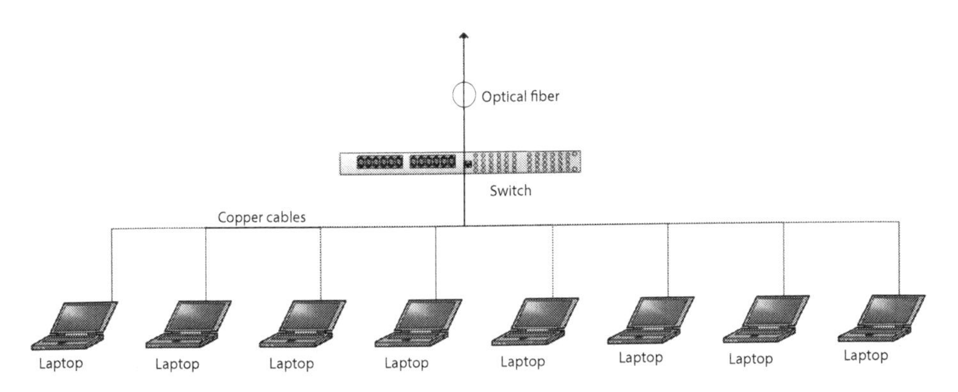

Figure 2.8: Aggregating network traffic from copper to fiber

How Optical Fiber Works—Total Internal Reflection

Optical fiber works on the principle of total internal reflection. You may recall, from experience or other classes, that light bends inward when it passes from air to water due to a phenomenon called refraction. The result is shown in Figure 2.10.

This is the conventional view of refraction because we generally care about the view of the world from air. This helps us understand why spoons appear bent in water, for example. To understand how optical fiber works, though, it is more useful to see the reverse path—what happens when light tries to leave water, or in the case of optical fiber, the glass. The glass used in optical fiber has a very high refractive index and bends the light inward very sharply. If light is bent very sharply at the glass edge going in, light cannot leave the glass edge going out, unless it hits the edge almost exactly at a right angle. The rest of the light is

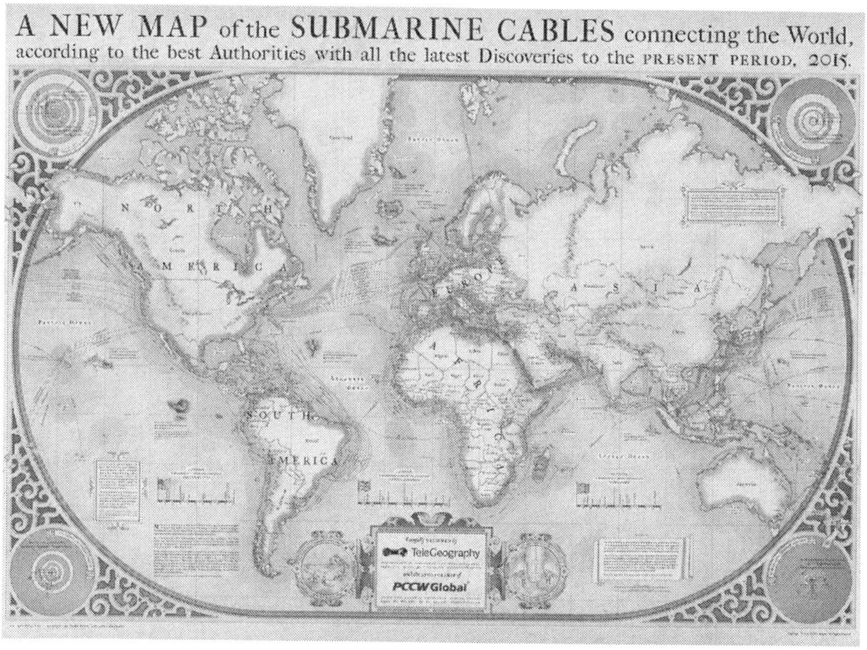

A NEW MAP of the SUBMARINE CABLES connecting the World, according to the best Authorities with all the latest Discoveries to the PRESENT PERIOD, 2015.

Figure 2.9: Map of major submarine optical cables
(Source: Telegeography)

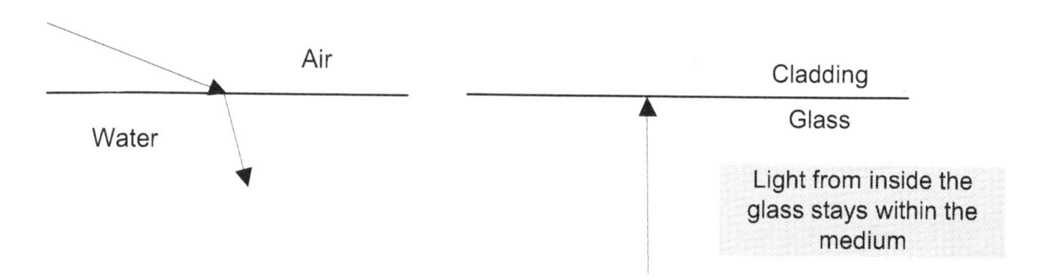

Figure 2.10 (*left*): Total internal reflection (conventional view)
Figure 2.11 (*right*): Total internal reflection (view from inside the fiber)

reflected back. This is called total internal reflection. It is shown in Figure 2.11. Seen from within water, this creates sunbeams as shown in Figure 2.12.

To summarize, total internal reflection makes glass very efficient at carrying signals over large distances because energy (in the form of light) sent into the glass medium is tightly contained within the glass. Unlike copper, which dissipates energy in the signals as heat, total internal reflection makes optical fiber a very efficient carrier of signals as light energy.

Types of Optical Fiber

There are two types of optical fiber in common use—multimode and single mode. Multimode fiber uses LEDs as the light source and can carry signals over shorter distances, about 2 kilometers (1.2 miles). Multimode fiber is usually 62.5 microns in diameter (1 micron is a thousandth of a millimeter). Single-mode fiber is about 10 microns in diameter and can carry signals over distances of tens of miles. Both single-mode and multimode fibers cost approximately the same. Multimode is preferred for shorter distances because the LED light source used in multimode fiber costs less than the laser light source necessary for single-mode fiber.

Figure 2.12: Sunbeam created from total internal reflection
(Source: © istock.com/AH Design Concepts)

Optical fiber and the 2009 Nobel Prize in Physics

Charles Kuen Kao, a Chinese-born Hong Kong, American, and British citizen, shared the 2009 Nobel Prize in Physics for "groundbreaking achievements concerning the transmission of light in fibers for optical communication." Kao, who pioneered the development and use of fiber optics in telecommunications and identified the significance of single-mode in optical fiber communication, is considered the "Godfather of Broadband," or "Father of Fiber Optic Communications." In 2015, there were 180 terabytes per second of Internet bandwidth globally, most of it optical, and it is growing approximately 30% annually.[6] The 2009 physics prize awardees were called the masters of light. The two other recipients created the digital image sensor at AT&T Bell Labs.

Construction of Optical Fiber

Optical fiber used in communication has a central light path surrounded by reinforcements to prevent wear and tear. A typical cross-section is shown in Figure 2.13. At the center of the fiber is the core. This is the glass with a very high refractive index through which the light travels. The core is surrounded by a special glass coating called the cladding. The cladding has a very low refractive index. The high refractive index of the core together with the low refractive index of the cladding restricts the light signal so that it stays within the core and moves along the fiber. The coating and fibers cushion the glass. The jacket makes the fiber easy to handle, adds weatherproofing, and encloses the fibers.

Now that we have seen the media over which signals travel, let us turn our attention to the signals themselves.

6 Global Internet geography, https://www.telegeography.com/research-services/global-internet-geography/ (accessed Dec. 2015).

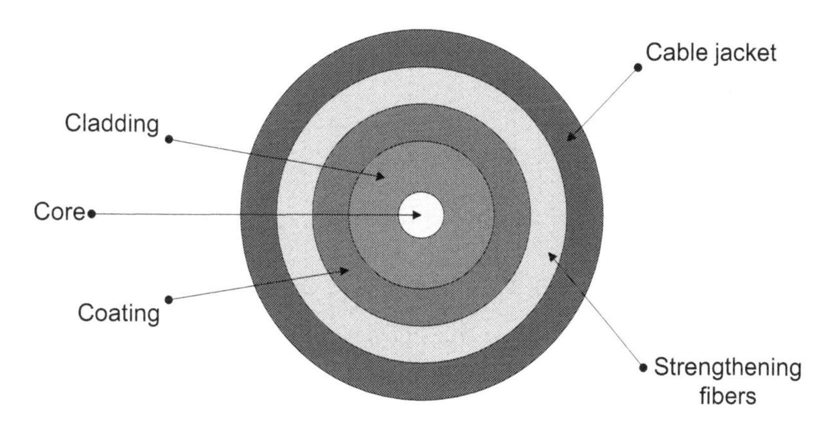

Figure 2.13: Cross-section of optical fiber

Gotchas

Students frequently write on exams that optical fiber is faster than copper because fiber sends data at the speed of light. These students mean to say that since the speed of light is the maximum possible speed in the universe, signal speeds in copper wire are probably much slower than signal speeds in fiber, and as a result optical fiber has higher data rates than copper cables.

However, the electronic signal in the copper wire is also an electro-magnetic signal, just as light is. Both travel at the speed of light. There are only very marginal differences in signal speeds between copper and fiber.

Why then do we say that fiber is faster than copper? The more correct statement would be something like, fiber can carry more data than copper in the same amount of time.

Think of it this way: Consider a two-lane expressway and a four-lane expressway, both with speed limits of 70 mph. Traffic in each lane moves at 70 mph in either expressway. But the four-lane expressway can carry much more traffic than the two-lane expressway in a given amount of time.

Think of copper as the two-lane expressway and fiber as the four-lane expressway. Fiber can transfer much more data in a given amount of time, though signals in both media travel at similar speeds. In fact, the analogy between the number of lanes on a highway and the data-carrying capacity of physical media is quite close. The data-carrying capacity of a medium is proportional to its bandwidth just as the traffic-carrying capacity of an interstate is proportional to its width in number of lanes. The bandwidth of optical fiber is thousands of times greater than the bandwidth of copper, allowing optical fiber to have data rates that are hundreds of times greater than data rates in copper.

Data vs. Signals

Information is the resolution of uncertainty. Claude Shannon, the pioneer of information theory, said that information is surprise, difficulty, and entropy. If only one message is possible in some context, then there is no uncertainty and there's no information in the received message. The physical layer converts data (which codify information) to signals for transmission over physical media such as copper wires or optical fiber. *Data are numbers, letters, or other symbols that can be processed by people or machines*. For example, the text that you are reading right now is data. Our primary interest in using computer networks is to exchange data with other computer users. For example, we may want to inform colleagues about an upcoming meeting. This can be done by sending an e-mail to all meeting participants.

Unfortunately, data cannot be transmitted over wires. Take the e-mail example above. Consider for a minute how you might send the e-mail over a wire. Perhaps you could print the e-mail. But how would you send the printed paper over a wire?

Consider another example that was introduced earlier. Thoughts and ideas are the data in human communication. How do we exchange thoughts and ideas with colleagues? As we all know, mind reading is just not possible (though wouldn't it be great if it were). If it were indeed possible, mind reading would be an example of transmission of raw data. Since such raw transmission of data is not possible, we use speech to share thoughts and ideas with listeners. Speech is a signal used in human communication.

Signals are detectable transmitted energy that can be used to carry information. Data is sent over wires as signals. Senders create changes in energy depending upon the data to be sent. When these changes in energy reach the receiver, the receiver interprets these changes as data. Signals are necessary because data cannot be transmitted directly.

> *How do troops march?*
> *In formation*
>
> Source: *Boys' Life* magazine, May 2010

Earlier in this chapter, we saw that the specific forms of energy to be used as signals depend upon the medium. When creating signals, the physical layer tries to best exploit the properties of the transmission medium used. Electrical energy is used for copper because copper is good at transmitting electrical energy. Light is used with optical fiber because glass is effective at carrying light but not electrical energy. We can now look at some common signals.

Signals and Their Properties

Claude Shannon, about whom you will read shortly, stated that "the fundamental problem of communication is that of reproducing at one point either exactly or approximately a message selected at another point. Frequently the messages have meaning."[7] Communication

---

7 Claude Shannon, "A Mathematical Theory of Communication," *Bell System Technical Journal*, July and Oct. 1948.

is achieved through signals. *Signals are changes in energy that can be detected at the receiver.* We saw an example of the change with the telegraph. Turning the switch on or off created a change in voltage that could be detected at the receiver. Good signals have four desirable properties—easy reception, noise resistance, efficient utilization of bandwidth, and multiplexing.

The first desirable property, of course, is that a signal must be easily detectable at the receiver. Signals that require complicated detectors at the receiver increase system costs. Signals that are easier to detect also require less energy for transmission.

Traffic signals and detectability[8]

Traffic signals use red, amber, and green lights. These colors have been chosen because the eye is especially well designed to detect these colors, perhaps for evolutionary reasons. The eye contains two types of light receptors—rods and cones. There are 120 million rods, which are more sensitive than cones. However, rods are not sensitive to color. There are 6 to 7 million cones, which detect color. It is believed that 64% of the cones detect red, 32% detect green, and 2% detect blue. Using red and green colors therefore makes it easy for drivers to detect the status of traffic lights even from a distance.

Cleveland, Ohio, is credited with installing the world's first electric traffic signal, on August 5, 1914. It had four pairs of red and green lights operated manually but configured such that conflicting signals would not be possible.[9] The first "no left turn" sign emerged later, in Buffalo, New York, in 1916.[10]

The second property of good signals is that they should be good at resisting noise that gets added during transmission. Signals that are affected by noise are difficult for the receiver to detect and at increased risk of transmission errors. Third, good signals are efficient at using bandwidth so that more information can be sent within the available bandwidth of a transmission medium. Finally, good signals make it easy for signals to be multiplexed so that multiple channels of transmission can be created within the same medium.

Just as a surfaced road minimizes road bumps compared to a dirt road, good media such as copper and optical fiber help minimize noise in communication signals. Also, just as well-built cars resist damage from wear and tear, good signaling schemes resist the impact of noise on the signal. In this manner, good media and good signaling schemes combine to counter the impact of noise to the greatest extent possible and increase the data rates and distances over which signals can be transmitted.

Each of these four properties of good signals—easy reception, noise resistance, efficient utilization of bandwidth, and multiplexing—affects our daily lives. Some examples of their impact are provided below to help you understand the vital role of signals in modern technology.

---

8 http://hyperphysics.phy-astr.gsu.edu/hbase/vision/rodcone.html (accessed Dec. 2015).

9 The system was patented by inventor J.B. Hoge as US Patent 1251666.

10 http://www.history.com/this-day-in-history/first-electric-traffic-signal-installed (accessed Jan. 2016).

Examples of the Impact of Improved Signaling on Our Daily Lives

Let us start with the impact of developments in signaling that have improved their detectability. Improvements in cell phones are almost entirely driven by improvements in signaling. The signals used by the earliest generation of cell phones were difficult to detect. As a result, cell phones had to transmit signals at very high power to be able to reach nearby cell phone towers. These cell phones needed large batteries to deliver this power and were therefore not very handy. CDMA signals used in modern phones are much easier to detect and cell phones can use smaller batteries. This has led to the development of extremely sleek cell phones in the last decade.

Batteries are themselves an important constraint in many environments. About a third of the 90 pounds of gear carried by US soldiers are batteries, required to power various viewers and communicators.

Now consider an example of the differences in noise resistance of signals. AM/FM radios illustrate the contrast in noise-resisting capabilities of different signals. FM is better at resisting noise than AM. As a result, you may have experienced that the sound quality on FM channels is far superior to the sound quality on AM channels. You may also have noticed that most music stations broadcast on FM. This is because music stations want to provide the best possible listening experience and prefer the superior noise resistance of FM. By contrast, AM channels are frequented by talk and news programs because on these stations users don't mind an occasional streak of noise.

Improved efficiency in bandwidth utilization by newer signals has led to major changes in traditional media. The recent move by the government to mandate high-definition transmission is motivated by the development of signals that are very efficient at utilizing bandwidth. These signals can pack greater amounts of data than the signals currently used in broadcast TV. Modern signaling techniques combined with digital compression allow almost eight high-definition TV channels to be transmitted in the bandwidth used by one analog TV channel. The freed bandwidth can be sold to cell phone operators for other communication purposes.

Radio and TV illustrate the utility of multiplexing. Multiplexing allows all radio and TV stations in one geographical location to broadcast over the same general area and gives the audience the opportunity to tune to the desired station without interference from other stations. A demonstration of multiplexing is provided later in this chapter.

Categories of Signaling Methods in Data Communications

All signals use energy to create change at the sending end that can be detected at the receiving end. As a simple example, consider the waves created on the surface of a lake when a pebble is dropped at one end of the lake. As the pebble is slowed down by the water, the kinetic energy of the pebble is transferred to the water, creating ripples on the lake. These ripples travel in all directions on the lake and can act as information-carrying signals.

FCC spectrum auctions

The Federal Communications Commission is responsible for managing the wireless spectrum in the US. It held its first spectrum auction in 1994, which raised more than $830 million. As of December 2015, the FCC has raised almost $121 billion from these spectrum auctions. Cell phone service providers have been the primary buyers of this spectrum, and these auctions have enabled the growth of the cell phone industry in the United States.[11, 12]

Contrary to what you may believe, many bidders at these auctions are small investors. For example, over the years bidders have included a retired marketing executive from AT&T, a mailman in Santa Monica, and an emergency room physician in LA.[13]

In data communications, two generic kinds of signals are in common use—digital and analog. Digital signals are created when we turn a switch on or off depending upon whether the data to be sent is a 0 or a 1. Analog signals are created when we take a sine wave and modify some property of the sine wave. Figure 2.14 shows a simple digital signal and an analog signal.

Digital Signals

Digital signals are signals in which discrete steps are used to represent information. Digital signals are intuitively easy to visualize. For example, a positive voltage could signal a 1 and a negative voltage a 0.

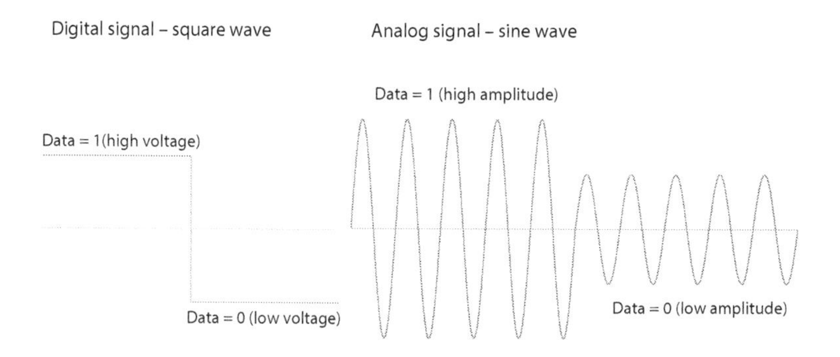

Figure 2.14: Digital and analog signals

11 See http://wireless.fcc.gov/auctions/default.htm?job=auctions_all for details.

12 Spectrum auctions are part of a more general trend towards privatizing spectrum. In the 1980s, the FCC experimented with un-licensing junk spectrum, or frequencies that were not commercially useful. This led to unintended benefits in the form of remote garage-door openers. Innovators used the freed spectrum to create what we now know as Wi-Fi. (Source: L. Gordon Crovitz, "Better broadband is no joke," *Wall Street Journal*, Mar. 22, 2010.

13 Thomas Gryta, "The gold rush hits wireless spectrum," *Wall Street Journal*, Dec. 10, 2013.

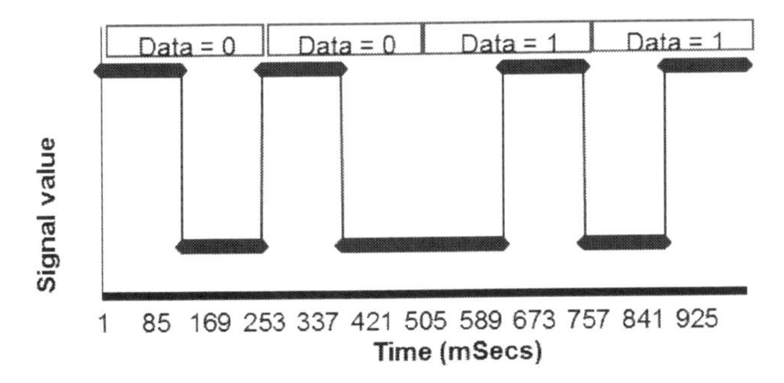

Figure 2.15: Common digital signal (Manchester encoding, used in Ethernet)

In a simple digital signal, a high or low voltage could be used to indicate a 1 or 0. However, such a signal is not used in practice because it becomes difficult to detect bit boundaries. For example, if the data has 50 1's and a high voltage indicates a 1, we would have a high voltage on the line for a long duration. How is the receiver to know whether the sender wanted to send forty-nine 1's or fifty 1's or fifty-one 1's? To avoid this confusion, when digital signals are used, a transition is forced in the signal for each data bit. Therefore, modern networks prefer to deal with changes in energy rather than with absolute energy levels. For example, a transition from a high voltage to a low voltage could mean a data bit of 0 and a transition from a low voltage to a high voltage could mean a 1. This is shown in Figure 2.15.

A bit period is the amount of time required to transmit one bit of data. When the signal includes transitions in every bit period, the sender and receiver do not have to synchronize clocks to count bits. The signals themselves identify when the data bit has been transmitted.

For this reason, modern signals force change in each bit period, as seen in Figure 2.15. The receiver does not have to count bits; it can simply see if the signal is going from high to low, or low to high, and use this to decide if the sender was trying to send a 1 or a 0. Another common change-based signaling technique is to send information over sine waves. Both these schemes are discussed later in the chapter.

Analog Signals

Analog signals are signals that have a continuous nature rather than a pulsed or discrete nature. Analog signals are a little more difficult to visualize and understand than digital signals. Analog signals are created by manipulating sine waves. To begin with, we will address why we use a complicated-looking signal such as a sine wave for signaling. We will see what is so special about sine waves and why they are used for data communications.

It turns out that sine waves are remarkably easy to create. Sine waves are associated in various ways with any spinning object and are the most elementary signals in nature. For example, the height of a point on a spinning wheel is described by a sine wave as shown in Figure 2.16. The power that reaches your home is a sine wave (at 60 Hz) because it is generated by a turbine spinning inside a constant magnetic field. Computers use the electronic equivalent of a spinning wheel to generate sine waves for data communication.

Sine waves have three properties—amplitude, frequency, and phase. Figure 2.17 shows these properties. The amplitude is the height of the wave. The frequency of a wave is the

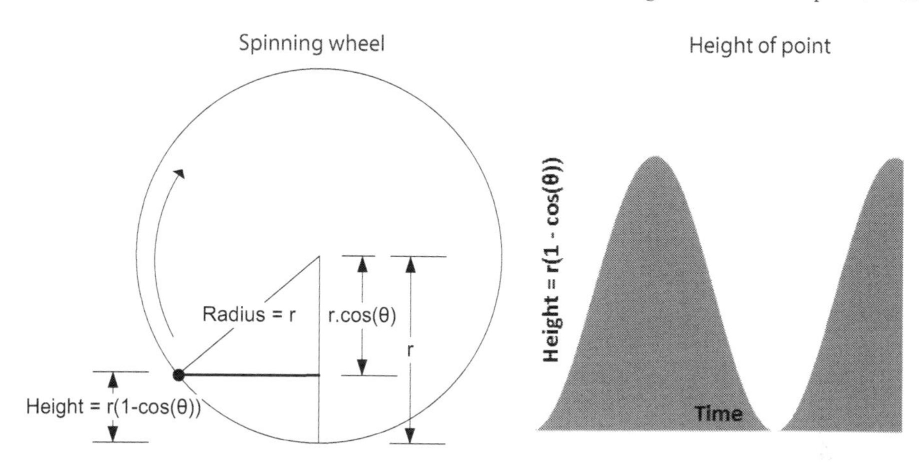

Figure 2.16: Sine wave generated from the height of a point on a spinning wheel

number of complete cycles made by the wave in one second. In Figure 2.17, the wave completes two cycles in one second and is said to have a frequency of 2 Hertz (Hz).[14] The phase of the wave is the position of the wave at the start time. The wave in the example has a phase of 0^0.

To create analog signals, we start by generating a sine wave at the sender end. This wave is called a carrier wave, and it travels from the sender to the receiver over the medium. To transfer information using the carrier wave, we change one or more properties of the carrier wave in response to data. These changes are called modulation. Modulation is detected at the receiving end and is interpreted as data.

Commonly used analog signals in data communication include amplitude modulation (AM) and phase modulation (PM). In amplitude modulation, a data bit of 1 may be sent as a wave of high amplitude and a 0 may be sent as a low-amplitude wave (see Figure 2.18).

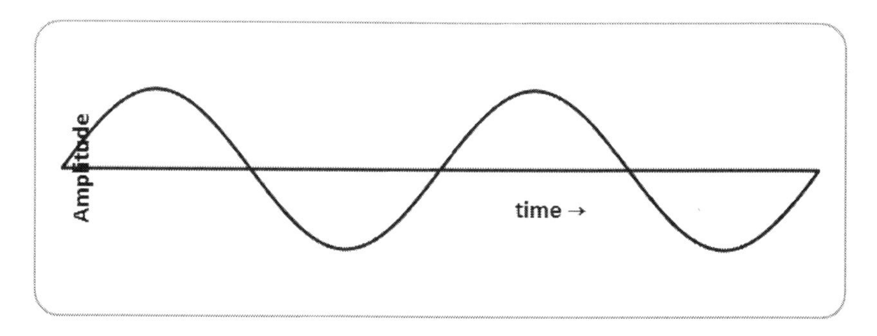

Figure 2.17: Properties of a sine wave

14 We detect frequency as pitch. Higher-pitched signals have higher frequencies than lower-pitched signals. Fun fact: Pitch is critical in Chinese and African languages. For example, alambaka boili [-_--_ _ _] = "he watched the riverbank"; alambaka boili [----_-_] = "he boiled his mother-in-law." Source: James Gleick: *The Information: A History, a Theory, a Flood* (Vintage, 2012).

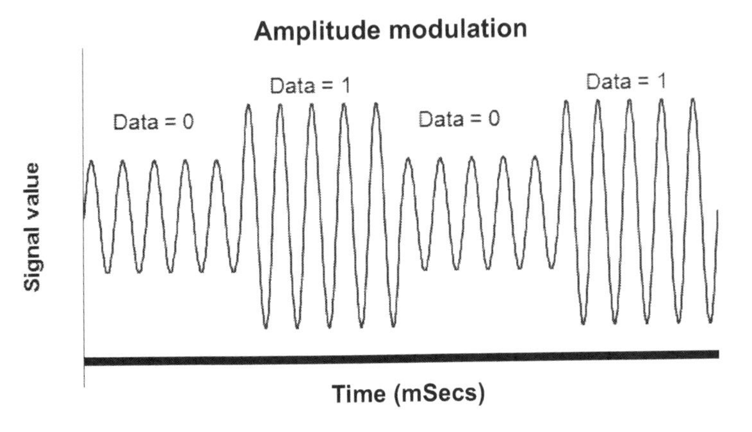

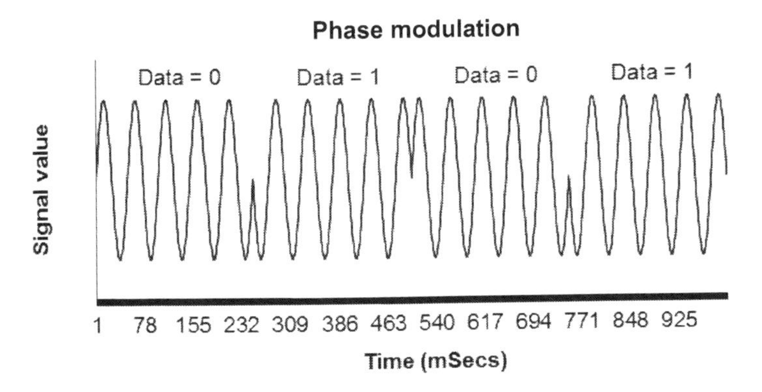

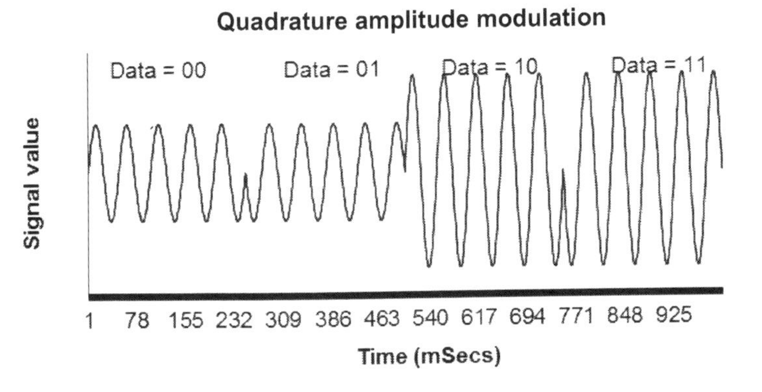

Figure 2.18: Common analog signals

In phase modulation, a wave starting with a phase of 0^0 may indicate a data bit of 0 while a wave starting with a phase of 180^0 could indicate a 1.

Another popular analog signaling technique is called quadrature amplitude modulation (QAM). QAM combines both amplitude modulation and phase modulation as seen in the final example in Figure 2.18. The figure shows how QAM could use two amplitudes and two initial phases to create four possible signaling levels. With four possible signaling levels, we can send 2 bits of data at a time. This is discussed in more detail in Chapter 4. In our example, a low-amplitude signal starting with a phase of 0^0 could indicate the bit combination 00. High amplitude and starting phase of 180^0 indicates the bit combination 11.

Take a look at the logo of the VAIO series of computers made by Sony. The V and A letters are designed to indicate an analog signal and the I and O represent 1 and 0. This logo was created to indicate that the computers were designed to be friendly for multimedia use.[15]

Though QAM is more complex than either AM or PM, its advantage is that it allows faster data transmission. In the examples shown, whereas AM or PM can only send 1 bit of data in one time period, QAM can send 2 bits of data per time period. With more amplitude and phase levels, QAM can send even more data in one time period. For reference, the 802.11ac Wi-Fi technology uses QAM to send 8 bits per time period whereas the QAM used by 802.11n only sends 6 bits per time period. This is one of the ways by which 802.11ac improves performance over 802.11n.[16]

The wave forms shown in Figure 2.18 use digital data to modulate an analog carrier. These modulations are more formally called amplitude shift keying and phase shift keying. The terms analog modulation and phase modulation more commonly refer to the transmission of analog data using analog signals. These were used in non-digital radio and TV broadcasts, and are also shown in the multiplexing example later in this chapter. The figures do not show analog modulation because it is visually more complex.

Impact of Noise and the Importance of Binary Signals

Before we leave the topic of signals, it is useful to spend a few moments considering the impact of noise on signals. *Noise is any disturbance that interferes with the normal operation of a device.* Noise is one of the most important constraints in signaling. As the signal travels through the medium, noise keeps getting added to the signal. Noise comes from various sources, such as cross-talk and heating of the wire from the signal flowing through it. The result is that, though the signal keeps getting weaker with distance, noise keeps accumulating. Eventually, at some distance from the source, it becomes very difficult to detect the signal from the background noise. You may have experienced this deterioration of the signal and

15 http://imjustcreative.com/the-meaning-behind-the-sony-vaio-logo/2011/08/11/.

16 Wi-Fi technologies are covered in the chapter on wireless (supplement).

the accumulation of noise while driving away from your home town while listening to one of your local stations on your car radio. Significant deterioration of the signal begins about 50 miles from your city.

As an example of the impact of noise, consider a simple digital signaling scheme where a high voltage represents a 1 and a low voltage represents a 0 (this is analogous to the signaling scheme used in the example of the telegraph). Let us now add noise to this signal. Figure 2.19 shows how the original digital signal from the sender might appear to a receiver after the addition of noise. The figure shows 4 bit periods with the sender sending a 1, followed by a 0, a 1, and then another 0. Observe how the sender has sent a clear signal but the received signal has only a slight resemblance to the transmitted signal.

Figure 2.19 shows the challenge receivers face when processing incoming signals. In the example, how will the receiver recover the data from the noisy signal it receives? A common technique is to average the signal value over the bit period. If the average is high, say any positive value, the receiver could interpret the signal as a 1 and if the average is negative, the receiver can interpret it as a 0. Clearly, when the noise level becomes very high, the means for both 0s and 1s will become similar and it will be difficult to detect 0s from 1s in the signal.

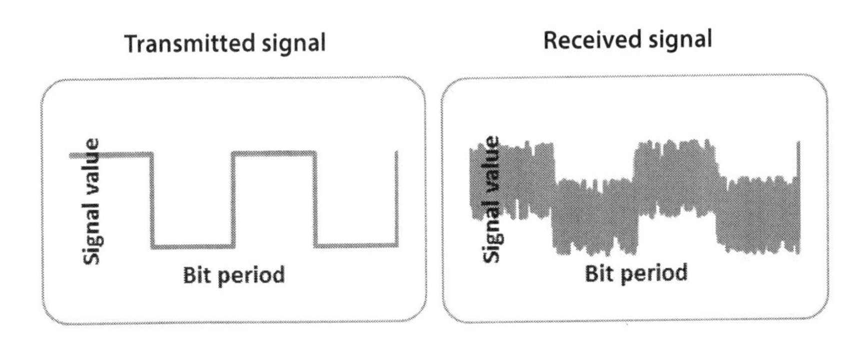

Figure 2.19: Impact of noise on a digital signal

The example also helps show why computers prefer to use binary numbers to represent data. Compared to other forms of representing data, such as decimal numbers, binary numbers help improve the reliability of reception in the presence of noise. To see why this is so, consider the example in Figure 2.20, this time using an AM signal.

In Figure 2.20, the left column shows the transmitted data and the right column shows the received data. The received data includes the effect of noise. The first row shows a signal with two possible values (binary signal). The second row shows a signal with three possible values (ternary signal).

To convert the received signals back to data, we would apply a rule similar to the one used in the digital example. In the binary signal we might decide to interpret the signal as a 0 if the mean amplitude in one bit period is less than some value (shown as a gridline in the top right cell of Figure 2.20) and 1 if the mean amplitude in a bit period is higher than the gridline.

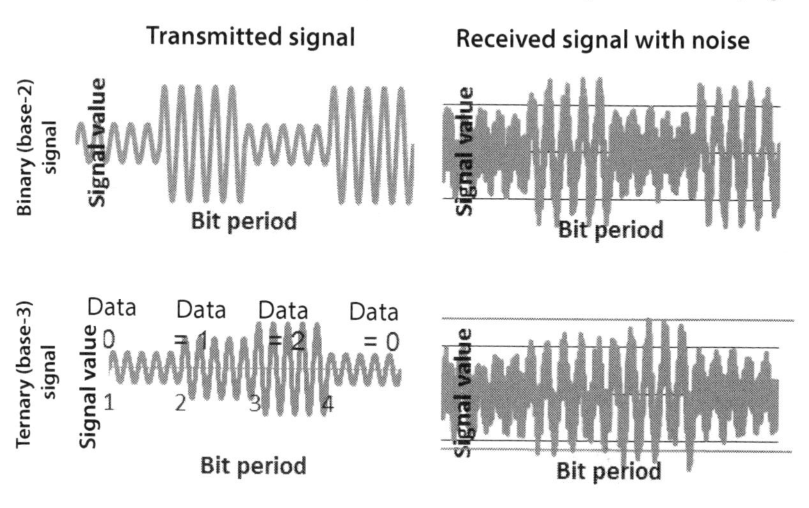

Figure 2.20: Binary and ternary signals

If we use base 3, each data bit can have three values—0, 1, and 2. To represent these, the signal needs to take three values as well. We can implement these three values as low, medium, and high. To convert a ternary signal back to data, the rules get more complex. We need two separators now. We would use a rule such as: If the amplitude is less than the first gridline, we interpret the signal to be a 0; if the amplitude exceeds the first gridline but is less than the second gridline, we will interpret the signal as a 1; and if the amplitude exceeds the second gridline, the signal represents a 2. This is clearly more complex than receiving a binary signal.

Signal conflict and noise

This section is extremely relevant to wireless communication. Any signal that is not associated with an ongoing communication acts as noise. For effective communication, therefore, it is necessary to ensure that there are no other (or as few as possible) nearby signals at the frequency being used for the communication. One way this is done is by allocating frequencies to specific user groups (Figure 2.23).

Which signal would you prefer to work with—the binary or the ternary signal? The advantage of the ternary signal is that it allows us to send more information in one bit period compared to the binary signal. However, it is also likely to lead to greater errors at the receiver end compared to the binary signal. This is because the differences between the signals representing the values of 0, 1, and 2 are less noticeable in the ternary signal than in the binary signal. For this reason, computers represent data in binary form. Binary representations are the most reliable form of representing data. The unit of information in a computer is called a binary digit or a bit. *A bit is a unit of information that designates one of two possible states of anything that conveys information.*[17] A bit usually carries too little information to be of practical use. As we will see later in this chapter, using 8 bits as a group allows us to represent all the letters in the English language. All computer systems therefore use 8-bit groups, called bytes, to represent data. A sequence of 8 contiguous bits, considered as a unit, is called a byte. A byte is usually written as B (in upper case), whereas bits are written as b (in lower case).

What is the data rate of your Internet service? Is the advertised data rate in bits/second, or bytes/second?

Shannon's Theorem

The discussion in the previous section is stated formally as the Shannon-Hartley theorem, which quantifies the maximum data-transmission rate over a communication channel of a specified bandwidth in the presence of a specified level of noise. The relationship is written as

$$C = B.log_2\left(1+\frac{S}{N}\right)$$

where C is the channel capacity in bits/second, B is the channel bandwidth in hertz, S and N are the signal and noise levels in watts. The ratio S/N is commonly known as the signal-to-noise ratio (SNR) and represents how much of the received information is useful.[18] The channel capacity of a medium is one of the most critical properties of a medium in determining its utility for data communication.

Data rates and bandwidths

It is true that some media are faster than others because the signals travel faster in some media than others. This is true when you compare, say, sound and light, but when we are talking electronic data rates, the signals travel at approximately the same speed in all media (the speed of light). The differences in data rates come from the greater parallelism of signals in a medium with a higher bandwidth.

The difference is akin to the differences in passenger capacity between a car and a bus. Both travel at the same speed (the car may actually drive faster since buses often have speed regulations), but a bus can carry 50 passengers whereas a car can carry perhaps 5. However, for a bus to operate between two locations requires wider roads throughout the span between the locations. Higher-bandwidth media such as optical fiber may be seen as bus lanes, and lower-capacity media may be seen as car lanes.

17 The bit is now widely recognized as a fundamental unit of measure, a unit for measuring information, relevant even to physicists in the discussion of black holes in the universe. It was introduced in Shannon's landmark 1948 paper, "A Mathematical Theory of Communication," cited in note 7 in this chapter.

18 Businesses use the term SNR more broadly to represent the utility of incoming information. For example, the book, *The Signal and the Noise: Why So Many Predictions Fail—But Some Don't*, by Nate Silver (Penguin, 2012), provides many examples of extracting useful information (signal) from the torrents of incoming information (noise).

Hertz

The unit of frequency is hertz (Hz), and it is defined at one cycle per second. Hertz represents the speed at which any repeating phenomenon recurs. For example, an indicator light blinking two times per second can be said to be operating at 2 Hz.

Communication media are usually capable of efficiently transmitting signals within specific ranges. For example, the phone wire can efficiently transmit signals in the frequency range of 0 Hz–3,400 Hz. The difference between the upper and lower frequency limits is called the bandwidth of the channel. For example, the bandwidth of the phone channel is 3,400 Hz - 0 Hz = 3,400 Hz. As seen from the channel capacity theorem, the ability of a communication medium to carry information increases in direct proportion to its bandwidth.

Relative channel capacities of common communication channels

The bandwidths of some popular communication channels are listed below. Assuming the same signal-to-noise ratio in all channels of 1,000, estimate the data rates of each of the channels below. For reference, log_2 (1+1,000)=9.967, which can be approximated as 10.

Channel	Bandwidth	Data rate
Phone line	3,000 Hz	?
Cat5 cable	100 MHz	?
Single-mode fiber	20 GHz	?
TV channel	6 MHz	?

Transmission and Reception of Data Using Signals

Before we close our discussion of signals, let us integrate all the ideas in this chapter to see how signals might be used in practice to send data. Let us consider a simple example. We will send the word "hello" using a simplified version of amplitude modulation. In this simplified version of AM, high amplitude will represent a 1 and a silent signal will represent a 0. This simplification is done to keep the plots as simple as possible.

Cody: How do you measure happiness?
Chuck: How?
Cody: In Giggle Bytes!

Source: *Boys' Life* magazine, March 2014

Transmission and reception of data is performed in four steps. At the sender end, (1) the data is coded as binary numbers representing the data, and (2) the carrier signal is modulated as specified by the binary representation of the data. These steps are inverted at the receiving end where (3) the incoming signal is demodulated into the corresponding binary numbers and (4) the binary numbers are decoded into the data.

Coding Data

Coding is the transformation of elements of one set to elements of another set. For example, representing characters as numbers is a form of coding. There are two widely used schemes to code data as binary numbers. The simpler scheme is called the American Standard Code for Information Interchange (ASCII). ASCII is used to encode the characters in the English alphabet.[19] In the ASCII scheme, the numbers 48–57 represent the numbers 0–9; 65–90 represent the upper case letters *A–Z*; and the numbers 97–122 represent the lower case characters *a–z*. Other numbers represent punctuation and other characters.

The second scheme is Unicode. Unicode represents characters in almost all languages. At last count, it defined over 100,000 characters from languages including Chinese and Korean.[20] English characters are represented in Unicode by the same codes as ASCII codes. Other Unicode characters include 945–969 for the Greek characters α–ω. Displaying these characters requires software capable of processing these characters.[21] Fortunately, most modern web browsers, mobile applications, and word processors are capable of doing so.

Business school fraternities in Unicode

Some of the popular business school fraternities are listed below. Look up their Unicode codes. Are you a member of any?

Fraternity: Pi Sigma Epsilon	Unicode: 928 931 917
Fraternity: Delta Sigma Pi	Unicode: ?
Fraternity: Phi Chi Theta	Unicode: ?

For our example, we use the simple ASCII coding scheme. In ASCII, the letters in the word "hello" are coded as shown in the center column of Table 2.3. In the third column of the table, the decimal numbers are converted to the corresponding binary representation. If you are not familiar with decimal-to-binary conversion, do not worry about it now. It is covered in the binary numbers overview section of Chapter 4, in connection with IP addresses. For now, simply accept column 3 of Table 2.3.

19 http://www.asciitable.com/.

20 A good site to search for Unicode characters is http://symbolcodes.tlt.psu.edu/bylanguage/greekchart.html.

21 This is a more general phenomenon prevalent in the modern world—we have created numerous symbols that we cannot understand without appropriate aids—compressed files and encrypted files, for example.

Table 2.3: Binary representation of example data

Letter	ASCII code	Binary representation
h	104	01101000
e	101	01100101
l	108	01101100
o	111	01101111

So, to send the word "hello," if both sides (sender and receiver) agree to use the ASCII code for encoding data, we would transmit the word "hello" as:

01101000	01100101	01101100	01101100	01101111
h	e	l	l	o

Modulating Carrier

After step 1 of the four-step transmission and reception process is completed, it is relatively straightforward to modulate a carrier signal according to this binary data. A signal corresponding to each bit of data is transmitted for one bit-period. The resulting signal is shown in Figure 2.21. Note how the carrier is sent with a high amplitude when 1 is to be sent and nothing is sent (low amplitude) when a 0 is to be sent.

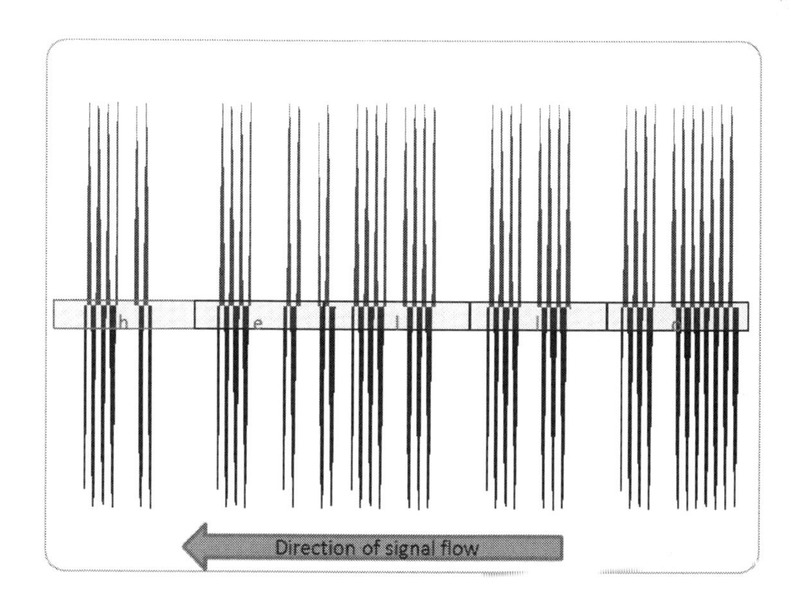

Figure 2.21: Amplitude modulated signal representing example data

In the figure, the oscillating signal is the transmitted signal. This is the signal that will be carried by the medium to the receiver. For reference, the characters being transmitted (*h*, *e*, *l*, *l*, *o*) and the bit-level breakdown of each of these characters are shown near the axis. For example, *h* is represented as 01101000, and, in the first bit period, we send a signal with no amplitude for one bit period. For the next two bit periods, we send a sine wave with high amplitude to send the two 1s, and then a signal with no amplitude, and so on. This completes step 2 of our four-step transmission and reception process.

Demodulating Signal

You probably can now anticipate what the receiver will do. As the signal arrives at the receiving end, the receiver interprets the signal amplitudes for each bit period as either a 0 or a 1. If no signal is received for a bit period, the receiver interprets that as a 0. If a wave is received during a bit period, the receiver interprets the signal as a 1. This completes step 3 of the transmission-reception process.

Decoding Data

Finally, in the last step, the receiver accumulates the bits in blocks of 8 bits each. When the first set of 8 bits is collected, the receiver compares the received bits 01101000 to the ASCII table and interprets the bits as the letter *h*. Similarly, it interprets the remaining bits as the characters *e*, *l*, *l*, and *o*. This gives us the word "hello" at the receiving end.

Again, please remember, signaling is used only because direct transfer of data is not possible and energy in the form of signals is necessary to transport data over the air, wire, or fiber. The complex transformation of data to signals and back would not be done if somehow it were possible to directly transfer data over the medium.

Multiplexing

Before we end the chapter, we consider an extremely useful property of signals—multiplexing. You may recall that multiplexing was introduced as a desirable property of signals. *Multiplexing is the combination of two or more information channels over a common medium.* Examples of multiplexing include the transmission of multiple TV channels over the same coaxial (TV) cable, or the transmission of multiple radio stations in the same air space over a metro area. As shown in Chapter 1, multiplexing is an extremely important milestone in data communication. Multiplexing is what made data communication economical because it allowed multiple phone, TV, and data signals to be combined for transmission over the same cable. This drastically reduced the need for laying out communication cables to homes and businesses. Since cabling costs are one of the biggest costs in data communication, a reduction in cabling costs significantly reduced the overall costs of setting up and maintaining a communication network.

Figure 2.22 provides a visual example of the usefulness of multiplexing and switching by showing a picture of the world before multiplexing and switching. The figure shows over 150 cables over Broadway in NYC around the year 1900. Each cable provided one telephone connection. As telephony and telegraphy gained popularity, all major streets

Figure 2.22: New York City, around 1900

in all metro cities were blanketed by such cables. When we use multiplexing and switching, trunk lines such as this set of 150 cables, can be replaced by one cable, vastly reducing installation and maintenance costs.

In general, there are two categories of multiplexing. The first is used with analog signals and is called frequency-division multiplexing (FDM). As the name suggests, in this form of multiplexing, signals from one channel are sent at one frequency and signals from another channel are sent at another frequency. Fortunately, as a result of the properties of sine waves, signals at different frequencies do not interfere with each other even when they overlap geographically. To receive signals in one channel, the receiver tunes into the sender's frequency. This is what you do when you tune to a station on a car radio.

Signals at different frequencies do not conflict with each other, but signals at the same frequencies do. The government coordinates the use of available frequencies by different organizations to eliminate conflicting uses of the same frequencies. In the US, many frequencies have been reallocated following spectrum auctions. Allocations as of 2011 (the most recent available) are shown in Figure 2.23.[22]

The second category of multiplexing is called time-division multiplexing (TDM) and is used with digital signals. In this form of multiplexing, signals from different streams are sent at different time slots. Receivers scan the medium at their respective time slots to receive their signal.

Though multiplexing sounds complex, it can be relatively easy to do. In the example that follows, we show an example of multiplexing two signals using amplitude modulation. For the rest of this section, it will be convenient to open the spreadsheet available at the link for "multiplexing" in Chapter 2 on the companion website. You can open the charts on the spreadsheet to see the graphs in larger scale.

22 The previous chart was for 2003, and can be found online for comparison.

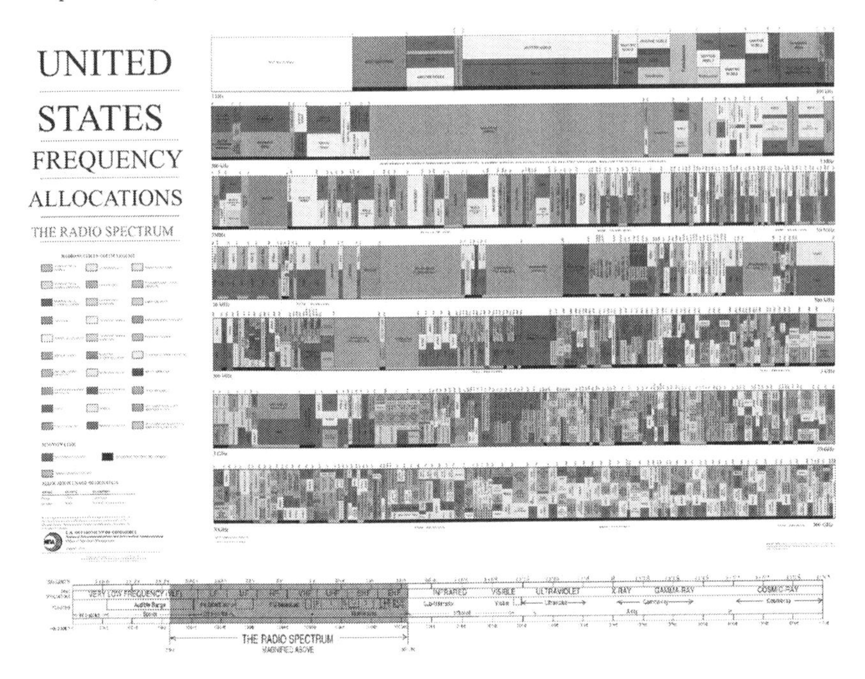

Figure 2.23: Frequency allocation chart (2011)

In the example, two stations wish to transmit signals. Each station has been allocated a carrier frequency. In Figure 2.24 below, the first column shows the activity at station 1 and the second column shows the activity at station 2. The rows show the signals, carriers, and modulated signals from the two stations. From the first row in the figure, observe that the two carriers are at different frequencies (the carrier wave in the first column has more waves than the carrier in the second column, though both cover the same amount of time). It is this difference in carrier frequencies that enables receivers to distinguish between the two signals.

Implications of carrier frequencies[23]

Carrier frequencies have implications. Signals at different frequencies behave differently when propagating through the atmosphere, propagating over the earth's surface, dealing with humidity, etc. For example, cordless phone signals travel about feet, Wi-Fi signals travel about 100 feet, cell tower signals travel about 20 miles, and amateur radio signals can travel around the globe. Besides, higher-frequency signals typically require shorter antennas, which can be very useful for consumer devices.

23 A good link is http://www.tele.soumu.go.jp/e/adm/freq/search/myuse/summary/ (principal uses and characteristics of radio wave, accessed Nov. 14, 2015).

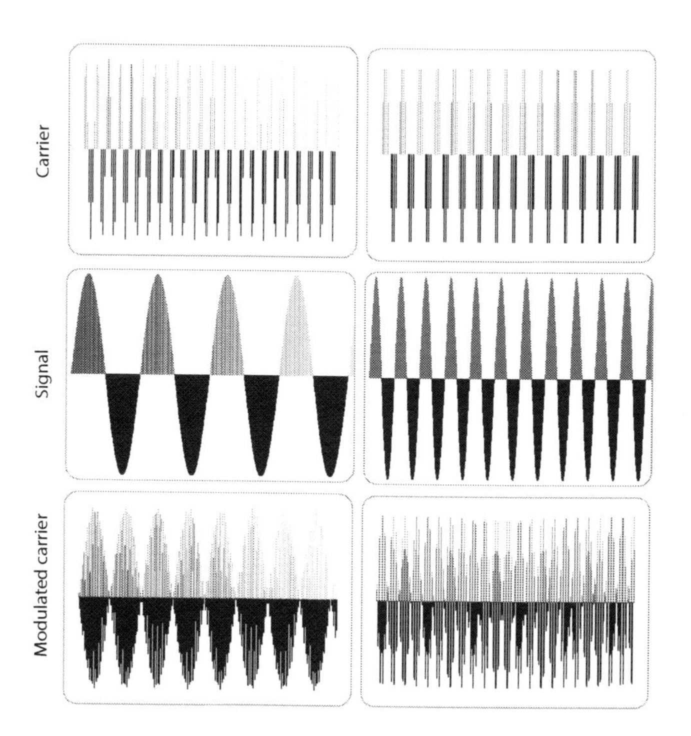

Figure 2.24: AM multiplexing example

In the second row, we show the signals to be carried. Stations can transmit any information; however, for the purposes of this example, each station has a plain sine wave to transmit. The signals from the two stations are at different frequencies for easy reference in the example.

The third row shows the modulated signals. This is an example of AM using analog data.

Comparing amplitude modulation of digital data and analog data

You may like to contrast the modulated carrier in Figure 2.24 to the amplitude modulated carrier in Figure 2.18. In Figure 2.18, digital data was used to modulate an analog carrier wave, and in Figure 2.24, analog information is used to modulate an analog carrier. Observe how the modulated signal in both cases is a sine wave with the frequency of the carrier wave, but whose amplitude varies with the amplitude of the transmitted signal.

Once each station has created the modulated signal, it simply transmits the signal into the shared medium. Here, all the signals get added up. The result for our example is shown in Figure 2.25. The signal in the figure is just the sum of the two signals in the last row of Figure 2.24.

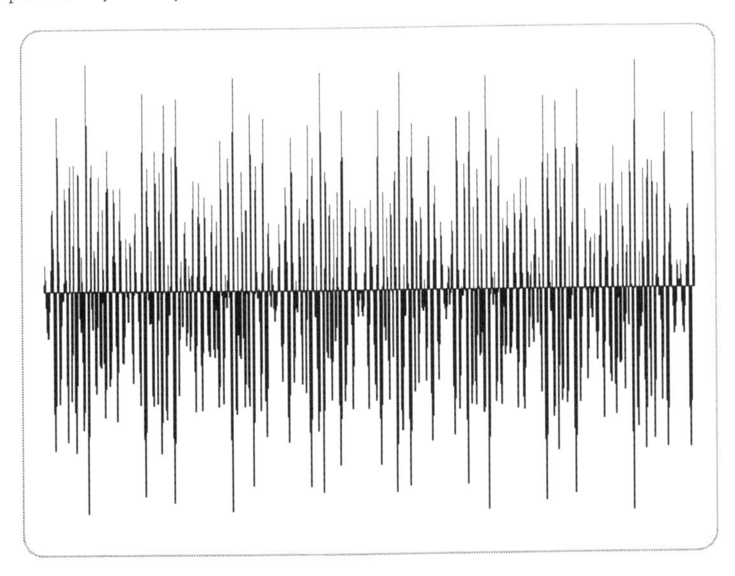

Figure 2.25: Resultant signal in the medium (sum of both modulated carriers)

So, the signal seen in Figure 2.25 is the result of our attempt to transmit two signals in a shared medium. Take a look at the signal. Can you discern signal 1 or signal 2? Probably not. Though the constituent signals are not readily apparent in this figure, we will now see how they can be recovered quite easily.

When using AM, the secret to recovering the original signals is to multiply the signal in the shared medium with a locally generated sine wave at the same frequency as the carrier. In a radio receiver, you do this when you tune your radio to a station. Your radio receiver generates a local sine wave at the selected frequency and multiplies it with the signal in the air. In the multiplexing spreadsheet, you see this in the columns titled "demodulated" of the "data" worksheet. As a result of the special properties of sine waves, this simple operation retains the modulated carrier from the selected station and eliminates all other signals. The results for the two stations are shown in Figure 2.26. Note how the result already approximates the signals sent by the two stations. The only task left is to remove the "rough edges" around the signals.

To remove these rough edges, the standard technique is to average out the signal.[24] The result is shown in Figure 2.27. We can see that we have been successful in recovering the signals sent by the two senders (row 2 of Figure 2.24). The rough edges in the output compared to the transmitted signal in Figure 2.24 are due to numerical approximations in spreadsheet calculations. These are similar to the effects of noise on signals.

This example shows how a few simple operations allow large numbers of senders to share a common medium to send data to their respective receivers without mutual interference.

24 This is formally called low-pass filtering.

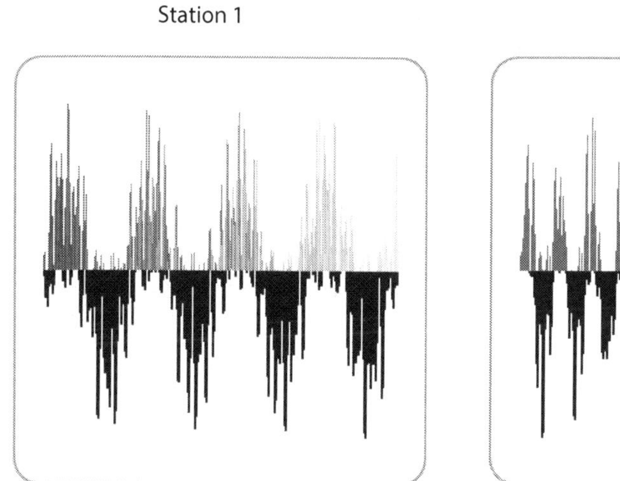

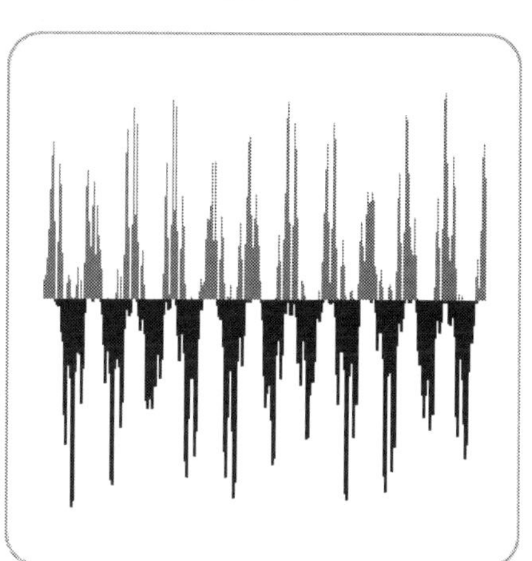

Figure 2.26: Result of demodulation at receiver end

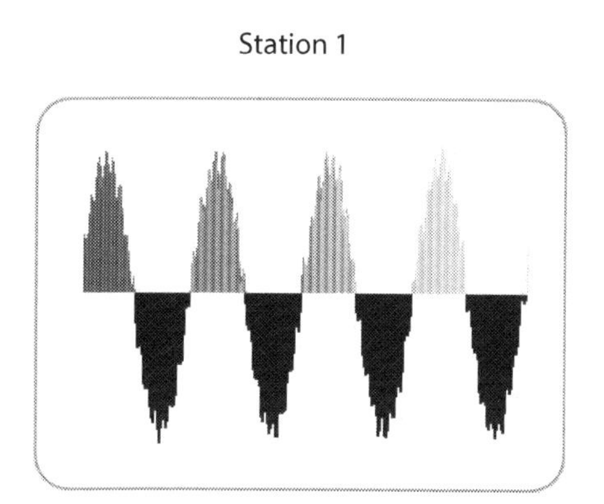

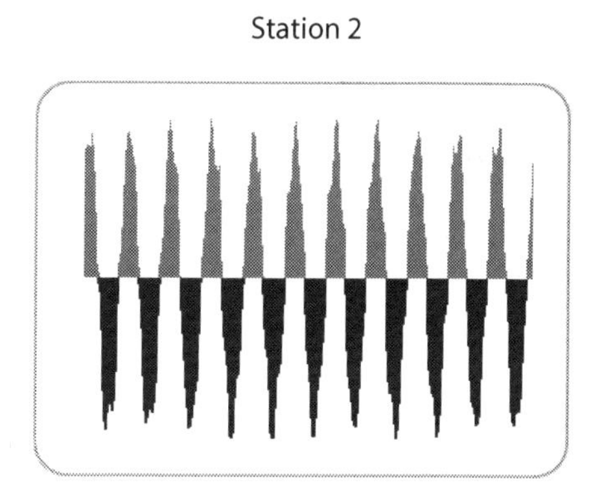

Figure 2.27: Recovered signals after removing noise by averaging

Multiplexing is used in one form or another in almost every electronic communication—telephone, television, cell phones, Internet access, etc.

Summary

In this chapter we described the role of the physical layer in data communication. The layer is responsible for sending data as signals between adjacent devices on a communication network. Signals travel over a transmission medium such as copper wire. Signals are necessary in data communication because data cannot be transferred over wires. We described some reasons why copper and optical fiber are popular transmission media. Common analog and digital signals were introduced. The effect of noise on these signals was described. An example of transmitting a word demonstrated how coding schemes and modulation help transmit data as signals. Finally, an example of multiplexing showed how different signals can be combined and separated.

The physical layer transmits data as best it can over the physical medium. Nevertheless, transmission is error-prone. One of the functions of the data-link layer is to remove these errors. In the next chapter, we look at the data-link layer to see how this is done.

About the Colophon

The Prussian statesman Otto von Bismarck defined politics as the art of achieving the possible. The 32nd president of the United States, Franklin D. Roosevelt, put this idea into words when he said that it was necessary to compromise between the ideal and the practical to accomplish almost anything worthwhile.

The physical layer seems to embody this idea of compromising between the ideal and the practical. The ideal method to accomplish the worthwhile goal of data communications would be to transfer data without the need for complex transformations between data and signals. However, the practical reality is that media cannot transfer data. Media can only transfer signals. And so, in data communications as in politics, we compromise between the ideal and the practical and convert data to signals for just as long as is necessary for data transfer over physical media.

For most students, learning about coding, signals, and modulation is not fun. But hopefully you appreciate that the practical realities of data communication force us to know at least the rudiments of these subjects.

EXAMPLE CASE—Smart Grids

Everyone is acutely aware of rising energy costs and the prospect that the world may soon run out of easy access to oil, no matter what contemporary prices indicate. One of the most promising solutions to the problem uses the power line as the physical medium of a computer network that links power producers and consumers in order to exchange information and optimize power utilization. It is estimated that the system has the potential to eliminate almost half of America's oil imports and hundreds of power plants. However, there are challenges to creating such a system. The challenges are centered primarily on the fact that the power grid was not designed to work as the physical layer of a computer network. Recent developments in physical layer technologies are helping overcome these challenges.

To celebrate the beginning of the 21st century, the US National Academy of Engineering carried out an exercise to identify the single most important engineering achievement of the 20th century. The Internet took 13th place on this list, and highways took 11th. At the top of the list, as the most significant engineering achievement of the 20th century, was electrification, as made possible by the grid. As summarized by the NAE, "Scores of times each day, with the merest flick of a finger, each one of us taps into vast sources of energy—deep veins of coal and great reservoirs of oil, sweeping winds and rushing waters, the hidden power of the atom and the radiance of the Sun itself—all transformed into electricity, the workhorse of the modern world."

The electric power infrastructure of a country is often called "the grid." The United States' grid consists of more than 9,000 electric generating units with more than 1,000,000 megawatts of generating capacity connected to more than 300,000 miles of transmission lines. With rising population, bigger homes, and more and bigger appliances, even this infrastructure is running up against its limitations. Efforts to upgrade the grid are based on the idea of "smart grids." *The smart grid is the system that delivers electricity from suppliers to consumers using digital technology to save energy, reduce cost, and increase reliability and transparency.* The smart grid allows informed participation by customers, integrates all generation and storage options, and optimizes asset utilization. The smart grid will be capable of monitoring power generation, customer preferences, and even individual appliances. Bi-directional flows of energy and two-way communication and control capabilities on the smart grid will enable an array of new functionalities and applications. Figure 2.28 is the conceptual model for information and power flows among the key players in the smart grid.

How can the smart grid and data communications lower energy costs? Between 30% and 50% of the typical power bill goes toward the transmission and distribution charge, which pays for the grid infrastructure. An estimated 10% of all generation infrastructure and 25% of distribution infrastructure is used less than 5% of the time, roughly 400 hours per year. By smoothing power demand, smart-grid technologies will eliminate the need for investments that are only marginally utilized. As an example, if all cars in the United States could run on batteries that were charged at night, the idle capacity in the power grid at night could meet 70% of the energy needs of cars and light trucks. This would improve asset utilization of generating stations and transmission lines. Improved asset utilization would lower energy costs for all consumers.

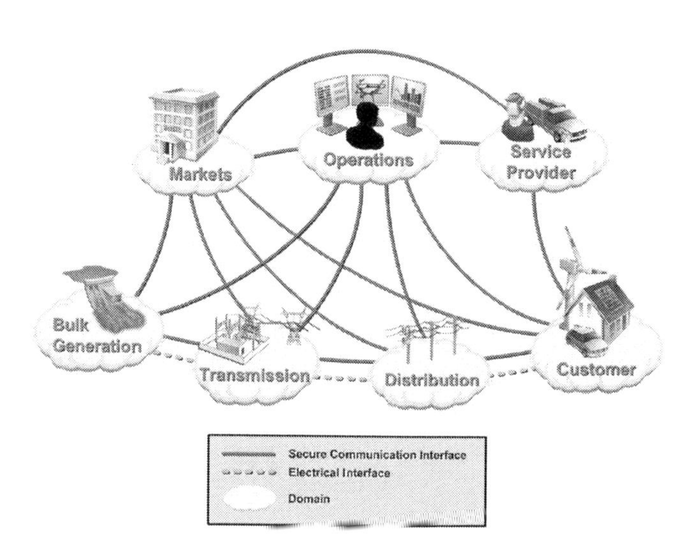

Figure 2.28: NIST smart-grid conceptual model

Smooth demand also means that power requirements can be met with lower-cost sources of power, such as nuclear power and coal. In addition, the smart grid will allow the integration of power-generating resources of every description: rooftop solar cells, fuel cells, and electric vehicles. In emergencies, communities will be able to generate sufficient electricity to keep essential services up and running even when the community is disconnected from the grid.

The Physical Layer as a Critical Component of the Smart Grid

One of the primary challenges in making the smart grid work is to reliably transmit demand and price signals over power lines in the presence of the extremely intense and noisy power signal. In fact, of the five technologies identified by the Department of Energy as drivers of the smart grid (communications, sensing, components, controls, and interfaces), all technologies except components focus on the capture, transmission, and processing of information in real time. All this information is transmitted over the power line, using the electrical wiring as the physical medium for data transmission. This takes advantage of the large installed base of electrical wiring in homes and businesses for signal transmission.

To accomplish these goals, physical layer technologies such as multiplexing and modulation that were introduced in this chapter are key areas of R&D for the smart grid. To multiplex data signals along with power, power line communication technologies have been allocated two frequency bands on power lines: 3 kHz–148 kHz for low-data-rate applications and 2 MHz–28 MHz for high-data-rate applications.

Since power signals are transmitted at considerably higher energy levels than data signals, electrical impulses (noise) generated when electrical appliances are switched on or off can overwhelm the low-power data signals. Further, the electrical meters and circuit breakers installed on power lines are not designed to facilitate data transmission and can degrade signals that pass through these devices. Therefore, as data-transmission functions are added to the power grid, these existing constraints have to be attended to. New modulation techniques, such as differential quadrature-phase shift keying (DQPSK) and orthogonal frequency-division multiplexing (OFDM) are helping to overcome these constraints.

The National Institute for Standards and Technology (NIST) has been assigned primary responsibility for coordinating development of protocols and standards for interoperability of smart-grid devices. NIST is coordinating with a number of groups that have been working on physical layer technology standards for home automation, home networking, and related technologies. These technologies include Homeplug (data rates up to approximately 200 Mbps), Homegrid/ITU G.hn (data rates up to 1 Gbps), IEEE 802.3 (Ethernet, covered in Chapter 3), and IEEE P1901 (data rates greater than 100 Mbps).

If successful, the smart-grid efforts will also lead to interoperability among home computing and entertainment devices. Unlike the media studied in this chapter (Cat5 cable, optical fiber, etc.), data transmission is not the primary function of electrical wiring, which is primarily designed to carry electrical power as 60 Hz (United States and Japan) or 50 Hz (Europe and Asia) signals. Yet, it is possible that as a result of developments in physical layer technologies, utility companies may someday become capable of offering high-speed Internet service using the same technologies that are being developed to help transmit power demand and price signals for the smart grid.

References

1. http://www.greatachievements.org.
2. Kintner-Meyer, Michael, Kevin Schneider, and Robert Pratt. "Impacts assessment of plug-in hybrid vehicles on electric utilities and regional U.S. power grids part 1: Technical analysis," https://www.ferc.gov/about/com-mem/5-24-07-technical-analy-wellinghoff.pdf.
3. National Institute of Standards and Technology, "NIST Framework and Roadmap for Smart Grid Interoperability Standards Release 1.0 (Draft)."
4. US Department of Energy, "Smart grid: An introduction."
5. US Department of Energy, "What the smart grid means to America's future."

REVIEW QUESTIONS

1. What is the ATIS telecom glossary? How can it be useful to you?
2. What is the primary responsibility of the physical layer in data communication?
3. Define *physical medium* in the context of computer networking. What are the common physical media used in computer networks?
4. What properties are required for a material to be suitable for use as a physical medium in computer networks?
5. What is *UTP*? Why is the copper cable commonly used in computer networks called UTP?
6. What are the common categories of copper cable used in networks? Under what conditions would you prefer to use each category of cable?
7. What factors favor the use of optical fiber as a physical medium over copper?
8. What is *total internal reflection*? How does it help optical fiber transmit light signals efficiently?
9. What are the two categories of optical fiber? Under what conditions is each category preferred?
10. What are the components of optical fiber? What is the role of each component?
11. Define *data.*
12. Define *signal.*
13. Why is there a need to convert data to signals?
14. What are the properties of a good signal?
15. What is *modulation*? How does modulation help in data transmission?
16. What is *amplitude modulation*?
17. How does noise affect signals? What happens if the level of noise becomes too high relative to the strength of the signal?
18. Given a communication channel with a bandwidth of 3,000 Hz, and a signal-to-noise ratio of 1,000: Use Shannon's theorem to calculate the maximum data rate that can be supported by this channel (this is close to the traditional phone line).
19. Why is binary representation preferred in computers over common representations such as decimal?
20. Briefly describe the standard procedure used by the physical layer to send and receive data as a signal.

21. What is the ASCII code? Why is it useful in data communication? What is the ASCII code for the letter *a*? For the letter *A*?
22. What is *multiplexing*? Why is multiplexing useful in data communication?
23. What are some examples of multiplexing in day-to-day life?
24. Describe how the Interstate system may be seen as a multiplexed transportation system.
25. What are the two categories of multiplexing?

EXAMPLE CASE QUESTIONS

1. Visit the website www.greatachievements.org. Which of the top technologies of the 20th century could be categorized as a networking technology, if a network is defined in more general terms as a collection of users sharing a resource?
2. For each of these technologies, briefly describe how networking (interaction) improves the utility of stand-alone components in the network.
3. Why should power be cheap at night and expensive during the day? You may find the Wikipedia article on capacity factors useful: http://en.wikipedia.org/wiki/Capacity_factor.
4. What changes in behaviors will be required of people to exploit the potential of the smart grid?
5. It is expected that smart meters will report the current price of power and smart appliances will be programmable to operate only when power prices fall below values you specify. How might you change your energy usage if smart meters and smart appliances were installed in your house?
6. It is expected that the smart grid will allow you to sell power stored in batteries (charged when power prices are low) or generated using solar, wind, and other means. What changes do you expect to see in your neighborhood if these technologies go mainstream?
7. Broadband over power line (BPL) may allow your local power company to compete as an Internet Service Provider (ISP). What advantages for the power company do you foresee over your current ISP if your power company decides to actually offer ISP service?
8. Using information from the Internet and other sources, write a short report (two to three paragraphs) about the smart-grid initiatives being undertaken by your local power company.

HANDS-ON EXERCISE—Amplitude Shift Keying

In this exercise, we will use Excel to simulate amplitude modulation of an analog carrier wave using a digital signal. The process is called amplitude shift keying (ASK). We will consider one of the simplest cases of ASK, called on-off keying, in which the presence of a carrier wave indicates a 1 and its absence indicates a 0. A common usage scenario for on-off keying was the transmission of Morse code. When the operator pressed a switch on, the signal would get transmitted, and when the operator turned the switch off, the signal would not be transmitted. A receiver could listen to these beeping signals and recover the text that was being sent.

The example in Figure 2.29 shows an on-off-keyed ASK signal for 1 second. During this time, the signal sends out 20 bits of data. The bit pattern sent is 01101100101101001100. The times on the x-axis are in milliseconds and the values on the y-axis are the signal amplitudes. Intervals of silence represent 0s being transmitted and the periods with the signals represent 1s being transmitted. For example, since the first bit transmitted is 0, the signal has 0 amplitude, from 0 ms to 50 ms.

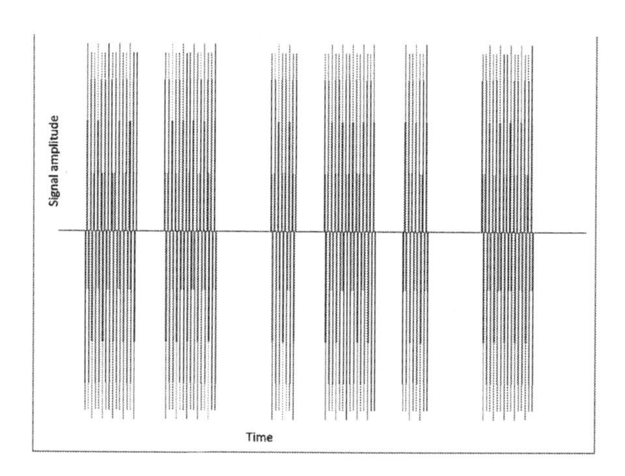

Figure 2.29: Amplitude phase shift keying example

Answer the following questions using the "Exercise" worksheet in the: Ch2_AM_multiplexing_example.xlsx spreadsheet, available on the companion website. Column A in the spreadsheet is the time, Column B is the data to be transmitted, Column C is the carrier signal, and Column D will show the modulated signal. Each row is 1/1,000 of a second.

1. If the signal sends out data at the rate of 20 bits per second, how long does it take for the signal to send each bit? This time is called the bit interval or bit period.
2. If we send 25 bits per second, what will be the bit interval of the new signal?
3. Using a bit rate of 25 bits per second, modify the contents of Column B so that it represents the bit pattern 0011100101111000101001011. Show a graph of the data in Column B. If you prefer, you may use column F for this bit pattern instead of modifying column B. For convenience, column E has a time series counting up to 1 second in 1 millisecond intervals.
4. Sine waves have values Sin(2πft) where f is the frequency of the wave. The carrier in column C of the example has a frequency of 150 Hz. Create a new column where the frequency of the carrier is 100 Hz. Plot both carrier signals on a graph. (The line chart in Excel is the best option for this graph. You may find it easier to plot a chart showing all the values in the two columns. However, if you prefer and are comfortable doing so, it will be more informative if you only plot both carrier waves for approximately 100 milliseconds.)

5. The amplitude-modulated signal using on-off keying is calculated as M = D*C, where M is the modulated signal, D is the digital data, and C is the carrier signal. Compute the value of the modulated signal in the column titled "modulated signal." (Hint: you can look at columns F and G in the "Data" worksheet to see how to compute the modulated signal.) Do not report anything for this question, you will do that in the next question.
6. Plot the modulated signal as a function of time as in Figure 2.29.
7. Assume you are explaining your plot to a lay person. Describe how to interpret the signal in the question above to recover the data sequence of Question 3.

CRITICAL-THINKING EXERCISE—The Value of Carriers

Signals have no direct use. Their only use is as carriers of information. Some of the most important quantities we deal with have the same property. Money and electricity come to mind. Money has no utility other than as a carrier of value. Electricity has no utility other than as a carrier of energy. To help you think this through, answer the following.

1. You have been given a truckload of $100 bills but are not allowed to use them as currency. What are three useful things you can do with the currency bills?
2. You have access to unlimited electricity for a month from your power company, with the condition that this electricity must be used directly, that is, it cannot be converted to other forms of energy by conventional electrical devices (e.g. bulbs, computers, air conditioners, etc). What are some (say, three) useful things you can do with this electricity?

IT INFRASTRUCTURE DESIGN EXERCISE—Media Selection

Using the information about TrendyWidgets provided with the Firm Details in the IT infrastructure design exercise of Chapter 1, answer the following questions:

1. For each location (Tampa, Amsterdam, Mumbai, and Singapore), identify the most suitable physical medium for building the network. Make a rough estimate of the total quantity of the media that will be needed for each location. Use online or other resources to estimate the cost of purchasing enough quantities of the medium for each location.
2. Identify the most suitable physical medium to connect the different locations to each other (the long-distance links).

CHAPTER 3

Data-link Layer

We may say that according to the general theory of relativity space is endowed with physical qualities; in this sense, therefore, there exists an ether.

Albert Einstein

Overview

This chapter describes the functions of the data-link layer. The data-link layer interfaces with the network layer above it to transfer data reliably across one hop. To accomplish this goal, the data-link layer depends upon the physical layer below it for signaling. At the end of this chapter you should know:

- the functions of the data-link layer
- an overview of Ethernet, the primary technology used to connect corporate desktops
- error detection using cyclic redundancy check (CRC)
- Ethernet frame structure
- MAC addresses and their organization
- switching

Functions of the Data-link Layer

The physical layer sends signals between neighboring devices on a network. Using this signaling service, the data-link layer sends data between neighboring devices on the network. To do this, the data-link layer has two primary functions: addressing and error-detection. Addressing ensures that signals from the physical layer reach the correct device on the network. Once the signals are received, the error-detection function of the data-link layer detects if any errors were introduced during signal transmission. If any errors are detected, the data-link layer discards the data.

Error detection and error correction

The technology we cover in this text can only detect errors. However, other technologies can also correct errors.[1]

You might wonder what happens when you use a technology that discards data with errors. In these cases, the transport layer (layer 4 of the TCP/IP stack) recognizes the missing data and fixes the errors. However, every data-link layer technology ensures that errors are detected and acted upon. Data with errors is never passed on to the network layer from the data-link layer.

Data-link layer analogy in real life

This is quite analogous to what happens when you make a long-distance trip. You walk to your car (transportation technology 1). You then drive your car to the airport (transportation technology 2), then switch back to walking on foot to the aircraft (transportation technology 3). Finally, you take the plane (transportation technology 4). We build on this idea in the next chapter.

The data-link layer is the building block of the Internet. Many technologies are available that provide the functions of the data-link layer. When you use the Internet, information generally passes through multiple instances of the data-link layer before it reaches its final destination. Each instance of the data-link layer ensures that the data safely hops to the next device in the network. For example, referring to Figure 1.1, the wireless link from the home PC to the home router is an instance of the data-link layer. The optical-fiber link from the home router to the carrier's router provides another instance of the data-link layer. Each link between the routers in the carrier's network is served by an instance of the data-link layer. In each case, the local instance of the data-link layer ensures that the data reaches the next device safely. It is common for large networks to mix and match different data-link technologies within the overall network, depending upon network needs.

Figure 3.1: Ethernet connector in the rear panel of a PC

Thus, the physical and data-link layers together ensure that the data sent by one device on the network safely reaches the next device on the network without errors.

In the rest of this chapter, we focus on a very common data-link layer technology—Ethernet. Almost every corporate desktop is connected to the network using Ethernet as

1 A good page for error-correction codes is http://www.eccpage.com/ (accessed Nov. 14, 2015).

the data-link layer technology. Anytime a computer uses a connector like the one shown in Figure 3.1, it is using Ethernet. Therefore, Ethernet is a very relevant technology for IT professionals.

For classroom purposes, Ethernet has another advantage over other data-link layer technologies. It is extensively documented and the documentation is publicly available.[2] Therefore, every aspect of Ethernet can be understood by referring to the formal technology specifications.

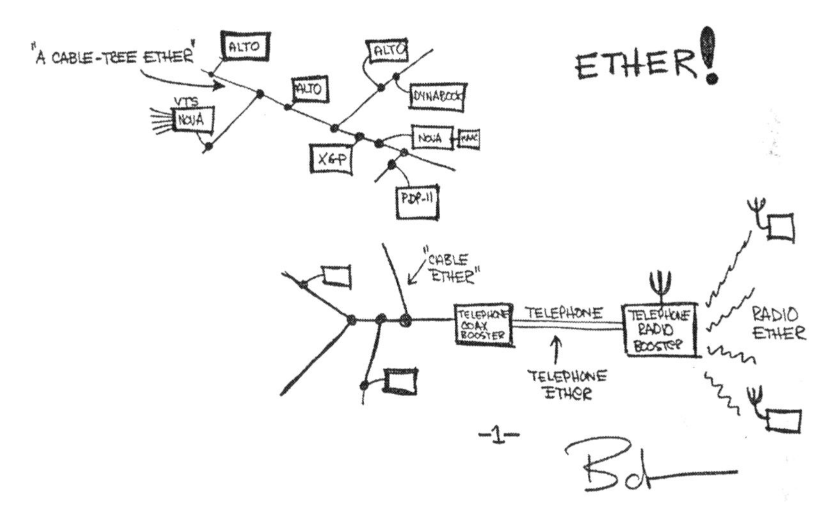

Figure 3.2: Early Ethernet vision
(Source: Bob Metcalfe's 1972 sketch of his original "ethernet" vision. Image provided courtesy of Palo Alto Research Center, Inc., a Xerox company.)

Ethernet

Ethernet is a standard protocol for local area networks (LANs) that uses carrier-sense multiple access with collision detection (CSMA/CD) as the access method. Ethernet is a technology for low-cost, high-speed communication in small networks.

A network using Ethernet can be up to 100 meters in diameter and have up to 250 devices. Because it covers a small territory, Ethernet networks are commonly called local area networks (LANs). Most departmental and small business networks use Ethernet for network connectivity. The technology was patented in 1977. The patent filing specified a maximum data rate of 3 million bits per second (3 Mbps). Technology has evolved considerably since then and most computers sold today support Ethernet data rates of up to 1 gigabit per second (1 Gbps = 1,000 Mbps), a speed increase of more than 300 times. Figure 3.2 shows how the creator of Ethernet, Bob Metcalfe, visualized the technology.

2 http://standards.ieee.org/getieee802/portfolio.html.

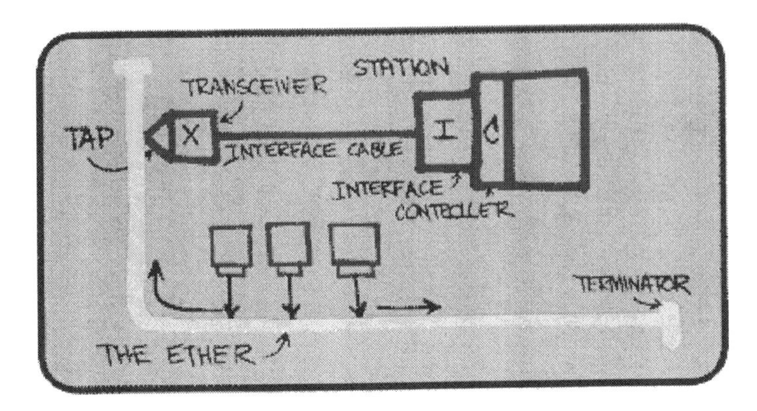

Figure 3.3: Early diagram of Ethernet
(Source: Robert Metcalfe and David Boggs, "Ethernet: Distributed Packet Switching for Local Computer Networks," *CACM* 19(7), pp. 395–404)

Figure 3.3 is an early sketch of the Ethernet drawn by Bob Metcalfe[3] and it provides an excellent visual overview of the technology.[4] As seen in this figure, the core of Ethernet was a cable available throughout a building. This cable was called ether, after the substance that was once believed to permeate the entire universe. Analogous to ether, Ethernet was visualized as a technology that would be available anywhere in a building wired for Ethernet. Computers, called stations in this figure, obtained network connectivity by tapping into the Ether. Figure 3.4 shows the network interface used for Ethernet in the early 1980s. It shows the tap on the cable serving as the Ether and the transceiver used to communicate between the computer and the other computers connected to the Ether.

When a station wanted to send data, it would simply send the data out into the Ether where the data would reach every other computer connected to the Ether. This mechanism ensured that the data always reached its destination, though the simplicity came at the cost of privacy of data, since all users could listen to all the traffic on the network.

Ethernet traded privacy in favor of simplicity. Since any data sent on Ethernet is accessible to all other computers on the network, there is no privacy in Ethernet. However, the benefit is technical simplicity. No complex mechanism is needed to route data to specific computers. There are no controllers or other equipment to regulate which computer may transmit at any time. These simplifications greatly reduce the cost and complexity of networking, making Ethernet affordable for organizations of any size and budget.

---

3 Bob Metcalfe was also responsible for the initial contact between Charles Simonyi, the creator of MS Word, and Bill Gates, the founder of Microsoft. Simonyi developed his editor for the computer his employer, Xerox, was building, but sensed no support for his project at Xerox. Metcalfe was a friend of Simonyi and kept pushing him to meet Bill Gates—"this crazy guy in Seattle"—to explore possibilities of bringing the product to the market. They met in fall 1980. Until then, Microsoft had no plans to develop applications. Source: Stephen Manes and Paul Andrews, *Gates* (Doubleday, 1993), p. 166. Charles Simonyi is also the originator of the Hungarian notation for naming variables in programs, in honor of his country of origin. In the Hungarian notation, variable names indicate their type or intended use, e.g. txtFirstName may indicate a text field for the first name.

4 Robert Metcalfe and David Boggs, "Ethernet: Distributed packet switching for local computer networks," *Communications of the ACM*, 19(7) (1976): 395–404.

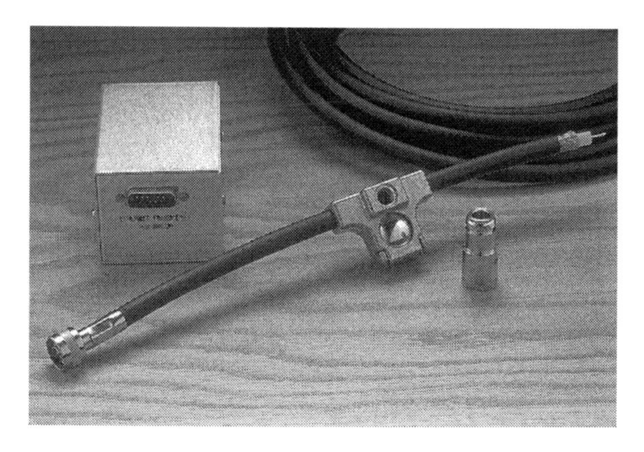

Figure 3.4: Ethernet transmitter-receiver, early 1980s

Fortunately, the lack of privacy is not a major concern for the contexts in which Ethernet is used.[5] Recall that Ethernet is limited to short distances, such as departments and small businesses. Within these small networks, it is safe to assume that all users have similar privileges to access and share data. Therefore, the loss of privacy is not a major concern in Ethernet and is an acceptable trade-off to simplify and economize the technology.

The Ethernet of Figure 3.3 has a major limitation. If the Ether failed, the entire network failed. And it was surprisingly easy for the Ether to fail. The Ether was a cable that usually ran through attics and closets in buildings. Here animals and workers could accidentally cut the cable in remote nooks that were difficult to locate. Therefore, later Ethernets placed the Ether in a box called the hub, with interface cables connecting individual computers to the hub from up to 100 meters away. Figure 3.5 shows such an arrangement.

There is no difference in the basic operation of a hub-based Ethernet (Figure 3.5) and the early Ethernet shown in Figure 3.3 although they look very different. The hub simply repeats all incoming data to all devices so that it acts like a transparent cable. Therefore, modern network diagramming software draws Ethernet as a long cable as in Figure 3.6.

The advantage of the newer hub-based Ethernet design is that damage to a cable only hurts connectivity at the computer connected by the cable.

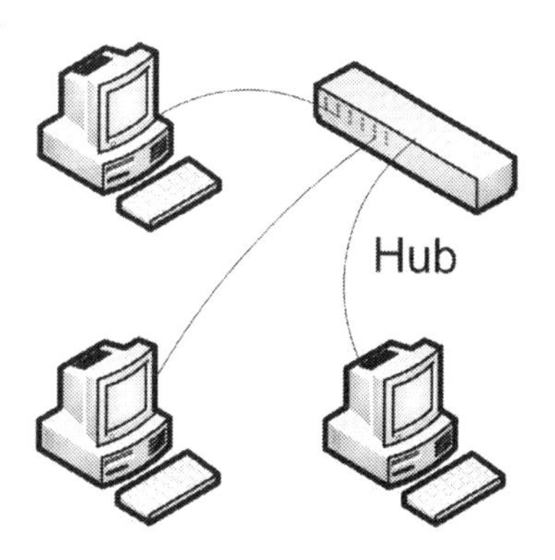

Figure 3.5: Hub based Ethernet

Ethernet Operation

We now look at the operation of Ethernet in a little more detail. A typical departmental Ethernet is shown in Figure 3.6. It includes the usual network elements such as a file server for shared files, a printer, and a few

5 To address situations where privacy is a matter of concern, newer technologies such as 802.1ae have been created. They are considered beyond the scope of this text.

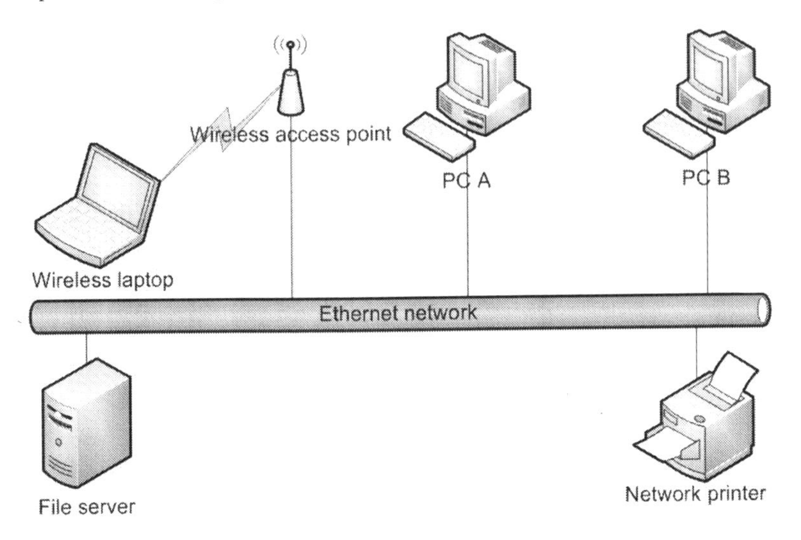

Figure 3.6: Typical Ethernet

desktops and laptops accessing these resources wirelessly. We will consider the example where PC A wishes to send some data to PC B.

To accomplish the transfer, A can transmit the data on the wire. The transmission will reach all the devices on the network, including B. The basic elements of data transfer are therefore quite simple. There is just one hitch. When the data reaches B, how will it know that the data was being sent to it, and not to some other device on the network?

Privacy and small networks

The treatment of privacy in Ethernet is very similar to the way we treat privacy in our daily lives. For example, social conversations are typically conducted among a well-connected small group of friends. Those engaged in conversation are quite comfortable with others in the group moving in and out of the conversation. There is an implicit assumption that there are no secrets among friends.

College classroom as an Ethernet

A college classroom operates a lot like an Ethernet. As in Ethernet, anyone in the class can speak at any time. Every remark made by any speaker is heard by all other students in the class, even if the remarks are addressed to the instructor or a specific person. No formal rules are necessary to decide the sequence in which speakers are permitted to make comments in the class. When a participant has something to say, he simply waits for the appropriate moment when no one else is talking and speaks, and everyone can hear every comment. Just as in Ethernet, there is no privacy in the classroom, but this does not diminish the functionality of the classroom.

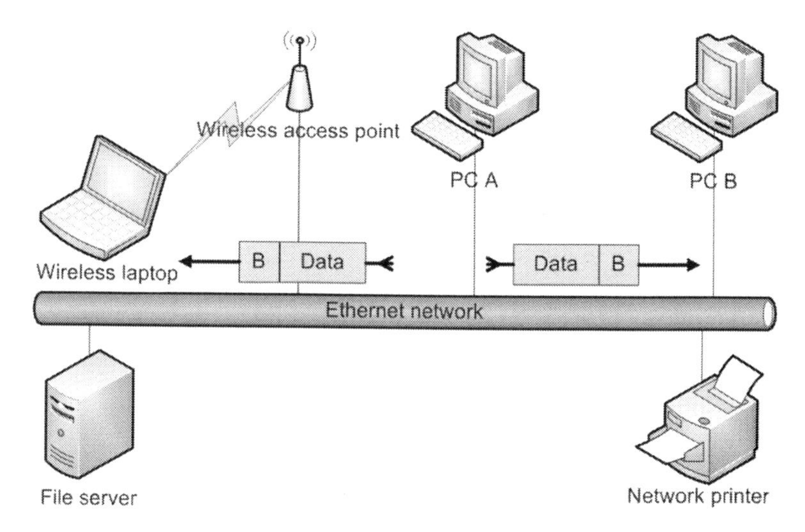

Figure 3.7: Packet in the medium

To overcome this problem, when A wants to send the data to B, it can add B's name to the data. This is analogous to putting a "To:" addressee on a letter before dropping it in the mailbox. To continue the analogy with a small group discussion used before, this is also like addressing a specific participant in the group before making a remark. The address alerts the named member that the comments that follow are addressed to him, though everybody in the group can listen to those comments. With this refinement of adding an address to the data, Figure 3.7 shows a data packet sent from A and addressed to B.

This idea of adding a "To:" addressee to every block of data sent on the network is our introduction to packet headers. A packet of data contains data and some overhead, such as names and addresses, necessary for the data transfer to be successful. The "To:" address discussed here is called the data-link destination address of the packet. As we proceed in this book, more overhead items will be introduced. With one exception, all the overhead in a packet is located in one place before the data. This location is called the packet header. Placing the overhead before the data facilitates packet processing. For example, as in Figure 3.7, if the receiver sees that a packet is not addressed to it, it does not have to worry about processing the rest of the packet. The only exception is error-detecting information, which is located at the end, as seen later in this chapter.

Broadcast in Ethernet

We have seen how any data sent in Ethernet is received by all other computers on the network. This mechanism is fundamental to the operation of Ethernet. Ethernet is therefore called a broadcast network. *Broadcasting is the transmission of signals that may be simultaneously received by stations that usually make no acknowledgement.* Broadcast is a very important technique

Figure 3.8: Mail broadcast

for simplifying communication in small-scale networks.[6] We have already seen the example of broadcast in a classroom. In fact, all forms of shouting use broadcast to send messages.

As another familiar example of broadcast, consider the ECRWSS service offered by the Postal Service. Local advertisers can use ECRWSS (Extended Carrier Route Walking Sequence Saturation) to cover the local area. You may have received such advertisements addressed to "ECRWSS postal customer." The mail carrier delivers such mail to every address along his route. Figure 3.8 shows an example of an ECRWSS letter.

Names and addresses in communications

Names and addresses are both used to get information to the right person. What is the difference between the two? What makes an identifier a name, and what makes it an address?

It is instructive to think about this distinction and its potential impact. Imagine a country where locations are identified by name and mail is addressed to locations such as "the old fig tree." Yet, Costa Rica and large parts of India use such mechanisms because street names are not fully defined and labeled. While such location identifiers are homely and quaint, they have costs. In Costa Rice, about 25% of the mail is undelivered, in spite of the best efforts of the postal system.[7]

In Chapter 4, we will see that data-link layer addresses are better seen as names than addresses. However, the industry calls them MAC addresses, and so we will also refer to them as MAC addresses to be consistent with industry.

---

6 Broadcast is so easy, even newborn babies know how to use it. Every time a baby cries, everybody within listening range knows that attention is required. From a data communications perspective, a newborn baby could potentially be seen to communicate exactly one bit of information—1 (crying) and 0 (not crying).

7 Leslie Josephs, "When Getting Directions, It Helps To Know Where the Fig Tree Was; Costa Rica Addresses Its Lack of Street Names; 'By Pizza Hut,'" *Wall Street Journal*, June 28, 2012.

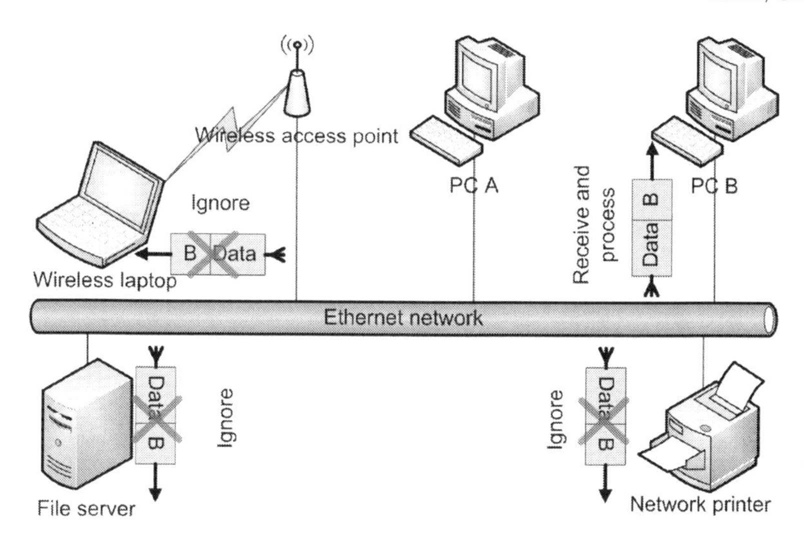

Figure 3.9: Receipt of packet in Ethernet

ECRWSS simplifies work for everyone. Local advertisers do not have to address mail to individual customers and the postal service does not have to sort the mail. The trade-off is that the mail reaches a number of customers who are not interested in the specific advertisement. As a result, a lot of the ECRWSS broadcast mail you receive every day falls in the category of junk mail.[8] The same thing happens in Ethernet as well. Most of the data a station receives in Ethernet is not of interest to the station. This suggests the major limitation of broadcast as a communication mechanism—it is simple, but also very inefficient.

Broadcast is great for small networks, but unsuitable for large networks. After all, how large can a network be if everyone were to shout at each other?

Packet Receipt in Ethernet

We have discussed the transmission of packets in Ethernet. Reception is a fairly straightforward process. When a packet reaches a networked device, the device compares its own address with the destination address of the packet. If they are not the same, the device ignores the rest of the packet. But if the addresses match, the device knows that the packet is addressed to it, and the device receives the packet. This process is shown in Figure 3.9.

CSMA/CD

Broadcast is simple, but broadcast networks require courteous behavior from their users for efficient operation. This is because broadcast networks have a major limitation—they can only handle one transmission at a time. What happens to the data sent by A if B also

8 In 2011, the postal service eased rules to reduce ECRWSS costs. Earlier, such mail was required to have the address of the recipient, forcing many small businesses to buy mailing lists. Now, the addressee name is not needed.

Broadcast mechanism vs. broadcast address

Expert users will make a distinction between broadcast as a transmission mechanism and broadcast as a deliberate transmission to all hosts on the network. Broadcast in this chapter refers to the transmission mechanism as shown in Figure 3.9, where frames are addressed to specific users, but sent to all users, and ignored by all other users.

Ethernet also defines a broadcast address, which is all 1s, which indicates to receivers that the frame is meant to be processed by all hosts on the network. This address is typically used during search operations. We will see this address in Chapter 7.

transmits some data at the same time? The result is shown in Figure 3.10. In the figure, a packet is being sent to B at the same time that a packet is being sent to A. The result is called a collision. *A collision is the situation that occurs when two or more demands are made simultaneously on a system that can only handle one demand at any given instant.*

Figure 3.10 shows that for efficient operation, Ethernet requires a mechanism that minimizes the possibility that two or more computers might send data over the medium at the same time. This is called medium access control (MAC). *Medium access control is the method used to determine who gets to send data over a shared medium.* The MAC mechanism used in Ethernet is called CSMA/CD, which is an acronym for Carrier Sense Multiple Access with Collision Detection. Though the term sounds complex, its operation is quite straightforward. Here is what each term in the acronym means:

Multiple Access A scheme that gives more than one computer access to the network for the purpose of transmitting information.

Carrier Sense An ongoing activity of a data station in a multiple access network to detect whether another station is transmitting.

Collision Detection The requirement that a transmitting computer that detects another signal while transmitting data, stops transmitting that data.

Multiple access refers to the ability of Ethernet to serve multiple computers. Carrier sense requires computers that wish to send data on a multiple access network to first sense the medium for the absence of other data-carrying signals. Computers transmit only when they sense that the medium is silent. If the medium is busy, a computer with data to send waits for the current transmission to end before transmitting data.

Carrier sense greatly improves the efficiency of multiple access networks such as Ethernet. Unfortunately, collisions are possible even if transmitting computers confirm that the medium is silent before transmitting data. Consider what can happen if more than one computer gets ready to send data while a transmission is going on. As soon as the transmission ends, each waiting computer will sense that the medium is silent and decide that it is acceptable to transmit data. When all these waiting computers transmit data, the packets will collide with each other.

Therefore, to improve the efficiency of Ethernet, transmitting computers do more than just carrier sense. They continue to listen to the medium after beginning transmission. This

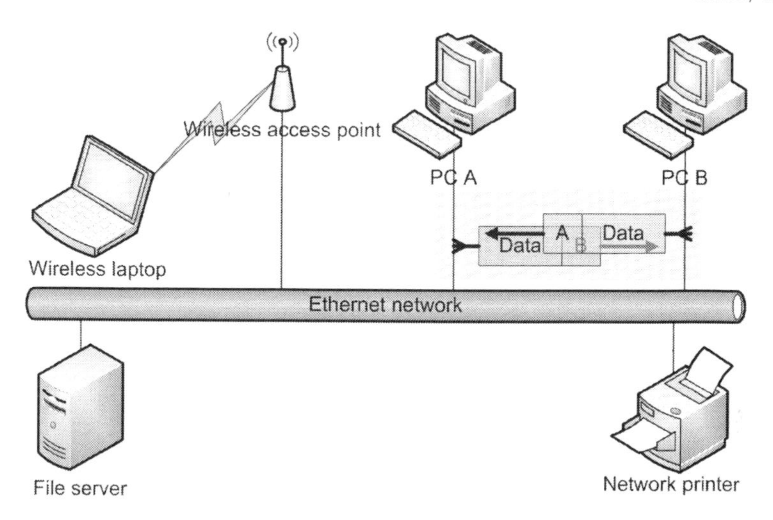

Figure 3.10: Collision in Ethernet

is called collision detection. If all goes well, there is only one station transmitting data, and the signal this sender senses in the medium will be identical to the signal this sender transmitted into the medium. However, if another computer is also transmitting at the same time, a collision will occur. If a collision occurs, the signal the sender senses will be different from the signal it is transmitting. Once a computer detects a collision, it knows that further transmission is futile because the receiver will not be able to successfully decode the signal. The computer therefore immediately stops transmission.

One final challenge remains. Once a computer detects a collision and stops transmission, it still has data to transmit. The computer also knows that there is at least one other computer on the network that is also waiting to transmit, though it does not know the identities of the other waiting computers. Therefore, a mechanism is required to enable all these waiting computers to send data over the shared network, without colliding with other waiting senders.

The mechanism that is used to resolve collisions is called random back-off. Each computer that detects a collision waits for a random amount of time before sensing the medium again. If the medium is clear after the waiting period is over, the computer begins transmission; otherwise it waits for the ongoing transmission to end before sending again. If it detects a collision a second time, it waits even longer. Eventually, the transmission succeeds.

CSMA/CD is just a formal way of saying: do not interrupt. Bringing up the analogy with a classroom again, it is easy to see that a classroom is a multiple access (MA) channel. The instructor and students share the same transmission medium. In this environment, it is efficient to allow speakers to finish what they are saying without interruption. This is analogous to carrier sense. When two or more people start speaking at the same time, they immediately detect the collision and stop. This is analogous to collision detection. The colliding speakers then find a courteous way to decide who goes first, which is functionally equivalent to random back-off.

Random back-offs are a very social gesture

Random back-off sounds complicated, but it is actually common in social conversations. When two or more people start speaking at the same time, they typically figure out how to proceed using random back-off, exactly as in Ethernet. They look at each other or use some other social process to decide who goes first. If the group is conscious of hierarchy, the senior-most person in the group may decide how to break the deadlock, or may simply be allowed to go first. Ethernet is very egalitarian in that sense. A supercomputer does not get priority over a cell phone if they are both contending for network access.

Advantages and Disadvantages of Using CSMA/CD in Ethernet

We have repeatedly drawn analogies with social groups, and Ethernets share similar advantages and disadvantages as social groups. The primary advantage of CSMA/CD is its simplicity. CSMA/CD does not require any additional equipment to work. It therefore greatly reduces the cost of networking. Further, since no configuration is required to make CSMA/CD work, Ethernet does not require any technical skills to set up. Additional computers can simply be plugged into the network. Home and small business networks are almost always Ethernets, and CSMA/CD is one of the reasons why it is so easy to add additional computers to home networks.

It may be useful to think of alternatives to CSMA/CD. How else would we let multiple users share a medium? We may allow users with short messages, or greatest seniority or some other property to go first. But this would introduce the complexity of first determining who has the shortest message or greatest seniority or another metric. All this introduces complexity, which generally translates to more expensive technology. By eliminating these complexities, Ethernet remains one of the most cost-effective technologies for LANs.

CSMA/CD does have some limitations. The primary limitation comes from the broadcast nature of a CSMA/CD network. Only one station can transmit at a time in a broadcast network. Collisions increase as the number of computers in a CSMA/CD network increases. Therefore, these networks are not scalable. Experience suggests that delays caused by collisions become unacceptably large when about 250 users join the network. Also, since computers need to be able to detect possible collisions before completing transmitting, all stations have to be in close proximity, about 100 meters.

Another potential disadvantage of CSMA/CD networks is that the maximum wait time to resolve a collision is undefined. There is no assurance that collisions will be resolved in a specified amount of time. As a result, CSMA/CD networks may not be suitable for applications that require guaranteed service. In its early days, this was considered a fatal defect of CSMA/CD networks. Modern networks resolve this limitation by over-provisioning—increasing the speed of the network so much that the likelihood of collisions is minimized. Also, as we will see later in this chapter, modern Ethernets have switching capabilities. Switching attempts to minimize the number of packets that reach devices other than their intended destinations. It therefore minimizes the use of broadcasts and virtually eliminates collisions.

At this point, we have completed our discussion of the first function of the data-link layer—addressing and delivering data to the correct destination. The discussion was based on Ethernet. Other data-link layer technologies may operate differently, but they share many features with Ethernet. They all deliver data across one hop and they all use addressing to locate the destination.

We now begin our discussion of the second function of the data-link layer—error detection. Again, the discussion will use the example of Ethernet, but virtually all data-link layer technologies use the error-detection technique used in Ethernet—cyclic redundancy check (CRC).

Error Detection

It is easy to see that as the signal passes through the medium, it is likely to get corrupted due to noise and factors such as power outages and power spikes. When the corrupted signal is decoded, errors are introduced in the received data. For reliable operation of the network, these errors must be detected and corrected.

The standard technique used to detect errors at the data-link layer is called CRC. To explain the need for a relatively complex error-detecting procedure such as CRC, it is useful to begin by seeing how errors are corrected in human communication. We will then see why error-correcting methods that work in human communication are unsatisfactory for computer communication. CRC overcomes these limitations in error-detection techniques that work so well in human communications.

To begin with, most human communication is social in nature with no great need for correct reception and therefore requires no error detection or correction. Social conversation is also quite redundant. In most cases, we can tune out a conversation for long periods and still understand the information being conveyed. Where error detection or correction becomes necessary, contextual cues often help us to fix errors.

Information redundancy and multitasking

Information redundancy is why most people can allow the TV to run in the background while they go about their business—they usually already know what is going on. There is no need to catch every phrase and move to figure out the plot. And most of the time, the work we do is quite familiar to us, so we can go about it without having to think too much.[9] When either of these two assumptions fails, multitasking is not advised.

Where error correction is important in human communication, various techniques are popular. Three techniques are considered here. The first is echo, or reading back the information. If you have purchased goods over the phone, generally the customer service agent wants to make sure that the credit card information you have provided is correct. To do this, the agent reads back the credit card number and other details to you to confirm that he correctly copied the information you provided. The second technique is to provide

9 Claude Shannon estimated that English has a redundancy of about 50%, and the typical English text can be shortened by half without loss of information (if u cn rd ths ...). Source: James Gleick, *The Information: A History, a Theory, a Flood* (Vintage, 2012).

redundant data to help reduce errors. For example, while fixing appointments, it is very useful to specify dates in detail. For example, instead of saying tomorrow, it is very useful to specify the day as: tomorrow, Friday, December 4, 2015. Finally, the third technique is for the receiver to contact the sender in case he is in doubt about any piece of information.

The last two of the three techniques discussed above are obviously not suitable for computer communication. Computers cannot yet figure out how to use redundant information to fix errors. Or, be in doubt that some received information such as a name or address might possibly be incorrect. The first technique, echo, seems promising. The receiver can echo back the information it receives from the sender. If the echo does not match what was sent, the sender knows there is a problem. However, echo has a major flaw. As Figure 3.11 shows, errors may get cancelled in the two-way exchange. In the example, the sender sent the word "hello" to the receiver. However, due to a transmission error in 1 bit of data, this was received as "gello." When the receiver echoed it back for confirmation, unfortunately, a second transmission error caused another 1-bit error to reverse the effect of the error introduced during onward transmission. This caused the word to be received back as "hello," leading the sender to believe that the transmission was successful. This kind of error cancellation may not happen every time, but the example shows that echo is not a very reliable method for error detection.

The above discussion suggests that informal methods that succeed in removing errors in human communication are inadequate for removing errors in computer communication. Whereas human communication is quite tolerant of errors, modern networks are used for shopping, banking, and other commercial transactions that are highly intolerant of errors. Therefore, extremely reliable methods of error detection are necessary in data communication networks.

The general approach used for error detection in computer communication is to help the receiver determine on its own that the data it has received has errors, without seeking confirmation from the sender. This is done by adding some meta-data[10] to the transmitted data. The sender computes this meta-data from the transmitted data and adds it to the

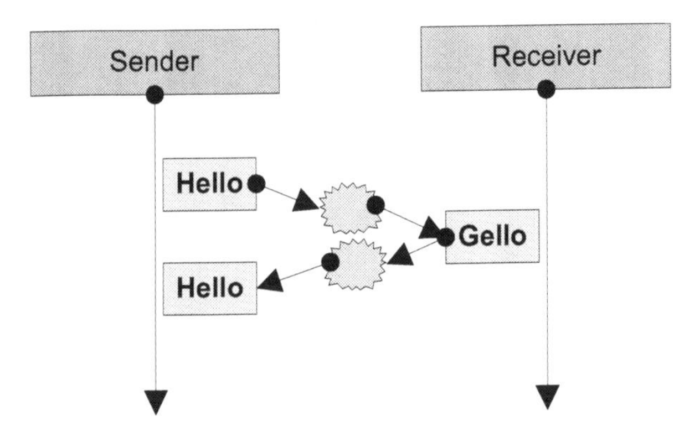

Figure 3.11: Error cancellation in echo

10 Meta-data is any data about data, usually used to provide definition or documentation.

packet. At the receiving end, the receiver re-computes the meta-data from the received data and compares its own computation of the meta-data with the meta-data sent by the sender. If the two meta-data match, the receiver accepts the data as correct. Otherwise, it rejects the data.

The ubiquity of meta-data

Meta-data is data about data. While this sounds esoteric, it is everywhere, commonly facilitating discovery of relevant information. Exif information about photographs is meta-data that gives information about the exposure and camera used to take the picture; and database meta-data provides information on the tables available in the database.

Simple Example of Error Correction

Let us consider a simple example of using meta-data for error-detection. The method used in the example is too simple to be useful in practice. However, it illustrates the basic ideas behind using meta-data for error-detection.[11]

Say we wish to transmit the word HELLO. To keep things simple, instead of using ASCII, we choose a simple coding scheme where we transmit each letter as its position in the alphabet. Thus, the letter *A* is coded as 1, *B* as 2, and so on. With this coding scheme, the data to be transmitted is: 8 5 12 12 15.

Let us calculate our error-detecting meta-data as follows: We will add up the digits to be transmitted until we get a single digit. Thus, we get $8 + 5 + 12 + 12 + 15 = 52$, we repeat to get our meta-data $5 + 2 = 7$. We can then transmit 8 5 12 12 15 7.

If there are no transmission errors, the receiver will get 8 5 12 12 15 7. The receiver knows that the last number received is the error-detecting meta-data and 8 5 12 12 15 is the data. It can calculate $8 + 5 + 12 + 12 + 15 = 52$ and $5 + 2 = 7$. Since the calculated meta-data matches the received meta-data, the receiver can conclude that the data was received without errors.

Let us see what happens if an error is introduced during transmission and the receiver gets 7 5 12 12 15 7. When the receiver repeats the error-detection computation, it obtains $7 + 5 + 12 + 12 + 15 = 51$ and $5 + 1 = 6$. Since $6 \neq 7$, the receiver can discard the received data as defective (note that 7 5 12 12 15 in our scheme was the code for GELLO).

So far, so good. The scheme will detect the error when one error is introduced during transmission. However, the proposed scheme is too naïve and error-prone. For example, this scheme will not detect errors in the above example if the receiver gets 10 5 11 11 15 7 (JELLO) or 2 5 10 11 15 7 (BEKLO). Clearly, there is a need for an error-detecting scheme that can detect arbitrary kinds of errors. This is provided by CRC, which is why commercial data-link layer technologies use CRC for error detection.

11 The scheme presented here is a variant of the parity scheme used on hard drives.

Cyclic Redundancy Check (CRC)

CRC is an error-checking algorithm that checks data integrity by computing a polynomial algorithm-based checksum. CRC used in Ethernet can detect all errors in data that affect 32 or fewer bits.[12]

CRC can also detect any errors affecting any combination of an odd number of bits. Errors affecting some combination of an even number of bits are also highly likely to be detected. The overall detection rate is 99.99999998% of all errors that affect 33 or more bits in the data. This is considered satisfactory for commercial use. For these reasons, CRC is preferred over simpler error-detection methods.

At a high level, CRC uses modulo-2 division, which may be called lazy division. The dividend in the division is the data for which the meta-data is to be calculated, along with a few zeros added as required by the procedure. The divisor is specified by the technology being used. When the division is completed, the remainder is the error-correcting meta-data.

The etymology of "algorithm"[13]

CRC is an algorithm. The word algorithm comes to us from al-Khowârizmî, who created the earliest known work in Arabic arithmetic. Al-Khowârizmî lived around 800 A.D. He was the first mathematician to publish rules for addition, subtraction, multiplication, and division using the (then) newly developed Hindu numerals. Speak the word al-Khowârizmî fast. It sounds a lot like algorithm, and the word algorithm means "rules for computing," in reference to the rules for computing developed by al-Khowârizmî.

Modulo-2 division is done the same way as regular division. The only difference is that while subtracting numbers, we use the rules in Table 3.1 instead of conventional subtraction.

Table 3.1: Modulo-2 subtraction rules

0 - 0 = 0	1 - 1 = 0	0 - 1 = 1	1 - 0 = 1

Only the third rule above is different from conventional division. In conventional division, it is not possible to subtract a smaller number from a larger number. However, this is possible in modulo-2 division. These rules may be called lazy division because you do not really have to subtract. If the numbers are the same, the difference is 0, otherwise it is 1.

At the sender end, the CRC meta-data is computed using a four-step process. These steps are summarized here before being explained by example (Figure 3.12).

- Step 1: The technology specifies the divisor. The data to be transmitted is the dividend.
- Step 2: Add specified number of 0s to the tail end of the dividend. The number of 0s to be added is one less than the number of bits in the divisor.

12 Recall from Chapter 2 that a bit is a unit of information that designates one of two possible states of anything that conveys information.

13 Peter L. Bernstein, *Against the Gods: The Remarkable Story of Risk* (Wiley, 1996).

- Step 3: Complete the modulo-2 division till the remainder has fewer bits than the divisor.
- Step 4: The remainder is the CRC meta-data. Add 0s to the head of the remainder till it has one less number of bits than the divisor. The resulting number is called the frame-check sequence (FCS).

The receiver's operation reverses the sender's operations. It performs a modulo-2 division using the data as the dividend with the same divisor as used by the sender (this divisor is specified by the technology). In place of the 0s added by the sender, the receiver places the FCS received from the sender. If the remainder is 0, the receiver accepts the data as correct, otherwise it rejects the data.

Let us now consider a simple example. In the example, the divisor specified by the technology is 1101. The data to be transmitted is 101010. The calculations are shown in Figure 3.12.

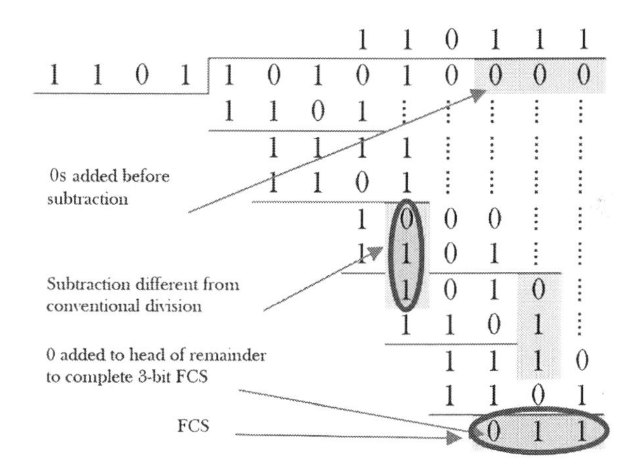

Figure 3.12: CRC—Sender operation

CRC Example

- Step 1: The divisor is 1101, the dividend is 101010.
- Step 2: Since the divisor has 4 bits, we add 3 0s to the tail of the dividend; this gives us 101010 000 to use as the dividend.
- Step 3: Perform modulo-2 division. This is shown in Figure 3.12. It is quite similar to regular division. The only difference is that the subtraction operation uses modulo-2 rules instead of regular subtraction. The instances where this happens in the example have been highlighted in the figure.
- Step 4: When the division ends, we have a remainder of 11. However, since the divisor has four digits, we need a three-digit remainder for the FCS. We therefore add a 0 to the head of the remainder to get an FCS of 011.

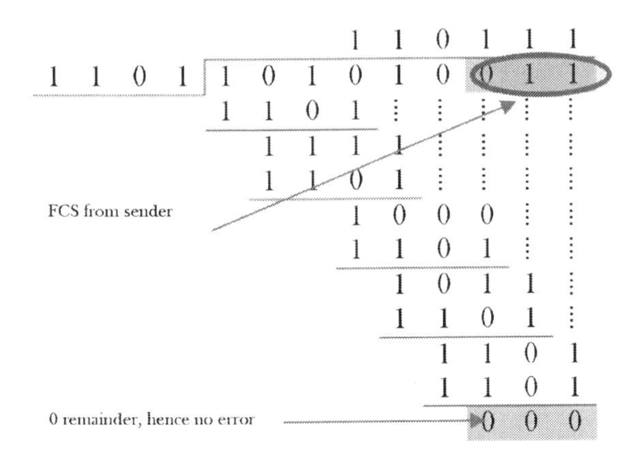

Figure 3.13: CRC—Receiver operation

The sender will now send the data (1010101) and the FCS (011) to the receiver. The receiver performs a modulo-2 division using the data and FCS as dividend with the technology-specified divisor. The operation is shown in Figure 3.13.

Since the remainder is 0, the receiver accepts the data as error-free. The receiver is confident that there are no errors affecting any combination of three or fewer bits in the data or any combination of odd number of bits. It is also convinced that there is only an extremely small chance there are any errors affecting some combination of more than four even numbers of bits.

The error-detection capability of CRC depends upon the choice of divisor and the length of the divisor. Ethernet uses a divisor of 1 00000100 11000001 00011101 10110111. This is 33 bits long and is judged as providing the required level of error-detecting capabilities.

So, how do you fix the detected errors?

If an error is detected, how does the receiver fix these errors? It doesn't because it doesn't have to.

Rather, most current Ethernet equipment simply discards the frame. Upper layers (specifically TCP) are capable of detecting this lost frame and having it re-transmitted. Therefore, Ethernet equipment can simply focus on making sure that no bit-level errors occur during transmission. Most current Ethernet equipment uses the store-and-forward mode of transmission, where each frame is received, processed, and forwarded only if it passes the CRC check. The alternate mode is cut-through, where the incoming bits are forwarded without processing.

In some sense, this disturbs the purity of the layered model, since in an ideal world, the data-link layer should do its job fully, without depending upon an upper layer. (What happens if TCP is changed and no longer detects lost frames?)

However, the speed benefits of this impure operation are so significant that it is considered an acceptable engineering practice.

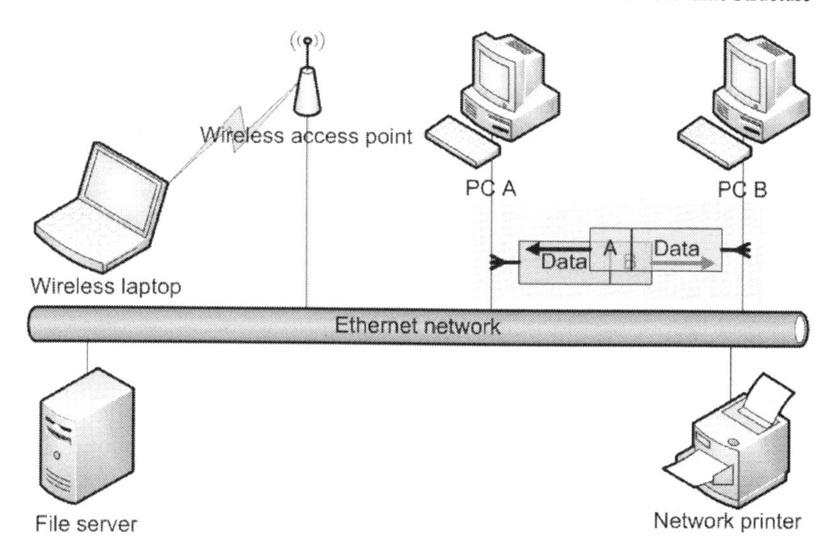

Figure 3.14: Packet with FCS in the medium

Whereas the destination address field helps the data-link layer perform its first function—delivery—the FCS field introduced in this section helps the data-link layer perform its second function—error detection. The FCS field is placed in the data packet at the tail end of the data. This is done because the FCS is appended to the tail end of the data in the CRC computation at the receiving end (see Figure 3.13). Locating the FCS at the tail end (called the packet trailer) eliminates the need for the receiver to rearrange data bits from different parts of the packet before performing the CRC verification. The data packet of Figure 3.7 can now be modified to also include the FCS, and it is shown in Figure 3.14. The FCS is the only overhead item in a packet among all layers in the TCP/IP stack that is placed in the trailer.

Just like Ethernet, all data-link layer technologies in use today ensure that errors are detected, and most make no attempt to fix these errors. In Chapter 5, we will see why it is perfectly fine for data-link layer technologies to not attempt to correct errors.

Ethernet Frame Structure

The packet at the data link is called a frame. This is because the preamble and FCS fields are seen as framing the packet from both ends. This is a unique feature of the structure of the data-link layer packet in Ethernet. Other layers (IP and TCP) have neither a preamble nor a field that tails the data. We can now look at the structure of an Ethernet frame. The complete Ethernet frame is shown in Figure 3.15.

Of the overhead items in an Ethernet frame, shown in Figure 3.15, we have already seen two—the destination address and the frame-check sequence (FCS). Another obvious required item of information is the source address. Just like the sender name in a letter, this field tells the receiver who to send the response to.

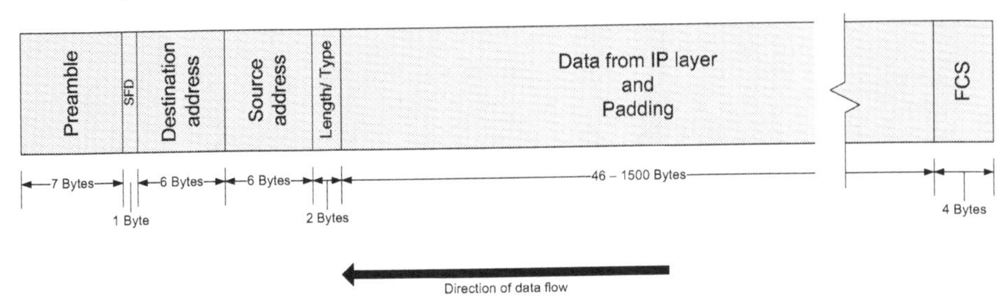

Figure 3.15: Ethernet frame structure

There are three fields in the Ethernet frame we have not discussed yet—the preamble, start-of-frame delimiter (SFD) and length/type. We can take them up now.

The preamble and start-of-frame delimiter (SFD) help the receiver locate the beginning of an incoming data-link layer frame, much like the whistle of an arriving train. Since the exact bit pattern is known in advance, these bits also help the receiver synchronize itself with the sender and know exactly how long it takes the sender to transmit each bit. These fields are necessary because in normal operation there are stray signals in the medium even if there is no network traffic. When signals from a new frame arrive at a receiver, it is extremely helpful to the receiver if it gets a clear indication that these signals represent data and are not random noise in the medium. Therefore, at the beginning of a frame, the data-link layer transmits a well-defined sequence of 7 bytes[14] (56 bits) called the preamble:

10101010 10101010 10101010 10101010 10101010 10101010 10101010

The bit sequence in the preamble is long enough and regular enough that it is extremely unlikely to be produced by random noise. At the end of the preamble, one more byte is sent to alert the receiver about the beginning of the frame. This is called the start-of-frame delimiter (SFD) and is the bit sequence 10101011. At the end of these 64 bits (56 in preamble and 8 in SFD), the useful information in the frame begins.

Since the exact bit pattern is known in advance, the preamble and SFD bits also help the receiver synchronize itself with the sender and know exactly how long it takes the sender to transmit each bit.

The field that immediately follows the SFD is the destination address, which, as described earlier, helps computers on the network determine whether the frame is addressed to them. After the destination address is the source address field. The source address field is required when creating a response to a frame. Upon processing a frame, if a computer determines that a response is required, it uses the source address of the incoming frame as the destination address of the response frame.

14 A sequence of 8 bits is called a byte. A bit is usually abbreviated using the lower case *b* (1 Mbps = 1 megabit per second), while a byte is usually abbreviated using an upper case *B* (1 MB = 1 megabyte).

After the address fields is the length/type field. The length field specifies the length of the data in the data-link layer frame. Ethernet specifies that the total length of the frame should not exceed 1,518 bytes. The length field is useful because, unlike the beginning of the frame, which is clearly marked by the preamble and SFD, there is no field that explicitly defines the end of the frame. Instead, the length field helps the receiver identify the end of the frame. The receiver counts bytes till it gets "length" number of bytes. It then knows that it has accounted for the entire frame. Once the receiver identifies the end of the packet from the length field, it knows that the last 4 bytes in the packet constitute the FCS. It then knows that the remaining information in the packet (between the length and FCS fields) is the data in the packet. As Ethernet developed, this field developed another interpretation—the type field. In this interpretation, the value of the field tells the receiver what type of frame this is. Examples include 2048 for IP, 2054 for ARP, and so on.[15] When the values of the field are greater than 1536, the field is interpreted as a type field, and when the values are less than 1536, the field is interpreted as length.[16]

After error-check, this data is passed to the receiver's network layer.

Ethernet Addresses

All the fields in the Ethernet header are automatically calculated by the host without any end-user administration or configuration. Of these fields, the address fields (source address and destination address) have a well-defined structure that may be of interest to network administrators. They are therefore described here.

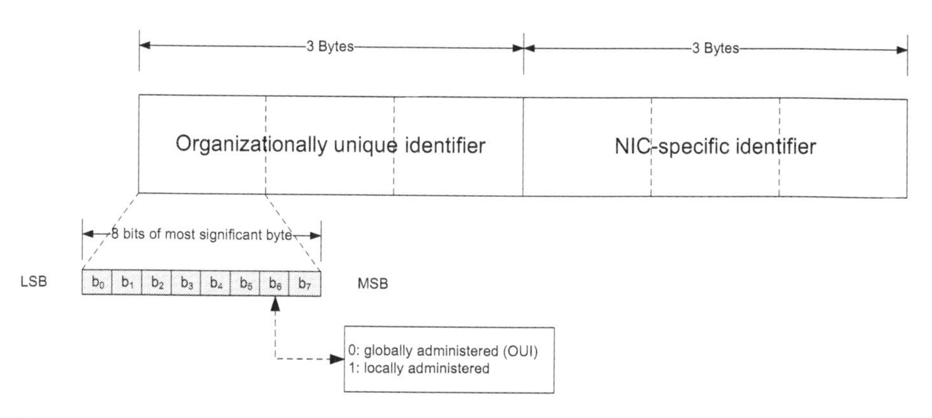

Figure 3.16: Ethernet address

15 A good reference is https://en.wikipedia.org/wiki/EtherType.

16 Modern enterprise networks use a concept called VLANs to organize their networks. A virtual LAN (VLAN) is a broadcast domain created by one or more switches. The administrator can configure one or more ports on a switch to be on the same VLAN, and broadcast packets are forwarded to ports on the same VLAN. When VLANs are used, VLAN information is included just before the length/type field. VLANs are considered to be beyond the scope of this text.

Every computer with a network interface, as in Figure 3.1, has an Ethernet address. Some computers even have two or more such interfaces for redundancy or to increase network capacity. Each such interface has its own Ethernet address. In the industry, Ethernet addresses are commonly called MAC addresses or physical addresses. The term *MAC address* comes from the fact that the address is associated with the data-link layer, which also performs medium access control (MAC). The term *physical address* comes from the fact that the data-link layer is commonly implemented in hardware called the network interface card (NIC).

Most modern computers have hardware for multiple network connections, including wired, wireless, and Bluetooth connections. In the industry, each of these connection points is commonly called an interface.

Ethernet defines the address field to be 48 bits in length. The 48 bits are split into two parts of 24 bits each as shown in Figure 3.16. The first 24 bits determine the organizationally unique identifier (OUI) or manufacturer ID. Manufacturers are assigned OUIs[17] by the IEEE.[18] The first 24 bits of every network interface card address are the OUI of the manufacturer. No two manufacturers have the same OUI. The remaining 24 bits are assigned by the manufacturer such that no two NICs made by a manufacturer have the same NIC-specific ID. Together, the OUI and NIC-specific ID ensure that every NIC card made in the world has a unique ID.

Virtualization and MAC addresses

Some modern virtualization technologies allow administrators to assign more than one physical address to a NIC. When this is done, the complete 48-bit MAC addresses are assigned by the local network administrator. As shown in Figure 3.16, Ethernet allows administrators to use one bit in the OUI part of the MAC address to indicate that the address is locally assigned. This is just mentioned here for reference. You don't have to worry about virtualization in this course.

You can view the physical address of a Windows computer by typing in the command `ipconfig/all` on your DOS prompt (Start → Type `cmd` in the Start search box). An example is shown in Figure 3.17. You may find it interesting to confirm the OUI allocation to the manufacturer of your NIC card at the IEEE registration authority website.

As seen in the figure, the physical address is usually displayed in hexadecimal format. Since most students are unfamiliar with the hexadecimal notation, an introduction is provided here. The hexadecimal notation is a way of representing binary numbers in 4-bit blocks. Thus, 48-bit MAC addresses are represented using 12 (48 ÷ 4) hexadecimal characters.

---

17 OUI records can be searched at http://standards.ieee.org/regauth/oui/index.shtml.

18 IEEE is the Institute for Electrical and Electronics Engineers.

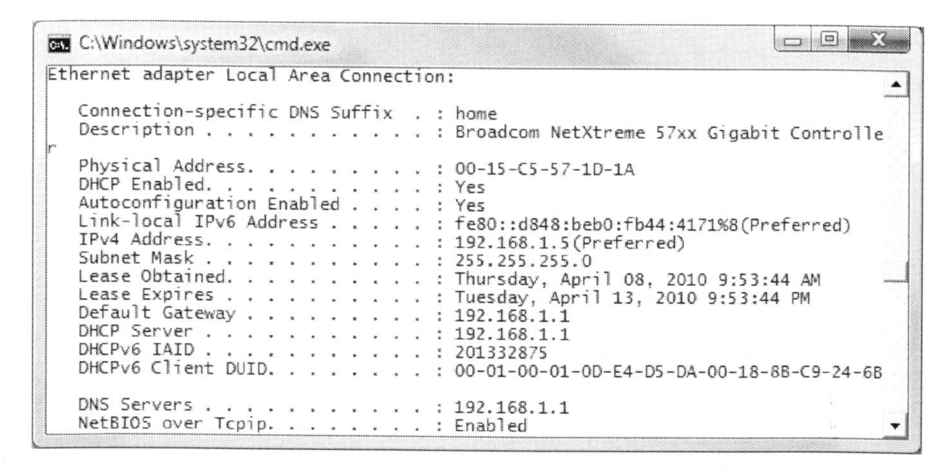

Figure 3.17: Ethernet address example

Table 3.2: Hexadecimal notation

Bits	Hex representation	Bits	Hex representation
0000	0	1000	8
0001	1	1001	9
0010	2	1010	A
0011	3	1011	B
0100	4	1100	C
0101	5	1101	D
0110	6	1110	E
0111	7	1111	F

Table 3.2 shows how each possible block of 4 bits is represented using the hexadecimal notation. For example, the bits 1100 are represented as C. The hexadecimal notation is very common in data communications. You can refer to this table for the hexadecimal representation when necessary in later chapters.

Based on the table, we can decode the binary representation of the physical address in Figure 3.17, as shown in Figure 3.18. The 48 bits of the MAC address are 0000000000010101110001010101011100011101000110 10. Looking up the OUI information, we find that the card was made by Dell.

Network arrangements—topologies[19]

Entities on a network can be connected to each other in various ways. *The general pattern of arrangement of devices on a network is called the topology of the network.* There are four standard topologies:

- Bus: All bodies on the network are connected by a single cable.
- Ring: All bodies on the network are connected in a ring.
- Mesh: A network in which at least two nodes have two or more paths between them.
- Star: A network in which peripheral nodes are connected to each other through a central node.

What is the topology of your Facebook network?

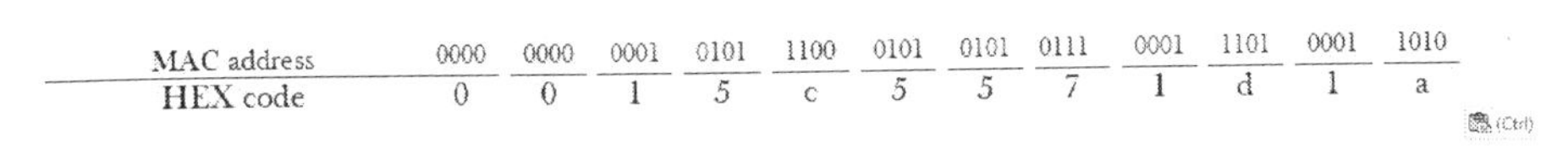

MAC address	0000	0000	0001	0101	1100	0101	0101	0111	0001	1101	0001	1010
HEX code	0	0	1	5	c	5	5	7	1	d	1	a

Figure 3.18: Binary representation of physical address shown in Figure 3.17

Switched Ethernet—State of the Market

Modern Ethernet networks largely follow the operational principles described in this chapter. However, there has been one important development that we have not covered yet. This is the replacement of hubs (shown in Figure 3.5) with switches. Switches greatly increase the data-transmission efficiency of hub-based CSMA/CD Ethernets. Almost all Ethernets designed in recent times use switches instead of hubs.

Whereas hubs broadcast all data to every computer on the network, modern high-speed electronics make it possible to create switches that read the destination addresses of incoming frames and send the frames only to the correct targets. By limiting broadcast, switches greatly reduce collisions.

With the evolution of technology, basic switches now cost about the same as hubs. There is therefore very little incentive today to use hubs instead of switches. Basic switches are very easy to use and require no configuration. Switches automatically discover the MAC addresses of computers connected to the different ports. Switches maintain this information about the computers connected to the different ports in a forwarding table. Switches look up the forwarding table to determine the port on which an incoming frame should be sent out.

Switches cannot eliminate broadcasts. Even in switched networks, a lot of support traffic continues to be broadcast. For example, if a switch hasn't yet learned about the location of a particular destination MAC address and receives a frame to be sent to this address, it floods the traffic out of all ports in the hope of finding the correct destination. This effectively makes the switch revert back to the standard hub operation for the frame. The important categories of this traffic that require broadcasts even in the presence of switches are covered in Chapter 7.

---

19 For more details and pictures, check out the discussion on "network topology" at the ATIS telecom glossary, http://www.atis.org/glossary/ (accessed Nov. 14, 2015).

Learning MAC address locations

Switches use a very interesting mechanism to discover MAC addresses. To populate their forwarding tables, switches monitor the source addresses of incoming packets on each port. For example, if the source MAC address of a frame coming into port 3 is 00:18:8B:c9:24:6B, the switch automatically knows that the computer with MAC address 00:18:8B:c9:24:6B is connected to port 3. The switch can add this information to its forwarding table. Now, if an incoming frame has destination address 00:18:8B:c9:24:6B, the switch knows that the frame should be forwarded on to port 3. This way, the forwarding table is automatically created and updated in real time with no manual intervention.

Another development is the widespread popularity of Wi-Fi networks. Many organizations are finding that their Ethernet ports have less than 50% utilization since users simply do not connect to the wired network. Wireless networks are discussed in the supplement, and share most of the core technology features with Ethernet.

Spanning Tree Protocol

Contemporary switches offer a very useful feature that protects against a common networking problem—loops in the network. When networks become large, and all connections are not visible at one place, administrators can accidentally create loops within networks (e.g. Figure 3.19). Network designers also like to design redundant paths in critical parts of the network to improve reliability (Chapter 11). These redundant paths also create loops. When this happens, frames that are being broadcast on the network get sent out by every switch on every port, creating broadcast storms that overwhelm the capacity of the network.[20]

Ethernet naming

Ethernet networks are known in the industry using a convention that summarizes the essential information about the network. The name has three parts—speed, modulation technology, and medium. Some examples are below:

- 10BaseT: Ethernet operating at 10 Mbps, using baseband modulation and twisted pair cables
- 100BaseT: Ethernet operating at 100 Mbps, using baseband modulation and twisted pair cables
- 1000BaseF: Ethernet operating at 1,000 Mbps, using baseband modulation and optical fiber cables

20 The Wikipedia article on the spanning tree protocol is a good read: https://en.wikipedia.org/wiki/Spanning_Tree_Protocol.

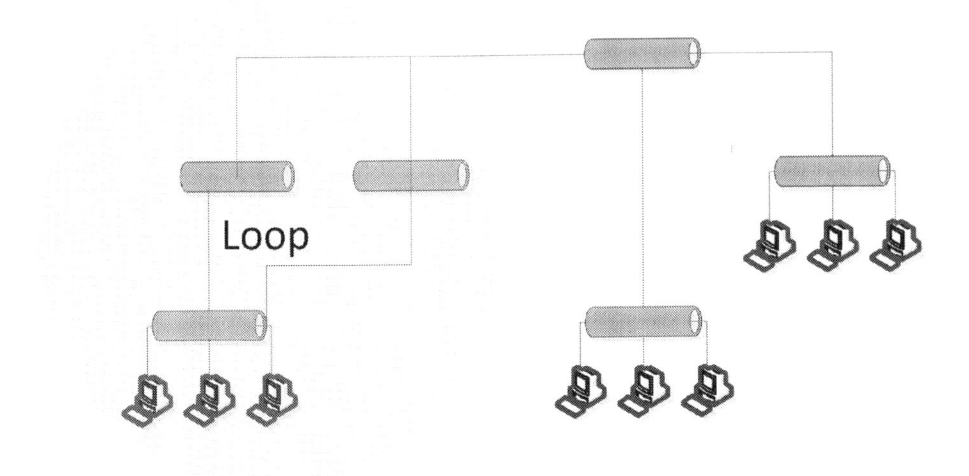

Figure 3.19: Loops in networks

Switches on a network can communicate with each other to eliminate loops by disabling the offending links. The protocol used today to ensure loop-free networks is called the spanning tree protocol. In the spanning tree protocol, switches on the network start by electing one switch as the root switch, then each switch determines the least-cost path to the root switch and disables all paths to the root that are not on the least-cost path. This eliminates loops and broadcast storms.

Spanning tree allows network administrators to maintain redundancies in networks. If one of the paths fails, the switch can immediately compute an alternate map of the network and forward frames along that path.

From Ethernet to the Outside World

Ethernet LANs are great for applications such as file and printer sharing on small networks. In the next chapter we will see how individual Ethernets are connected to other Ethernets and larger networks to create even more useful networks—Internets. At a very high level, Ethernets connect to Internets in the manner shown in Figure 3.20.

The shaded network is a departmental Ethernet LAN. The LAN is connected to the carrier network by a device called a router. Traffic directed to computers outside the LAN is sent to the router. The router determines the next hop for the packet among its neighboring routers. This process continues through the maze of other networks until the packet reaches its final destination on the other network.

Outside the Ethernet, traffic is no longer sent by broadcast for reasons of efficiency. The alternate mechanism is called point-to-point and is used everywhere outside LANs. This is covered in later chapters.

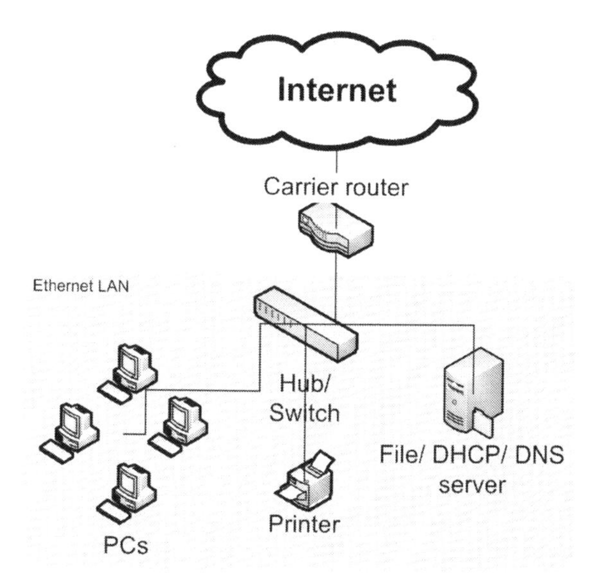

Figure 3.20: Ethernet as part of larger network

EXAMPLE CASE—Networks Helping Big Data Applications

We are all familiar with local area networks which we use every day at school, work, and home. In this example case let us see how computer networks help improve decisions and improve life. A common theme in these cases is that computer networks simplify the aggregation of data and decision rules in real time to improve productivity.

Agriculture is the cornerstone of civilization, tracing back almost 12,000 years. Many of the clay vessels dug up in archaeological excavations were made to store agricultural products. Agriculture allowed communities to get larger and eventually form cities. Once people settled down as a result of predictable food supplies, they began accumulating material goods and developing rules for self-governance (politics).

The earliest advances in agriculture were related to identifying edible plants, domesticating animals for agricultural work, building tools for agricultural work, and developing fertilizers to improve productivity. An early non-agricultural development that affected agriculture was the use of transportation networks to trade crops with distant lands. In the 20th century, high-yielding varieties of grains largely eliminated famine in most parts of the world.

While consumer-facing applications such as Google and Facebook capture much public imagination when technology is discussed, significant developments are happening behind the scenes that improve people's lives. In the 21st century, agriculture is yet another industry where significant productivity gains are likely to be achieved from the use of computer networks. Mobile applications are helping farmers determine the finest details of planting seeds to get the best harvests. Farm equipment now monitors yields and soil conditions in every part of a farm and allows farmers to use this information when planting seeds the next season. Farmers can view many layers of relevant details, such as soil type overlaid on farm maps, to make planting and fertilizing decisions. Depending upon soil conditions, optimal planting density can vary from 25,000 seeds per acre to 200,000 seeds per acre. At a cost of approximately $3 per 1,000 seeds, excess seeding can significantly hurt profits. Thus, removing the guesswork and optimizing planting can have significant impacts on farm viability.

Sellers of farm equipment are bundling these services with their equipment to generate more steady revenues after the initial sale of their equipment. These technology-driven improvements in planting and fertilization can improve corn production by about 10 bushels per acre where average production is about 150 bushels per acre.

Another modern use of computer networks is in monitoring manufacturing plants that involve the four Ds—dull, distant, dirty, dangerous. Many mines and oil rigs have these conditions and it is difficult to find qualified workers to monitor equipment in these settings. But now computer networks help operators in convenient locations to monitor these facilities and troubleshoot many problems remotely. In more complex cases, they can turn machines off until engineers are able to drive to the locations to fix problems.

Similar developments are taking place in mines. As the easy deposits have been harvested, newer mines are digging deeper and in more remote, unpopulated locations. In these hazardous locations, wired networks, satellite links, and GPS services are being used to run fully automated mines that are operated remotely from urban centers. The technologies help operators drive the machines and load and unload ore without human accidents.

These developments are raising interesting security challenges, however: Can providers of these data services gain undue advantage over competitors? For example, when a farm equipment manufacturer or seed provider is also the developer of a popular computer application to monitor farm yields, it can use the information collected by its own software to gain competitive information about the performance of its competitors. Or a monitor of plant facilities can provide information about plant performance to competitors.

References

1. Berry, Ian. "Farmers prepare for the data harvest." *Wall Street Journal*, June 4, 2012.
2. Hagerty, James. "Keeping plants running from afar." *Wall Street Journal*, Sept. 4, 2013.
3. Matthews, Robert. "Miner digs for ore in the outback with remote-controlled robots." *Wall Street Journal*, Mar. 2, 2010.
4. Zemlicka, Jack. "Writing a prescription for the future of farming." *No Till Farmer*, May 1, 2014.

Summary

This chapter covered the data-link layer. This layer transfers data over one network link or hop, such as between a laptop and the wireless router at home. The data-link layer helps senders and receivers identify themselves on the local network (addressing). The layer also ensures that any errors in data transmission over the physical medium are detected. We used the example of Ethernet to discuss the advantages and disadvantages of using broadcasts as a mechanism for data transfer. The use of CRC for error detection was described. The concept of headers was introduced for the first time in the book with a description of the Ethernet header. The two parts of Ethernet addresses were discussed. We also examined switches and how they improve the efficiency of Ethernets over hubs.

About the Colophon

For a long time physicists used the concept of ether to explain how light could travel through space where no physical medium existed. Ether was thought of as an omnipresent, undetectable medium that could propagate magnetic waves such as light waves.[21] The colophon shows how Einstein saw parallels between ether and physical matter.

When Bob Metcalfe and David Boggs were building the first network at PARC, their plan was to run a cable through every corridor in the building to create an omnipresent medium for the propagation of electronic signals carrying data packets. Seeing the parallels between ether and their network, Metcalfe and Boggs named their network "Ethernet" on May 22, 1973. Prior to getting this name, their network was called the Alto Aloha network because it connected Alto computers using ideas developed for a wireless network at the University of Hawaii.[22]

Though modern Ethernets are significantly different from Metcalfe and Boggs's network, the Ethernet name survives.

REVIEW QUESTIONS

1. What are the primary functions of the data-link layer?
2. Ethernet is the most popular end-user technology at the data-link layer. What is *ether* in the context of computer networking?
3. What are the components of a typical Ethernet? What are the functions of each component?
4. What is *broadcast* in the context of Ethernet?
5. What are the advantages of broadcasting data in Ethernet? What are some other examples of communication in day-to-day life that use broadcast?
6. What are the limitations of broadcast as a method of sending data to the intended receiver of communication?
7. What is *carrier sensing* in Ethernet? What is *multiple access*? What is a collision and what is collision detection? How are collisions detected in Ethernet?
8. Describe some techniques you have used in the past to ensure error-free communication over the telephone.
9. Provide a lay person's overview of CRC.
10. Why is CRC preferred over simpler computational techniques?
11. With a divisor of 1101, perform the sender-side computation and calculate the CRC when the data is 1001010.

---

21 Albert Einstein, "Ether and the Theory of Relativity," address delivered on May 5, 1920, at the University of [illegible], http://www.tu-harburg.de/rzt/rzt/it/Ether.html (accessed Feb. 27, 2010).

22 Cade Metz, "Ethernet—a name for the ages," *The Register*, Mar. 13, 2009, http://www.theregister.co.uk/2009/03/13/metcalfe_remembers/print.html (accessed Feb. 27, 2010).

12. Check your computation by performing the receiver-side computation.
13. What is the size of the smallest Ethernet frame? The largest frame?
14. List the fields in the Ethernet header. What are the roles of each of these fields?
15. The start-of-frame and preamble fields are unique to the data-link layer in that they do not carry any useful information. What are the roles of these fields?
16. The SFD field alerts the receiver about the beginning of a data frame. How does the receiver know when the frame ends?
17. What is the structure of a MAC address? What information can be gathered from a MAC address?
18. What is the MAC address of your computer? You can get this information by typing `ipconfig/all` in Windows, or `ifconfig` on Mac/Linux (Figure 3.21).
19. What is the hexadecimal notation? How is the number 14 represented in hexadecimal?
20. Write the binary number 01010000 in hex (hint: break the number into two 4-bit blocks and represent each 4-bit block in hexadecimal notation).
21. What are *hubs*?
22. What are *switches*?
23. What are the advantages of switches over hubs in Ethernets? Under what conditions might you prefer to use a hub instead of a switch?
24. What is the spanning tree protocol? What is it used for?
25. What are the common data-transfer speeds in Ethernet? What is the maximum possible speed of the network card on your computer? In Windows, you can right-click on the network adapter to check its speed (Control panel → Network connections → <select adapter>).

EXAMPLE CASE QUESTIONS

1. Check out the websites of Field View and Seed Sense. What are some of the key features of these applications?
2. If you had to add a feature to each of these apps, what would they be?
3. Mining has historically been an important source of well-paid jobs for unskilled workers. Automation eliminates thousands of these jobs. You are the local elected representative in an area where copper has just been discovered and a miner is planning to invest in a heavily automated mine that will bring in few local jobs. Will you support this investment? Why, or why not?

HANDS-ON EXERCISE—OUI Lookup

We saw in this chapter that Ethernet specifies a 48-bit address for network cards. This address is popularly called the MAC address or the physical address of the interface. Recent laptop computers usually have three such interfaces—one for the wired network card, another for the wireless LAN card, and a third for the Bluetooth interface. In this exercise, you will identify the MAC addresses of the interfaces on a computer, convert the hexadecimal representations to the 48-bit binary addresses, and identify the manufacturers of the network cards.

```
C:\Windows\system32\cmd.exe

C:\>ipconfig /all | more

Windows IP Configuration

   Host Name . . . . . . . . . . . . : U252406
   Primary Dns Suffix  . . . . . . . : forest.usf.edu
   Node Type . . . . . . . . . . . . : Peer-Peer
   IP Routing Enabled. . . . . . . . : No
   WINS Proxy Enabled. . . . . . . . : No
   DNS Suffix Search List. . . . . . : forest.usf.edu
                                       home
                                       usf.edu

Ethernet adapter Bluetooth Network Connection 3:

   Media State . . . . . . . . . . . : Media disconnected
   Connection-specific DNS Suffix  . :
   Description . . . . . . . . . . . : Bluetooth Device (Personal Area Network)
#3
   Physical Address. . . . . . . . . : 00-1A-6B-30-C1-C8
   DHCP Enabled. . . . . . . . . . . : Yes
   Autoconfiguration Enabled . . . . : Yes

Wireless LAN adapter Wireless Network Connection:

   Connection-specific DNS Suffix  . : home
   Description . . . . . . . . . . . : Dell Wireless 1490 Dual Band WLAN Mini-Ca
rd
   Physical Address. . . . . . . . . : 00-19-7E-30-0A-2B
   DHCP Enabled. . . . . . . . . . . : Yes
   Autoconfiguration Enabled . . . . : Yes
   Link-local IPv6 Address . . . . . : fe80::d0b0:fb8f:a087:ebe8%9(Preferred)
   IPv4 Address. . . . . . . . . . . : 192.168.1.152(Preferred)
   Subnet Mask . . . . . . . . . . . : 255.255.255.0
   Lease Obtained. . . . . . . . . . : Wednesday, March 03, 2010 10:19:28 PM
   Lease Expires . . . . . . . . . . : Thursday, March 04, 2010 10:19:27 PM
   Default Gateway . . . . . . . . . : 192.168.1.1
   DHCP Server . . . . . . . . . . . : 192.168.1.1
   DHCPv6 IAID . . . . . . . . . . . : 151001470
   DHCPv6 Client DUID. . . . . . . . : 00-01-00-01-0D-E4-D5-DA-00-18-8B-C9-24-6B
   DNS Servers . . . . . . . . . . . : 192.168.1.1
   NetBIOS over Tcpip. . . . . . . . : Enabled

Ethernet adapter Local Area Connection:

   Connection-specific DNS Suffix  . : home
   Description . . . . . . . . . . . : Broadcom NetXtreme 57xx Gigabit Controlle
r
   Physical Address. . . . . . . . . : 00-15-C5-57-1D-1A
   DHCP Enabled. . . . . . . . . . . : Yes
   Autoconfiguration Enabled . . . . : Yes
   Link-local IPv6 Address . . . . . : fe80::d848:beb0:fb44:4171%8(Preferred)
   IPv4 Address. . . . . . . . . . . : 192.168.1.5(Preferred)
   Subnet Mask . . . . . . . . . . . : 255.255.255.0
   Lease Obtained. . . . . . . . . . : Wednesday, March 03, 2010 11:00:39 PM
   Lease Expires . . . . . . . . . . : Thursday, March 04, 2010 11:00:39 PM
   Default Gateway . . . . . . . . . : 192.168.1.1
   DHCP Server . . . . . . . . . . . : 192.168.1.1
   DHCPv6 IAID . . . . . . . . . . . : 201332875
   DHCPv6 Client DUID. . . . . . . . : 00-01-00-01-0D-E4-D5-DA-00-18-8B-C9-24-6B
   DNS Servers . . . . . . . . . . . : 192.168.1.1
   NetBIOS over Tcpip. . . . . . . . : Enabled

Tunnel adapter Local Area Connection* 6:

   Media State . . . . . . . . . . . : Media disconnected
   Connection-specific DNS Suffix  . :
   Description . . . . . . . . . . . : isatap.{EC25907B-A190 [illegible] [illegible]-40CF2E2DB
995}
```

Figure 3.21: Viewing configuration of network interfaces

The ipconfig utility is a convenient way to see the configuration information of all these interfaces on Windows computers. On Macs and Linux machines, the equivalent utility is ifconfig. To use ipconfig, you need to open up the command prompt (Start → Run → type `cmd` on the search field). At the command prompt, you can type the command `ipconfig /all` to view the configuration of all network information. If you have a number of such interfaces, you can paginate the output using the command `ipconfig /all | more` as shown in Figure 3.21.

The output in Figure 3.21 shows that the MAC address of the Bluetooth interface is 00-1A-6B-30-C1-C8; of the wireless adapter is 00-19-7E-30-0A-2B; and of the wired Ethernet card is 00-15-C5-57-1D-1A. We can look up the hexadecimal table in the chapter to convert each hexadecimal digit to binary and obtain the 48-bit MAC address of the Bluetooth interface as 0000 0000 0001 1010 0110 1011 0011 0000 1100 0001 1100 1000.

Looking up the OUI 00-1A-6B of the Bluetooth adapter card in the public OUI listing[23, 24] we find that the Bluetooth adapter is manufactured by USI, located in Taiwan, Republic of China. Similarly, the wireless card is manufactured by Hon Hai Precision Ind. Co., Ltd, Taiwan, Republic of China; and the OUI of the wired Ethernet interface card is owned by Dell.

Do the following on a computer you use at work or at home.

1. Show the output of the command ipconfig /all or ipconfig /all | more (if there are numerous adapters on your computer).
2. From Question 1 above, what are the MAC addresses of the different interfaces on your computer?
3. Express each of these MAC addresses as 48-bit binary addresses.
4. Look up the OUIs of each of these MAC addresses and list the names and locations of the manufacturers of these MAC cards.

CRITICAL-THINKING EXERCISE—Broadcast and Search

In this chapter we saw how broadcast is useful in search. Broadcast ensures that information reaches the intended destination. However, driverless cars, the much-anticipated technology, do not use broadcast for obstacle detection, even though the technology needs to detect all obstacles in the vicinity. Since broadcast by its very nature is directionless, broadcast would not help the cars determine where a detected obstacle is located. Instead, the cars currently use LIDAR (light radar), which sends highly directed laser beams in all directions around the car, and makes more than one million observations each second. When the beam hits an object, it is reflected back, and this reflection can accurately locate the obstacle.

1. What are some situations in which we do use broadcast as a mechanism for locating targets in our daily lives?

23 http://standards.ieee.org/regauth/oui/index.shtml.
24 http://standards.ieee.org/regauth/oui/oui.txt.

2. Processing the million observations each second generates computational complexity. Imagine a future where all vehicles and pedestrians on the road could be guaranteed to have responders to queries from nearby cars. Can this simplify obstacle detection by autonomous vehicles? If yes, suggest some ways.

IT INFRASTRUCTURE DESIGN EXERCISE—Ethernet Diagram

Answer the following question for TrendyWidgets, based on the details about the firm provided at the end of Chapter 1.

1. Assume that all locations use Ethernet for local connectivity within the buildings. Typically, each floor in each building will have its own Ethernet, and these Ethernets will be connected to other Ethernets in the same building through a switch. Draw the Ethernet diagram for the second floor of the AP service center.

CHAPTER 4

Network Layer

The journey of a thousand miles begins with a single step.

Chinese proverb

Overview

This chapter covers the network layer. The network layer transfers data from a source computer to a destination computer via one or more networks. Each network in the path uses the data-link layer technology of its choice to transfer data to the next network. The standard protocol used by all networks at this layer is the Internet protocol (IP). IP assigns an address to every device on the network, and every IT professional and most computer users need a working knowledge of IP addresses. At the end of this chapter you should know:

- the functions of the network layer
- an overview of the Internet protocol (IP)
- the IP header
- IP addresses
- the CIDR notation for IP addresses
- obtaining IP addresses
- IP version 6

Functions of the Network Layer

The network layer is responsible for transferring packets of data from the source computer to the destination computer via one or more networks. The data-link layer is responsible for transferring data within a network. In general, though, the source and destination of a data packet can be located on different networks. For example, when you visit your school's home page from your home PC, the source is on your home network and the destination is on the university network. The data-link layer cannot exchange data between these two computers, and the network layer becomes necessary to transfer data across networks. This function of the network layer may be summarized in one word—routing. *Routing is the process of selecting a path on the Internet that can be used to deliver data to a destination.*

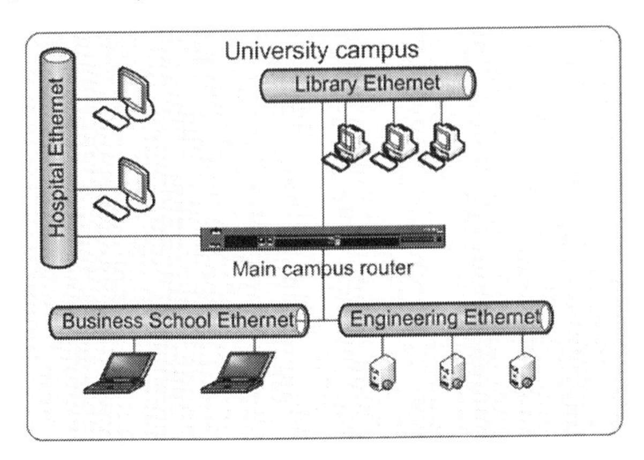

Figure 4.1: Routers connect networks.

Routing is performed in networks by devices called routers. *Routers are devices that connect two or more networks and forward incoming packets to the appropriate connected network.* Routing is examined in Chapter 8. This chapter focuses on the tasks performed by the source and destination computers to help routers perform their routing function.

Figure 4.1 provides an overview of the role of routers. When a network grows to the point where it is too large for an Ethernet, the network can be divided into multiple Ethernets. A router can be used to connect the Ethernets, creating a network of networks.

To help routers perform their routing function, every device on the network is assigned a unique network layer identifier. Since this address is defined by IP, the network layer protocol, this address is popularly called the IP address of the computer. *IP addresses are identifiers used as addresses of computer resources on the Internet.* Before sending a data packet out on the network, the sender adds the destination's IP address to the header of every data packet.

In Chapter 3, we saw that networked computers also have MAC addresses. You might wonder how data-link layer addresses are related to IP addresses. If you simply want to identify a computer on the network, why do you need two different addresses for every computer? The relationship between data-link layer addresses and IP addresses is clarified in Figure 4.2. The path from source to destination comprises many hops. A hop is a link from a router to the next neighboring router. The data-link layer address allows the packet to be delivered correctly within each individual network (i.e. over one hop) on the path to the destination. The IP address retains the address of the actual source and the final destination. Note in Figure 4.2 that the destination IP address does not change throughout the packet's journey from the source (PC) to the destination (web server). However, the data-link layer addresses keep changing as the packet gets transferred from network to network. Thus, immediately as the packet leaves the PC, the data-link destination address points the packet to the nearest router. The data-link layer technology on this hop uses the data-link layer destination address to deliver the packet to router 1. On the second hop, the data-link layer destination address is router 2, and so on. All this time, the destination IP address maintains the address of the final destination.

Political impacts of networking

In recent years, some significant events have highlighted the political impacts of a free and open Internet. Here is an example. Select the context in which the quote below was made (answer in the footnote[1]):

"*This could literally make the difference in terms of what happens in that country.*"

A. The US Defense Secretary at a congressional hearing on the defense budget.
B. A State Department official requesting Twitter to delay an upgrade that would have shut down the network while riots were going on in Iran.
C. The Central Banker of the European Union at a policy meeting discussing economic recovery.
D. The Chinese Premier discussing the annual budget.

On its way to the destination, the packet encounters its first router (R1) when it reaches the outer edge of its local Ethernet. This router is connected to one or more other routers. The router looks at the destination address of the packet and determines the best neighboring router to hand the packet over to. In this example, it is R2. With each successive handover, the packet moves closer to its target until it is finally delivered to the correct destination (web server).

It is amazing that this process works. But it is one of the miracles of modern technology that billions of data packets are delivered in this manner on the Internet every day and they all reach their destinations almost instantaneously without incident.

There is a very direct analogy between routing and a typical road trip. Say you want to travel from your home in Arlington, Texas, to Busch Gardens, Tampa, Florida. The final destination address is Busch Gardens, Tampa. Busch Gardens, therefore, corresponds to your IP destination address. However, your route map will point to several intermediate destinations along the way. For example, exit 53B on I-30E to US-80, left on N 56th Street, and so on. Each of these intermediate destinations is analogous to a data-link layer destination address. All the while, you never forget your final destination address—Busch Gardens. However, at each point on the trip, your goal is always to get to the next exit. As long as you get all the intermediate destinations right, you are guaranteed to get to the final destination.

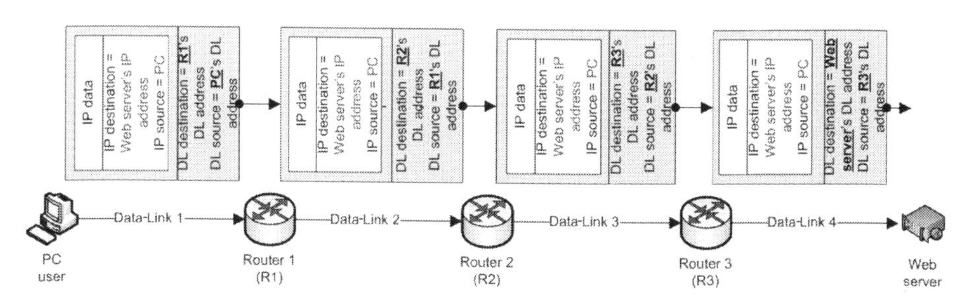

Figure 4.2: IP addresses and their relationship to data-link layer addresses

1 Answer: B; Source: Nick Bilton, *Hatching Twitter* (Portfolio, 2013); also see http://www.reuters.com/article/2009/06/16/us-iran-election-twitter-usa-idUSWBT01137420090616.

Virtual trip, destination: ?

Think of an international tourist destination you would like to go to on your next vacation. Write down the major intermediate hops and technologies used to reach each hop. Each Internet packet goes through almost the same process as it hops from router to router on its path to the destination.

Technology: Car	destination: Airport
Technology: ?	destination: ?
Technology: ?	destination: ?
Technology: ?	destination: ?
Technology: ?	destination: ?
Technology: ?	destination: ?

A related issue is: Why is it necessary to use a different address format at the data link layer, such as the MAC address, to identify intermediate destinations? Why can't intermediate destinations also be identified by IP addresses? One reason is that some of these data-link layer technologies (e.g. Ethernet) were developed independently of IP, and already had their own address formats before being integrated with IP. Another reason is that allowing data-link layer addresses to be independent of IP addresses provides great flexibility in technology design at the data-link layer. It also allows the data-link layer to be independent of the network layer. Upgrades to IP, whenever necessary, will not require simultaneous upgrades to data-link layer technologies. For example, the transition to IPv6 has not required new Ethernet cards because Ethernet does not use IPv4 addresses.

The designers of the Internet deliberately designed the network layer in such a way that all the complexity of routing is located in specialized devices called routers. The source and destination hosts do not have any routing responsibilities. The only thing senders have to do is to label packets with the correct source and destination addresses. Routers take care of the rest to deliver the packets to the correct destination. This design allows even simple devices, such as handheld computers and inexpensive security cameras with limited processing capability, to be connected to the Internet. These devices do not have to be capable of performing the complex operations required to route packets. Any device that can label packets with source and destination IP addresses can access the wonders of the Internet, without worrying about routing.

Overview of the Internet Protocol (IP)

The most common protocol used at the network layer is the Internet protocol, also known by its acronym, IP. A number of technologies have competed for dominance at the network layer in the past. However, IP has emerged as the clear winner at the network layer with virtually no viable competition at this time. Therefore, this book focuses on IP when discussing the functioning of the network layer.

IP was specified in 1981 in RFC 791. A major factor that helped popularize IP was its use to enable networking in the BSD UNIX distribution. At that time, though few large communities of users had access to computers, the computer science academic community was already using BSD UNIX for its day-to-day computing needs. The incorporation of TCP and IP in BSD UNIX dispersed these protocols among the leading-edge users of the time, paving the way for further adoption of TCP and IP among other communities and, eventually, by all major operating systems.[2]

RFCs

RFCs stand for requests for comments. *RFCs are a series of documents that began in 1969, describing Internet technologies and related experiments.* RFCs capture the research and engineering thought underlying the Internet. They also served as an early version of a publicly available Facebook wall of the group of experts who were developing networking technologies. A good introduction to RFCs is RFC 2555, summarizing the first 30 years of RFCs.

RFCs reflect the "perpetual beta" heritage of the Internet. All current Internet technologies are working drafts, serving as starting points for collaboration instead of being the final word on a technology. RFCs also reflect the informal nature of the Internet. Steve Crocker, then a graduate student at UCLA, typed the first RFC on the night of April 7, 1969, in a bathroom so his friends sleeping in the apartment were not disturbed.[3]

IP is highly adaptable. This adaptability has helped IP maintain its popularity for more than a quarter of a century, even as the applications being used on the Internet have changed dramatically since the introduction of IP in the early 1980s. Most applications that make the Internet popular today were not even envisioned when IP was first introduced. The web was created almost 10 years after IP, in the early 1990s, bringing network interactivity to the masses. Real-time applications such as instant messaging emerged even later. Early this century, bandwidth-hungry media applications such as video became popular. IP has successfully handled the routing needs of all these new applications. The success of IP at meeting the evolving needs of the Internet so far gives us confidence that IP will continue to be successful in the future at the network layer.

When defining the functions of IP, its designers deliberately limited the capabilities of IP in one important respect. IP does not provide end-to-end reliability. In other words, if there are problems such as defective routers along the path from source to destination, IP will not worry about ensuring that data packets will actually reach their destination. Instead, reliability is provided by the transport layer. In other words, once a router dispatches a data packet on to a neighboring router, the packet is left to its own fate and the router does not worry about the packet anymore. If the neighboring router is unable to process the data packet for any reason, the packet can get lost, but routers do not worry about recovering lost packets. Instead, routers focus on routing the next packet. For this reason, the service provided by IP is also called best-effort delivery. *Best-effort delivery is a network service in which the network does not provide any guarantee that data will be delivered.* Limiting IP functions to best effort greatly simplifies the design of routers.

2 http://www.isoc.org/internet/history/brief.shtml.

3 Johnny Ryan, *A History of the Internet and the Digital Future* (Reaktion Books, 2013).

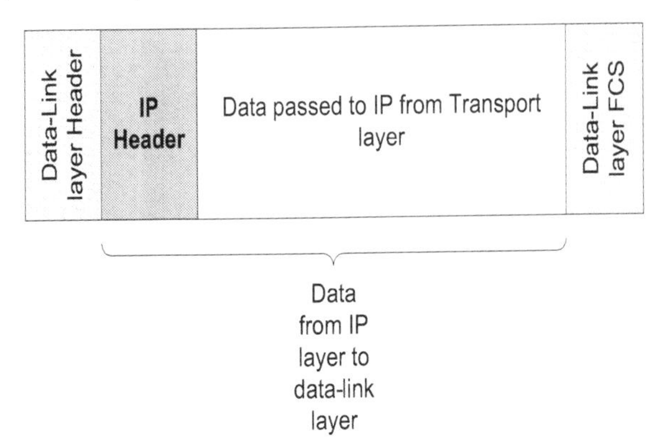

Figure 4.3: Data-link and IP headers in relation to packet

The postal service is an example of best-effort delivery. There are no schedule guarantees, and in fact, not even a guarantee that mail will actually be delivered. However, eliminating all the tracking associated with more reliable package delivery dramatically reduces costs, and the service provided is deemed acceptable.

IP Header

Before proceeding further to examine the IP header, it is useful to integrate what we have learned so far. When a packet is in the medium, the location of the data-link layer header, the IP header, and the data passed to IP from the transport layer is as shown in Figure 4.3. Leading the packet is the data-link layer header. Immediately behind it is the IP header. The data-link layer FCS is at the trailing end of the packet. The IP header and transport layer data together constitute the IP data for the data-link layer. A comparison between Figure 4.3 and Figure 3.15 would help to further clarify how IP packets fit into the data-link layer frame.

The IP header is shown in Figure 4.4. The data bits follow immediately after the header. The IP header adds information that enables IP to do its job. The information in the various fields of the IP header is used primarily by routers to perform routing. The information in the fields and their functions are briefly described below.

Version The version field tells routers the version of IP being used. Having a separate field to identify software versions simplifies upgrades.[4] Most of the Internet currently runs version 4 of IP. The upgrade to IP is numbered version 6. The primary motivation for upgrading IP is to increase the potential size of the Internet. More details on version 6 are provided in a later section of this chapter.

---

4 Ethernet does not have a version field. Therefore, as new features such as VLANs got added to Ethernet, the interpretation of the length field was hacked to mean length/type to provide information analogous to the IP version field. Had Ethernet had a version field, it would have been much easier to define newer versions of Ethernet, with different features, without resorting to these hacks.

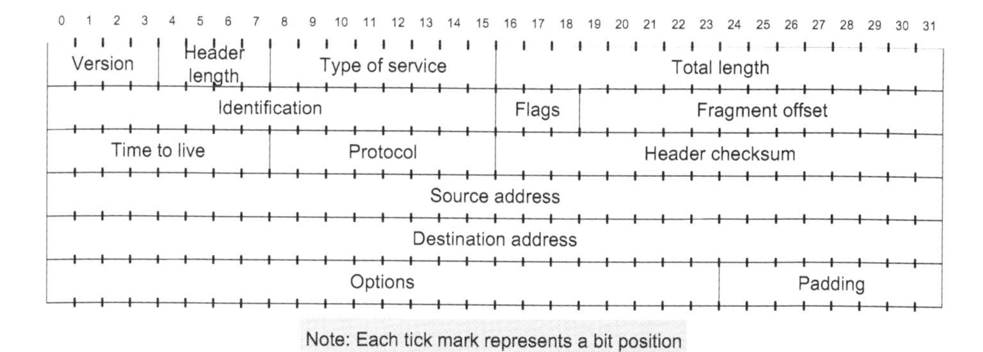

Figure 4.4: IP header

Header length This field specifies the length of the IP header. This field is necessary because IP allows the header to include various options, which means that the length of the header can be variable. The IP header specifies where the header ends and the data begins. Note that the header length of IP only specifies the length of the IP header, not the length of the entire frame. This field would not be needed if the IP header had a fixed size.

Type of service *Type of service is a specification of the desired priority.* The designers of IP allowed the source to specify the desired service priority. This field is colloquially called the TOS field. A higher number in the TOS field signifies a higher desired priority. Implementing TOS requires that routers be able to maintain different queues for packets with different TOS values. This is analogous to the efforts necessary to maintain high-occupancy-vehicle (HOV) lanes in some busy parts of the interstate system, or in separate lines for priority passengers while boarding aircraft. Carriers implementing TOS may want to charge more for higher-priority delivery. Since implementing multiple queues is difficult, at this time networks are not required to honor the desired service priority.

Total length This field specifies the size of the packet, including the header information and data. The maximum possible value of this field is 65,535 (the discussion on binary numbers later in the chapter will show why this is so). This field, therefore, limits the size of IP packets to approximately 65 KB. Note that this field is analogous to the frame-length field of Ethernet. Interestingly, initially the recommended size limit was 576 bytes to accommodate the limited processing capability of computers.

Identification IP allows packets to be fragmented if necessary. Fragmentation allows packets to be broken down into smaller fragments if an intermediate network in the path is unable to handle large packets. If a packet is fragmented, all fragments of a packet have the same value in the identification field, enabling reassembly by the receiver. In this book, we do not focus on the fragmentation function of IP.

Flags This field indicates whether a packet may be fragmented, and whether it has, in fact, been fragmented.

Fragment offset If an IP packet is fragmented, this field specifies the position of the current packet with respect to all other fragments with the same identification.

Time to live As the name suggests, this field specifies the remaining life of the packet on the network. When a packet is sent out, the source puts some value in this field, usually 128 or 256. Each time a packet passes a router, the router decrements the value of this field by 1. If the value of the TTL field of a packet ever reaches 0, the router discards the packet. This mechanism ensures that no packet can live on forever on the Internet even if poorly configured routers send packets on endless circular routes. Thus, this field acts as a safety valve on the Internet.

Note that this field also implies that there can be no more than 256 routers between any source and destination on the Internet. In practice today, packets usually pass through a maximum of about 15–20 routers.

Protocol Many transport layer technologies have been defined as potential users of IP. The protocol field in IP identifies the transport layer technology that is sending and receiving the packet. The values in the protocol field for familiar transport layer technologies include 6 (TCP) and 17 (UDP). These numbers are defined in RFC 1700.

Header checksum This field carries error-detection information for the packet header. The IP header checksum is only calculated over the IP header, not over the packet data. Unlike the CRC checksum that is used to ensure data integrity in the data-link layer, the error-detection procedure used in the IP header is very limited in its error-detection capability. However, the most vital information in the IP header is the source and destination IP addresses, and the IP header checksum provides some protection against errors to these addresses. The IP header checksum therefore gives a certain amount of assurance that IP packets will only get delivered to the correct destination. The header checksum is recalculated at each node.

A very effective way to learn about a protocol is to look at the protocol header and examine the functions of each field in the header.[5]

Thus, there are two major differences between the IP header checksum and Ethernet CRC. CRC is for the entire frame whereas the IP checksum is only computed over the header. Also, while CRC is very robust at error detection, the IP header checksum is less reliable. Since the Ethernet data also includes the IP header, the Ethernet CRC also checks for errors in the IP header. In fact, in the newest version of IP, IP version 6, the header checksum has been eliminated since the Ethernet CRC was deemed sufficient.

5 RFC 791 is a great resource for this section. A reading of the RFC is highly recommended for readers of this chapter. The RFC is available from many sources, e.g. http://www.rfc-editor.org/rfc/rfc791.txt. Figure 4.4 is from RFC 791.

Source and destination addresses As indicated in Figure 4.2, these fields identify the originating source and ultimate destination of the packet. The next section of this chapter focuses on these two fields. In fact, these two fields are a major focus of this chapter.

At the end of this course, even if you are not involved in networking, you are likely to find yourself repeatedly using two items of information from this course extensively—the names and functions of the OSI layers (Chapter 1) and IP addresses (this chapter).

Options This field allows the source of the packet to specify various kinds of optional information for use in routing. For example, one early option was to define source routing, whereby the source could specify the path from source to destination and insist that routers only send packets along that route. This option was considered necessary for military users of the Internet who would wish to specify a route that avoided routers in hostile countries. As the name suggests, options are optional, a source is not required to specify any options. Options are generally not used any more.

Padding This field consists of a string of 0s to ensure that the IP header is a multiple of 32 bits in size. Padding becomes necessary if options are used.

What is data? What is overhead? Depends on who you ask

Thus far, we have seen the protocol headers at two layers—the data-link layer and the network layer. This introduction to protocol headers demonstrates a very interesting feature of protocols. End users care only about the data in the packet, and treat the protocol header as overhead. But the hardware and software implementing the protocol only know how to interpret the fields in the header. The data is merely payload that makes the headers necessary.

On the receiving side, as soon as a layer interprets the header information inserted by its counterpart on the sender side, it runs into information that it does not understand. For example, see Figure 4.3. As soon as the data-link layer processes the data-link header, it encounters the IP header. The data-link layer software has no idea how to interpret the IP header. So, the data-link layer passes all this information to the IP protocol at the network layer for processing. This process is repeated at every layer until eventually the application layer passes usable data to the end user.

This can also be compared to mail. Say you mail a legal contract. To you, the address label is merely overhead necessary for mail delivery. To the mailman, however, the overhead is precisely the information that lets him do his job. The mailman knows how to interpret zip codes, street addresses, and everything else that appears on the envelope. But even if the mailman opened the letter, he would have little idea what the contract was about. To the mailman, the mail contents are incomprehensible.

So, the network knows how to process the header, while the sender and receiver know how to process the data.

IP Addresses

Of all the fields in the IP header, the source and destination address fields get the most attention. As a matter of fact, these two fields are probably the best-known fields among all protocol headers. In this section, we examine the IP address fields in more detail. Given the attention paid to IP addresses, it is important that every IT professional has a good working knowledge of IP addresses and their organization. IP addresses may be the most important networking-related operational detail for any IT professional.

An address is a unique label that helps locate an entity on a network. As Figure 4.4 shows, IP address fields are 32 bits in length. Every network connection of a computer connected to the Internet is assigned a unique 32-bit number as an IP address.[6] For example, a laptop with a wireless connection and a wired connection will have one IP address for the wireless connection and one IP address for the wired connection. This number serves as the network layer source address for all packets leaving the computer from the connection. This is also the destination address for all packets arriving at the connection.

Names and addresses

We mentioned in the context of Ethernet addresses that MAC addresses were more like names than addresses. The addresses discussed in this chapter, IP addresses, function like addresses. So, what is the difference between a name and an address?

What makes addresses different from names is that whereas names have no directionality, addresses give users an indication of the direction in which to move to get closer to the destination. Street numbers play this role on roads. For example, in Figure 4.5, a driver knows that he needs to go left if the street address is below 16900 and to the right if the street address is above 17000. Names identify individuals and can be used for addressing within a small group, say, for example, a family, a classroom, or a birthday party. Typically, when the recipient is known to be in the vicinity and communication uses broadcast, names are more convenient for locating the recipient. But when the recipient has to be located within a wider area, the directionality information embedded in addresses becomes very useful.

The importance of IP addresses in packet routing is reason enough to learn about IP addresses. There is, however, an even more important reason for you to spend time learning about IP addresses. By design, network administrators have tremendous flexibility and discretion in allocating IP addresses to computers within their organizations. Therefore, an understanding of the organization of IP addresses will help you leverage this flexibility when required.

A working knowledge of binary numbers is essential when working with IP addresses. The following overview provides the information necessary to be comfortable working with binary numbers involved in computer networking in general, and IP addresses in particular.

6 This statement is not strictly true. As we will see in Chapter 7, Network Address Port Translation (NAPT) allows a set of IP addresses to be reused.

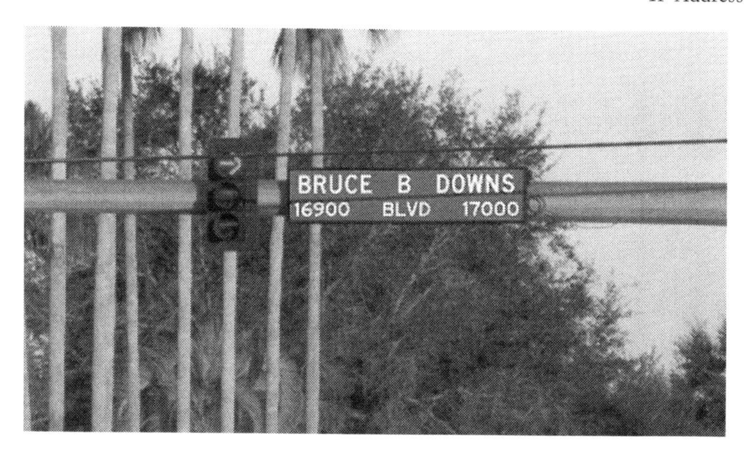

Figure 4.5: Street numbers help direct users to their destination
(Source: Author photo)

If you are comfortable working with binary numbers, you can skip the overview of binary numbers below.

Binary Numbers Overview

Binary numbers represent numerical quantities using two digits, 0 and 1, where each digit is a coefficient of the corresponding power of 2. This overview provides everything you need to know about working with binary numbers for computer networking. We know from Chapter 2 that all computer data is stored as binary numbers. It becomes particularly important to be comfortable with binary numbers when working with computer addresses. IP addresses, like all other protocol header fields, are binary numbers. Working knowledge of binary numbers helps network administrators compute the best possible organization of IP addresses, and calculate the maximum possible sizes of computer networks and other network details.

Table 4.1: Labeling with binary numbers

Bits	Labels	Number of labels	Formula for number of labels
1	ID: 0 ID: 1	2	2^1
2	ID: 00 ID: 01 ID: 10 ID: 11	4	2^2
3	000, 001, 010, 011, 100, 101, 110, 111	8	2^3
n			2^n

One particularly important skill is the ability to convert from binary numbers to decimal and vice versa. Fortunately, in most cases, you will only deal with binary numbers that are 8 bits in length. You should therefore be very comfortable working with 8-bit binary numbers.

Our primary interest in binary numbers in this book is to use binary numbers to assign computer addresses. An address is essentially a unique label that locates a computer on the network. Larger numbers allow more computers to be identified. Therefore, our primary goal in this overview is to be able to determine how many computers can be identified, given the size of binary numbers used as identifiers.

There are 10 types of people in this world.
Those who know binary, and those who don't.

Before using binary numbers to calculate the number of possible addresses, let us use decimal numbers and a simple example to understand what we are trying to do. Say you are asked to use one-digit decimal numbers to label computers in a student lab. How can you proceed? You can label the first computer as 0, the second computer as 1, and so on, until you label the 10th computer as 9. At this point you will run out of unique labels. If the labels are not unique, they cannot serve as addresses. For example, if two computers have the label 0, there would be conflicts and confusion in determining which computer really had the label 0. Therefore, even if you have more computers in the lab, if you can only assign one-digit labels, you will only be able to use 10 of the computers. If you use two-digit labels, you will be able to assign labels to 100 computers (0–99) before you run out of labels.

Car lights and binary numbers

A student group in my class at USF presented a great example of binary numbers—a car's turn indicator lights. Each light takes on two values—on and off. The two lights constitute a 2-bit number. When both lights are off, it indicates that everything is normal. When the left (or right) light is on, it indicates that the driver intends to turn left (or right). When both lights are on, it indicates that the driver is slowing down.

Therefore, we see that the number of possible labels depends upon the size of numbers used to assign the labels. In general, using n-digit decimal numbers, we can assign 10^n labels.

Binary digits are called bits. Each bit can take two values—0 and 1. Continuing the idea in the previous paragraph, when using binary numbers to assign labels, if we have 1-bit binary numbers, we can assign two labels. The first label will be a 0; the second label will be a 1. If we have 2-bit binary numbers, we can assign four labels—00, 01, 10, and 11. With 3-bit binary numbers, there can be 8 labels—000, 001, 010, 011, 100, 101, 110, and 111. With n-digit binary numbers, we can have 2^n labels.

These relationships are shown in Table 4.1. The first row of the table shows the labels that are possible with 1-bit binary numbers. We can have two such labels as shown in the first row. The table shows one computer being assigned label 0 and the other computer being assigned label 1.

Table 4.2: Example of decimal number

3	5	8	Digit
100	10	1	Place value
(102)	(101)	(100)	

The second row uses 2-bit labels. We can have four computers with 2-bit labels. Computers with these labels are shown. The labels are 00, 01, 10, and 11. Similarly, if we have 3-bit labels, we can label eight computers. For brevity, these computers are not shown in the table, though the labels that would be used are shown. In general, with n-bit labels, we can label 2^n computers. This means that if we have n bits to label computers in a network, the network can accommodate at most 2^n computers if every computer on the network requires a unique label to serve as its address on the network. More address bits imply more possible computers on the network. Every additional address bit doubles the potential size of the network.

Table 4.3: Example of binary number

1	0	1	Digit
4	2	1	Place value
(2^2)	(2^1)	(2^0)	

Network administrators specify the number of address bits available on a network in a standard format called the subnet mask (subnet masks are discussed in Chapter 9). The subnet mask expresses the number of address bits in a network as a binary number. To deal with these representations as a network administrator, you will need to be able to convert numbers from decimal form to binary, as well as in the reverse direction—from binary form to decimal.

Conversions between Binary and Decimal Representations

As a network administrator, you will need to be comfortable with numbers in the range 0–255. These correspond to binary numbers that are up to 8 bits long. Before discussing the conversion process between binary and decimal, let us quickly recapitulate how decimal numbers are represented. A number such as 358 represents the number 3*100 + 5*10 + 8*1. We say that the place value of 3 is 100, the place value of 5 is 10 and the place value of 8 is 1. This is shown in Table 4.2. Since the place values are increasing powers of 10, decimal numbers are called base-10 numbers.

Table 4.4: Template for conversions between decimal and binary

128	64	32	16	8	4	2	1
(2^7)	(2^6)	(2^5)	(2^4)	(2^3)	(2^2)	(2^1)	(2^0)

The same idea is used in binary numbers. The only difference is that the place values are powers of 2 and the possible digits are 0 or 1. Thus, the place value of the right-most digit is 1; the place value of the next digit is $2^1 = 2$, followed by $2^2 = 4$, and so on. An example is shown in Table 4.3. We can use a procedure similar to the procedure used with decimal numbers in Table 4.2 to calculate the number represented by the binary number 101 as shown in Table 4.3. The binary number 101 represents 1*4 + 0*2 + 1*1 = 5, as can be verified from Table 4.3. Table 4.4 extends Table 4.3 to 8 bits.

With this background, we can work through some examples. To do the conversions, I find it convenient to start with the template shown in Table 4.5 (which is the same as Table 4.4, it just removes the references to the powers of 2).

Table 4.5: Template for conversions between decimal and binary (simplified)

128	64	32	16	8	4	2	1

To convert from binary to decimal, we start by writing the digits of the binary number in the appropriate places in Table 4.5, one digit per place value, and add up the place values with 1. For example, if we are interested in finding the decimal equivalent of the binary number 11100011, we start by filling the template of Table 4.5 with the digits of the binary number as shown in Table 4.6. The decimal representation of 11100011 is then the sum of all place values with 1, giving 1*128 + 1*64 + 1*32 + 1*2 + 1*1 = 227.

Table 4.6: Example to convert from binary to decimal—11100011

1	1	1	0	0	0	1	1
128	64	32	16	8	4	2	1

We can use the same template to convert from decimal to binary, following the three-step procedure below to convert a number from decimal to binary:

1. Starting from the left, put 1 in the largest place value that is less than or equal to the number
2. Subtract the place value from the number. The remainder is the new number
3. Repeat until the remainder is 0. Fill all remaining places with 0.

We can use an example to see how this works by converting 133 to binary. We start with the template in Table 4.5. The largest number in the template that is less than or equal to 133 is 128. So we place a 1 above 128. The result is shown in Table 4.7.

Table 4.7: Converting 133 to binary—using 128

1							
128	64	32	16	8	4	2	1

The remainder is 133 - 128 = 5. The largest number in the template that is less than or equal to 5 is 4. So, we place a 1 above 4. The result is shown in Table 4.8.

Table 4.8: Converting 133 to binary—using 4

1					1		
128	64	32	16	8	4	2	1

The remainder is 5 - 4 = 1, so we place a 1 over 1 and fill the remaining positions with 0, giving the binary representation of 133 as 10000101 as shown in Table 4.9.

Table 4.9: Converting 133 to binary—using 1 and completing

1	0	0	0	0	1	0	1
128	64	32	16	8	4	2	1

A good value to remember is the largest possible 8-bit binary number. This number is 128 + 64 + 32 + 16 + 8 + 4 + 2 + 1 = 255. You will run into this number quite often when you work with IP addresses, particularly with subnet masks.

Table 4.10: Value of $100 invested every month

	Annual growth		
Years	**6%**	**8%**	**Contribution**
1	$1,233.56	$1,244.99	$1,200
5	$6,977.00	$7,347.69	$6,000
10	$16,387.93	$18,294.60	$12,000
15	$29,081.87	$34,603.82	$18,000
20	$46,204.09	$58,902.04	$24,000
25	$69,299.40	$95,102.64	$30,000
30	$100,451.50	$149,035.94	$36,000
35	$142,471.03	$229,388.25	$42,000
40	$199,149.07	$349,100.78	$48,000
45	$275,599.26	$527,453.99	$54,000
50	$378,719.11	$793,172.75	$60,000

Number systems have implications[7]

The decimal and binary numbers used above are called number systems. *Number systems are methods for representing numbers, together with operations that can be performed on those numbers.* Number systems have implications. The earliest numbering system was developed by the Greeks in about 450 B.C. The Greek system used alphabets to represent numbers, much as Roman numerals, that are popular even today (these are the numbers such as i, v, x that are often used for page numbering). Though these numbering systems were useful, they had many limitations. Most importantly, though these numbers could be used to easily represent the results of computations, they could almost never be used to perform computations in the head (for example, try subtracting iv from xiii using mental math). Therefore, these numbers were generally used to record results computed by other methods, usually an abacus.

For traders trying to compute sale prices (unit price * quantity), profits, losses, etc., the computation itself posed challenges. Besides, the Greek and Roman numbers did not include 0, and could not represent negative numbers. This made it difficult to use these numbers to represent losses in commercial transactions.

Almost a thousand years later, in about 500 A.D., the Hindu number system we use today was developed.[8] This system defined the concept of place values and included a 0, and used only 10 symbols, from 0 to 9, to represent all conceivable numbers using the concept of place values. This created the required regularity so that commercial computations could be performed mentally without the use of an abacus. The introduction of this regularity had huge commercial implications. Sale prices, profits, losses, etc., could now be performed mentally. This greatly simplified commerce. Until this system became popular in the western world in around 1,000 A.D., all but the most elementary computations required the use of an abacus, seriously constraining commerce.

Compounding and personal finances

In Table 4.4, we see that the powers of 2 grow dramatically as we move left. The same concept applies to personal finances. Savings invested at reasonable rates of return grow slowly at first, but the growth increases dramatically over time. Table 4.10 shows what happens if you invest $100 each month in securities that return 6% and 8% each year, which are normal rates of return. After about 25 years, growth is driven largely by the increase in value of existing investments. Over these longer durations, small increases in growth rates also have very significant impacts. Most financial bubbles happen when people do not have the patience to wait so long and rush into speculative schemes to make money quickly. If you are young, and have time on your side, Table 4.10 may be the most useful information in this book.

7 This remark is drawn from Peter L. Bernstein, *Against the Gods: The Remarkable Story of Risk* (Wiley, 1998).

8 In his book *Predictably Irrational*, Dan Ariely describes Indian astronomer Aryabhata's remarks at this eureka moment as "sthanam sthanam dasa gunam," which is translated as, "Place to place is 10 times in value" (Harper Perennial, 2010).

Dotted Decimal Notation for IP Addresses

With this background of binary numbers, we can look at the standard manner in which IP addresses are written. Recall that IP addresses are 32-bit binary numbers. For convenience of representation, the 32-bit IP addresses are broken down into four blocks of 8 bits each. (This is why we emphasized 8-bit binary numbers in the introduction to binary numbers.) Each of these 8-bit blocks is called an octet. For user display, each octet is converted to decimal and the four decimal numbers are separated by dots. An example of an IP address written in this manner is 192.168.1.5 as shown in Figure 4.6. It can be verified that the IP address 192.168.1.5 represents the 32-bit binary address 110000001010100000000001000 00101.

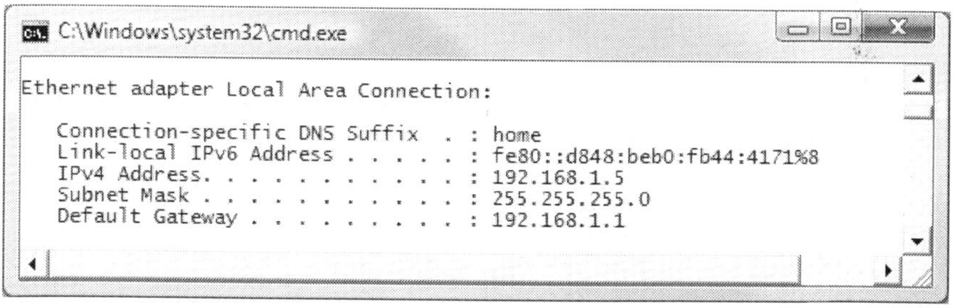

```
C:\Windows\system32\cmd.exe

Ethernet adapter Local Area Connection:

   Connection-specific DNS Suffix  . : home
   Link-local IPv6 Address . . . . . : fe80::d848:beb0:fb44:4171%8
   IPv4 Address. . . . . . . . . . . : 192.168.1.5
   Subnet Mask . . . . . . . . . . . : 255.255.255.0
   Default Gateway . . . . . . . . . : 192.168.1.1
```

Figure 4.6: Example of dotted decimal representation of IP addresses

You can view the IP address of your computer by opening up a command shell (Start → `cmd` in Windows, or Applications → Utilities → Terminal on a Mac) and viewing your network configuration (`ipconfig /all` in Windows or `ifconfig` on a Mac). You will see information similar to Figure 4.7. Every time you see an IP address in dotted decimal notation, you are seeing a human-readable representation of an underlying 32-bit binary number.

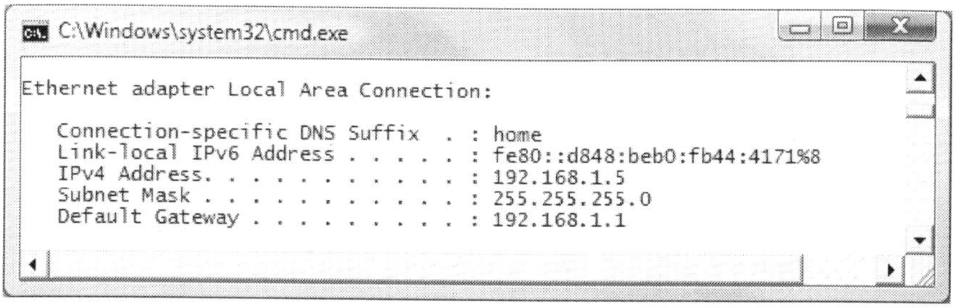

```
C:\Windows\system32\cmd.exe

Ethernet adapter Local Area Connection:

   Connection-specific DNS Suffix  . : home
   Link-local IPv6 Address . . . . . : fe80::d848:beb0:fb44:4171%8
   IPv4 Address. . . . . . . . . . . : 192.168.1.5
   Subnet Mask . . . . . . . . . . . : 255.255.255.0
   Default Gateway . . . . . . . . . : 192.168.1.1
```

Figure 4.7: Network configuration of your computer

Structure of IP Addresses

IP addresses are organized in a manner that makes them very friendly toward network administrators. This is extremely helpful because network administrators deal with IP addresses all day long. Unlike MAC addresses that are assigned in a factory, IP addresses are assigned by network administrators. This is why it is very important to understand the structure of IP addresses so that you can leverage the potential of IP addresses to your benefit.

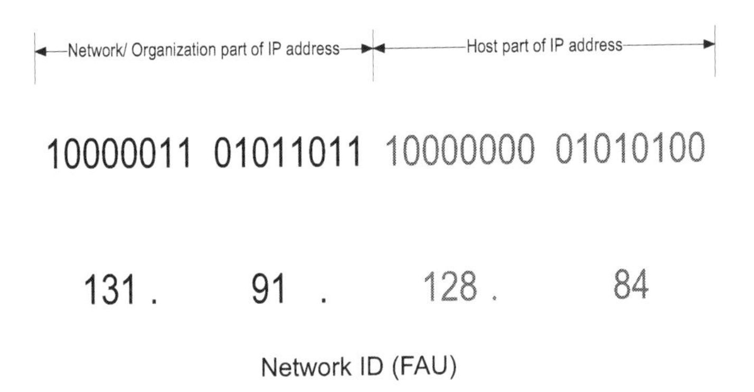

Figure 4.8: Two-part view of IP addresses

Broadly speaking, IP addresses are broken down into two parts. The left part of the IP address represents the network (organization or ISP) to which the IP address belongs. This is called the network part of the IP address. The remaining bits of the IP address identify the computer within this network. This is often called the host part of the IP address. For example, in the IP address 131.91.128.84, the first two octets (131.91) are the network part (Figure 4.8). Later you will see that these numbers indicate that this address belongs to the network of the Florida Atlantic University. The remaining two octets (128.84) identify a computer within the FAU network. This is analogous to the way we write conventional street addresses as shown in Figure 4.9. Homes on a street typically only have their street numbers on the mailbox. We complete their addresses by adding the street informaation to this number. Similarly, a computer's address can be viewed as an aggregation of its network address and host address.

Special IP Addresses

There are a few IP addresses that are defined to have special meanings. These are listed below:

- 255.255.255.255: The broadcast address on the local network to which the host is connected
- 0.0.0.0: The default route
- 127.0.0.1: This computer, or localhost. On any computer, the IP address 127.0.0.1 can be used to address itself.

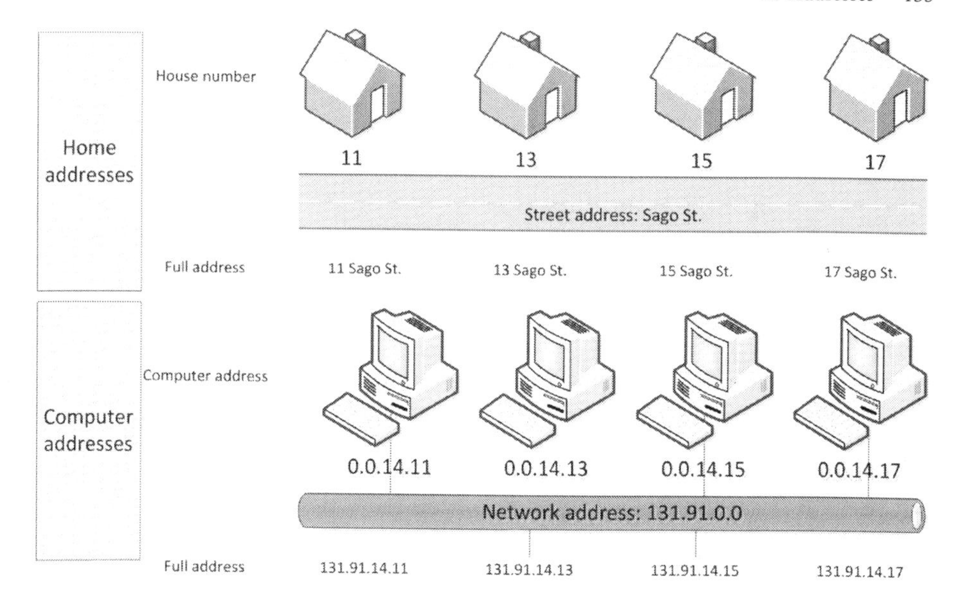

Figure 4.9: Analogy between home addresses and computer addresses

- All 0s in the host part: The network address
- All 1s in the host part: Broadcast to all computers on a network to which the host is not connected. For example, a packet addressed to 10.255.255.255 is a broadcast to all hosts on the 10.0.0.0 network. This is also called a directed broadcast. However, since this privilege can be easily abused, most routers are configured to block these packets.
- Reusable addresses: Defined in RFC 1918 (discussed in Chapter 7)

Multi-part Addressing Examples

Multi-part addresses combine different address components into a single field. IP addresses are interpreted as multi-part addresses and students often take time to understand the multi-part addressing scheme used in IP. However, this method of breaking down addresses into two parts where the left part identifies a larger organization and the right part identifies a user within the organization is very common. Three other analogues come to mind and are briefly described below.

Phone Numbers

The most familiar multi-part addressing scheme is the numbering scheme used in phone numbers (Figure 4.10). The phone network uses a three-part addressing scheme where the first part identifies the geographical area; the second part identifies the local exchange within the geographical area; and the remaining digits identify the specific user within the exchange. For example, in the phone number (813) 974-6716, the digits 813 indicate that the number

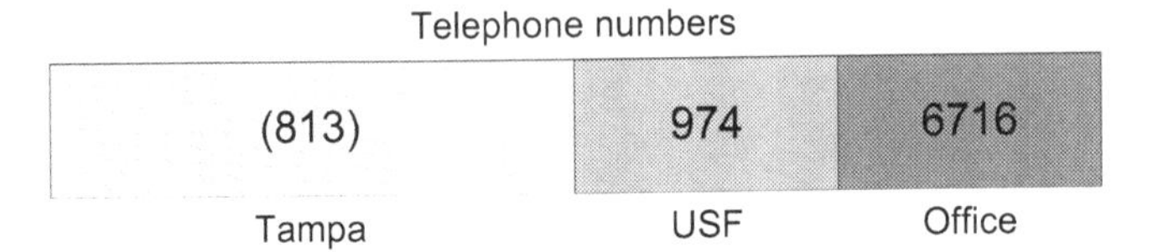

Figure 4.10: Multi-part addressing in phone numbers

is in Tampa, Florida. The next set of three digits, 974, indicates USF within Tampa, and finally the remaining digits, 6716, are the phone number within the USF network in Tampa.

While the organization of phone numbers into multiple parts is very familiar, there are two other examples of multi-part addressing that we all use every day but that are less familiar—credit card numbers and zip codes.

Credit Card Numbers

Credit cards also use a multi-part addressing scheme. You may have noticed that most Discover credit cards begin in 6011; all Visa cards begin with 4; all MasterCards begin with 51, 52, 53, 54, or 55, and so on. It turns out that credit cards and debit cards have a three-part structure, called the payment card number or the bank card number. The left-most six digits of a credit card number identify the issuer—the network and bank that issued the card. (Therefore the issuer ID is itself a two-part number.) The remaining digits identify the account number within the issuer.[9] Interestingly, the last digit of the credit card account number is actually an error-detection digit that uses the Luhn algorithm[10] for error detection (analogous to the FCS in the data-link trailer).

Is 123456789098765 a valid credit card number, as verified by Luhn's algorithm? Verify your answer from the footnote.[11]

Zip Codes

Zip codes are also multi-part numbers. As seen in Figure 4.11, the left-most digit of a zip code identifies the national region. All zip codes beginning in 1 are in New York or Pennsylvania. The second digit further narrows down the location of a zip code as shown in Figure 4.12.[12] For example, zip codes beginning in 10 are in Manhattan, New York City, New York.

All these instances of multi-part addressing—phone numbers, credit card numbers, and zip codes—serve the same purpose, which is to simplify delivery. Multi-part addresses are not just labels assigned to objects; these labels specify *where* the object is located to increasing degrees of precision.

---

9 Wikipedia has a very nice description of the credit card numbering system at http://en.wikipedia.org/wiki/Credit_card_numbers.

10 https://en.wikipedia.org/wiki/Luhn_algorithm.

11 No, the correct checksum is 8, so the number could be 123456789098768.

12 Wikipedia also has a good description of the structure of zip codes at http://en.wikipedia.org/wiki/Zip_codes.

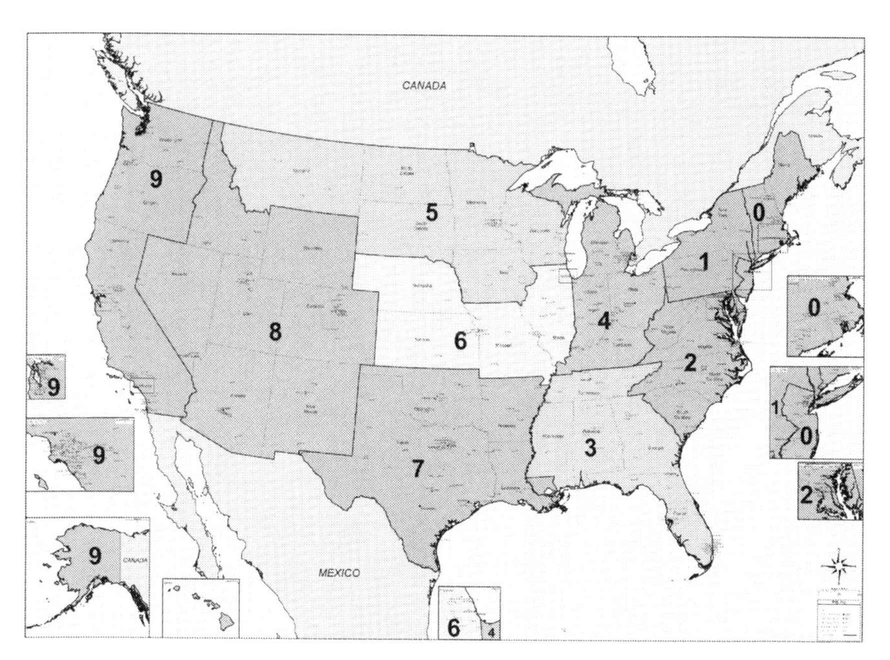

Figure 4.11: Zip codes—by left-most digit
(Source: MarketMaps, http://www.marketmaps.com)

This is particularly useful in networks which contain a large number of objects that need to be located. For example, when you dial a phone number, the local exchange does not need to figure out the exact location of the dialed number. All it needs to do is look at the first three digits and transfer the call to the appropriate metro area. At the metro exchange, the next three digits are used to transfer the call to the appropriate local exchange. Only the local exchange needs to know the exact location of the phone, and it only has to do so by looking at the last four digits.

Similarly, when you swipe a credit card at the grocery store, the card reader looks at the first few digits and selects the appropriate card network—Visa, MasterCard, Discover, American Express, etc. The selected network uses the remaining digits of the issuer ID to send the customer request to the appropriate bank. The bank uses the remaining digits of the credit card number to identify the account within its network and sends back the appropriate credit authorization. Finally, when you drop a letter in the mailbox, the local post office does not have to locate the exact street address of the addressee on the map. All it needs to do is create 10 bins and drop each letter into the appropriate bin by looking at the left-most digit of the zip code. Each of these 10 bins can then be sent to the distribution center in the appropriate geographical area. For example, the bin for zip codes beginning with 1 can be sent to the NY/PA area. At the distribution center, the mail can be sorted further by the remaining digits of the zip code until it reaches the mail carrier at the addressee's post office. The mail carrier of, course, is expected to know all street addresses on his route, so he can deliver the letters to the correct home. But the mail carrier only needs to know

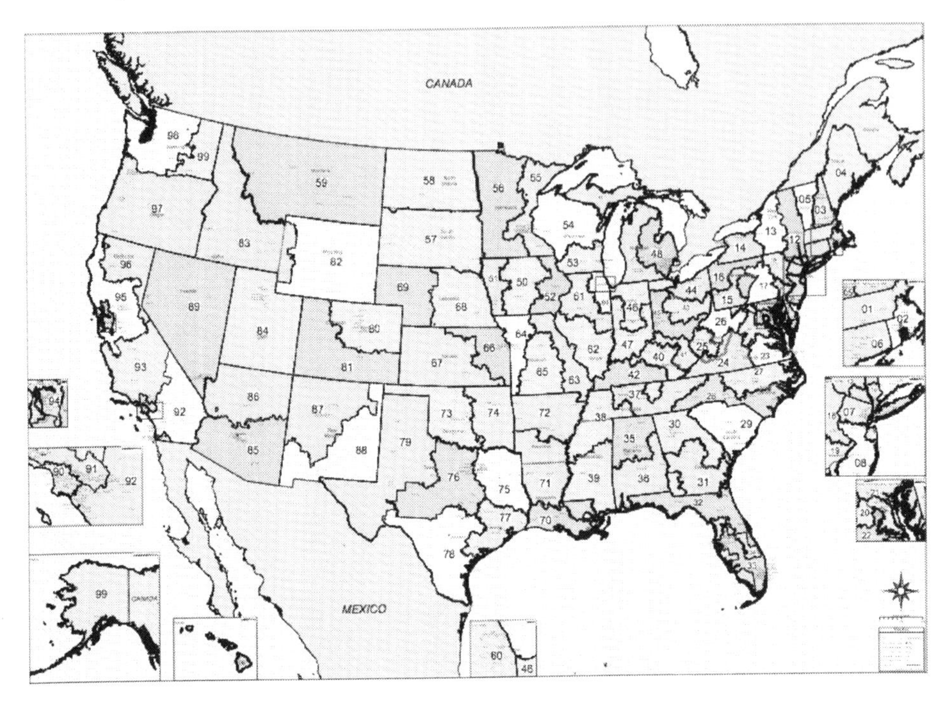

Figure 4.12: Zip codes—by two left-most digits
(Source: MarketMaps, http://www.marketmaps.com)

exact locations on his delivery route. And, other than the mail carrier, nobody in the USPS is required to locate a street address.

Multi-part addressing simplifies the work at each location of the network. With three digits in the area code, we can have 1,000 (10^3) exchanges. Each exchange only needs to know how to connect to these 1,000 exchanges in order to be able to connect all its users to any destination within North America. This is a much simpler task than knowing how to connect to each of the 10 billion possible phone numbers. (With 10 digits, we can have 10^{10}, or 10 billion, phone numbers.)[13]

An excellent animation of multi-part addressing in zip codes is available at http://benfry.com/zipdecode/.[14]

As in these examples, IP addresses are broken into two major parts. The left part is the network part and the remaining bits are the host part. The reason for this partition is also the same as in the examples above—the partition facilitates delivery of data packets. The network part identifies the network to which the address belongs. Usually, this is a telecom

13 In 1956, George Miller wrote a very influential article, "The Magical Number Seven, Plus or Minus Two" (*Psychological Review Review* 63 [2]: 81–97), which reported that most people can only hold up to seven items in working memory at any time. It may be noted that phone numbers are seven digits long, to be easily recalled by most people.
14 Ben Fry is also the creator of processing.org, a language for artists to create animations.

carrier such as AT&T or BellSouth. But it could also be a large organization such as a metropolitan university. The larger Internet only looks at the network part of the destination IP address of incoming packets and transfers packets to the correct networks. When the packet reaches the destination network, the destination network looks at the remaining bits of the IP address to locate the computer within its own network and delivers the packet to the computer.

IP Address Classes

Looking at the Structure of IP addresses and recalling the relationship between the size of a number and the number of addresses it supports, it is easy to see that organizations with more bits available in the host part of the IP address can address more computers and thereby have larger networks. Since the size of the network part + the size of the organization part = 32 bits, we can restate this as follows: organizations with smaller network parts can have larger networks because they have larger host parts in their IP addresses.

Recognizing that networks come in different sizes, and that one size would not fit all organizations, in the early days of the Internet, the network parts of IP addresses were classified into three address classes—A, B, and C. Class-A addresses were for the largest organizations, Class B was for medium-sized organizations, and Class C was for the smallest organizations. Organizations could request an address block of the size that best suited their needs. The address classes are shown in Figure 4.13.

In Class-A networks, the first 8 bits identified the network ID and the remaining 24 bits identified hosts within the network. Of the 8 bits identifying the network, the first bit identified the fact that this was a Class-A network. Therefore, 7 (8 - 1) bits uniquely identified each Class-A network. Recalling from Table 4.1 that n bits allow 2^n labels, with 7 bits we could label 2^7 = 128 Class-A networks. Therefore, we could have 128 Class-A

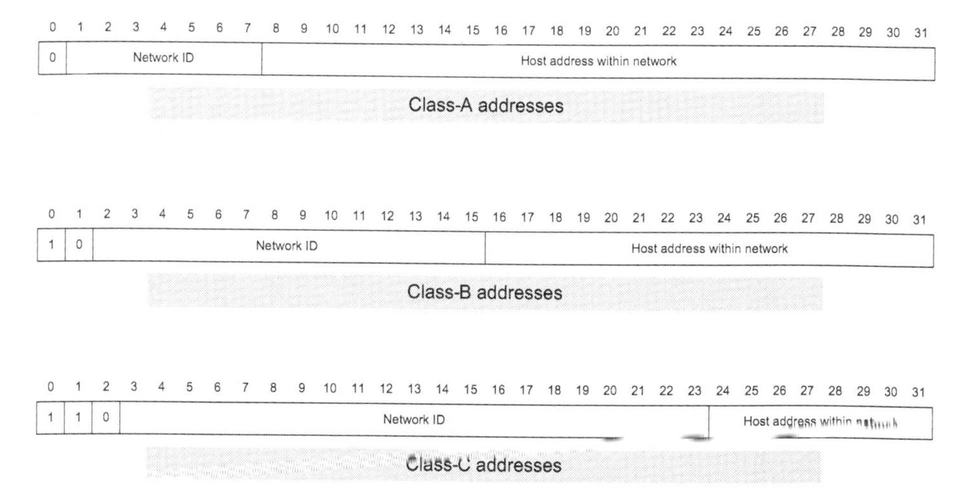

Figure 4.13: IP address classes

networks, and each Class-A network could have 2^{24} = 16,777,216 hosts. The first octet in all Class-A networks was in the range 0–127, giving 128 possible Class-A networks.

In Class-B networks, the first 16 bits of IP addresses identified the network ID and the remaining 16 bits identified hosts within each network. Within the network part, the first 2 bits identified the address as a Class-B network, leaving 14 bits to identify each network. We could therefore have 2^{14} = 16,384 Class-B networks and each Class-B network could have 2^{16} = 65,536 hosts. The first octet of all Class-B networks was in the range 128–191.

Finally, in Class C, the first 24 bits were the network part, leaving 8 bits for the host part within each network. Each Class-C network could, therefore, have 2^8 = 256 hosts. Of the 24 bits identifying the network, 3 bits were used to label the network as a Class-C network, leaving 21 bits to identify each Class-C network. Therefore 2^{21} = 2,097,152 Class-C networks were possible. The first octet of IP addresses in Class-C networks was in the range 192–223.

Figure 4.13 shows the organization of IP addresses in this scheme. By convention, bit positions start from 0, so the 32 bits go from position 0 to 31. The figure shows how the 32 bits of an IP address are assigned to network and organization parts in the three address classes. The figure shows the boundaries of the network and organization parts of the different address classes. Observe that as we go from Class A to Class B to Class C, the host part of the IP address class gets smaller, leading to smaller supported networks.

In the addressing scheme described above, the class of an address is identifiable from the first octet of the address. If the first octet is in the range 0–127, we know that the IP address belongs to a Class-A network. If the first octet is in the range 128–191, we know that the IP address belongs to a Class-B network; and if the first octet of an IP address is in the range 192–223, we know that the IP address belongs to a Class-C network. For example, 97.166.26.237 is a Class-A address and 219.23.46.78 is a Class-C address.

An interesting exercise is to look at the distribution of all available IP addresses among the different address classes. This is shown in Figure 4.14. The entire figure represents all available IP addresses (2^{32} = 4,294,967,296 IP addresses). Of these, half are Class-A

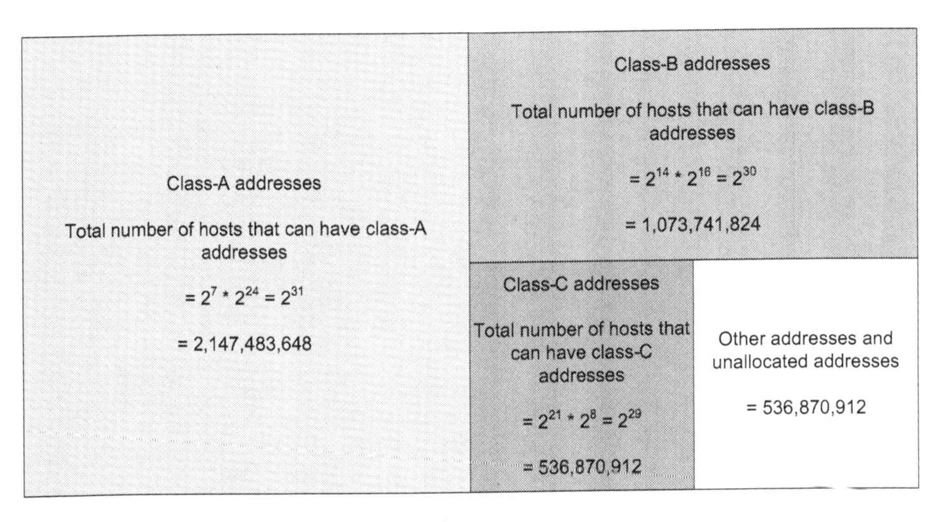

Figure 4.14: Available addresses in each class

addresses ($2^7 * 2^{24} = 2{,}147{,}483{,}648$). One-fourth are Class-B addresses, and one-eighth are Class-C addresses. The remaining one-eighth of the address space is either unallocated or consists of other address types.

If you observe Figure 4.14, the defects of the IP address allocation scheme discussed so far and used in the early days of the Internet become apparent. One hundred twenty-eight networks own half the addresses on the Internet. Surely this is unsustainable. Which will be these 128 lucky networks? What happens when an organization with a Class-A address block shrinks in size, for example, if the owner goes out of business? Or, what happens when an organization with a Class-C address block outgrows its allocated block? It would need to request additional blocks, which would most likely be non-contiguous, and inconvenient to use.[15] CIDR was created to address these concerns.

Classless Inter-domain Routing (CIDR)

The addressing scheme described in the previous section is called a class-based addressing scheme because there are three distinct sizes of address pools—Classes A, B, and C, as shown in Figure 4.13. This figure also indicates a significant problem with the class-based addressing scheme: there are only three possible sizes of address blocks. If you need 2,000 addresses, you can either get eight Class-C blocks or one Class-B block. Handling eight blocks is complex, and the Class-B block wastes a lot of IP addresses. What if you are a global bank and need 500,000 addresses? Should you be given a Class-A address block? What happens to the 15.5 million IP addresses you will not use? Neither of these solutions is satisfactory. Managing multiple Class-C address blocks adds to network complexity, and using a larger address block is inefficient.

CIDR (classless inter-domain routing) was introduced in 1993 to address the problem of unavailability of address blocks of reasonable size. As the name suggests, *CIDR is an address allocation scheme that eliminates the concept of address classes and allows address blocks of arbitrary length.* Whereas the network part of class-based addresses could only be 8 bits (Class A), 16 bits (Class B), or 24 bits (Class C), in classless addressing the network part can be of any length. Thus address-block sizes of any power of 2 are possible with CIDR.

For example, say we need 2,000 IP addresses. We can get the required number of addresses if we have 11 bits in the host part of the IP address ($2^{11} = 2{,}048$); 10 bits would be too few (1,024 addresses) and 12 bits would be too many (4,096 addresses). Since IP addresses are 32 bits in length, using 11 bits in the host part requires that the remaining 21 bits be used as the network ID. CIDR enables this. Using CIDR, we can specify that the network part of the address in this organization should be 21 bits in length ($32 - 11 = 21$). Thus, CIDR enables near-surgical precision in allocating IP addresses.

The primary beneficiaries of CIDR are medium-sized organizations such as universities. These organizations are too big for Class-C address blocks and too small for Class-B addresses. CIDR enables the aggregation of multiple Class-C address blocks into larger, more useful blocks. CIDR also enables the splitting up of an unused or reclaimed Class-A address block into multiple smaller address blocks that are more suitable for medium-sized

15 This is analogous to what happens in the large metros that have different non-contiguous area codes for phone numbers. For example, NYC has the area codes 212, 347, 646, 718, 917, and 929.

organizations. Most large ISPs have CIDR address blocks larger than Class B but smaller than Class A.

The flexibility of CIDR introduces a fresh complexity. In the class-based addressing scheme, the size of the network part of an IP address was obvious from the first octet of the address, as we have discussed in the previous section. If the first octet was in the range 0–127, it was a Class-A network, 128–191 was Class B, and 192–223 was Class C. There is no such regularity in CIDR. Therefore, it becomes necessary to specify the size of the network part of CIDR addresses. Accordingly, CIDR addresses are specified by two numbers—the network part of the address and the size of the network part. Routers need to know the network part of an IP address because they use this information to direct the packet to the correct destination network.

In CIDR, the size of the network part is specified by a number that accompanies all network addresses in CIDR notation. The number specifies the number of bits in the network part of the IP address. For example, the CIDR address 73.5.0.0/17 (Figure 4.15) indicates that the network ID for this organization has 17 bits. This leaves 32 - 17 = 15 bits for the host ID in the organization.

To complete the implications of CIDR, once we know the CIDR network address, we can calculate the number of computers that can be supported by the network. In our example, if 17 bits are taken up by the network part, the remaining 15 bits (32 - 17) can be used to identify hosts within the network. This means that a /17 network can have at most 2^{15} = 32,768 computers.

You can view examples of network addresses at any network registry (discussed in the next section). As an exercise, visit https://www.arin.net/ and type in the following IP addresses in the WHOIS search box on the page. Confirm that the IP addresses belong to

Network	
Net Range	73.5.0.0 - 73.5.127.255
CIDR	73.5.0.0/17
Name	MEMPHIS-10
Handle	NET-73-5-0-0-1
Parent	CABLE-1 (NET-73-0-0-0-1)
Net Type	Reassigned
Origin AS	
Customer	Comcast IP Services, L.L.C. (C05262175)
Registration Date	2014-08-26
Last Updated	2014-08-26
Comments	
RESTful Link	https://whois.arin.net/rest/net/NET-73-5-0-0-1
See Also	Upstream network's resource POC records.
See Also	Upstream organization's POC records.
See Also	Related delegations.

Figure 4.15: Network ID example

the organizations listed. What are the CIDR addresses of these organizations? How many computers can each of these networks support?

129.107.56.31 (University of Texas at Arlington)
204.154.83.103 (Shenandoah University)

Obtaining IP Addresses

Now that you know about IP addresses, you are probably curious about the procedure for getting an IP address for yourself or your organization. Initially, IP addresses were allocated directly by Jon Postel, who managed number assignments on the Internet. But as the Internet has evolved and taken on commercial importance, formal procedures have been developed for allocating network addresses.

Obtaining IP addresses in the early days of the Internet

To see how far we have come, it is instructive to read the first paragraph of RFC 790 to see how easy it was to get network addresses in the early days. The RFC states: "... the assignment of numbers is also handled by Jon. If you are developing a protocol or application that will require the use of a link, socket, port, protocol, or network number please contact Jon to receive a number assignment."

At the top level, the Internet Assigned Numbers Authority (IANA) is responsible for managing all the available IP addresses. There are 4 billion IP addresses as we have calculated earlier. IANA distributes this pool of IP addresses among regional registries. There are five such registries at this time: ARIN for North America; RIPE for Europe, Middle East, and Central Asia; APNIC for the Far East; LACNIC for Latin America; and AfriNIC for Africa. Users contact the registries within their respective continents to obtain IP addresses. Organizations can use their IP address blocks to assign IP addresses to hosts on their network anywhere in the world.

The distribution of the overall IP address space among these registries is available from the IANA website.[16, 17] Distributing IP addresses among registries has played a major role in democratizing the Internet by allowing all countries to help contribute to operating the Internet.

For administrative convenience, the registries have adopted a policy of allocating addresses only to large telecom carriers. This limits the number of organizations the registries have to deal with. It also provides the registries with some assurance that organizations with address allocations will have suitable technical expertise to manage these address allocations. Finally, allocating IP addresses in large blocks also simplifies the global routing of packets by limiting the growth in the size of routing tables. This is discussed in Chapter 8.

---

16 A very good overview of IANA's role is at http://www.iana.org/numbers/.
17 The distribution of IP addresses is available at http://www.iana.org/assignments/ipv4-address-space/.

Formally, the policy adopted by the registries for address allocation is specified in RFC 2050. Only networks that are connected to two or more networks and that route packets between these networks are allowed to apply to the registries to obtain a network address.[18, 19] Most organizations that need IP addresses will therefore contract with their local ISPs to obtain network addresses. The registries actively try not to allocate IP addresses directly to individuals and very small organizations.

You may note that unlike phone numbers, IP addresses have limited geographic significance. An IP address block will identify the network, but the network can expand globally. This is because the phone networks in most countries were government monopolies or state-regulated utilities. As a result, phone networks were organized along national boundaries. However, the Internet has developed in a more entrepreneurial environment where intellectual-property-based boundaries are better recognized than national boundaries. As a result, IP network addresses identify networks (organizations), but not geographies.[20]

IPv6

The current version of IP has served network needs very well thus far. The primary focus of the book is therefore IPv4, the current version of IP. However, as the Internet gains popularity, one bottleneck of IP has come to the fore. This is the lack of availability of IP addresses. Many short-term solutions have been developed to deal with this problem. These are addressed in Chapter 7. The long-term solution to this problem is an upgrade of IP from version 4 to IP version 6. A brief discussion of IPv6 follows.

When IP was originally designed, the IP address space was expected to meet the needs of the Internet for the foreseeable future. The 32-bit IP address fields allowed more than 4 billion (2^{32}) IP addresses. This equated to about one IP address per human being on earth at the time. Given the relatively low penetration of telephones a generation ago and the exorbitant costs of computers in those days, allocating one IP address per human being was considered quite generous and adequate to meet the needs of the Internet.

The allocation of IP addresses has turned out to be quite inefficient, however. Some experts have expressed fears that the highest efficiency that will be attained in using IP addresses will be about 15%, which suggests that we will run out of IP addresses once about 600 million computers are connected to the Internet. We are getting quite close to that number and there is increasing concern about the shortage of IP addresses. At last count, there were about 625 million computers connected to the Internet.[21] Others are optimistic about the availability of IPv4 addresses in the future.[22]

This problem of lack of availability of IP addresses is particularly acute outside the United States. The Internet originated in the US, and organizations in the US were quicker in realizing the potential of the Internet. They also had relatively easier access to the Internet

---

18 The guidelines for obtaining an initial address allocation from ARIN are at https://www.arin.net/resources/request/ipv4_initial_alloc.html.

19 The prerequisites for such allocation are at http://www.arin.net/policy/nrpm.html#four.

20 An excellent read that brings out many of these interesting insights is John D. Day, *Patterns in Network Architecture: A Return to Fundamentals* (Prentice Hall, 2008).

21 http://ftp.isc.org/www/survey/reports/current/ (accessed Nov. 15, 2015).

22 http://www.cisco.com/web/about/ac123/ac147/archived_issues/ipj_6-4/ipv4.html.

registry, which was located in California in its early days. Most large American organizations and telecom carriers, therefore, secured the required blocks of IP addresses.

As organizations and telecom carriers in other countries come online, they have to make do with the limited blocks of IP addresses that remain available. For example, most state universities in the United States have /16 IP addresses (65,536 addresses). By contrast, Tsinghua University, one of the top-ranked universities in China, the world's largest country by population, shared a /16 address pool with its ISP (211.151.0.0/16). As another example, BSNL, the national Internet backbone of India, the world's second-largest country by population, had 131,072 IP addresses (61.0.0.0–61.1.255.255). By contrast, one of the many address pools available to RoadRunner, a mid-sized ISP in the United States, has 712,704 IP addresses (97.96.0.0/13, 97.104.0.0/15, 97.106.0.0/17, 97.106.128.0/18, 97.106.192.0/19).[23] This is more than five times the size of the BSNL address pool that covers all of India.

As a result, many networks outside the United States are being forced into using IPv6 due to their huge populations and growing demand. In fact, more IPv6 address blocks have been allocated outside the US than within the US.[24] IPv6 is also being used on a growing number of cellular networks to provide per-handset addressing because IPv4 simply doesn't have enough available addresses.

IP version 6 was defined in RFC 2460 in 1998 to expand the IP address space. The source and destination address fields in the upgraded version of IP are 128 bits in length, giving 2^{128} possible IP addresses. During the upgrade, the IP header was also simplified to take advantage of the highly robust network infrastructure available today. Recall that in the current version of IP, the header checksum of every packet is recalculated by every router, since each router decrements the time to live by 1. This adds computational complexity to routers, which is difficult to manage since many routers handle tens of thousands of packets each second. IPv6 relies on the robustness of the data-link layer and the inherently low error rates of modern networks. It eliminates the header checksum to greatly simplify the task of routers because there is no checksum to be computed by routers. Finally, IPv6 also adds header fields that can potentially allow telecom carriers to offer value-added services to customers.

The IPv6 address space is very large. With 128-bit IP addresses, IPv6 has 2^{128} addresses. 2^{128} is 340,282,366,920,938,463,463,374,607,431,768,211,456 ($340 * 10^{36}$). At first sight, this number seems large but not particularly impressive. To get an idea of the magnitude of the number of IP addresses available in IPv6, consider that the surface area of the earth is 510 million square kilometers ($510*10^6*10^6$ m^2). Do the math and you get 600 billion trillion IP addresses for each square meter of the earth's surface. Since 1 square meter is approximately 10 square feet, this is 60 billion trillion IP addresses per square foot of the earth's surface—including the oceans, deserts, and other inhospitable areas. A home with about 1,000 square feet can therefore have 60 trillion trillion IPv6 addresses. It would be very surprising if we ever ran out of IP addresses in IPv6.

23 RoadRunner also has other address pools including 24.24.0.0/14, 24.28.0.0/15, 24.92.160.0/19, 24.92.192.0/18, 24.93.0.0/16, 24.94.0.0/15.

24 http://www.ipv6actnow.org/info/statistics/.

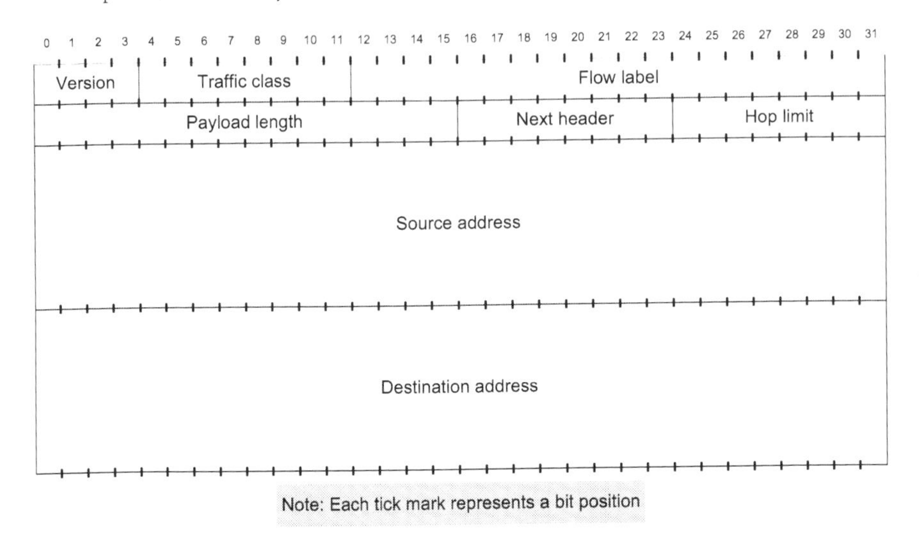

Figure 4.16: The IPv6 header

Figure 4.16 shows the IPv6 header. Whereas the IPv4 header has 14 fields, the IPv6 header only has 8 fields, leading to a simpler protocol. The main change is the noticeable increase in the size of the address fields. The functions of the other fields are summarized below.

Version This number is 6 for this version of IP.

Traffic class This field is similar to the type-of-service field in IPv4. It allows senders to specify a class of service for the packet. This field is useful if routers are capable of offering different classes of service for different packets.

Flow label This field is similar to the traffic-class field. It allows senders to designate a few packets for special handling. Again, this field is useful only if routers are capable of handling different packets differently.

Payload length This field is similar to the total-length field in IPv4 and specifies the size of the packet data.

Next header Again, this field is like the protocol field of IPv4 and specifies the transport layer protocol that is using IP to deliver this packet. As in IPv4, the protocol numbers are defined in RFC 1700.

Hop limit Finally, the hop-limit field is like the time-to-live field in IPv4. Each router decrements this field by one. When the hop limit reaches 0, the packet is discarded. As in IPv4, this field acts as a safety valve, removing damaged packets from the network.

Traffic classes in practice

Though it sounds very promising, differentiating between service classes is quite complicated. As a familiar example of what it takes to offer differentiated services, consider high-occupancy vehicle (HOV) lanes on highways. These lanes enforce different classes of service to different users of the highway network. Since HOV lanes are typically empty, cars with more than one occupant can move faster. But implementing HOV lanes adds the complexity of lane designation, driver education, fine enforcement, etc.[25, 26] Similarly, implementing traffic classes on computer networks requires that routers be capable of selecting packets belonging to each traffic class and rearranging their sequence in outgoing traffic according to priority. This is clearly a very complex task. Therefore, traffic class is quite difficult to implement in practice.

IPv6 Addressing Example

IPv6 makes a large number of addresses possible, but then system administrators have to determine how to best allocate the available addresses. Figure 4.17 shows how the network operations team at USF decided to allocate the available 96 bits across the campus. For future-proofing purposes, 8 bits are currently reserved (i.e. unused). The remaining bits are first distributed among the different institutions sharing the address block, and then each institution partitions the available bits according to its needs.

IPv6 Address Notation[27]

IPv6 addresses provide all the addresses we need. RFC 4291 describes the architecture of IPv6 addresses. However, 128-bit addresses are unwieldy, e.g. 01001100 00000111 11010101 01010101 01000000 01111111 11010101 01010101 00011100 11010100 01000111 00101100 01101100 00001100 11001010 11000110. So, a more concise representation is needed, which is specified in RFC 5952.

Whereas IPv4 used the dotted decimal notation for representation, IPv6 uses the hex notation for compaction and representation. Since each hex character represents 4 bits, the 128-bit IPv6 address is represented using 32 hex characters (32 * 4 = 128) and is written as 8 blocks of 4 hex characters each, with the blocks separated using the : character, e.g. 2001:abcd:2346:1234:a1b5:fedc:0011:35ac.

However, even this can be unwieldy, so RFC 5952 has specified some compaction rules so that there is exactly one possible representation of a given IP address. These rules are:

---

25 Emergency vehicles are another example of a differentiated multiple class. Again, it is not enough for these vehicles to flash lights and blow horns, it is also necessary that the rest of the traffic on the road allows quick passage to emergency vehicles.

26 Fast-Pass offerings at theme parks are yet another example of differentiating between traffic classes. As you may have experienced, it is not enough to issue fast passes (traffic labeling), it is also necessary to create a separate fast-pass queue and enforce pass restrictions on this queue.

27 The US National Institute for Standards and Technology (NIST) has a good guide on deploying IPv6, *Guidelines for the secure deployment of IPv6*, NIST publication 800-119, http://csrc.nist.gov/publications/nistpubs/800-119/sp800-119.pdf (accessed Dec. 2015).

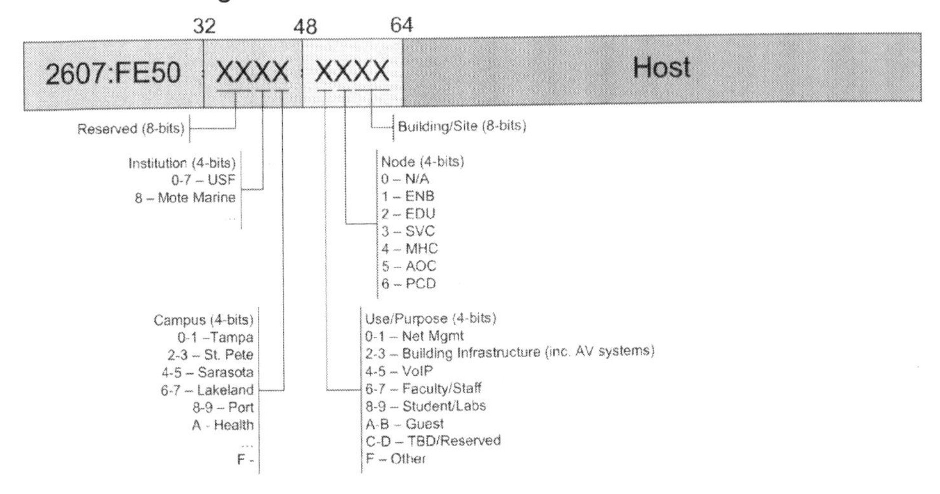

Figure 4.17: IPv6 allocation at the University of South Florida

1. Leading zeros in an individual 16-bit field must be omitted. The IP address 2001:0db8:aaaa:0000:0000:0000:eeee:0001 must be written as 2001:db8:aaaa:0000:0000:0000:eeee:1.
2. A single 0000 field must be written as 0. Thus, the above IP address must be written as 2001:db8:aaaa:0:0:0:eeee:1.
3. :: indicates a group of successive blocks of 0 fields. It can only appear once in an address, and must be used to its maximum capability. The above address must be further compressed as 2001:db8:aaaa::eeee:1. The address 2001:0:0:1:0:0:0:1 must be written as 2001:0:0:1::1 since it compresses three fields.
4. When there is a tie for possible uses of ::, the first possible occurrence of :: must be used.
5. The hex characters a, b, c, d, e, f must be written in lower case.

EXAMPLE CASE—Retailing

Supply-chain management is the flow of information from cash registers to an order-entry system to ensure that product replenishment matches store needs. Wal-Mart trucks are a familiar sight on America's highways, replenishing more than 75,000 different items from more than 145 distribution centers in more than 4,000 stores around the country. What is not so visible is the computer network that carries sales and inventory data from point-of-sale systems back to headquarters, and which gives the company its unique competitive advantage. This data and its strategic use have been the key to the company's success in keeping costs low, while ensuring that the products are always stocked. While Wal-Mart is the best-known user of computer networks in the retailing industry, every large retailer now depends upon computer networks to manage inventory and distribution. Retailers that try to compete based on the expertise of merchandising managers have a high performance bar.

Sam Walton was driven from a young age, becoming the youngest boy in the state's history at that time to become an Eagle Scout. After running franchised variety stores in his early career, Sam and his younger brother James L. (Bud) Walton opened their first Discount City store in Rogers, Arkansas, in 1962. Sam Walton invested 95%, his brother 3%, and the store manager 2% in the store. The same year, Sebastian S. Kresge opened his first Kmart store in Garden City, Michigan. Interestingly, Target started this same year, opening its first four branches around Minneapolis in 1962. Even more interestingly, Carrefour, the leading European grocer, also opened its first hypermarket in 1962.

Kmart expanded rapidly at first. The second Wal-Mart opened two years later in 1964, and five years after opening their first stores, Wal-Mart had 18 stores with annual sales of $9 million. By this time, Kmart had 250 stores with total annual sales exceeding $800 million. Thus, in the 1960s, each Kmart store had more than six times the sales of each Wal-Mart store. During the 1960s, Sam Walton made many changes in the way he operated his stores, inspired largely by what Kmart was doing.

Move forward to the year 2000. Both Wal-Mart and Kmart had slightly more than 2,000 stores each. But Wal-Mart's revenue for the year exceeded $200 billion, compared to less than $40 billion for Kmart. Not only did Wal-Mart have higher revenues, it also earned more than 10 times the profit margins that Kmart earned. How did the two chains reach these widely divergent outcomes from similar origins and goals? How did Kmart lose its huge early lead?

A big differentiator between the two companies is the way they use computer networks to manage information. By the mid-1960s, when he had just five to six stores running, Sam Walton was finding it difficult to track the inventory in his stores from the reports he received from his store managers. He understood that better record-keeping would help him figure out how much merchandise he had in each store, what was selling, what was not selling, what to mark down to push out the store, and what to reorder. If shelves were empty less frequently, store sales would go up.

Based on this insight, Wal-Mart first entered the computer age in 1969, by leasing an IBM 360 computer to manage inventory at its distribution center. Stores increasingly began to use electronic cash registers to transmit sales data back to distribution centers and headquarters for prompt inventory refills. The company built its first computer network using phone lines in 1977. The network was used to transmit information, such as orders, messages from buyers, and payroll data, speeding up restocking of merchandise in the stores. In 1979, the company built its first data center. In 1987, Wal-Mart created its own satellite network. The increased bandwidth of this network compared to the telephone network speeded up credit card authorizations and cut processing time in half from 14 seconds to 7 seconds, speeding checkout and reducing the number of cashiers needed at checkout counters.

Once the network and computer systems were in place, Wal-Mart adopted technologies such as EDI, and developed a system called RetailLink to computerize the creation and exchange of documents, such as purchase orders, invoices, and receipts, between Wal-Mart and its suppliers. The result was that suppliers such as Procter and Gamble could check on store inventory levels and sales data at Wal-Mart in real time to determine their own manufacturing and delivery schedules. The system would aggregate sales of products from each vendor throughout the entire store chain and streamline payments to suppliers.

Eventually, these suppliers forced Kmart to also develop a system similar to RetailLink, called Kmart Information Network.

Another component of Wal-Mart's success was its process of cross-docking at its distribution centers. In this procedure, manufacturers send full trucks of merchandise to distribution centers. At these centers, goods are quickly repacked and loaded to trucks for delivery to stores, often within 48 hours of arrival. There is no storage of goods at warehouses. For this system to be successful without inventory accumulation or stock-outs, it was critically necessary that information about store demand be transmitted as far back in the supply chain as possible. Wal-Mart's investments in its computer network enabled this information transmission.

The company's founder, Sam Walton, was famously frugal, flying coach class and sharing hotel rooms while traveling even when he was wealthy enough to afford a personal jet. Given this frugality, he was believed to hate technology because of its high costs, and to consider IT a very expensive overhead. The reality, however, was that he scouted nationwide, pursuing experts in distribution, logistics, and technology to work for him. He even spent time at an IBM training center in 1966 to find talented executives to automate his company. He spent more than two years wooing David Glass, who led the installation of the company's first computer network in 1977, and who later succeeded Sam Walton as the CEO of the company. Sam Walton has been quoted as saying, "It's only because of information technology that our store managers have a really clear sense of how they're doing … they get all kinds of information transmitted to them over the satellite on an amazingly timely basis: … up-to-the-minute point-of-sale data that tells them what is selling in their own store.…"

Wal-Mart's computer network and distribution system began to give it an important cost advantage. By the mid-1980s, compared to industry average distribution costs of 5%, Wal-Mart's distribution costs were 2%, giving it a 3% cost advantage over all other retailers, including Kmart.

Meanwhile, at Kmart

Whereas Wal-Mart invested heavily in computer networks and other information technologies to identify what was selling at each store and what was not, Kmart always believed that the store manager had the best knowledge about the store's neighborhood. It was part of the company's belief system that a good store manager could accomplish anything. In 1973, when the almost-700 Kmart stores were sending almost 40,000 invoices daily to headquarters, which took weeks or even months to process, Kmart considered setting up a computerized system to collect and centralize all this information. But the plan was met with fierce opposition from store managers and merchandisers at the company who did not want to give up control. The computerization plan was dropped as a result of this opposition. Such was the faith of Kmart in the ability of its store managers and merchandisers that in the early 1990s, when IT developed storewide reports identifying which products were not selling, store managers and senior managers chose to ignore the reports and ultimately asked IT to stop producing those reports. As late as 2002, Kmart gave store managers the authority to tailor the store's inventory.

Without a computer network and technology to provide a clear picture of what was selling and what was not selling, eventually it was common at Kmart stores to see piles of unsold merchandise alongside hundreds of bare shelves of hot-selling and basic items. We know from personal experience that once a busy customer leaves empty-handed from a store, it is difficult to motivate the customer to return to the store. Kmart was beginning to fall behind.

When new management took over in 2000, they located and disposed of 15,000 trucks of unsellable inventory gathering dust over the years behind stores, more than seven trucks per store. Even more interestingly, because of the poor information systems at Kmart, all this merchandise was unaccounted for. It was later discovered that what was happening was that when the company found good deals, they would buy large quantities of the product and the unsold units were pushed into trucks behind stores.

By contrast to Kmart, which depended upon store managers to decide what to stock in a store, Wal-Mart depended largely on combining sales data with publicly available information, such as zip code profiling, to decide what to stock in a store.

Without a clear vision for the role of IT in the company, between 1994 and 2002, Kmart had six CIOs. Each CEO/CIO threw out parts of systems developed by their predecessors and started building a new system from scratch. The systems that resulted were often incompatible with each other, limiting opportunities for integrating reports and data. When Kmart first installed computers in stores, it did not link cash registers to the computers, so it could not collect sales data in real time as Wal-Mart could do. In 1987, the same year that Wal-Mart invested $20 million in its satellite network, Kmart announced a $1 billion investment in IT. Even with this investment, Kmart developed one system for apparel and another system for the rest of the store, with no possibility of information exchange between the systems. In the meantime, Wal-Mart systematically built a network that gave its executives a complete picture at any point in time of where goods were and how fast they were moving in the entire chain.

Even in the 1990s, Kmart analysts did not have automated systems to generate chain-wide reports on supply and demand and had to use spreadsheets to integrate information from multiple data sources to gather this information. Most Kmart managers did not know how to use computers to track sales and orders.

In 2002, Kmart filed for bankruptcy protection. That same year, for the first time, Wal-Mart topped the Fortune 500 list as the biggest company in America, measured by annual revenue, earning $219.8 billion in 2001. The magazine noted that Wal-Mart was the first service company to top the list.

Recent Developments

Online sales are becoming increasingly important, hitting 5% of all consumer-goods sales, reaching more than $225 billion in 2012. Wal-Mart is behind market leader Amazon in this space, with online revenues of approximately $10 billion, compared to Amazon's $65 billion. Wal-Mart is trying to figure out how to process orders most efficiently. Currently, its shipping and handling costs are almost twice Amazon's costs. In 2012, Wal-Mart invested more than $400 million in its e-commerce investments.

Wal-Mart is also trying to develop its own mobile payment system to reduce dependence on intermediaries and hence processing fees. Called Wal-Mart pay, the feature on its mobile app will allow customers to use any payment method stored in their online account to make payments for purchases at Wal-Mart stores or online.

References

1. Banjo, Shelly. "Wal-Mart's e-stumble with Amazon." *Wall Street Journal*, June 19, 2003.
2. "How Kmart fell behind." www.baselinemagazine.com, Dec. 10, 2001.
3. "Is Wal-Mart good for America? The rise of Wal-Mart." Public Broadcasting Service, Nov. 16, 2004.
4. Nassauer, Sarah. "Wal-Mart to offer smartphone payments in stores." *Wall Street Journal*, Dec. 10, 2015.
5. Ortega, Bob. *In Sam We Trust: The Untold Story of Sam Walton and How Wal-Mart Is Devouring the World*. Times Business, 1998.
6. Piliouras, Teresa C. Mann. *Network Design: Management and Technical Perspectives*. Auerbach Publishers, 2004.
7. Turner, Marcia Layton. *Kmart's Ten Deadly Sins*. John Wiley and Sons, 2003.
8. Vance, Sandra S., and Roy V. Scott. *Wal-Mart: A History of Sam Walton's Retail Phenomenon*. Twayne Publishers, 1994.
9. Wal-Mart timeline, http://corporate.walmart.com/our-story/our-history.

Summary

The network layer routes packets between networks. IP is used almost universally at the network layer and we briefly described the roles of the different fields of the IP header. Given their importance to network administrators, the chapter paid particular attention to IP addresses and the breakdown of IP addresses into a network part and an organization part. The relationship between the length of an address field and the number of computers that can be connected on the network was described. The CIDR notation to improve IP addressing flexibility was introduced. Finally, a brief description of IPv6 was provided.

About the Colophon

What is good for people can be good for networks. The quote in the colophon is attributed to Chinese philosopher Laozi in the 6th century B.C. It is widely used as a motivational instrument in books and movies to help people move progressively towards their goals. Many significant goals seem unattainable at first. However, if people take even small steps that bring them successively closer to their goals, the goals can often be achieved.

As we saw in this chapter, this quote is an excellent visualization aid for describing how packet-switched networks route packets. Most routers are not directly connected to the final destination. Yet, they can pass each packet to a neighboring router that is one step closer to the destination. Eventually, all packets reach their destinations.

Maybe the next time you see a seemingly difficult goal, pause and think of how routers are successfully able to deliver billions of packets each day. They do this simply by helping packets take one well-directed step at a time. You may be able to do the same with the tasks required to help you reach your goals in life.

REVIEW QUESTIONS

1. Briefly describe routing—the primary function of the network layer.
2. How are IP addresses similar to MAC addresses? In what ways are the two addresses different?
3. What is the need for a computer address at the IP layer when computers also have a MAC address?
4. What are the advantages of designing the Internet in such a way that specialized devices called routers handle all the details of routing? What may be the possible disadvantages?
5. What are the advantages of designing IP as a best-effort protocol?
6. What are RFCs? Read RFC 791 that defines IP, and briefly describe one thing in the RFC that caught your attention.
7. Which, in your opinion, are the three most important fields in the IP header? Briefly describe the functions of these fields.
8. What is the need for the time-to-live field in the IP header?
9. What is the size of the largest possible IP packet?
10. How many objects can be uniquely labeled with 10-bit address labels?
11. You wish to assign unique labels to 200 objects using binary numbers. What is the minimum number of bits needed?
12. How would you represent 217 in binary? 168?
13. What decimal number does the binary number 10001101 represent? 11010101?
14. What is *dotted decimal notation*?
15. What information is conveyed by each part of a two-part IP address?
16. How are the 32 bits of an IP address organized in a typical large network?
17. Find the IP addresses of any five department websites at your school. Do you observe any patterns in the IP addresses of the websites? To find the IP address of a website, you can open a command prompt or terminal window and use nslookup; for example, nslookup www.msu.edu.
18. In what way are the 32 bits of an IP address organized similarly to the 10 digits of phone numbers? In what way are they different?
19. What were the three address classes in early IP networks? How many hosts (computers) could be accommodated in a network in each address class?
20. What are the disadvantages of using address classes? How does CIDR overcome these disadvantages?
21. What is *registry* in the context of IP addresses? What are *regional registries*? What is the need for regional registries?
22. What are the requirements that an organization must satisfy in order to obtain IP addresses directly from a registry?
23. What is the correct representation of the IPv6 address 2607:FE50:0010:0000:0000:1000:0101:abcd, as specified by RFC 5952?
24. On your home computer, what is the IP address reported by ipconfig /all (Windows) or ifconfig (Mac/Linux)?
25. From your home computer, go to www.whatismyip.com and make a note of your IP address. Type this address into the search box at www.arin.net. Who is the owner of that address block?

EXAMPLE CASE QUESTIONS

1. Based on the case, what computer networking technologies do retailers like Wal-Mart use?
2. Based on the case, how do retailers like Wal-Mart use computer networks to lower the costs of goods sold?
3. What is *cross-docking*?
4. Among the processes at a retailer that can be managed using IT are merchandise planning, sourcing, distribution, and store operations. Briefly define each process using any information source available to you (such as Wikipedia or Google).
5. For each of these processes, give an example of how IT can be used to manage the process. If your example uses computer networks, highlight the role of the network.
6. Based on the case, what are some reasons that Kmart did not achieve the same benefits from IT investments as Wal-Mart?

HANDS-ON EXERCISE—ipconfig and ping

This chapter introduced IP addresses. The hands-on exercise in this chapter will use two handy utilities that are very useful to gather network layer information—ipconfig and ping. While ipconfig was introduced in Chapter 3, ping will be introduced here.

Ipconfig

In the hands-on exercise in Chapter 3, you used ipconfig (or ifconfig) to obtain information on the MAC addresses of the network interfaces on your computer. In this exercise, you will continue the same exercise to get information about the IP addresses assigned to these interfaces. You have already seen a preview of this information in Chapter 3 because ipconfig provides information on both the MAC addresses and IP addresses. Figure 4.18 shows an example ipconfig output on Windows. The IP address of the Ethernet interface in the example is 131.247.95.118.

Use ipconfig (or ifconfig on the Mac/Linux) to answer the following questions:

1. What are the IP addresses assigned to the interfaces on your computer?
2. We have seen in this chapter that the network parts of IP addresses identify the owners of IP address blocks. Use the WHOIS search facility at the American Registry for Internet numbers (http://www.arin.net) to search the WHOIS database and identify the owner of the address block to which your IP address belongs. You can do this by visiting http://www.arin.net → typing in your IP address into the search WHOIS field on the site → press enter or click the button next to the search field. (Please see Figure 4.19 for reference.) Follow the link to the "related organization's POC records" just below the table to gather information on the following fields associated with your IP address:
    a. OrgName
    b. NetRange
    c. NetType

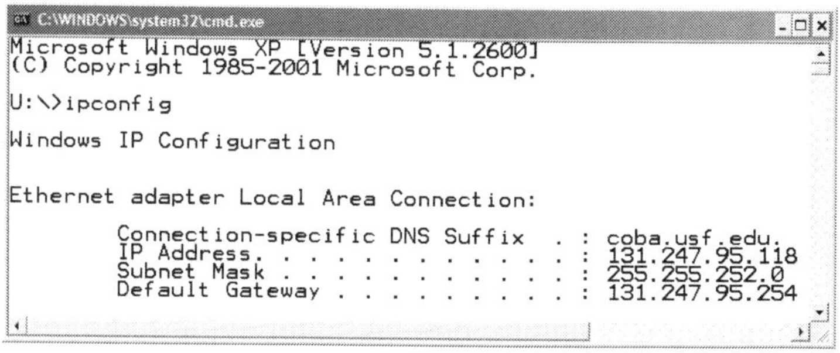

```
C:\WINDOWS\system32\cmd.exe
Microsoft Windows XP [Version 5.1.2600]
(C) Copyright 1985-2001 Microsoft Corp.

U:\>ipconfig

Windows IP Configuration


Ethernet adapter Local Area Connection:

        Connection-specific DNS Suffix  . : coba.usf.edu.
        IP Address. . . . . . . . . . . . : 131.247.95.118
        Subnet Mask . . . . . . . . . . . : 255.255.252.0
        Default Gateway . . . . . . . . . : 131.247.95.254
```

Figure 4.18: ipconfig output showing IP address of Ethernet interface

d. NameServer
e. OrgTechName
f. OrgTechPhone

3. Is the OrgName the same as your ISP's name? (If not, we will explore the main reason for this in Chapter 7.)

Ping

The network layer is responsible for host-level routing, i.e. the layer makes sure that packets are delivered to a remote computer if the remote computer is connected to the Internet. This role of IP can be reverse-engineered to detect whether a remote computer is connected to the Internet and networking has been properly configured on the remote computer. If the IP software on your computer can deliver a packet to the remote computer, we know that the remote computer must be connected to the Internet and its networking software must be properly configured. If IP cannot deliver packets to the remote computer, there must be some problem in network connectivity between the two computers.

Ping is the utility that is used to check for network connectivity. Ping sends a special form of IP packets called ICMP ECHO packets to the target host. When the target host receives these IP packets, it sends back a reply. Ping reports the round-trip time for the packet request-response cycle, giving an indication of the network connectivity between the two computers. Ping is extremely simple to use as can be seen from the examples in Figure 4.20. Here we use the ping command to check for connectivity to the web server at Purdue University and the University of Tokyo.

The ping output also reports on the average packet round-trip times. The two examples in the figure show that packet round-trip times depend upon the destinations. The ping requests originated from Tampa, Florida. It takes four times as long for a reply to be received from Tokyo (205 milliseconds) as it takes for a reply from Purdue (55 milliseconds). This is understandable since Tokyo is much farther from Tampa than is Indiana (where Purdue is located).

The ping utility can also be used to confirm network connectivity and configuration at your own computer. If you encounter network connectivity issues, you can ping a remote computer that you know is connected to the Internet. If you do not hear a ping response,

Figure 4.19: Searching the ARIN database for IP address block ownership

it is likely that the problem is at your end. You may need to check a few sites because many sites do not respond to ping requests for security reasons.

Answer the following questions.

1. How does ping work? Use Wikipedia or another resource to write a brief description of how ping works.
2. Use Wikipedia or another resource to write a brief description of the Internet Control Message Protocol (ICMP) and its use.
3. Ping the website of your university. Show the output. What is the average round-trip time? (If your university is locked down, i.e. does not respond to ping queries, try your ISP's website.)
4. Ping the website of a university in a neighboring city or town. Show the output. What is the average round-trip time? (Again, if the website is locked down, try the website of a company in another state.)

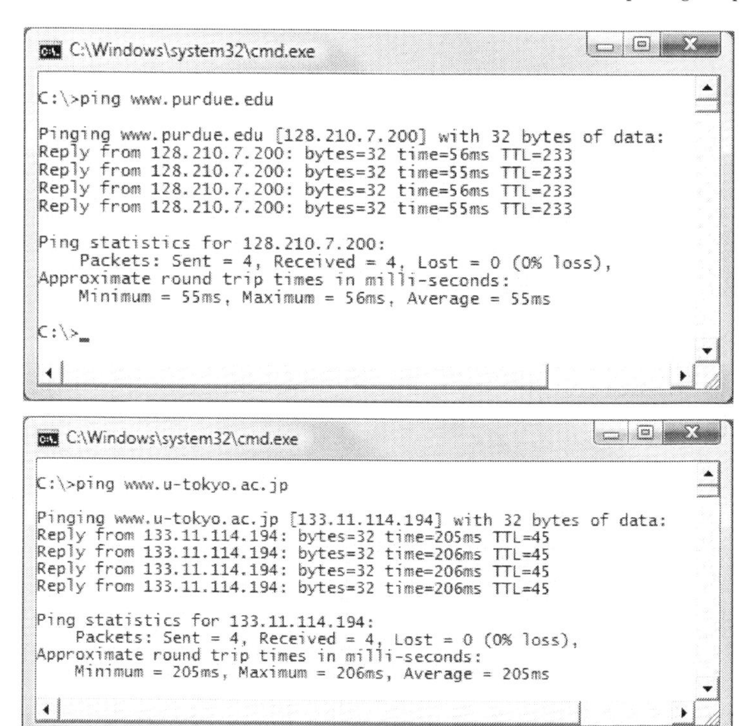

Figure 4.20: Using ping

5. Ping the website of a company, university, or organization located abroad. Show the output. What is the average round-trip time?
6. You wake up one day to find that you are unable to connect to the Internet from home. You make some calls to your friends and neighbors and find that they have no issues connecting to the Internet. You conclude that the fault is within your home. As seen in Figure 4.21, there are two networking components within your home—your computer and your home router. As seen in Figure 4.22, the IP address of the home router is the IP address of the default gateway in your ipconfig output, and home routers typically respond to ping requests. How can you use ping to determine whether the network connectivity problem is at the home router or at your computer?
7. The creator of ping has an interesting description of its creation at http://ftp.arl.army.mil/~mike/ping.html. What is the most interesting story narrated by the author of the use of ping for network troubleshooting?
8. (optional) A related utility is pathping. It combines the features of ping (reachability), tracert (hop visibility), and packet loss statistics per hop. An example use is pathping –n www.google.com. The most useful information in the output is the packet loss information.

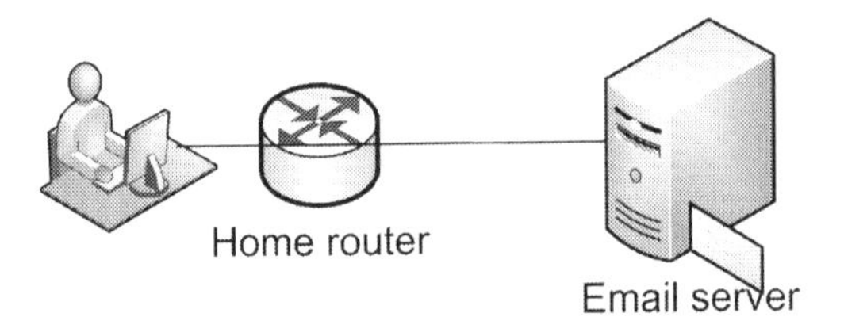

Figure 4.21: Home network connection

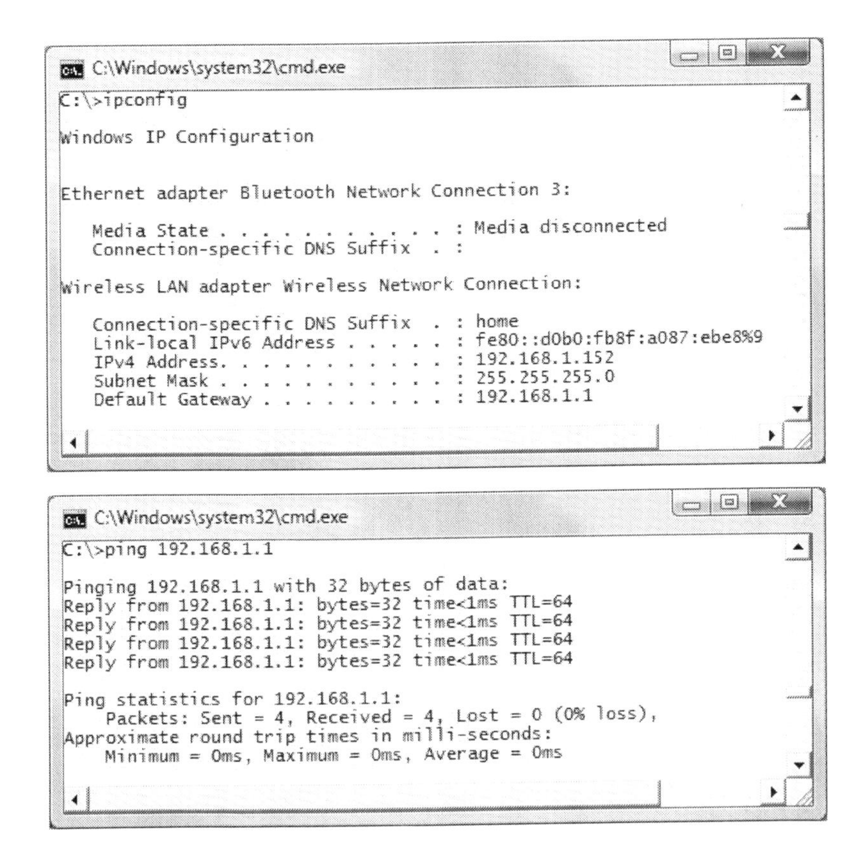

Figure 4.22: Pinging the local home router

CRITICAL-THINKING EXERCISE—Genetic Code

The genetic code uses three-letter DNA-words, where each letter can take four values—A, C, G, and T. How many words are possible in the DNA dictionary?[28]

IT INFRASTRUCTURE DESIGN EXERCISE—Estimating CIDR Requirements

In this exercise, we estimate the CIDR address block requirements for TrendyWidgets. For example, if TrendyWidgets needs about 8,000 IP addresses, you would say that TrendyWidgets needs a /19 CIDR block. This is because a /19 address block uses 19 bits for the network part of the IP address, leaving 32 - 19 = 13 bits available for the host part. This allows unique labeling for 2^{13} = 8,192 hosts.

Answer the following questions:

1. If each computer on TrendyWidgets's network is allocated an IP address, how many IP addresses is TrendyWidgets likely to need?
2. Based on this estimate, what is a CIDR address-block size suitable for TrendyWidgets?

---

28 This is from Richard Dawkins, *The Greatest Show on Earth: The Evidence for Evolution* (Free Press, 2009). The genetic code is universal, all but identical across animals, plants, fungi, bacteria, archaea, and viruses, and is believed to provide strong evidence of evolution. The 64 DNA words are translated into 20 amino acids and one punctuation mark, which toggles reading. The 64-word dictionary is universal across the living kingdoms.

CHAPTER 5

Transport Layer

I view the fundamental problem of resource sharing to be the problem of inter-process communication.

D.C. Walden, RFC 62

Overview

This chapter covers the transport layer. The network and lower layers deliver packets between computers across networks to the best of their capability; the transport layer provides all the remaining bookkeeping functions necessary for reliable data transfer between applications on these computers. At the end of this chapter you should know about:

- the need for a transport layer
- Transmission Control Protocol (TCP), the popular transport layer protocol
- TCP functions, including segmentation, reliability, flow control, multiplexing, and connection establishment
- the fields of the TCP header
- UDP, a simpler protocol at the transport layer

The Need for a Transport Layer

We saw in Chapter 4 that the network layer does best-effort delivery of packets between any two computers located anywhere on the Internet. Since computer networks are used for data transfer, and the network layer already performs the task of data delivery, what is the need for yet another layer of software for data communication? What value can the transport layer add?

It turns out that some additional processing is required before the data-transfer capability offered by the network layer can become useful for computer applications. Three tasks are vital: (1) The network layer can accept at most 65,535 bytes of data per packet. If an application has a bigger block of data to transfer (audio and video files come to mind), some entity needs to chop the larger block of data into smaller segments before handing the segments to the network layer. This task is called segmentation. If the transport layer did not perform segmentation, the application developers would have to do it while developing their applications; (2) The network layer only provides best-effort delivery and may drop packets or duplicate packets. Some entity needs to resolve these losses and duplications. Again, if the transport layer did not do this, the applications would have to do it. This task is called reliability; (3) The network layer does not distinguish between applications on the computer. It simply sends data from one computer to the other. Once the packets reach the destination computer, some entity needs to distribute incoming data packets to the appropriate application on the computer. This task is called multiplexing because it allows multiple applications on a computer to share a common network link.

Therefore, the transport layer has to perform many functions to ensure reliable data delivery. At the sending end, the transport layer receives data from the application layer and breaks down the application data into segments. The transport layer keeps track of these segments to account for packet loss or duplication by the network layer. The transport layer also provides a mechanism to distinguish between data segments created by each individual application on the computer. Finally, the transport layer provides for graceful use of network resources by allowing receiving computers to specify the data transfer rate.

Figure 5.1 shows the placement of the transport layer relative to the network layer (implemented as the IP protocol) and the application layer. The transport layer receives data from applications and performs all the bookkeeping functions described above to make the inherently unreliable computer networks (network layer) appear reliable to applications. The result is that, as far as computer applications are concerned, apart from network delays, there is no difference between accessing data from the local hard drive and from across the network.

The most popular transport layer protocol is the Transmission Control Protocol (TCP). The TCP/IP stack derives its name from this protocol. *TCP is a highly reliable host-to-host transport layer protocol over packet-switched networks.* Most end-user applications such as e-mail and the web use TCP. However, many applications do not need the segmentation and reliability

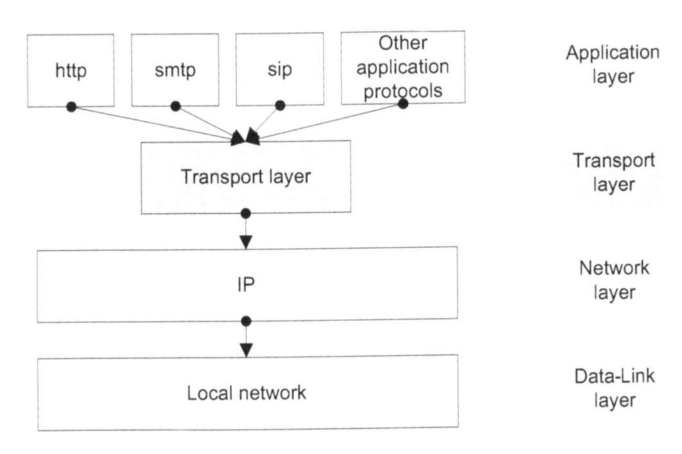

Figure 5.1: Transport layer relative to applications and the network layer

offered by TCP. For these applications a simpler transport layer protocol called the User Datagram Protocol (UDP) has been defined. UDP eliminates almost all the computational processing associated with TCP. Most of this chapter focuses on TCP. The last section of the chapter looks at UDP, in particular the various scenarios where UDP is preferable to TCP.

UDP was defined in RFC 768 in 1980. TCP was defined in RFC 793 in 1981.

Transmission Control Protocol (TCP)

At a very high level the operation of TCP may be explained as follows. At the sending end, computer applications generate the data that needs to be transmitted over the network. Examples of such data include e-mail, a web page from a web server, or a movie stream by a media server. The application hands this data off to TCP for transmission over the network. If the data to be transmitted is large, TCP first breaks the data into segments smaller than 65 KB (the data limit for IP). TCP then assigns sequence numbers to each segment and hands over the numbered segments to IP. IP transmits these segments over to the receiving computer.

At the receiving end, when TCP gets the segments from IP, it uses the sequence numbers assigned by the sender TCP to reassemble the data into its original form. TCP then passes the reconstructed data to the application. For example, if a web server sends an image that is about 100 KB in size, the sender TCP could break the image into two segments and label the segments as segment 1 and segment 2. At the receiving end, TCP would use the segment numbers 1 and 2 to reassemble the fragments in the correct order. The reconstructed image would then be passed to the browser. Neither the sending web server nor the receiving web browser (user applications) would be aware of the segmentation of the image into fragments by TCP during transmission.

To summarize, TCP provides a reliable communication service between applications over networks of arbitrary complexity and any level of unreliability. The only thing computer applications have to do to use the network is to pass data to TCP. TCP takes care of all communication complexities on behalf of the applications. TCP delegates the task of best-effort routing to IP, but performs all other bookkeeping required for reliable communication. TCP may be viewed as the data communications project manager for each network connection.

TCP Functions

We have already seen that TCP has to perform three essential functions to resolve the imperfections of IP—segmentation, reliability, and multiplexing. In addition, as you will see, TCP also performs two more useful tasks—flow control and connection establishment. We now look at each of these tasks in detail.

Segmentation

Segmentation is the process of dividing data blocks into smaller units. Segmentation is commonly used to facilitate transmission, and the segmentation function of TCP allows IP, which has a maximum packet size of 65,535 bytes, to transfer application data of arbitrary size.

A note on packet names: We have seen three names for data packets so far—frames, packets, and datagrams. The data created at the data-link layer is called a frame. At the network layer, IP creates packets, and at the transport layer, TCP and UDP create datagrams. Using these names consistently instead of the generic name—packet—facilitates professional discussion. If you are talking about frames, everyone understands that you are talking about the data-link layer.

TCP assists IP by breaking application data into segments of manageable size for IP. These segments are called datagrams.

TCP adds a sequence number to every datagram before transmission. Sequence numbers help the receiving TCP to reorder datagrams even if they are received out of order from the network. TCP assigns a sequence number to each byte of data and places the sequence number of the first byte of data in a datagram as the sequence number of the datagram.

Figure 5.2 shows an example of segmentation by TCP. In the example, the application is trying to send 930 bytes of data over the network. In real life, such a small data block would not be segmented; however, using small numbers keeps our example simple.

Say the transport layer decides to segment the data into three segments of similar sizes—300 bytes, 320 bytes, and 310 bytes. The first segment is numbered 1 as expected. The sequence number of the second segment also accounts for the length of the first segment and is assigned sequence number 301. This number is calculated as the sequence number of the first segment + the length of the first segment. Since the second segment has a length

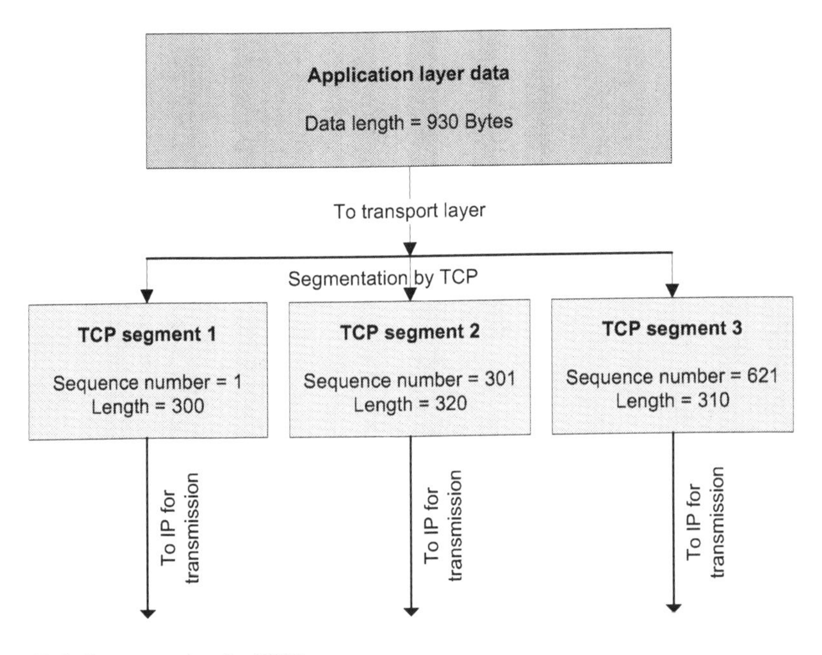

Figure 5.2: Segmentation by TCP

of 320 bytes, the third segment has a sequence number of 621 (301 + 320). This is shown in Figure 5.2.

What is the advantage of using a scheme that assigns sequence numbers by data bytes and not by datagram? In other words, do we have any advantage in using the numbering scheme shown in Figure 5.2 instead of the simpler scheme where the segments would be numbered 1, 2, and 3? The primary advantage of the numbering scheme shown in Figure 5.2 and used by TCP is that the scheme makes acknowledgments easier. Since every data byte in TCP gets a sequence number, the receiver has the flexibility to acknowledge by data byte rather than by segment boundary. The receiver is not constrained to acknowledge complete segments. In other words, in our example, it is possible for the receiver to acknowledge that 400 bytes or 600 bytes have been received successfully. TCP generally acknowledges datagrams by datagram boundary, but TCP can acknowledge partial datagrams because of the sequence numbering scheme used in TCP.

Therefore in general:

Sequence number of a TCP segment	=	sequence number of the previous segment + length of previous segment

Segmentation has both advantages and disadvantages. Let us start with the advantages. If every data byte has equal likelihood of being delivered in error, smaller segments are less likely to encounter errors during transmission. Segmentation reduces the chances of error. Further, in case a segment does get corrupted during transmission, only the corrupted segment needs to be retransmitted. Without segmentation, the entire data block would have to be retransmitted if even a single byte of data encountered a transmission error. Therefore, segmentation reduces the need for reprocessing in case of error. Finally, with segmentation, it is easier for routers to handle segments. Imagine sending an entire DVD movie as one data segment. A regular DVD has about 4 GB of data. If we were to send out a 4 GB data segment, every router along the path would have to dedicate 4 GB of memory to store the packet before forwarding it to the next router. Clearly, this would make routers extremely expensive. Thus, segmentation simplifies packet handling.

You may now better appreciate the analogy between packetization and knock-down kits that was introduced in Chapter 1. A lot of furniture is shipped as knock-down kits. These kits are easy to handle, take up less space, and can be transported in any general-purpose truck, lowering costs. The trade-off is that customers need to reassemble the furniture at home. The knocking down and labeling of each furniture part is akin to what the sending TCP does. Reassembly of furniture based on part labels by end users at home is akin to the TCP receiving function. In the DVD example above, the DVD is like the table that is difficult to ship. The data segments of the DVD are like the kits that are easy to ship, though slightly complex to reassemble.

The primary disadvantage of segmentation is the additional complexity introduced by segmentation and reassembly. The sending TCP needs to compute sequence numbers and the receiving TCP needs to reassemble the segments based on sequence numbers. It is possible for segments to get lost or duplicated and these duplicates and losses need to be identified and processed appropriately.

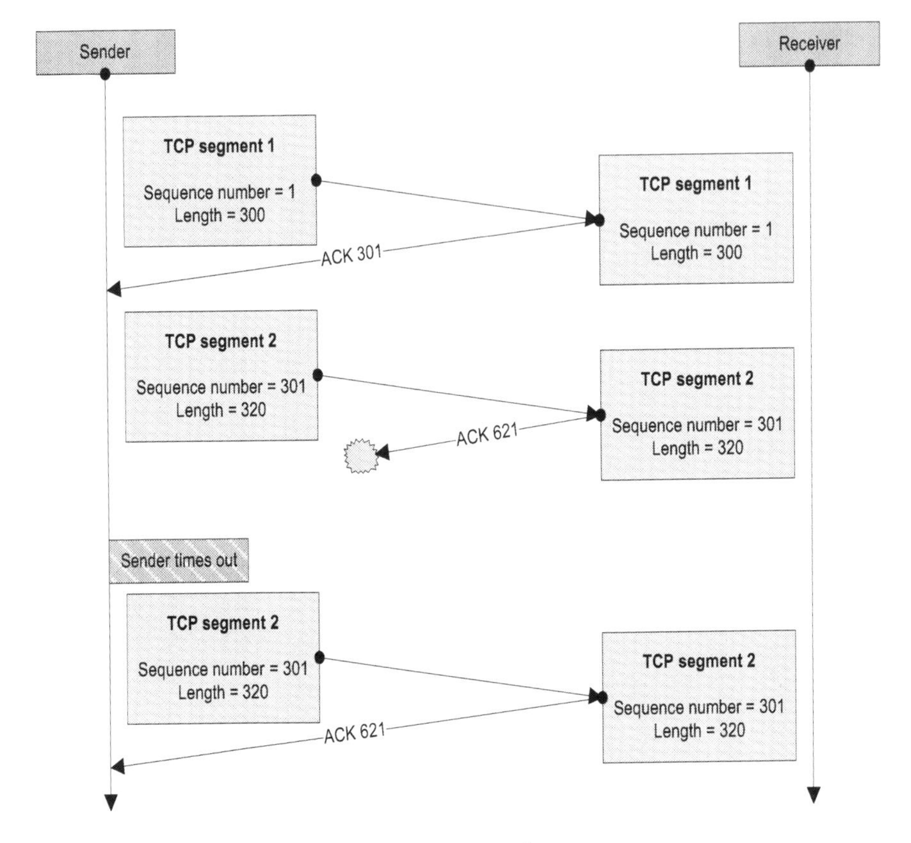

Figure 5.3: TCP reliability based on sequence numbers

Reliability

TCP makes the network reliable. The basic mechanism for getting reliability is for the receiver to periodically acknowledge received datagrams. If an acknowledgment is not received within a reasonable amount of time, the sender assumes the worst and resends the datagram. This basic mechanism is illustrated in Figure 5.3.

In the example, the sender begins transmission and the first segment is acknowledged successfully. When the first segment is acknowledged, the second segment is transmitted. Unfortunately, a router on the return path loses power and the acknowledgment is lost, even though the segment was received successfully by the receiver. After waiting for some time (called the time-out interval), the sender resends the segment and this time the acknowledgment successfully reaches the sender.

As the example shows, it is possible for segments to be duplicated. Note how the receiver gets two segments with sequence number 301, indicating that these segments have duplicate data and the receiver can discard one of them. The TCP software takes care of handling duplicate datagrams.

In addition to keeping track of datagrams, there is a second dimension of reliability offered by TCP. This is in ensuring that the received datagrams are complete. Recall that the maximum frame size in Ethernet is about 1,500 bytes, but the maximum IP packet size is 65,536 bytes. The fragmentation function of IP allows further fragmentation of packets to fit Ethernet frames. It is possible that some Ethernet frames may get lost during transmission. Since IP does not detect packet-data errors (it only detects header errors), when the receiving TCP attempts to reassemble such a segment, it needs a mechanism to know that the segment is incomplete. For this purpose, TCP also computes a checksum to each datagram. This checksum is not as robust as the data-link layer CRC, but it is good enough to detect damaged segments.

When a TCP segment is identified as being damaged, the receiving TCP does not have to do anything. It can simply discard the segment and wait for the sender to timeout and resend the segment.

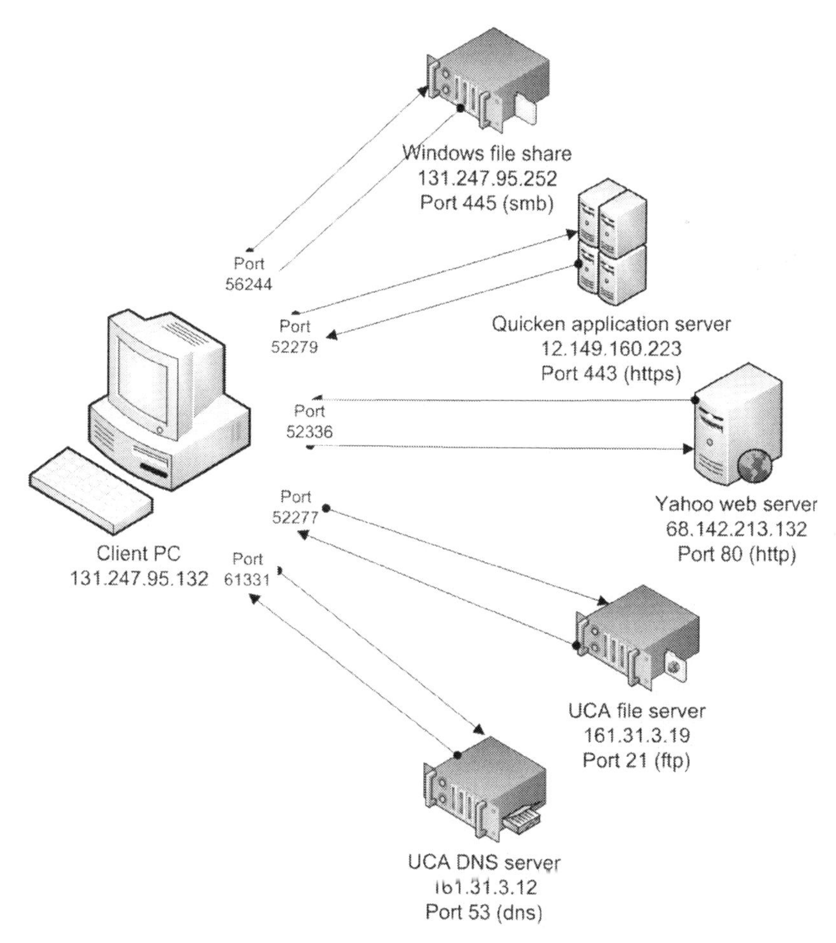

Figure 5.4: TCP ports and multiplexing

Multiplexing

The third essential function of TCP is multiplexing. *Multiplexing is the combining of two or more information channels onto a common transmission medium.* Multiplexing at the transport layer allows multiple applications to share the same network interface card and network link. This is a good place to highlight an important capability of computers as communication devices. You may have noticed this difference between computers and telephones. Telephones can only help in one (albeit, highly useful) communication task—talk. You pick up the phone, dial a number, and talk. More interestingly from the point of view of multiplexing, you can only carry out one telephone conversation at a time. But computers can multitask. You can browse the web (with multiple browser tabs open simultaneously, even streaming different video channels on each tab), chat on an instant messaging client, listen to streaming music, monitor a stock quote service, and watch live traffic updates. And, you can do all these things simultaneously on your computer. This is where multiplexing comes in.

Each of these applications is sending and receiving data segments. All these data segments go out of the same Ethernet or other network link. Yet, we never see computers displaying information in wrong windows. So, clearly the computer has a mechanism to distinguish between packets coming to each application on the common link. This is accomplished by the multiplexing function of TCP. In the context of TCP, the information flowing to and from each individual application is an information channel and the common transmission medium is the network connection of the computer.

Sockets

If you have developed network applications, you may be familiar with the concept of a "socket." *A network (IP) address and a port address together constitute a communication socket.* The application can send data over the network using the socket. Each network connection is then identified by a pair of sockets: the port and IP addresses of the sender are the socket at one end of the connection, and the port and IP addresses of the receiver are the socket at the other end of the connection. Together, the two sockets make up the network connection.

As we have seen in Chapter 2 and Chapter 3, the standard way to uniquely identify an entity is to assign it a unique identifier (ID). *Port numbers are identifiers assigned to each application process by the transport layer.* TCP enables multiplexing by providing multiple port addresses within each host. Each application that needs a network connection—a browser tab, for example—is assigned a port number. Each port, therefore, identifies a communication channel on the host.

Figure 5.4 shows an example of various applications on a client PC simultaneously connecting to applications on various computers on the Internet. Each application on the PC is assigned a separate port by the client operating system. Network traffic related to the application leaves from and arrives at the assigned ports. When packets come to the client PC, the computer can use the port numbers to identify what application to pass the packets to.

In the example, the web browser is using port 52336 to connect to the Yahoo web server. The Yahoo web server is running on port 80. Packets leaving the client PC have a source port address of 52336 and a destination port address of 80. Similarly, packets going to the Windows file sharing have a source port address of 56244 and a destination port address of 445. These port numbers uniquely identify the applications running on the respective computers.

From a networking point of view, there are two very interesting questions here: (1) How does the client PC know the IP address of the UCA file server because, after all, no end user ever types in IP addresses into any application? (2) How does the client know that the UCA DNS server is running on port 53 because again, nobody ever specifies a port number while using the Internet?

We will address the first question (obtaining receiver IP addresses) in Chapter 7 when we discuss DNS. DNS is a service that translates site names such as www.uca.edu to IP addresses such as 161.31.3.35. The second question (obtaining receiver port addresses) is addressed in later chapters.

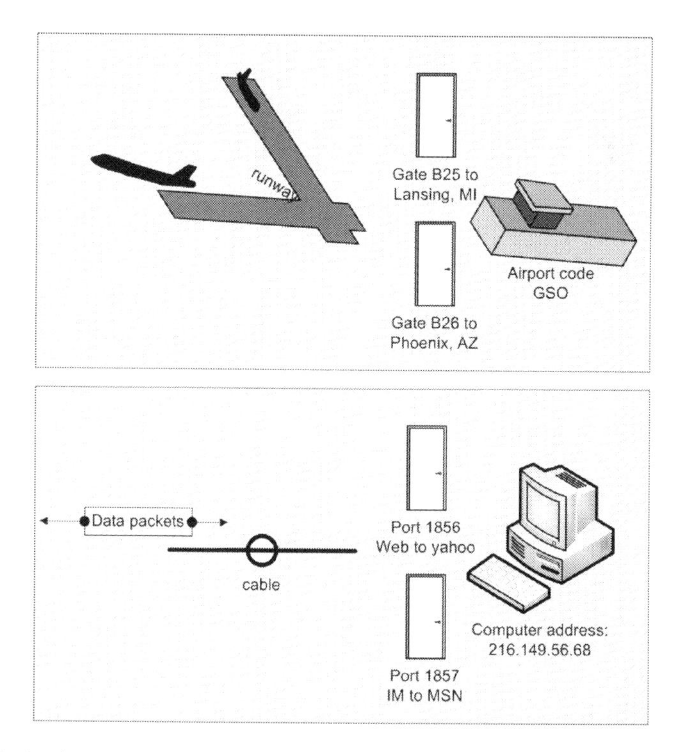

Figure 5.5: Analogy between TCP ports and airport gates

Computer port numbers are a lot like airport gates. The runway at the airport is analogous to the computer's network cable. Just as all aircraft landing at an airport share the runway, all network traffic related to the computer is carried by the cable. The airport code is analogous to the computer's IP address—just as aircraft identify an airport on the air-traffic network by its airport code, packets identify a computer on the computer network by its IP address. Port numbers are like airport gate numbers. A single airport can serve multiple destinations and airlines simultaneously by parking different aircraft at different gates. Similarly, a single computer can serve multiple applications connecting to multiple destinations. Figure 5.5 shows this analogy.

In Figure 5.5, the aircraft at gate B25 might be a USAIR Airbus A320 leaving for Lansing, Michigan, arriving at gate 7 of the Capital Region International Airport in Lansing. The aircraft at gate B26 might be a Southwest Boeing 737 going to Phoenix, Arizona, arriving at gate D7 of Phoenix Sky Harbor airport. Similarly, packets leaving port 1856 on the computer might carry web traffic to port 80 at yahoo.com, while packets leaving port 1857 might carry IM traffic to port 1863 at MSN.

The analogy between airports and routers is pretty close. Routers transfer traffic across data networks. Airports transfer traffic across traffic networks—the road network and the air network. Figure 5.6 is a photo of Tampa International Airport. Passengers arrive at the airport from the road network and depart over the air network. The airport provides all services necessary to interface between the two networks.

Figure 5.6: Tampa International Airport
(Source: Photo courtesy Google)

We can now turn our attention to the second question raised earlier. And this is the last issue that needs to be resolved to make port addressing work—how do we know what port to connect to? The client's port number is assigned by the local operating system and this information is available to the receiver in the incoming datagram. But how do we know what number to put in the destination port address field? How do we know that the web server is listening on port 80? If we want to access a Windows file server, how do we know what port to connect to?

There are some parallels to this problem of locating services, and what happens when you enter a retail store. First, note that each section of an aisle in the store is analogous to a server port since it serves up one type of goods. Now, as you enter the store, how do you know where to find what you are looking for? Milk? Bread? Cereals? The solution adopted in the store problem is to have labels high above the aisles and visible across the store to announce the location of each section as shown in Figure 5.7.

Most stores also adopt a variant of the standard port solution by standardizing on-aisle locations to the extent possible, so visitors know exactly where to find a product even if they are visiting a different branch of the store.

Figure 5.7: Store aisle directions are analogous to server port directions
(Source: Author photo)

In the absence of an elegant solution to port discovery, the IANA (Internet Assigned Numbers Authority)[1] has defined standard ports. The standard port for web servers is 80, the standard port for e-mail is 25, and the standard port for Windows file sharing is 445. If you connect to any of these standard services and do not specify a port, your application automatically connects to the standard port defined for the service.

You can see the list of standard ports known to your computer. In Windows, they are defined in the file C:\Windows\System32\drivers\etc\services. On UNIX derivatives, they are in the file \etc\services. By convention, ports 1–1023 are reserved for defined services such as web (80), e-mail (25), and secure web (port 443). User applications may use the remaining ports (ports 1024–65535). Figure 5.8 shows a screenshot of the etc\services file on Windows. The file shows the ports for common services such as e-mail (SMTP) and web (HTTP).

1 This is the same agency that has ultimate responsibility for IP addresses.

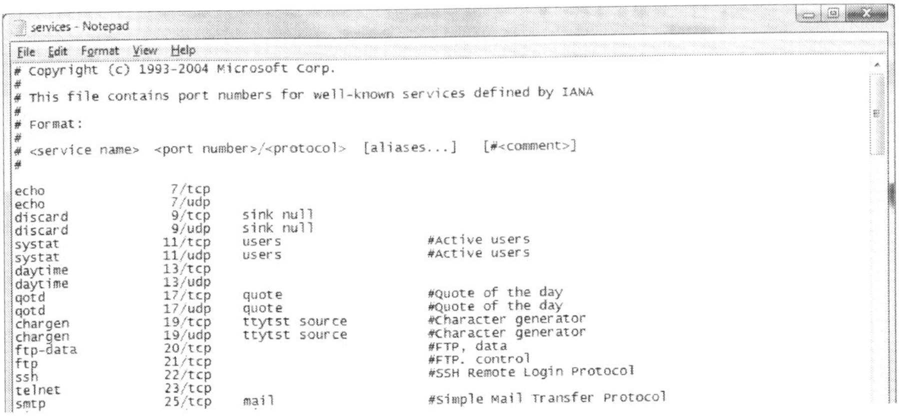

```
services - Notepad
File Edit Format View Help
# Copyright (c) 1993-2004 Microsoft Corp.
#
# This file contains port numbers for well-known services defined by IANA
#
# Format:
#
# <service name>  <port number>/<protocol>  [aliases...]   [#<comment>]
#

echo                7/tcp
echo                7/udp
discard             9/tcp    sink null
discard             9/udp    sink null
systat             11/tcp    users                  #Active users
systat             11/udp    users                  #Active users
daytime            13/tcp
daytime            13/udp
qotd               17/tcp    quote                  #Quote of the day
qotd               17/udp    quote                  #Quote of the day
chargen            19/tcp    ttytst source          #Character generator
chargen            19/udp    ttytst source          #Character generator
ftp-data           20/tcp                           #FTP, data
ftp                21/tcp                           #FTP. control
ssh                22/tcp                           #SSH Remote Login Protocol
telnet             23/tcp
smtp               25/tcp    mail                   #Simple Mail Transfer Protocol
```

Figure 5.8: The etc\services file on Windows

Port conflicts

What happens if two different applications listen on the same port? Typically, the operating system will pass incoming packets on this port to one of the listening applications. If you are an application developer and you develop an application that listens on port 80, in most cases the application will run fine because client PCs rarely run web servers. But if your application is deployed on a server that is also running a web server provided by the OS vendor (say IIS on Windows), chances are that you will receive an installation error and even if there are no install errors, packets sent to port 80 will be directed to the web server and not to your application. Your application will appear unresponsive as a result. As an application developer, therefore, it is a good idea to avoid the reserved ports (0–1023).[2]

All operating systems provide a utility—netstat—to view port usage. On Windows, you can run netstat by typing `netstat` on the command prompt (Start → Programs → Accessories → Command prompt → `netstat`). Some useful options require netstat to run with administrative privileges. To run the command prompt with administrative privileges, right-click the command prompt application in the Accessories program group and select "Run as administrator." Figure 5.9 and Figure 5.10 show the output of the netstat utility with some useful options.

Security use of netstat

The netstat –b utility is an excellent tool for identifying potential network security weaknesses on a computer. This utility identifies all applications on your computer that are communicating on the network (ESTABLISHED state) or have the potential to communicate (LISTEN state), and the ports on which they are expecting packets. When you look at the output of netstat –b, all applications you see listed should be known

2 Also, in Linux and UNIX systems, root privileges are sometimes necessary to listen on the reserved ports.

or expected applications. Any unexpected application needs to be investigated as a potential security hazard. You can use this information to stop unnecessary applications or use a firewall to block packets coming from unsafe locations on the Internet to these open applications.

Another useful option is –e. netstat –e displays errors in sent and received packets. We expect to see 0 errors. If the number of errors is large, it can indicate problems with a network card or cable or a configuration error.

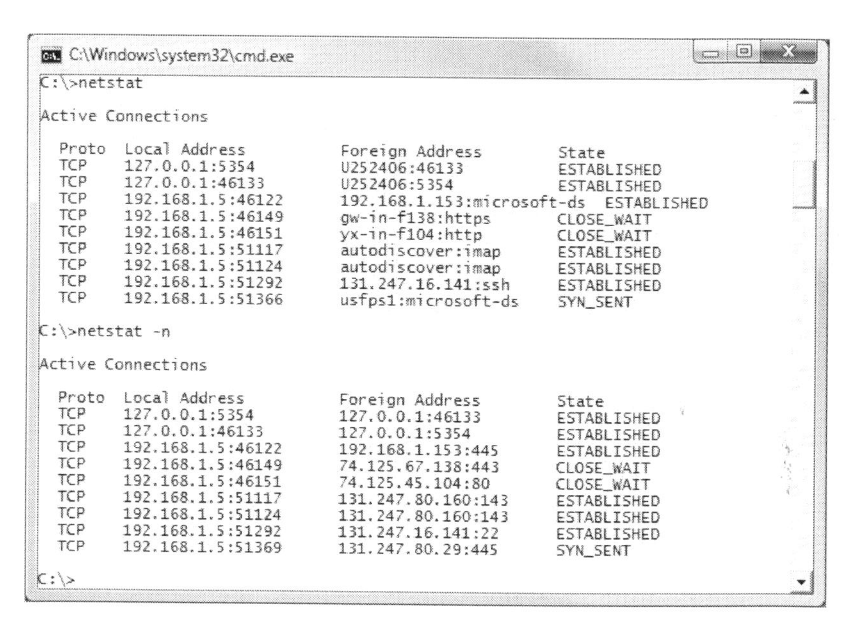

Figure 5.9: Viewing used ports with the netstat utility

```
Administrator: Command Prompt
C:\>netstat -b

Active Connections

  Proto  Local Address          Foreign Address             State
  TCP    192.168.1.5:46122      192.168.1.153:microsoft-ds  ESTABLISHED

 Can not obtain ownership information

x: Windows Sockets initialization failed: 5
  TCP    192.168.1.5:46149      gw-in-f138:https            CLOSE_WAIT
 [GoogleCalendarSync.exe]
  TCP    192.168.1.5:46151      yx-in-f104:http             CLOSE_WAIT
 [GoogleToolbarNotifier.exe]
  TCP    192.168.1.5:51117      autodiscover:imap           ESTABLISHED
 [opera.exe]
  TCP    192.168.1.5:51124      autodiscover:imap           ESTABLISHED
 [opera.exe]
  TCP    192.168.1.5:51418      mangrove:epmap              SYN_SENT
 [lsass.exe]
```

Figure 5.10: netstat –b shows the executables that are using ports

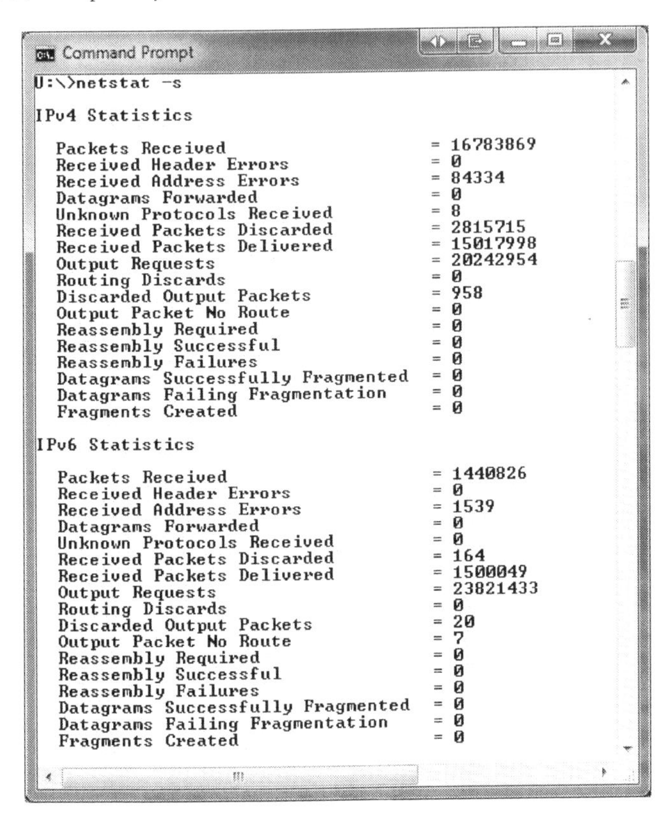

Figure 5.11: netstat –s option

Another useful option with netstat is –s. netstat –s provides a summary of packet statistics (Figure 5.11).

> ### Multiplexing at the movies
>
> The term multiplex is also used in reference to large entertainment venues with multiple screens showing many movies at the same time. The usage is analogous to our airport example—you go to one venue and the movie you see depends upon the door you take within the complex.

Flow Control

In the last three parts of this section, we have looked at the three critical tasks performed by TCP—segmentation, reliability, and multiplexing. We now look at two additional useful functions of TCP. These functions allow graceful network operation (flow control) and one support function (connection establishment) required for TCP sequencing to work.

Flow control is the control of the rate at which data are transmitted from a terminal so that the data can be received by another terminal. This becomes useful if a receiver is too slow to process data at the rate at which it is being delivered by the sender. Colloquially, we use the term "drinking from a fire hose" to describe this situation. A fire hose does serve water, and you could drink from it if absolutely necessary, but most of the water would get wasted. To improve efficiency, therefore, it is useful to allow the receiver to slow down the sender based on the receiver's capabilities, just as an adult might prefer a faster water flow than a child. Similarly, in computer networks, newer computers may be capable of handling higher data rates than older computers. TCP allows the receiver to specify the amount of data it is capable of handling. The sender uses this information to regulate the rate at which it transmits data. This is the flow-control function of TCP.

Flow control in the classroom

Flow control is one of the greatest challenges for any instructor. Every class has students with a wide range of prior experience in the subject. When the instructor goes too slowly, students who are familiar with the subject lose interest in the class. When the instructor speeds up, students with limited prior experience in the subject feel lost.

Most instructors prefer accommodating the latter group more than the former group and adapt their pace of instruction so that these students are able to keep up.[3]

A primitive mechanism to accomplish flow control is shown in Figure 5.12. In this mechanism, the sender sends a datagram and waits for the datagram to be acknowledged before sending the next datagram. This is an extremely courteous flow-control mechanism in that no data is sent unless the previous data has been acknowledged. Since this is the basic model for flow control, this mechanism even has a name—"stop-and-wait."

Why does flow control only try to slow down the sender? Why is no attempt made to speed up the sender? This is because it is not necessary. Senders always attempt to send data at the highest rate possible. If a receiver is faster than the sender, any command from the receiver to the sender to speed up is not meaningful if the sender is incapable of speeding up any further. Efficiency gains in terms of reduced loss of data are possible only by allowing the receiver to reduce the sender's speed. Hence, flow control only slows down the sender, it does not speed it up.

The limitations of stop-and-wait are obvious—it is very slow. Once the sender sends a datagram, it waits for an acknowledgment before sending the next datagram. Similarly, while the acknowledgment is making its way back to the sender and the next datagram is

3 It may be noticed that TCP is more efficient than most classrooms. In a classroom, students are the receivers of instruction. All these receivers get information at the same rate because the classroom does not facilitate individualized instruction. TCP, on the other hand, is able to adapt the transfer rate according to the capabilities of each receiver. Each receiver gets data at the highest rate it specifies.

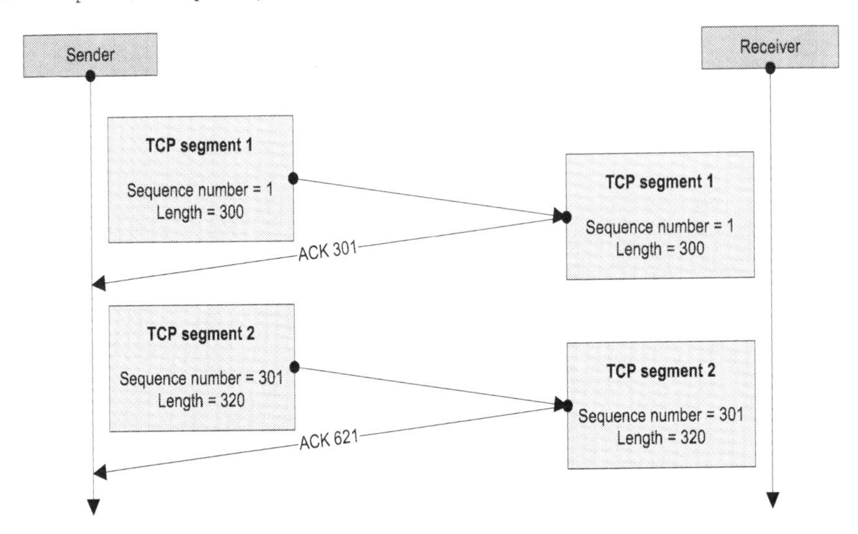

Figure 5.12: Stop-and-wait flow control, a very simple flow-control mechanism

reaching the receiver across the network, the receiver sits idle. Therefore, this flow-control mechanism is not particularly useful, even though it does ensure that all transmitted data can be processed by the receiver. Clearly, there is room to improve efficiency in terms of increasing data-transfer speeds.

TCP improves upon stop-and-wait flow control by allowing the receiver to indicate a "window size" with every acknowledgment. *The window size is the amount of data the receiver is capable of processing.* This is the amount of data that the sender may send without waiting for an acknowledgment from the receiver. Using the window-size information, the stop-and-wait flow control of Figure 5.12 may be improved as shown in Figure 5.13. In the figure, the receiver begins by announcing a window size of 700 bytes. Assuming that the sender has 620 bytes of data to send, since 620 < 700, it sends all the data at once, without waiting for an acknowledgment from the receiver for the first datagram. Please note that it is unlikely that such a small amount of data would be split into two datagrams, but the figure illustrates how multiple datagrams might be sent without waiting for an acknowledgment. TCP acknowledgments are cumulative. The receiver acknowledges all this data by sending a datagram, labeled ACK 621 in Figure 5.13. ACK 621 (value of 621 in the acknowledgement number field) means that all data up to byte 620 has been received successfully.

Figure 5.13 is the basic mechanism used by TCP to implement flow control. Figure 5.14 shows how this mechanism works when the sender has larger quantities of data to send. When the sender receives the window size from the receiver, it transmits the permitted quantity of data. The transmitted data may be said to be in the "sent window" of the sender. The data in the "sent window" has been transmitted but has not yet been acknowledged by the receiver. The "width" of the "sent window" is equal to or less than the advertised window size. The sender holds on to the data in the "sent window" because it is possible

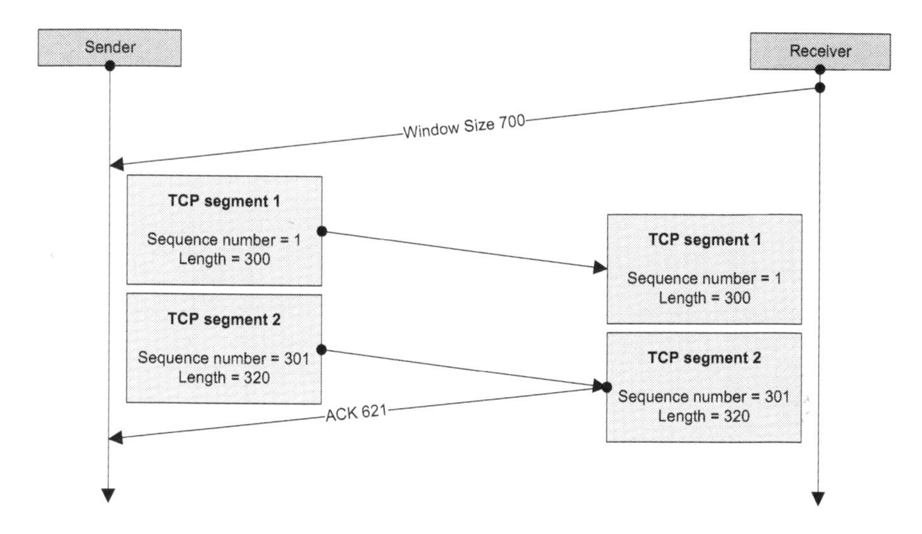

Figure 5.13: Using TCP window size to refine stop-and-wait flow control

that there could be transmission errors which may require some data within the "sent window" to be resent. If the sender sends more data, it might overwhelm the receiver, leading to lost packets and wasteful use of network resources. When the sender reaches the limit on transmission permissions, it waits for an acknowledgment from the receiver.

When an acknowledgment is received (datagram with ACK 601; window size 900 in Figure 5.14), the sender is free to send more data. In our example, the receiver has advertised a higher window size and so the sender can now send up to 900 bytes of data starting from sequence number 601. The example shows how the sender sends this data, places the data in the "sent window," and waits for acknowledgments before sending more data.[4]

Since the "sent window" keeps sliding over the data to be sent, the flow-control mechanism is called sliding-window flow control. *Sliding-window flow control uses a variable-length window that allows a sender to transmit a specified number of data units before an acknowledgement is received.*

Connection Establishment

We now come to the last function of TCP. This relates to the initialization of sequence numbers for a TCP connection and exchange of this initialization information between sender and receiver.

We have thus far used a sequence number of 1 for the first byte of data in all our examples. While this is intuitive to understand, the problem with this simple scheme is that it becomes difficult to identify duplicate datagrams, particularly if a user revisits a network destination in quick succession from the same client. If a receiver gets two datagrams with sequence numbers of 1 from the same computer, it becomes difficult to determine whether the datagrams are duplicates or two different connections.

4 There are a number of good online animations to demonstrate TCP.

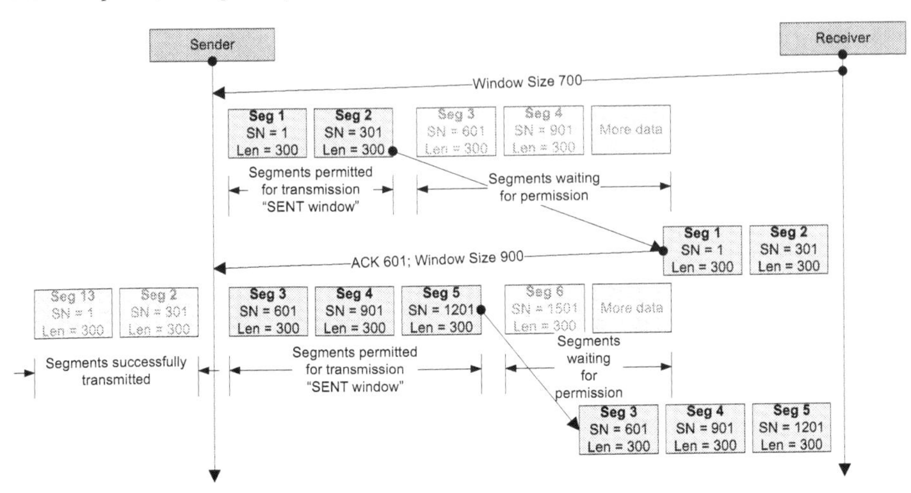

Figure 5.14: Sliding-window flow control

To overcome this confusion, TCP ensures that TCP numbers are not repeated in quick succession. This is done by generating the initial sequence number (ISN) for connections from a continuously advancing number generator. Once the initial sequence number has been generated for a connection, sequence numbers for all subsequent datagrams in the connection are sequentially numbered by data bytes as discussed before. With this refinement, Figure 5.12 can be modified as in Figure 5.15. In the figure, the ISN is 1620789.

Since data communication is usually bidirectional, TCP always creates a bidirectional connection to facilitate bidirectional transfer of data. For example, when you visit a website, you send the name of the web page you are interested in browsing and the website sends

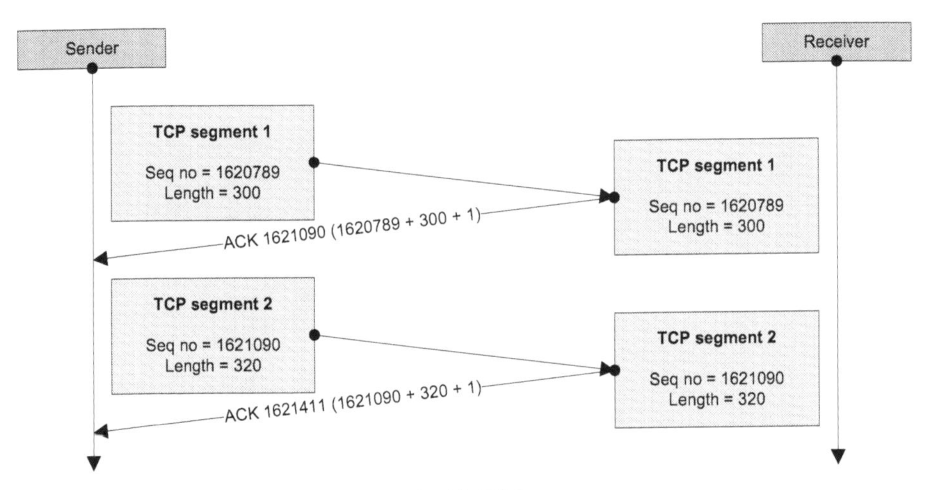

Figure 5.15: Stop-and-wait flow control with ISN

the contents of the page to you. In a bidirectional connection, each time two computers communicate, two TCP streams of data are generated—one from the sender to the receiver and the other from the receiver to the sender. Therefore, two ISNs are generated. Each side generates an initial sequence number for outgoing data from its side.

One final task now remains. If the two sides are not going to start with a sequence number of 1, each side has to inform the other side of the initial sequence number it will use for the connection. For this task, TCP defines a procedure in which three special packets are exchanged between the two computers before any data exchange takes place. This is called the three-way handshake and is shown in Figure 5.16.

The three-way handshake is really a simple procedure. Say the sender generates an initial sequence number (ISN) of 83441 and the receiver generates an ISN of 2713867. The sender begins by sending its ISN to the receiver in a blank datagram (no data). This is the transfer labeled 1 in Figure 5.16. The receiver acknowledges the sender ISN by putting the 83442 in the ACK field of its response as expected.[5] In addition, it also inserts its own ISN of

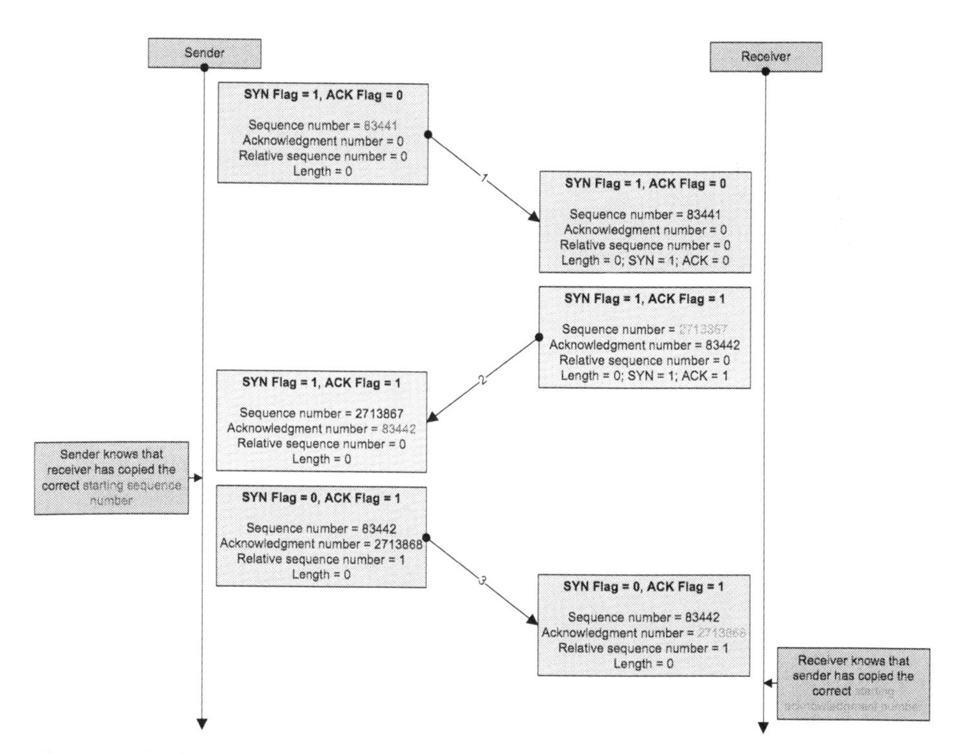

Figure 5.16: Three-way handshake to exchange initial sequence numbers

5 The careful student may note that the formula for calculating the ACK field isn't quite followed in the three-way handshake. The ACK is usually sequence + length. However, in the handshake, the ACK is actually sequence + length + 1 (since length is zero). In both cases, the ACK tells the other end "this is the byte I want next."

2713867 in the sequence number field of the acknowledgment. This transaction is labeled 2 in Figure 5.16. When the sender receives this acknowledgment from the receiver, it knows that its ISN has been copied correctly by the receiver. It also knows the ISN selected by the receiver. In the final datagram of the three-way handshake, the sender acknowledges the receiver's ISN and places the expected acknowledgment number of 2713868 in the acknowledgment number field of its response. This is labeled 3 in Figure 5.16. When the receiver gets this datagram it knows that its own ISN has also been correctly copied by the sender.

Once the two ISNs have been successfully copied, data transmission can proceed. The sender and receiver use the sequence numbers they just exchanged for segmentation, reliability, and flow control.

Initial sequence number—1 or a random number?

Previously, it was mentioned that the sequence number of the first segment is 1, but we see here that the initial sequence number is selected at random. Which one is correct?

The earlier example used an initial sequence number of 1 to simplify the discussion while introducing the topic. In practice, the initial sequence number is chosen at random as described in this section.

Multi-path TCP

Devices such as smart phones have multiple network connections that are active simultaneously, including Wi-Fi, 3G/4G, and Bluetooth. Recent updates to the TCP protocol (RFC 6824) allow devices to distribute the traffic across all these network paths. The latest versions of smart-phone operating systems have begun to use these extensions. An important benefit is that phone calls can seamlessly increase or decrease the extent to which voice and data traffic is routed through Wi-Fi networks as they become available.

TCP Header

We have completed our discussion of the functions of TCP. We can now look at the TCP header and identify the fields in the TCP header that support the TCP tasks discussed in this chapter. Figure 5.17 shows the TCP header.

The roles of the important fields in the TCP header have already been discussed. For reference, a quick summary of each field is provided below:

Source and destination port addresses These are the port numbers on the two sides of the data transfer.

A useful point to note in the TCP header is that the port address fields are 16 bits in length. This means that 65,536 (2^{16}) port numbers are possible on each computer. Since a connection is identified by both the IP addresses and port address, if the computer has sufficient processing power to manage the data on all these connections, a computer running TCP can uniquely label 65,536 separate communication channels simultaneously

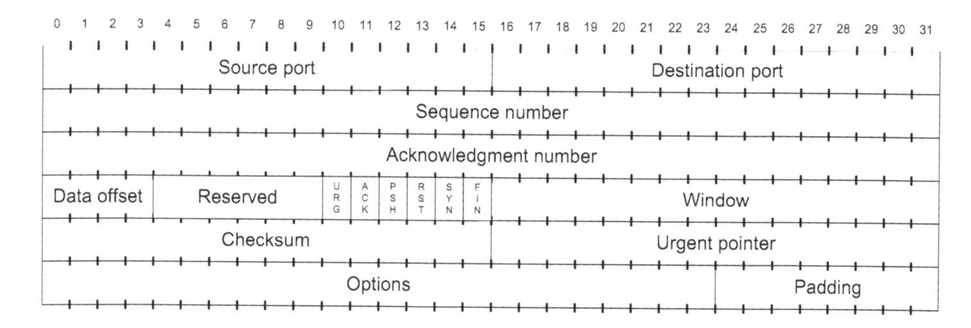

Figure 5.17: TCP header

to a given remote machine on a fixed port.[6] As has been pointed out before, it is useful to compare this capability of TCP with the limited capability of the telephone, which can only support one connection at a time.

Sequence number The sequence number of the first data byte in the datagram

Acknowledgment number The next sequence number expected by the sender of the datagram

Data offset Size of the TCP header. This field identifies where the data begins in the datagram. This is necessary because a TCP datagram may have variable-sized options.

Reserved These bits have no current use, but may be used to add new functionality in the future.

Control bits Six bits indicating special information. Control bits indicate changes in TCP state, such as whether a new connection is being opened, or an existing connection is being closed.

Window This is the window size, and it indicates the number of data bytes the sender of the datagram is willing to accept beyond the acknowledgment number field of the datagram.

Checksum A checksum on the entire datagram and the IP header. This is used for error detection.

Urgent pointer Used to indicate that the data in the datagram must be processed immediately. This feature is not generally used anymore but was popular in earlier applications.

6 The only limitation to the maximum possible number of connections from a computer is processing power. The same port number can be used to communicate with different remote computers.

Options Potential options that can be defined, such as maximum receivable datagram size

Padding 0 or more bits to ensure that the TCP header length is a multiple of 32 bits. This field is useful if options are used.

TCP checksum and layering purity

The TCP checksum is a very interesting field. This field includes information from the IP header in its computations. Layering requires that each layer be independent of other layers.[7] The TCP checksum is therefore the one field in the technologies we have discussed so far that explicitly violates the principles of layering by being dependent upon the value of fields in another layer.

UDP

This chapter has shown how TCP is very reliable. But the reliability comes at the cost of a fairly complex protocol. TCP requires the generation and exchange of ISNs, checksum, sequence numbers, and reassembly of datagrams. TCP is very useful when ordered and reliable data delivery is required for an application. Examples of such applications include the web, e-mail, and any application carrying financially sensitive information. However, there are many applications where reliable and ordered delivery is not required. In these cases, much of the complexity of TCP can be eliminated in favor of a simpler transport protocol.

An example of a situation where reliable, ordered delivery is not required is sensor data. Sensors usually send out very small amounts of data (e.g. time = 9:30 a.m., or temperature = 43^{0}F). Such small transmissions do not require segmentation. Sensors also generally send data at frequent intervals, say, every minute, which creates a lot of redundancy to handle lost datagrams. Thus, even if an occasional sensor reading is lost, it does not significantly affect applications using this data. These applications therefore do not need the reliability of TCP.

Another situation where reliable, ordered delivery is not useful is real-time communication, such as voice or video. If a data packet from a video frame is lost, there is no point in receiving a retransmitted datagram once the video frame has advanced and the TV is displaying the next scene. It is therefore best to ignore lost datagrams in video feeds, tolerate the occasional loss, and move on.

For situations such as these, or where application developers might be interested in enforcing their own flow control, an alternate transport layer protocol has been defined. This protocol is called UDP, which stands for User Datagram Protocol. UDP was defined in RFC 768 in 1980. *UDP is a transport layer protocol without the reliability provided by TCP.*

The UDP header is shown in Figure 5.18. Comparing the UDP and TCP headers, we see that UDP eliminates all sequencing and acknowledgments, but preserves port numbers for the multiplexing function. UDP essentially adds port addresses to the IP header. Since the function of UDP (multiplexing) is a subset of the functions of TCP, there is no additional description provided here to understand the functioning of UDP.

7 What happens to the checksum when IPv6 is used? See section 8.1 of RFC 2460 (IPv6).

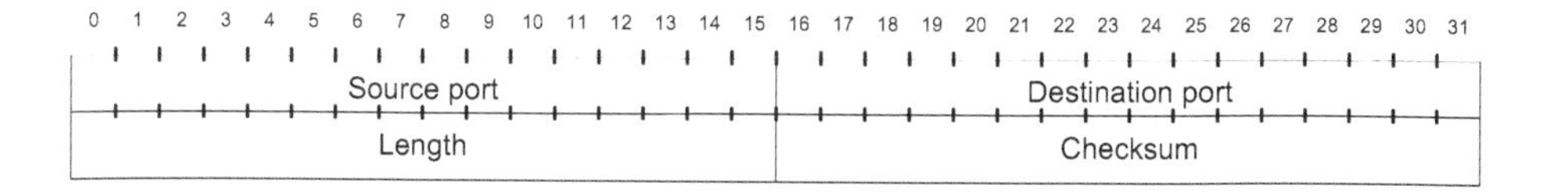

Note: Each tick mark represents a bit position

Figure 5.18: UDP header

For reference, the fields of the UDP header are described below:

Source and destination ports These fields perform the same functions as TCP ports.

Length This field defines the length of the entire datagram.

Checksum This field is computed in a manner similar to the TCP checksum, though it is often not used in computer-intensive applications such as video streaming.

EXAMPLE CASE—Financial Industry

What is money? Is it the cash in your pocket? Or, perhaps, the cash in all the pockets of the world combined? Facts suggest otherwise. As a result of fractional reserve banking, most of the world's money exists as accounting entries in the databases of financial institutions worldwide. In fact, it may not be an exaggeration to say that almost all the money in the world is just data in a distributed database.

The numbers bear this out as well. The total volume of US currency in circulation at the end of September 2015 was $1,390 billion. A more general measure of money is called M1 and includes deposits in checking accounts and travelers checks. At the end of September 2015, M1 was $3,064 billion. This quantity of money supports bank assets worth $11,413.1 billion (reported as H.8 by the Federal Reserve)—more than four times the amount of cash in circulation. More interestingly, the total volume of transactions on the securities markets in 2015 was approximately $300 billion each business day.[8] In other words, each business day the value of securities transactions in the United States is worth about one-quarter of the total amount of currency in circulation.

If you have worked in any job, chances are that you received your salary in the form of a check that you deposited in your bank account. In fact, other than the occasional fundraiser for charitable causes, there was probably no cash transaction between you and your employer. If the job was in a corporate setup, your salary was probably deposited electronically into your bank account, without even the paper check.

For most of your shopping, you probably use a credit card or debit card. When the monthly payments on these cards come due, you probably mail a paper check, or increasingly, ask the card company to withdraw payments directly from a designated bank account.

8 https://www.batstrading.com/market_summary/.

Just as you don't see cash for most of your income sources, you probably don't use cash for most of your household expenses either. In fact, the most common use for cash may be expenses such as payments at toll booths, charges that are too small for electronic payment systems to process.

You may notice that a common feature of both income and expenses in the typical household is that the transactions take the form of debits and matching credits in the databases of your employer's bank, your bank, and your credit card company. All of these databases are networked using extremely reliable computer networks.

Financial transactions take the form of data exchange between networked databases in all parts of the financial system. For example, if you buy stocks in a brokerage account, your brokerage firm debits the cost of the transaction from your account and transfers the money to the brokerage firm of the seller of the stock, which then credits the payment to the seller's account. No paper cash changes hands. In fact, no paper stock certificates change hands either. Since the early 1970s, stock ownership in the United States is recorded exclusively in the form of bookkeeping entries at a company called the Depository Trust and Clearing Corporation (DTCC). When the stock sale is settled, DTCC records a change in the ownership of the stock in its databases.

Indeed, transactions could not happen at the required speeds if money in the form of cash and securities in the form of paper certificates needed to physically change hands in each transaction. Without computer networks helping automate the settlement of trades, securities markets could not exist in their current form.

Latency in the Financial Services Industry

A special feature of computer networks in the financial industry is their need for low latency. Almost 50% of all trades in many securities markets are now electronic, where computerized platforms match orders received from computerized algorithmic engines at buyers and sellers. Trading algorithms on these exchanges try to profit from split-second price imperfections. Buyers and sellers who can trade faster have an advantage in these markets. As a result, improvements in computer networking technologies now play a central role at the core of the financial services industry. In response, technology vendors in the financial services sector are engaged in what is called a "low-latency arms race."

When latency becomes critical, the reliability functions of TCP become expensive because they can slow the transmission of data. Among the measures taken to reduce latency are minimizing router hops, optimizing the TCP/IP stack, disabling Nagle algorithm, and tuning TCP Window and datagram sizes. Some architectures even avoid TCP in favor of protocols such as sockets-direct protocol (SDP) to reduce latency. Improvements in data storage are also helping to reduce latency. Solid-state devices can provide access times that are hundreds or thousands of times faster than hard disks.

Even the speed of light can be a limiting factor in reducing latency. Data travels at about 100,000 miles per second in optical fiber, leading to latency of about 10 microseconds per mile. Algorithmic traders cannot afford to wait that long to execute their trades. In response, trading exchanges are increasingly offering proximity services that allow trading firms to locate their electronic trading systems physically close to the exchange order-matching systems, sometimes even within the premises of stock exchanges.

$300m to save 6 ms[9, 10]

In 2011, installation began for the Hibernian Express submarine cable. After careful planning, and following the shallow continental shelf instead of the deeper ocean, the route reduced the distance by 310 miles, saving 6 milliseconds in transit time for electronic signals from London to New York, a saving of 10% over the existing transit time of 65 milliseconds. This allows traders to race to the head of the queue when executing trades across continents.[11]

Another company in this business, Spread Networks (http://spreadnetworks.com), dug a direct route between financial capitals New York and Chicago, including drilling tunnels through the Allegheny mountains to create an almost straight-line route, at an estimated cost of $300 million, resulting in a round-trip time of 13.10 milliseconds, a saving of 100 miles and 3 milliseconds.[12] Analysts estimate its low latency can let the company charge almost 8-10 times the normal transmission price. The construction was financed largely by Jim Barksdale, one of the earliest CEOs of the Netscape browser.

References

1. Board of Governors of the Federal Reserve System. "H.8—Assets and Liabilities of Commercial Banks in the United States," http://www.federalreserve.gov/Releases/h8/Current.
2. Depository Trust and Clearing Corporation. "Following a trade."
3. Depository Trust and Clearing Corporation. "Transaction statistics and performance."
4. Federal Reserve Bank of New York. "The money supply," http://www.newyorkfed.org/aboutthefed/fedpoint/fed49.html.
5. Federal Reserve statistical release. "H.6—Money stock measures," http://www.federalreserve.gov/releases/H6/Current.
6. Wesley, Daniel. "The value of US currency in circulation," http://visualeconomics.creditloan.com/the-value-of-united-states-currency-in-circulation/.

Summary

This chapter discussed the transport layer. The transport layer provides segmentation, reliability, and multiplexing functions in data communications. Where necessary, it also provides flow control. The layer receives raw data from the end-user application and delivers it to the receiving computer reliably, without errors. For purposes of network transport, the layer may segment the data into datagrams. The segments are labeled with sequence numbers before transmission to facilitate reassembly of the data at the receiving end. Before delivery

9 http://www.telegraph.co.uk/technology/news/8753784/The-300m-cable-that-will-save-traders-milliseconds.html (accessed D[illegible] 2015).

10 Doug Cameron and Jacob Bunge, "Underwater Options? Trans-Atlantic Cable Targets High-Frequency Traders," *Wall Street Journal*, Oct. 1, 2010.

11 Andrew Blum, *Tubes: A Journey to the Center of the Internet* (HarperCollins, 2012).

12 Christopher Steiner, "Wall Street's Speed War," *Forbes*, Sept. 9, 2010.

to the application at the receiving end, the data is reassembled in the correct sequence. The end-user application is completely unaware of the segmentation and reassembly performed by the transport layer for data transfer.

Data segments at the transport layer are called datagrams. The primary protocol at the transport layer is TCP. For applications that do not need reliability or ordered delivery of data, an alternate protocol, UDP, has also been defined. Application developers are often interested in specifying an appropriate value for the port address fields of TCP or UDP to ensure that their applications do not interfere with other applications on computers where their applications are installed.

An interesting pattern may be observed in the placement of protocol layers. The physical layer does best-effort information transmission across a single link. This transmission can introduce bit-level errors. These single-link errors are removed by the data-link layer. The network layer does best-effort routing across networks. Some packets can be lost during routing. The transport layer removes these errors. Thus, protocol layers alternate best-effort transmission and error correction.[13]

With this chapter we have completed discussion of the technologies that reliably transfer any amount of data across any two computers located anywhere on the Internet and interconnected by networks of arbitrary complexity and unreliability.

In the next chapter we will see some standard network applications that make these network services useful for end users.

About the Colophon

This book has focused on data communication between applications on different computers. However, the problem is more general. Anytime two or more applications wish to share any computing resource, even if the applications are running on the same computer, the sharing mechanism uses most of the principles discussed in this book. Computer scientists use the term *process* to refer to running programs. Processes need to define rules for cooperation with each other. These rules are called protocols. To send and receive data, operating systems allocate ports to processes. These are the same ports that we have seen in detail in this chapter. This idea is described very well in RFC 62, from which the quote has been taken. Thus, the ideas introduced in this chapter describe the standard method by which computers share resources, even among processes located on the same computer.

REVIEW QUESTIONS

1. What are the functions of the transport layer?
2. Why are two protocols, TCP and UDP, defined at the transport layer instead of just TCP?
3. What is *segmentation*?
4. Why is segmentation useful?

13 This is yet another useful insight from John D. Day, *Patterns in Network Architecture: A Return to Fundamentals* (Prentice Hall, 2008).

5. How is reliability at the transport layer different from the reliability provided by the data-link layer?
6. What are the important potential problems with reliability that are handled by the transport layer?
7. How does TCP provide reliability?
8. What is *flow control*?
9. Why is it more useful to allow the receiver to control flow speed, rather than the sender?
10. How is flow control implemented?
11. What is *sliding window* in the context of flow control? Describe its operation.
12. What is *multiplexing* in the context of the transport layer?
13. How is the multiplexing at the transport layer different from the multiplexing at the physical layer?
14. What is a socket?
15. What is a port?
16. Why is it necessary to define port numbers at the receiving end for network services (such as web, e-mail, etc.)?
17. From the \etc\services file on your computer, list any five standard ports not listed in the text.
18. If you were developing an application that provided services over the network, could you have your application listen to client requests on port 80 (the port for web servers)? If yes, do you think it would be a good idea? Why, or why not?
19. What is *connection establishment* in TCP? Why is it necessary?
20. Describe the three-way handshake used in TCP.
21. What is the maximum number of possible TCP ports on a machine?
22. Why is the initial sequence number for a connection chosen at random?
23. A receiver sends an acknowledgment packet with the number 2817 in the acknowledgment number field. What inference can the sender draw from this packet?
24. What is the function of the window-size field in TCP?
25. What are some application scenarios where UDP may be more useful than TCP? Why?

EXAMPLE CASE QUESTIONS

1. Watch your personal expenses for a week. What fraction of your total expenses is in the form of cash? What fraction is in the form of network data exchange (credit and debit cards, online payments)? For privacy reasons, please do not report actual amounts, just report fractions.
2. Do a similar exercise for your net worth. What fraction of your net worth is in the form of cash or cash equivalents (checking and savings accounts)? What fraction is in the form of networked data (retirement assets, brokerage account assets)? Again, please only report fractions. Hint: you may find websites such as mint (www.mint.com) helpful in gathering this information from multiple financial institutions in which you have accounts.

3. What is the sequence of actions in settling a trade in securities markets? How many of these actions are completed over computer networks? How many by manual transfer? Hint: look at the broker-to-broker trade in the DTCC publication, *Following a Trade.*

For the following questions, please use online sources to gather the required information. Answer each question in not more than four to five sentences each.

4. What is the Nagle algorithm? Why is it useful in TCP? Why might it be a good idea to disable the use of the algorithm when TCP is used in the financial services industry?
5. Specialist firms have emerged to help organizations, especially financial institutions, reduce latency. Write a brief report on the services offered by one such firm.
6. What is an Internet socket? What is socket-oriented programming or network programming? Why do you think financial services firms are interested in experts in computer network programming?

HANDS-ON EXERCISE—netstat

The utility used to gather information about open ports on a computer is netstat. By default, netstat shows all the open ports on the local computer and the ports on remote computers that they are connected to. The first output in Figure 5.19 shows the use of netstat to list open ports on the local computer. We see three open ports.

The various options available with netstat can be seen using the DOS help for netstat using the command `netstat /?`. Without any options, netstat shows the network connections open on the computer. The –n option shows IP addresses instead of computer names.

netstat can be used to infer how networked applications work. For example, we can use netstat to see the connections opened by web browsers to display web pages. Figure 5.19 shows a second invocation of netstat a few seconds after the first call to netstat.

The main activity performed during this interval was a visit to the home page of St. Thomas University. From looking at the first two outputs of netstat, we see that the visit to the page opened five ports on the local computer. These ports allow the browser to quickly gather all the different pieces of information required to assemble the web page from different sources on the Internet. netstat has an option, -f, to view the URLs of the remote computers. This option is handy for identifying the remote computers. Output from netstat –f is also included in Figure 5.19.

Another netstat option you may find interesting is –r, which shows the IP routing table on your local computer. netstat –s shows various summary statistics.

Answer the following questions:

1. Show the output of netstat on your computer.
2. netstat has an option, -b, that shows the applications on the computer that open each port. Using the –b option requires that netstat be run with administrator privileges. You can do this by right-clicking the Command Prompt icon and selecting "Run as Administrator"). What applications have opened each open port on your computer?
3. Visit your university or college website. Then run netstat again. Show the output.

Figure 5.19: Output from netstat before and after connecting to www.stu.edu

4. How many new ports were opened by the browser to gather all the information on the page?
5. How does opening multiple ports simultaneously speed up the display of the web page on your browser?
6. Display the output from netstat –s. The output can be lengthy. A simple and elegant way to capture it is to save it to a text file, then copy and paste the file contents into your assignment. The "greater than" sign, ">" is the redirect operator in almost every major operating system and can be used to send the output of a command to any location, including networked locations. Try netstat –s > netstat_output.txt. Now open the netstat_output.txt file to view the results of executing the command.

CRITICAL-THINKING EXERCISE—Flow Control of Distractions

This is a bit of a stretch but let us try to see if we can use the concept of flow control to improve our personal productivity. Most of us know we spend too much time consuming media content, and too little time producing school or professional content. According to the American time use survey 2014, we spend 4 hours each day on work or education, but 5.3 hours on leisure and sports. And this survey does not even have an item for web browsing or social media consumption! We know that for our own good, in the long run we need to focus more on work and education. But in the short run, the gratification from TV and social media consumption is so overwhelming for most of us that these entertainment channels easily get past the flow-control mechanisms (e.g. calendars, reminders, to-do lists) we have in place to take control of our time and productivity.

1. Looking ahead at your life and career, what should you be spending most of your time on these days?
2. What flow-control mechanisms/tools (e.g. calendars, lists, etc.) do you use to direct your time to the most productive uses you identified in question 1?
3. To what extent do distractions (e.g. TV and social media) foil your flow-control measures to manage your time (you may use metrics such as how many unscheduled hours or how many incomplete tasks etc., you lose each day due to these distractions)?
4. Why are these distractions so attractive to you?
5. Based on your responses above, how can you improve your flow-control mechanisms to improve how you use your time to maximize personal productivity?

IT INFRASTRUCTURE DESIGN EXERCISE—Estimating Data Requirements

There isn't much configuration that network administrators have to do at the transport layer. However, since the transport layer is the primary location for segmentation, this is a good place to estimate data-rate requirements for the long-haul network links. In this exercise you will estimate the data-rate requirements for the link from Amsterdam to Mumbai.

Assume that the traffic on the link from Amsterdam to Mumbai primarily consists of data required to support the call center in Mumbai. Traffic from the Mumbai call center has two components:

- Database transactions to pull customer data from the data center in Amsterdam. Each customer transaction takes an average of 3 minutes, during which time the agent typically makes three queries to the database. Each query generates about 3,000 bytes of data.
- Customer support calls to a 1-800 number routed to Mumbai from Tampa through Amsterdam. The calls use a G.728 codec and each call requires a 31.5 Kbps bit rate.

Answer the following questions:

1. What is the average data rate in bits per second required to support the database queries with no unnecessary delays? To do this, you would like to have enough capacity so that all agents can run queries simultaneously. Remember that 1 byte = 8 bits and carriers report data rates in bits per second because this gives a larger number, which is useful for marketing purposes (10 megabits per second is more marketable than 1.25 megabytes per second).
2. What is the data rate required to support the voice traffic?
3. Adding both the above, what is the total data rate required on the Amsterdam–Mumbai link (in bits/sec.)? What fraction of this traffic is data, and what fraction is voice?

CHAPTER 6

Application Layer

Ben Franklin wouldn't be impressed by our pace of innovation. He invented the post office and showed us electricity, and it still took us 200 years to come up with email. We're not good at connecting the dots.

Scott Adams

Overview

This chapter covers the application layer. The application layer helps end-user applications to use the data communications service provided by TCP and UDP. At the end of this chapter you should know about:

- the services provided by the application layer
- the Hypertext Transfer Protocol (HTTP), used to retrieve web pages
- the Simple Mail Transfer Protocol (SMTP), used to exchange e-mails
- the File Transfer Protocol (FTP), used for transferring large amounts of data
- Instant Messaging and presence services

Application Layer Overview

The application layer enables end-user applications to use TCP and UDP in meaningful ways, for example to send e-mail or to download web pages. The application layer hides TCP and IP from the end user. The application layer provides functions that greatly simplify the development of networked end-user applications. For example, the HTTP application layer protocol retrieves web pages. Developers of web browsers can focus on creating a pleasant end-user interface and use the HTTP protocol to retrieve content to populate the web pages.

The application layer enables a critical feature that makes the Internet so useful and different from earlier information technologies—utility for groups of all sizes. For example, the telephone is convenient for one-on-one communication, but extremely inconvenient for mass communication. TV and print media (such as newspapers and magazines) are convenient for communicating to large groups, but not for individuals. But websites (HTTP) allow broadcast to large groups, mailing lists for communication to smaller groups, and e-mail (SMTP) to individuals.

The application layer deals with all the idiosyncratic requirements of each individual application. For each specified end-user activity, there is a specific application layer protocol suitable for the activity. For example, there is HTTP for web transfers and SMTP for e-mail. Application layer protocols define commands that are appropriate for accomplishing the specified end-user activity and are, therefore, unique to the activity. For example, the HTTP application layer protocol has commands to retrieve web pages. But since e-mail is not used to retrieve web pages, the SMTP application layer protocol does not define commands to retrieve web pages. Instead, SMTP has commands to locate the receiver on the target mail server.

There is a very clear analogy between the ways in which end users interact with computer networks and transportation networks, such as the US Postal Service (USPS). The USPS can transport any cargo for end users. All this cargo is transported in the same USPS trucks. However, different items are treated differently. Books are shipped in boxes, get media mail rates, and are transported with the lowest priority. Liquids are packaged in watertight containers. Fragile objects are mailed in specially labeled containers and receive special handling. The key point to observe is that different items may be packaged differently though they are all shipped using the same trucks. The Internet uses the same principle. The application layer provides the required support for each application. However, all traffic is transported over the same TCP/IP network.

The application layer is therefore unique among the five layers of the TCP/IP stack. The lower layers do not distinguish between end-user applications. Any physical layer and data-link layer technology can be used in combination with IP and either TCP or UDP to transport data for any application. However, the application layer protocols are customized to each application. This is shown in Figure 6.1. All differences between applications are reflected in the differences among the various application layer protocols.

There is yet another difference between application layer protocols and the lower-layer protocols. All application layer protocols interface with computer resources external to the communication system, typically the file system. The file system on the computer is where you store all your documents and programs. For example, the HTTP protocol reads files from the web server's file system and writes the contents of the file on the web browser's local file system. The file system is outside the communication system. The lower-layer protocols deal only with communication entities and have nothing to do with the rest of the computer that is not involved in communication.

The TCP/IP design of using a common data-transport mechanism to support any arbitrary application has greatly contributed to the popularization of the Internet. The architecture greatly improves efficiency in terms of lower costs. End users just have to buy one computer capable of running TCP/IP and can be assured that the computer will be capable of running all network applications, including those that haven't yet been developed. For example, voice-over IP (VoIP) is a relatively new application. Yet, using VoIP on a computer only requires installing a VoIP application such as Skype. No upgrades to TCP/IP, Ethernet, or the physical layer are necessary to add VoIP functionality to a computer.

The architecture of using common transport protocols is particularly economical for network carriers. Carriers can deploy one common TCP/IP network infrastructure to

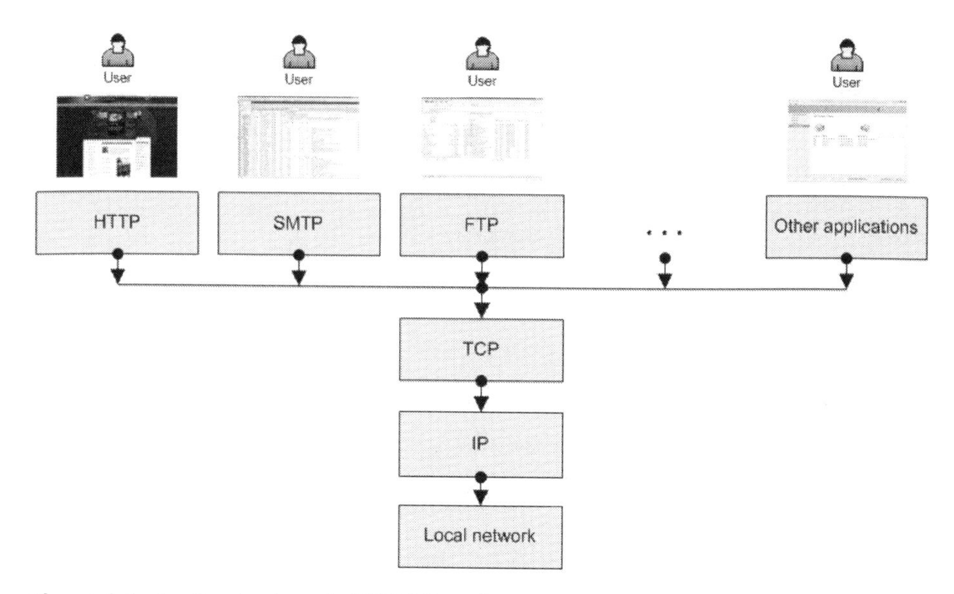

Figure 6.1: Application layer in TCP/IP stack

support all networking applications, including voice, video, and data. Contrast this with the earlier generation of networks where each city had one network for telephones and another network for television.[1]

An initial set of application layer protocols (SMTP, FTP, and Telnet) for using TCP/IP was defined in RFC 1123 in 1989. You may note that this list does not include one of the most popular application layer protocols today, HTTP, which is used for retrieving web pages. HTTP was only defined in 1996 in RFC 1945. Many other interesting applications have been defined in recent years, including instant messaging, file sharing, and even video streaming. The ability of the Internet to accommodate all these diverse applications with no changes to TCP/IP demonstrates the versatility of TCP/IP. It also reinforces our belief that TCP/IP should be capable of handling other network applications that may emerge in the future.

In the following sections of this chapter, we will discuss some common application layer protocols in more detail. Where possible, we will also relate these protocols to contemporary business practices.

HTTP

The simplest protocol to understand is the Hypertext Transfer Protocol (HTTP). *HTTP is a protocol that facilitates the transfer of files between local and remote systems on the World Wide Web*

1 Of course, unified networks were not possible 50 years ago because the cables available then were incapable of handling the large volumes of network traffic in a unified network. Modern fiber-optic cables can handle virtually unlimited quantities of data.

The World Wide Web (web for short) is an information system that displays pages containing hypertext, graphics, and audio and video content from computers located anywhere around the globe. The web is only a small fraction of the Internet, yet it is so important that lay users treat the web and the Internet as synonyms (interchangeable terms).

Internet, internet, and the web

Lay users often use these three terms interchangeably. But there are slight differences in meanings that you should be aware of. Figure 6.2 shows the relationships between the three terms. The term internet (spelled with a lower case *i*) is distinguished from the Internet (spelled with an upper case *I*). *An internet is any interconnection among or between computer networks.*

The Internet is a worldwide interconnection of individual computer networks that provide access to all other users on the Internet.

The web is the part of the Internet where information is accessed using HTTP. Information on the web is typically accessed using web browsers.

If you set up LANs in two neighboring homes and connect the LANs together, you get an "internet." This internet may be used to share images, music, etc. If you connect one of these LANs to "The Internet," this internet becomes part of the Internet. If you run a web server on one of the computers in the LANs, the information on your LANs, accessible through the web server, becomes part of the web.

Over the last few years, users have found innumerable uses of the web. One of the earliest exciting developments was the ability to shop online. This was soon followed by comparison shopping online. Today, most people do not buy any expensive item before researching it online.

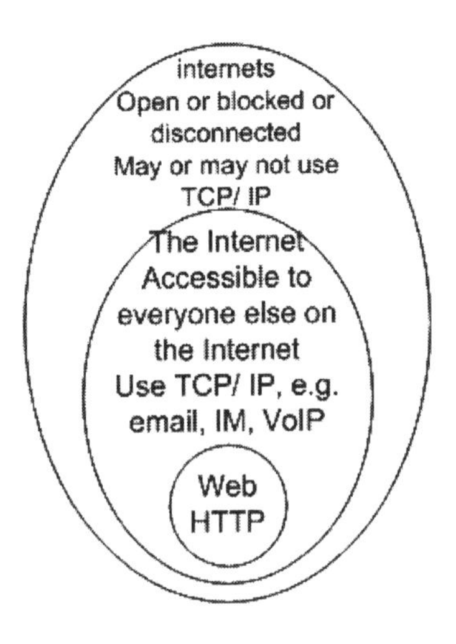

Figure 6.2: Internet vs. internet vs. web

The web has transformed traditional business models in a wide range of industries. Stockbrokers have been impacted because investors can place their own trades online and do not have to call share brokers to trade on their behalf. As a result, brokerage fees have come down from more than $70/trade to as low as $7/trade. The travel agent industry has been almost decimated because most travelers now make their own vacation arrangements. Amazon has become a formidable competitor in many retail categories. Wikipedia has emerged as a popular online reference encyclopedia. A recent class of applications is social networking, where people with shared interests can interact over the Internet. You have probably registered a profile on at least one of these social networks. You probably also check your assignments and grades from home using a web-based system.

Evolution of the Web

The web is useful because web servers allow many of these users to make information on their computers available on the web. People have different motivations for making information publicly available. Some are interested in selling things, others like to share personal interests, and many are just interested in expressing themselves. A common feature, however, is that most people making information available online have a great interest in providing the right information and in a form that engages viewers. Therefore, the web has become an effective and useful information filter that is attractive to viewers.

> *Why was the fish scared of the computer?*
> *It didn't want to get caught in the Internet.*
>
> Source: *Boys' Life* magazine, November 2011

The popularity of the web began around 1993. At this time, commercial ISPs were beginning to sell Internet connections to individuals. The ability to browse the web was a major motivation for individuals to pay a monthly subscription fee for Internet access. In 1992, AOL went public with about 150,000 paying subscribers. In its last public filing before merging with Time Warner, AOL had 26.7 million ISP customers. These numbers mean that AOL sustained an annual subscription growth rate of almost 80% for more than 9 years (Figure 6.3)! There are very few contemporary examples of businesses that sustained such high growth rates for such a long time.

	A	B	C
1		AOL subscriber growth rate	
2			
3	Date	Subscribers	Growth rate
4	33604	150000	=(B5/B4)^(1/((A5-A4)/365))-1
5	36891	26700000	
6			
7			
8	Date	Subscribers	Growth rate
9	January 1 1992	150000	77.78%
10	December 31 2000	26700000	

Figure 6.3: Calculating AOL's subscription growth rate

Tim Berners-Lee[2]

Tim Berners-Lee is credited with inventing the World Wide Web in 1990. His contribution was in combining two existing technologies—hypertext and the Internet. Each of these has existed since the 1960s. But Tim Berners-Lee was the first to create software to deliver hypertext over the Internet (i.e. a web server), including the associated protocols such as HTTP, HTML, and URLs. He deployed it at his employer, CERN, the European laboratory for Particle Physics, in Geneva, Switzerland.

The web browser was a major driver of the growth in popularity of the Internet. Netscape Navigator was released to the public in 1994. For the first time, non-technical users could access the wealth of information on the Internet. The most captivating feature of the browser was its ability to display images alongside text. It should be remembered that at this time (1994), very few computer applications were capable of displaying text alongside images on the same page. Therefore, the multimedia web page was itself a novelty. Further, the blue hyperlinks had almost magical powers and could instantly take users to web pages in other countries and even other continents.[3] By 1998, however, Internet Explorer became the dominant web browser, and on January 23, 1998, the Netscape source code was released to the open-source community, where the Mozilla project was born. We now see Mozilla as the Firefox browser.

Estimated size of the Internet

The current size of the Internet can be estimated using various metrics. We will look at three—number of hosts, web servers, and web pages. According to the Internet Systems Consortium (ISC),[4] there are more than 1 billion hosts (computers) on the Internet. Netcraft[5] estimates that there are approximately 900 million websites on the Internet as of 2015, of which about 170 million are active. Public websites have approximately 50 billion pages.[6] If we assume that each web page has approximately 10,000 characters (for a size of 10 KB/page), the text content of the public web would be approximately 500 TB in size.

---

2 Source: Abhijit Chaudhury and Jean-Pierre Kuilboer, *E-Business & E-Commerce Infrastructure: Technologies Supporting the E-Business Initiative* (McGraw-Hill/Irwin, 2001). A very interesting related project was the Internet Mapping Project, started by blogger Kevin Kelly in 2009. Readers were asked to submit pictures of their view of the Internet. A summary is by Lic. Mara Vanina Oses at https://psiytecnologia.files.wordpress.com/2009/06/the-internet-mapping-project2.pdf. It is an interesting view of how people 12–72 years of age view the Internet.

3 The website www.dejavu.org re-creates the web experience of the time, including the user interfaces of the early web browsers.

4 ISC offers the authoritative count of the number of hosts connected to the Internet.

5 Netcraft has been conducting surveys of websites for a long time and is the standard resource for this statistic.

6 http://www.worldwidewebsize.com.

Netcraft also reports on the market shares of web servers. As of 2015, the open-source Apache web server had the highest market share, accounting for almost 37% of all web hosts. Microsoft's Internet Information Server (IIS) was second with almost 27% of web hosts. In third place, with an almost 17% share, was nginx, a web server that focuses on low memory needs per connection.

blogs

The web has enabled some very interesting forms of communication. One of these, a form of intensely personal communication on the web, is a blog. Any person with an opinion can express it on a blog and allow anybody on the web to critique the opinion. The term weblog was coined by Jorn Barger in December 1997. The abbreviation, blog, was coined by Peter Merholz in May 1999.[7]

Web Pages

Let us now turn to the technology behind the web. The web is a collection of web pages. Web pages are special in that they are connected to each other through hyperlinks on the pages. Figuratively, web pages linked to each other through links create a web of pages. Each of these pages can be located anywhere in the world. The two ideas are combined to get the name World Wide Web, a map of which is shown in Figure 6.4.

Figure 6.4: Map of World Wide Web
(Source: http://www.vlib.us/web/worldwideweb3d.html)

7 Elias Awad, *Electronic Commerce*, 3rd ed. (Pearson Education, 2002).

Web and marketing

A big part of the commercial importance of the web is its role in marketing. If a web surfer is searching for a product or service that a business offers, and is located in the coverage area of the business, he is a potential customer. It is useful to attract his attention. This makes the web a powerful marketing medium, and search engines are becoming a powerful intermediary in web marketing. Google earned 90% of its revenues in 2014 from online advertising[8] ($59bn out of $66bn).

About 40% of web searches have a commercial motivation—they are performed by people looking for products or services. Also, most web purchases begin with a web search.[9] Mobile advertising promises to be big, too.[10]

Google's search advertising product, AdWords, is named by employee #59, Douglas Edwards, who had the privilege of giving the product a name. Creation has its privileges.[11]

```
<html>
      <head>
      <title>HTML 101</title>
      </head>
      <body>
      <h1>Welcome to html page at
      <a href = "www.usf.edu">USF</a> </h1>
      <p>Please check back later</p>
      </body>
</html>
```

Figure 6.5: Example web page written in HTML

Authors of web pages link to other pages that they determine are relevant to the content on their pages. *Links on other web pages that point to a page are called inlinks (or backlinks).* Therefore, the more inlinks to a page, the more likely it is that the page contains relevant information. Information about inlinks can be useful. If you have your own page on the web, the inlinks to your page provide a measure of interest on the web in your page. Inlinks can help you identify other websites that link to your site.

Google provides an easy way to search for inlinks to a page. For example, a search for `link: http://www.washington.edu/` will list the inlinks to the home page of the University of Washington.

---

8 Google form 10-K for 2008.

9 Michael A. Cusumano, "Google: What it is and what it is not," *Communications of the ACM*, 48(2) (Feb. 2005).

10 You may like to read the entrepreneurial story of Omar Hamoui, who started AdMob in his dorm and sold it to Google for $750 million: "The new advertising age," *Entrepreneur Magazine*, http://www.entrepreneur.com/article/204976 (accessed Dec. 2015).

11 Doug Edwards maintains a blog to reminisce about the early days of Google, http://xooglers.blogspot.com (accessed Feb. 2016).

Web pages are created using a simple language called Hypertext Markup Language (HTML). Figure 6.5 shows an example of a simple web page written in HTML. Figure 6.6 shows this page as it appears on a web browser. USF is a hyperlink on the page. Clicking on USF on the page takes you to the home page of the University of South Florida.

Figure 6.6: Example web page as displayed in a browser

In October 2014, HTML was updated after 17 years. HTML 5 greatly simplifies web page development by introducing new elements and attributes that better describe web pages (e.g. ARTICLE and SECTION tags), simplify the display of audio and video, and eliminate the need for JavaScript for common tasks (e.g. displaying date and time pickers).

As indicated in the example, unlike programming languages, HTML does not have conditions, loops, or methods. It simply defines how a certain piece of text should be displayed on the screen and what the browser should do if the user clicks on a hyperlink.

HTML pages have some special features that are not present in book pages. The most important feature is the idea of hypertext. *Hypertext is text that includes navigable links to other hypertext.* This allows web-application developers to create web pages that provide viewers with immediate access to related information elsewhere. For example, on a merchant website, you can tempt customers with links to other items they may be interested in. (These items may be identified using prior shopping history.) On social networking sites you can provide links to the profiles of common friends. The result is that though the page structure of a merchant site may look the same to everybody, the contents of the page can be customized to each visitor.

Kaden: Where does a spider usually hang out?
Aiden: I don't know.
Kaden: A web site!

Source: *Boys' Life* magazine, June 2014

12 https://en.wikipedia.org/wiki/HTML5.

The HTTP Protocol

HTML pages may be viewed in many ways.[13] The most common method is to store them on web servers and transfer them to browsers using the HTTP protocol. The HTTP protocol is a simple request/response protocol used to transfer raw data. HTTP was first defined in RFC 1945 in 1996. The current version of HTTP was specified in RFC 2616 in 1999. After SMTP, FTP, and Telnet, which were defined in the early 1980s, HTTP was the first new application layer protocol that gained importance. Though HTTP is newer than FTP or SMTP, and newer technologies are usually more complex than older technologies, HTTP is the simplest application layer protocol to understand. This is why this chapter begins with HTTP.

In using HTTP, the client sends a request for a file to the web server. The requested file is usually an HTML page. The web server responds to the request by sending the requested file over TCP. In addition, the server provides a status code that indicates the extent to which the server was successful in fulfilling the client's request. The server also provides meta-data that indicates the nature of the content to the client. The client can use the meta-data to decide how to process and display the content. The HTTP transaction for the example web page (Figure 6.6) is shown in Figure 6.7.

```
GET /index.html HTTP/1.1
User-Agent: Opera/9.27 (Windows NT 6.0; U; en)
Host: www.ismlab.usf.edu
Accept: text/html, application/xml;q=0.9, application/xhtml+xml, image/png, image/jpeg, image/
gif, image/x-xbitmap, */*;q=0.1
Accept-Language: en-US,en;q=0.9
Accept-Charset: iso-8859-1, utf-8, utf-16, *;q=0.1
Accept-Encoding: deflate, gzip, x-gzip, identity, *;q=0
Cookie: __utmz=118619277.1227574991.6.4.utmcsr=coba.usf.edu|utmccn=(referral)|utmcmd=referral|
utmcct=/departments/isds/faculty/agrawal/index.html;
__utma=118619277.2614276856621331000.1212003453.1226937862.1227574991.6
Cookie2: $Version=1
Cache-Control: no-cache
Connection: Keep-Alive, TE
TE: deflate, gzip, chunked, identity, trailers

HTTP/1.1 200 OK
Date: Sun, 25 Jan 2009 19:58:26 GMT
Server: Apache/2.2.6 (Unix) PHP/5.2.4
Last-Modified: Sun, 25 Jan 2009 19:37:03 GMT
ETag: "3e021-b9-bf1755c0"
Accept-Ranges: bytes
Content-Length: 185
Keep-Alive: timeout=5, max=100
Connection: Keep-Alive
Content-Type: text/html

<html>
.<head>
.<title>HTML 101</title>
.</head>
.<body>
.<h1>Welcome to html page at
.<a href = .www.usf.edu.>USF</a> </h1>
.<p>Please check back later</p>
.</body>
</html>
```

Figure 6.7: HTTP transaction for example web page

13 As a coding mechanism, HTML is very interesting in that while it has been created to be interpreted by machines, it is quite readable by humans, unlike most other codes we see in the IT industry (e.g. binary).

We can now examine salient features of the HTTP request/response cycle shown in Figure 6.7.

The HTTP transaction began with the client's request:

```
GET /index.html HTTP/1.1
```

The request also included the following supplemental information:

```
User-Agent: Mozilla/5.0 (Windows NT 6.1; WOW64) AppleWebKit/537.36
(KHTML, like Gecko) Chrome/48.0.2564.109 Safari/537.36
Host: www.ismlab.usf.edu
Accept: text/html, application/xml;q=0.9, application/xhtml+xml,
image/png, image/jpeg, image/gif, image/x-xbitmap, */*;q=0.1
Accept-Language: en-US,en;q=0.9
Accept-Charset: iso-8859-1, utf-8, utf-16, *;q=0.1
Accept-Encoding: deflate, gzip, x-gzip, identity, *;q=0
Cache-Control: no-cache
Connection: Keep-Alive, TE
TE: deflate, gzip, chunked, identity, trailers
```

It is convenient to start looking at this request from the third line. In this line, the client specifies that it is trying to connect to the web server named www.ismlab.usf.edu. In the first line of the request, the client has asked for a file called /index.html. This is specified using the HTTP command: GET /index.html.

This command means that the client wants the file called index.html located at the top of the web server's directory for the site. In the second line, the client has specified that it is using the Chrome web browser. This is a useful piece of information in case the website customizes responses for specific browsers. In the fourth line, the client has specified that it is capable of processing HTML text and various image formats. In the remaining lines the client has specified some other capabilities, such as language and compression.

Get vs. POST

This example shows the use of the HTTP GET command. Get is used to retrieve documents. The other major HTTP command is POST. Post is used to send data from the user to the server, and is primarily used when a form is submitted from a web page. Some differences between GET and POST include:

Parameters for a GET request are sent in the URL (e.g. in our example, the file name index.html was sent as part of the URL). POST parameters are sent in the body of the page.

GET parameters are limited by the maximum URL limits, and can only be text. There are no such limitations on POST parameters, which can even include file attachments.

A GET request should not attempt to make any changes on the server. A POST request is intended to make changes on the server (for example, the uploaded data may be saved to a database).

When the server receives this request, it knows what file the client is requesting, and what its capabilities are. We see from Figure 6.7 that the web server sends the following in response:

```
HTTP/1.1 200 OK
Date: Sun, 25 Jan 2009 19:41:12 GMT
Server: Apache/2.2.6 (Unix) PHP/5.2.4
Last-Modified: Sun, 25 Jan 2009 19:37:03 GMT
Accept-Ranges: bytes
Content-Length: 185
Keep-Alive: timeout=5, max=100
Connection: Keep-Alive
Content-Type: text/html

<html>
        <head>
        <title>HTML 101</title>
        </head>
        <body>
        <h1>Welcome to html page at
        <a href = "www.usf.edu">USF</a> </h1>
        <p>Please check back later</p>
        </body>
</html>
```

Figure 6.8: Web pages are often zipped before transmission

In the very first line, the server provides a status code for the response. In this example, the status code is 200. We can look up the code in RFC 2616 and see that status code 200 means that "the request has succeeded." For human readability, HTTP also allows the server to include a reason phrase "OK" following the status code (i.e. that code 200 means "OK"). In line 3, the server identifies its version (Apache 2.2.6). In line 6, the server indicates that the response is 185 bytes in length. The content type is text/html, which means that the browser can display the text as regular HTML text. After providing this background information, the web server sends the file we saw in Figure 6.5, and which the browser can display as shown in Figure 6.6.

In the example in Figure 6.7, the web page was sent as plain text. Pages can also be compressed before transmission. Many web servers now do this by default to conserve network bandwidth. Figure 6.8 shows an example, with the search results in zipped format and an alert to the browser of the compression using the content-encoding: gzip. For reference, the page as it appeared on a browser is shown in Figure 6.9. This is one way of improving the performance of websites over slow links.

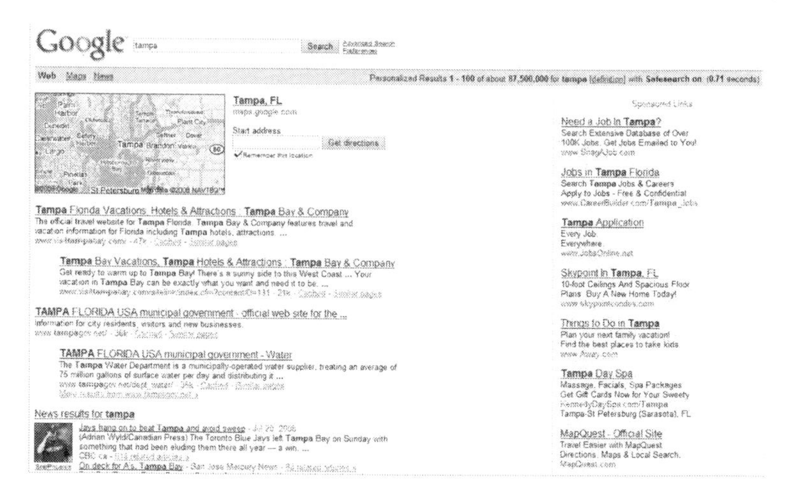

Figure 6.9: Google results (compressed above) displayed in a browser

Please note that though the HTTP application layer protocol facilitates web browsing, the protocol is not the web-browsing application. The web-browser application and web pages are not a part of the TCP/IP stack. The browser application uses commands from the HTTP application layer protocol to retrieve web pages from web servers.

Amazon and Yahoo[14]

Amazon was launched on July 16, 1995, just as the Internet was taking off. The site got about 5–6 orders per day. Three days after launch, Jeff Bezos, the founder of Amazon, got an e-mail from Jerry Yang, a co-founder of Yahoo, asking whether Amazon would like to be listed on the Yahoo's "what's cool" page. With the Yahoo listing, Amazon began receiving orders in excess of $12,000 each week.

URLs

Before leaving the section on the web, let us take a look at the information conveyed by familiar addresses, such as http://www.slu.edu, that are used to get pages in web browsers. These addresses are called URLs. *A URL, or uniform resource locator, is a character string describing the location and access method of a resource on the Internet.* URLs were defined in 1994 in RFC 1738 as a way to represent any resource available through the Internet.

Though URLs are mostly used to get web pages from web browsers, and we typically don't pay much attention to their structure, URLs have been designed to be a general method to access any resource on the Internet. It is useful to know about the components of URLs. All URLs use the following general form:

protocol://host [:port]/[abs_path]

14 Richard L. Brandt, *One Click: Jeff Bezos and the Rise of Amazon.com* (Portfolio, 2012).

The URL begins with the protocol used. The most familiar example is of course, http. When the receiver sees a URL beginning with http, it knows that the rest of the URL is to be interpreted in the manner specified for http. Http://www.slu.edu is an example of a familiar Internet resource being accessed using the HTTP protocol. Another example URL is ftp://128.197.27.121, which uses the FTP protocol to access a resource at the host 128.197.27.121.

The second part of the URL is the host. The host specifies the IP address or domain name of the computer on which the client is trying to access the Internet resource. Thus the URLs http://www.slu.edu or http://165.134.39.20 specify that the client is trying to access a resource using the HTTP protocol on the computer with IP address 165.134.39.20 (or domain name www.slu.edu).

The protocol and host are the only required parts of a URL. All other parts are optional.[15] When used, the third part of a URL is the port address. The port address is separated from the host address using a colon. An example could be ftp://ftp.bu.edu:21. This would indicate an FTP resource running on port 21 on the computer ftp.bu.edu. As we saw in the section on Transport layer Multiplexing in Chapter 5, port numbers are generally unnecessary because services typically run on default ports. We could have accessed the FTP resource above with the URL: ftp://ftp.bu.edu.

Once we access the Internet service running on a port on a computer, the last part of the URL allows navigation through the directory structure of the service. If a web server at www.example.com has the page today_prices.html at the top level, we can access it using the URL: http://www.example.com/today_prices.html. As another example, if the FTP server in the previous paragraph has a top-level directory called etc, its contents can be accessed using the URL: ftp://ftp.bu.edu/etc. A file in this directory, called group, can be accessed by FTP using the URL: ftp://ftp.bu.edu/etc/group.

Web defaults

Just as default ports have been defined at the transport layer for convenience, web browsers and web servers recognize some default values for the convenience of end users. For example, if you do not specify the HTML file name of the home page of a website, the server assumes by default that you are looking for a file called index.html or index.htm.

Web applications allow users to send application parameters through the URL in a GET request. For example, a URL such as http://www.example.org:8080/grades?fName=john could be used to send the value "john" as the value of the variable "fName" in web application "grades" listening on port 8080 on the host www.example.org. Google allows users to specify the search term as the value for the "q" variable to the search application. For example, you can directly type in the following URL in your browser to obtain the results of a Google search for "TCP": http://www.google.com/search?q=tcp.

15 In the industry, [] is a standard notation used to indicate optional parameters.

URLs can be used to access any resource on the Internet. A common use of URLs by application developers is to specify database connections. For example, a MySQL database called ism4220 running on host ismlab.usf.edu can be accessed using the URL: mysql://ismlab.usf.edu:3036/ism4220 ?user=testuser&password=testpass. This causes the client to use the MySQL protocol (using the MySQL driver installed locally) to connect to port 3036 on the host ismlab.usf.edu and attempt a connection to the ism4220 database using the username and password specified.

E-mail

E-mail is one of the oldest Internet applications. E-mail was the primary use of the Internet for more than two decades—the 1970s and 1980s. The web only became popular in the latter half of the 1990s. E-mail is widely considered to be the "killer app" on the Internet. Figuratively, a killer app is a computer application to die for. It is an application that is so useful that it causes people to buy a larger, more expensive system just to get the functionality of the killer app.

E-mail is an electronic means for communication in which information—including text, graphics, and sound—is sent, stored, processed, and received. Messages are held in storage until called for by the addressee.

At the time of the creation of the Internet in the late 1960s, e-mail and FTP were visualized as the two major Internet applications. E-mail was designed for short messages, and FTP was designed to transfer bulk information. Though HTTP has supplanted most of the functionality of FTP, e-mail continues to be one of the most popular applications on the Internet. Businesses have a high regard for e-mail because e-mail can be used in various ways to improve business productivity. As a measure of e-mail's integration into daily life, note how the contact information of almost every professional now includes an e-mail address along with a phone number.[16]

E-mail has changed how we communicate. E-mail has played such a dominant role in modern business that e-mail now has a place in financial history as the leading cause of one of the greatest financial manias of all time. In his famous book on market manias, Charles Kindleberger has attributed the dot-com bubble to e-mail and related technologies.[17] The book states that "events that lead to a [financial] crisis start with a 'displacement,' some exogenous, outside shock to the macroeconomic system." In other words, a financial mania begins with some unanticipated event that has great economic impact. Further, to describe the dot-com mania, the book states that "the shock in the United States in the 1990s was the revolution in information technology and new and lower-cost forms of communication and control that involved the computer, wireless communication and *e-mail*."[18] Though we now take e-mail for granted, this vignette is intended to show that the business community

---

16 Not all e-mail is useful. Irritating e-mail, called spam, is a perennial feature of e-mail. The term spam has a rather convoluted, geeky association with irritating e-mail. The food that goes by the same name is associated with austerity, and was used in Monty Python skits as a tedious repetitive joke. The pioneers working on Internet technologies were fond of Monty Python comedy and began using the word spam to refer to tedious content. Source: Finn Brunton, *Spam: A Shadow History of the Internet* (MIT Press, 2013).

17 C.P. Kindleberger and R. Aliber, *Manias, Panics, and Crashes: A History of Financial Crises*, 5th ed. (Wiley, 2005).

18 E-mail has been italicized here for emphasis.

was so wonderstruck about the possibilities created by e-mail when it first became available to a wider audience, that an entire financial bubble developed as a result.

The popularity of e-mail is based on its unique ability to meet the human need to communicate. Where HTTP allows information dissemination to a wide audience, e-mail allows communication with targeted individuals. E-mail should continue to be extremely popular in the near future because the human need to communicate has not diminished.

2015—Communication continues to be a killer app[19]

In the 1960s, communication was the killer app on the Internet. This trend has continued in the 2010s. As of 2015, 6 of the top-10 most-used applications globally are messaging apps.

There are, however, some changes happening on the e-mail front. Among young users, e-mail seems to be losing some ground to more fast-paced technologies such as text messaging, chat, and social networks. In response, e-mail providers have improved the technology to offer "push-e-mail" in which incoming mail arriving at a mailbox located on the server is "pushed out" almost instantaneously to a mobile device. Also, as we will see when we examine the technology behind it, e-mail is a very expensive technology to support. As a result, to manage costs, colleges and universities are now increasingly partnering with larger companies, such as Google, Microsoft, and Yahoo, to offer e-mail service to their users.

E-mail as a Communication Medium

Every communication medium has its own unique properties. If we know some of the unique properties of e-mail, we can use this information to identify situations where the use of e-mail is appropriate and where it is not.

A key property of e-mail as a communication medium is that e-mail uses a *push* form of communication. This means that the sender decides who to send the message to and when to send the message. This may be contrasted with the web's *pull* nature. On the web, it is the receiver who decides which web page to read and when. While there is no guarantee that the audience will ever visit a web page, e-mail is a mechanism to reach the recipient's mailbox.

The push nature of e-mail makes it an extremely cost-effective measure for organization leadership to instantly disseminate information to everybody in the organization. In the traditional method of pushing information down through direct reports, not only do managers have no control over whether the information is being disseminated at all, they also have no control over the actual message being passed on. Print flyers are useful but very expensive and time consuming to print and mail. Putting up a website is easy, but employees need a way to know what site to visit. Therefore, CEOs now routinely use e-mail to directly send vital information to all employees.

19 Mary Meeker, "Internet trends," a perennially fascinating read for over a decade, http://www.kpcb.com/internet-trends (accessed Dec. 2015).

What is push worth? $6 billion?[20]

It was reported in 2012 that Google offered almost $6 billion in 2010 to acquire Groupon, the firm that pushed discount offers for services.

E-mail has helped reduce communication barriers created by organizational hierarchies. E-mail also eliminates cues about age, gender, and appearance. These cues sometimes influence communication on other channels such as the telephone and face-to-face conversations.

For its convenience and capabilities, e-mail is now an essential part of project communication plans. Before projects begin, project managers develop communication plans that define the stages of a project when status updates will be sent out, and the people who will get these updates. In many project communication plans, e-mail is now the primary communication channel.

E-mail also has other advantages as a communication medium. Unlike the telephone, e-mail is an asynchronous[21] medium. This means that the sender and the receiver do not have to be on their computers at the same time to communicate. A sender can send the message at his convenience and receivers can reply at their convenience. E-mail therefore eliminates the game of phone tag. Within organizations, e-mail eliminates hierarchical barriers.

Human communication dynamics reflected in e-mail[22]

Stefan Wuchty and Brian Uzzi of Northwestern University looked at all 1.5 million e-mails sent by all 1,052 employees of a firm between July 2006 and Jan. 2007. They found that people got the most e-mail from non-associates and the least e-mail from social friends. However, they responded to social friends in approximately 7 hours, to professional colleagues in about 10 hours, and everybody else in about 50 hours. In their words, underlying behavioral patterns continue to operate in a classical fashion in e-mail interactions.

E-mail also makes it very easy to locate expertise in remote locations of the organization. E-mail is particularly useful in obtaining the opinions of introverted people or people who are hesitant to speak before large groups. Many of these people are quite comfortable expressing themselves over e-mail. In groups where some individuals have the tendency to dominate conversations, e-mail can be effective in extracting the opinions of the less-vocal group members.

---

20 Amir Efrati and Geoffrey Fowler, "Google plots move from search to sales," *Wall Street Journal*, Dec. 1, 2010.

21 Asynchronous = not required to occur at the same time.

22 Stefan Wuchty and Brian Uzzi, "Human communication dynamics in digital footsteps: A study of the agreement between self-reported ties and email networks," *PLOS One*, 6(11) (Nov. 2011).

From an economic perspective, e-mail eliminates the costs of communication. Once the Internet connection is paid for and the e-mail account is set up, there are no marginal costs to sending e-mail. This generally increases the supply of e-mail. An unfortunate consequence has been the rise in spam e-mail. *Spam is the use of IT infrastructure to exploit existing aggregations of human attention.*[23] It is just as easy to send a mass e-mail to 20 million recipients as it is to send an e-mail to one recipient. Even if just 0.001% of all the receivers make purchases following the e-mail, spamming can be profitable. As a result, it is a very challenging problem to stop spam.[24]

The low marginal cost of sending e-mail introduces other problems. E-mail is often copied to more people than necessary to keep them "in the loop." Each of these receivers must spend precious time deciding how to act on the mail. The low cost of sending e-mail encourages people to "shoot" e-mails before thinking enough. Also, unlike organized meetings, there is often no clear leader in e-mail discussions and discussions can drift toward unrelated topics.

Overall, the benefits of e-mail have made it the communication medium of choice in the professional workplace. E-mail is so important that even without realizing it, e-mail is central to how many professionals conduct their business. Some of the most important business communications are now sent and received only by e-mail. As a result, organizations are now required to develop formal retention policies for e-mail to comply with rule 34 and rule 26(f) of federal civil procedure.[25, 26] These rules require companies to make information that is relevant in legal proceedings available to courts of law. Many large companies now save all e-mail for at least one month. If an e-mail is required in a legal matter, it is retained until the matter is settled.

Zubulake vs. UBS Warburg, 2003–2005[27, 28]

A landmark case on the use of e-mail and the responsibility of corporations to produce e-mail in disputes was the gender discrimination lawsuit filed by Laura Zubulake, a senior executive with UBS Warburg, a NYC-based investment bank. UBS tried to stall the production of incriminating e-mails but was compelled by the court to do so at its own expense. Eventually, in 2005, Ms. Zubulake was awarded $29 million in damages. The rulings in the case are among the most highly cited rulings on the issue of electronic discovery.

23 Source: Finn Brunton, *Spam: A Shadow History of the Internet* ((MIT Press, 2013).

24 A very interesting early essay on spam filtering is Paul Graham's "A plan for spam," http://www.paulgraham.com/spam.html (accessed Jan. 2016).

25 http://www.law.cornell.edu/rules/frcp/Rule26.htm.

26 http://www.law.cornell.edu/rules/frcp/Rule34.htm.

27 Eduardo Porter, "UBS Ordered to Pay $29 Million in Sex Bias Lawsuit," *New York Times*, Apr. 7, 2005, http://www.nytimes.com/2005/04/07/business/ubs-ordered-to-pay-29-million-in-sex-bias-lawsuit.html (accessed Dec. 2015).

28 Zubulake vs. UBS Warburg. *Opinion and order* 02 Civ 1243, US District Court, Southern District of New York, May 13, 2003, http://www.jeffparmet.com/pdf/electronic_discovery.pdf (accessed Dec. 2015).

E-mail System Architecture

The technology components and protocols that make e-mail work are shown in Figure 6.10. Each e-mail user has access to an e-mail client and a mailbox on an e-mail server. The mailbox provides long-term storage for e-mail. In industry parlance, the e-mail client is called a mail-user agent (MUA). The e-mail server is called the mail-transfer agent (MTA). Common mail-user agents include Outlook and Thunderbird, and most users are familiar with their use. These clients provide a familiar interface for users to compose e-mail and to read and delete e-mail in their mailboxes. Common MTAs include Exchange, Sendmail, and Postfix. Since MTAs act behind the scenes and end users don't deal with MTAs, many e-mail users are not familiar with MTAs. However, MTAs are commercially very important, with Microsoft Exchange generating more than $1 billion in annual sales.

E-mail Protocols

E-mail is a relatively complex service to set up. Two sets of protocols are employed to use e-mail. *SMTP, or Simple Mail Transfer Protocol, is the protocol used to transfer e-mail between mail servers.* Figure 6.10 shows how SMTP is used to deliver e-mail between MTAs. Mail access protocols such as IMAP or POP are used to retrieve e-mail from mailboxes located on the mail servers. The figure also illustrates how users use POP/IMAP to manage mail boxes. It may be noted that outgoing e-mail does not go to the mailbox.

SMTP was one of the earliest application protocols created. It was defined along with FTP and Telnet in RFC 821 in 1982. At that time, users typically ran e-mail servers on their workstations and directly interacted with their mailbox. There were no mail clients or mail access protocols such as POP/IMAP. SMTP was used to send e-mail messages to other e-mail users running SMTP on their own workstations. This was fine as long as the

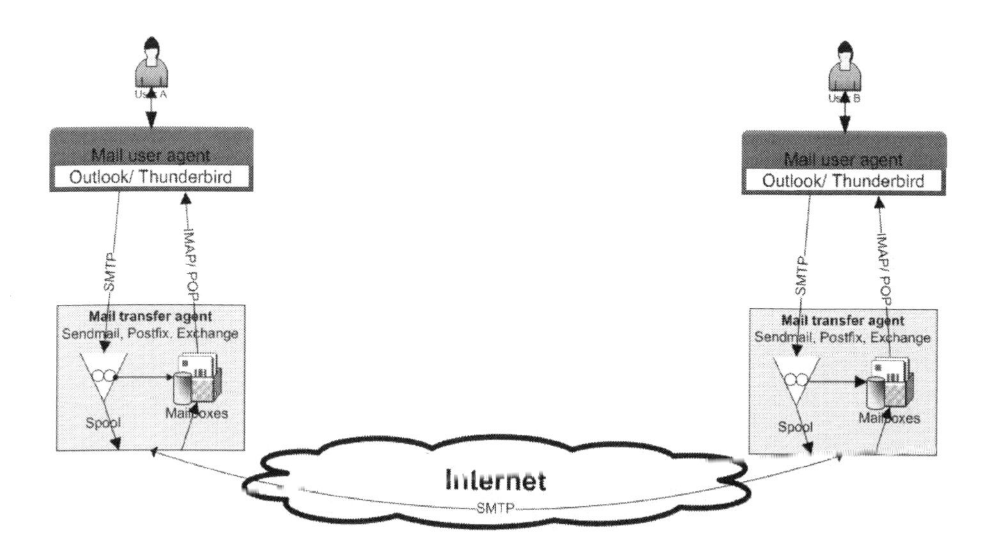

Figure 6.10: E-mail system architecture

workstations were located on university networks and could be left online 24/7, ready to receive incoming e-mail at any time of the day.

As e-mail proliferated, many users who joined the Internet only had dial-up connectivity and could not be expected to be online 24/7 to receive incoming e-mail. It then became necessary to create a mechanism so that e-mail servers running SMTP could be hosted on university or ISP networks where they would be online 24/7 under the watchful eyes of a trained system administrator. These e-mail servers would be ready to receive incoming e-mail at any time of the day. Whenever convenient, end users could access e-mail that had been delivered to their mailboxes since the last access. POP and IMAP were therefore developed as protocols for users to access e-mail from their mailboxes. POP was developed in 1996 and IMAP in 2003. IMAP added features that allowed multiple e-mail clients (possibly at work and home) to maintain synchronization with each other and the mailbox.

The older e-mail system where SMTP was the only e-mail protocol and all users ran mail servers directly on their PCs is analogous to the system used by carriers such as FedEx and UPS that require users to sign off on packages upon receipt. If there is no one to receive the package (which would be analogous to the mail server not being online when an e-mail message is to be delivered), these carriers return a second time, possibly even a third time. If the carriers fail to find the receiver after multiple attempts, delivery fails and the package is returned to the sender. SMTP operates almost exactly in this manner and makes multiple attempts to deliver e-mail; if it fails after repeated attempts, it informs the sender of the failure.

The newer e-mail system that separates mail delivery to the mailbox from mail clearance out of the mailbox is analogous to the way the postal system works. All users have a mailbox that is available 24/7 for a mail carrier to deliver mail. The mail carrier can deliver mail to this mailbox even when residents are not at home. The mail carrier's operation is analogous to the working of SMTP. Like the mail carrier, SMTP performs only one task—delivery of mail to the mailbox. Residents typically check their mailboxes when they return from work in the evenings. Residents' operations are analogous to the operations of POP/IMAP. Residents perform various actions on each mail: open mail, throw some mail in the trash, stack some mail for later action, and save some mail in special folders for later reference (e.g. bank statements). POP/IMAP actions such as read, delete, move, create folder, etc., are analogous to these actions.

Bill: Did you get email about the canned-meat diet?
Phil: No.
Bill: Oh, it probably got stuck in your spam folder.

Source: *Boys' Life* magazine, July 2013

SMTP

As illustrated in the postal system example, SMTP delivers e-mail messages to mailboxes. SMTP does this in three steps. In the first step, SMTP initiates a connection with the receiving SMTP server and informs the receiver of the sender's e-mail address. In the second step, the sender SMTP provides the receiving SMTP with a list of recipients of the message on the receiver's system. Finally, the sender sends the message. This sequence of steps may be seen in the sequence of SMTP transactions shown below. These transactions show a user, John, on the system *example.com*, using SMTP to send an e-mail to users Joe, Jane, and Jill on the system *example.net* about the wonders of business data communications. In the example, SMTP commands are in UPPER CASE.

- Step 1: Sender initiates the communication and informs the receiver of the sender's e-mail address.

```
Sender: MAIL FROM john@example.com
Receiver: OK
```

At this point, the receiving e-mail server knows that an e-mail message is coming in. It also knows the e-mail address of the sender. If necessary, this information is used to send notifications of delivery failures.

- Step 2: The sender provides the receiving SMTP (e-mail) server with a list of recipients on the server.

```
Sender: RCPT TO joe@example.net
Receiver: OK
Sender: RCPT TO jane@example.net
Receiver: FAILURE
Sender: RCPT TO jill@example.net
Receiver: OK
```

In this step, the receiver knows the mailboxes to which the message is to be delivered. Also, the sender has been informed of receivers who do not have mailboxes on the server. These could simply be typos, or they could be users whose accounts have been deleted.

- Step 3: The sender provides the data that will be delivered as e-mail messages to the two receivers (Joe and Jill) with mailboxes on the receiving SMTP server.

```
Sender: DATA
Receiver: START MAIL INPUT
Sender: Business Data Communication technologies are great
Sender: for business and personal communications
Sender: .
Receiver: OK
```

Figure 6.11: SMTP Wireshark capture

In this step, the sender sends a message of arbitrary length and ends it with a pre-specified end-of-file indicator (.). When the receiver sees the end-of-file indicator, it knows that the message has been completed. The message is then delivered to the INBOX of each user's mailbox. At this point, the responsibilities of SMTP are over.

The e-mail message will stay in the two inboxes until the receivers use POP/IMAP to delete or move them to other folders. Figure 6.11 shows a Wireshark capture of an SMTP mail transmission.

SMTP has historically used port 25. Ports 465 and 587 are also used, especially for encrypted e-mail.

POP (Post Office Protocol)

The Post Office Protocol (POP) is a protocol that allows a user to access a mailbox on an e-mail server and perform useful actions on the contents of the mailbox. The most common use of POP is to retrieve new e-mail messages that are stored in the mailbox since the last time the mailbox was accessed. After the download is successful, the mail is usually deleted from the mail server. IMAP (discussed in the next section) allows more complex operations on the mailbox.

The POP transaction is performed in a sequence of three states. The POP transaction begins with the AUTHORIZATION STATE where the user identifies himself to the mail server to get access to his mailbox. The AUTHORIZATION STATE is followed by the TRANSACTION STATE where the user retrieves messages that are waiting to be read. As each message is retrieved, it is marked for deletion. When the TRANSACTION STATE is completed, POP enters into the UPDATE STATE where all messages marked for deletion in the transaction state are deleted.

We can continue our e-mail example that was introduced in SMTP. User Jill now has at least one message in her mailbox. When she opens her POP e-mail client (such as Thunderbird or Outlook) to check her mail, the operations are as follows. POP commands are in UPPERCASE.

The transaction begins with Jill's e-mail client opening a connection to the server on a TCP port waiting for POP connections. By default, this is port 110. When the connection is established, the POP software sends an OK message to indicate that it is ready to accept POP commands. POP then enters the AUTHORIZATION STATE. In the example below, gthyf5675rder45srgafde5 is Jill's password.

POP server: +OK
Jill: APOP jill gthyf5675rder45srgafde5
POP server: +OK jill's mailbox has two messages (400 bytes)

At this point, Jill has provided her credentials (username and password) to the POP server. Satisfied with this information, the POP server provides a status report indicating that two messages with a total size of 400 bytes are waiting to be read. The AUTHORIZATION STATE ends and the TRANSACTION STATE begins.

Jill: RETR 1
POP server: +OK 200 Bytes <POP server sends message>
Jill: DELE 1
POP server: +OK message 1 marked for deletion
<repeat for message 2>

At this point, Jill has retrieved both her messages and these messages are marked for deletion on the POP server. The TRANSACTION STATE is over because the transaction is over as far as Jill is concerned. All that is left is some bookkeeping to prepare the mailbox to receive subsequent messages. To do this, the POP server enters the UPDATE STATE.

Jill: QUIT
POP server: +OK (mailbox empty)

When the UPDATE STATE ends, the mailbox is empty and ready to receive more e-mail. Jill can now read her downloaded messages locally.

The term "Post Office Protocol" makes a lot of sense when it is viewed in terms of the operations performed by a customer of a post office (PO) box. The postal service drops mail at the PO box just as SMTP drops e-mail in the inbox. Customers are given keys they can use to come into the post office at any time of the day, open the PO box, and retrieve the mail delivered since the last time the PO box was emptied. The PO box becomes empty once again when the mail is retrieved by the user. You may note that POP works exactly the same way.

IMAP (Internet Message Access Protocol)

As e-mail became popular and people began accessing e-mail from home and work, a major limitation of POP emerged. Since POP deletes e-mail upon retrieval, it was difficult to access the same e-mail from home and work. IMAP was developed to overcome this limitation. IMAP also allows multiple e-mail clients to synchronize with the contents on the server. *The Internet Message Access Protocol is a protocol that allows a client to access a mailbox on an e-mail server and manipulate messages located on the server as conveniently as they could be manipulated locally*. IMAP includes operations for creating, deleting, and renaming mailboxes, checking for new messages, permanently removing messages, setting and clearing flags, and searching for and selective fetching of messages and portions of messages.

IMAP differs from POP in that IMAP adds two information fields to every message on the server: flags and message IDs. Flags are a user-friendly feature that allows users to mark each message with 0 or more flags: seen, answered, flagged, deleted, draft, recent.

Message IDs are the key field that makes IMAP more capable than POP. Every message on the server has a unique 64-bit ID called a UID. No two messages on an IMAP server (across all mailboxes on the server) can have the same UID. With 64 bits, a mailbox can have 2^{64} (more than 18 billion billion) unique messages, which is sufficient today, even with the GB-sized mailboxes offered by many service providers. In addition, messages within a mailbox have message sequence numbers (MSN) that indicate the relative position of messages within the mailbox. The oldest message in the mailbox has an MSN of 1. IMAP messages are accessed using either the UID or the MSN. UIDs allow multiple clients to check the contents of a mailbox and resynchronize with the server upon connection to the server.

IMAP transactions pass through four states—NOT AUTHENTICATED, AUTHENTICATED, SELECTED, and LOGOUT. These states roughly correspond to the POP states with the IMAP, NOT AUTHENTICATED, and AUTHENTICATED states mapping to the POP AUTHORIZATION STATE, IMAP SELECTED to POP TRANSACTION, and IMAP LOGOUT to POP UPDATE states. As in the POP example, we can continue the SMTP example to show how Jill may read her e-mail using IMAP.

The e-mail client begins by establishing a TCP connection with the e-mail server listening on an IMAP port. (By default, this port is 143.) When the connection is successful, the IMAP server indicates that it is ready to accept commands, and the transaction begins:

```
IMAP Server: * OK IMAP4rev1 Service Ready
```

Note: The server is now in the NOT AUTHENTICATED STATE, and Jill has to provide her credentials to get access to her mailbox.

```
Jill: a001 login jill fgjf5656kjhfjhfg456jhgcv5654jhv
        IMAP Server: a001 OK LOGIN completed
```

At this point, the NOT AUTHENTICATED STATE is over, and Jill gets access to her mailbox with the server in the AUTHENTICATED STATE.

```
Jill: a002 select inbox
        IMAP Server: * 18 EXIST
        IMAP SERVER: * FLAGS (\Answered \Flagged \Deleted \Seen \Draft)
        IMAP SERVER: * 2 RECENT
        IMAP SERVER: * OK [UNSEEN 17] Message 17 is the first unseen message
        IMAP SERVER: * OK [UIDVALIDITY 3857529045] UIDs valid
        IMAP SERVER: a002 OK [READ-WRITE] SELECT completed
```

In the AUTHENTICATED STATE, Jill selects the folder she is interested in working with. The IMAP server responds with information about the contents of the selected folder, indicating that the folder has 18 messages, of which 2 (messages 17 and 18) are new. After providing this status update, the server enters the SELECTED state where Jill can manipulate the messages in the folder.

```
Jill: a003 fetch 17 full
        IMAP SERVER: * 17 FETCH (FLAGS (\Seen) INTERNALDATE "17-Jul-
2008." . . . . . . <mail contents>)
        IMAP SERVER: a003 OK FETCH completed
Jill: a005 logout
```

In this example, Jill is only interested in reading message 17. (We would like to think that this is the message delivered by John about the wonders of data communications.) In the SELECTED STATE, she asks for the message to be delivered to her e-mail client for offline reading. When she logs out, the IMAP server enters the LOGOUT STATE for graceful closure.

```
        IMAP SERVER: * BYE IMAP4rev1 server terminating connection
        IMAP SERVER: a005 OK LOGOUT completed
```

In the LOGOUT STATE, the IMAP server closes the connection, performing any background tasks on the contents of the mail folder.

As you can see, e-mail is a very complex service to provide. It requires multiple protocols (SMTP and POP/IMAP) in cooperation to provide end-user functionality. Provision of e-mail also requires technical skills and resources to maintain hardware that operates 24/7 indefinitely, even in the presence of disk failures. For this reason, many large organizations, particularly universities, are handing over e-mail functionality to service providers with the required expertise. This provides considerable savings in hardware and personnel costs. The service providers attempt to monetize the service through advertising, cross-selling, and other activities.

E-mail outages at USF[29]

Over Christmas 2005, as well as on Thanksgiving 2006, USF's e-mail servers experienced hardware failures, irretrievably destroying e-mails received within the prior 24-hour period, and causing system administrators to spend their vacations restoring e-mail contents for more than 50,000 users from backup tapes. These incidents were strong motivators for USF to seriously consider outsourcing e-mail.

At the beginning of this chapter, we mentioned that application layer protocols define appropriate end-user commands. We saw earlier how HTTP uses the GET command. In this section we saw that e-mail does not have a GET command. Instead, SMTP has MAIL FROM, POP has RETRIEVE, and IMAP has FETCH. All these commands transfer data using TCP/IP. Whereas it makes sense to have a GET command in HTTP to retrieve web pages, it does not make sense to have GET in SMTP because SMTP does not "get" anything. SMTP only sends e-mail. So SMTP does not have the GET command.

Web Mail

An increasing number of users use web mail (e.g. gmail) instead of traditional e-mail clients. Web mail retains all the components of the e-mail system shown in Figure 6.10. In addition, to provide the web interface, web mail introduces a web server between the mail server and the end user. An application running on the web server accepts user commands to manipulate e-mail and executes the commands on the user's behalf on the SMTP and POP/IMAP servers. This is shown in Figure 6.12.

Web mail drafts folder as a terrorist tactic[30]

David Headley, the spy whose reconnaissance for terrorist group Lashkar-e-Taiba was instrumental in the execution of the December 2008 attacks in Mumbai, India, that left 168 dead, used the drafts folder of several online e-mail accounts as a mechanism for secure communication. He would leave information in the drafts folder and fellow conspirators would periodically log into the same accounts to collect the information. This tactic is called electronic dead drops.

29 http://www.usforacle.com/news/view.php/829077/WebMail-on-recovery (accessed Nov. 29, 2015).

30 Scott Stewart, "Tactical implications of the Headley case," *Stratfor*, Dec. 31, 2009.

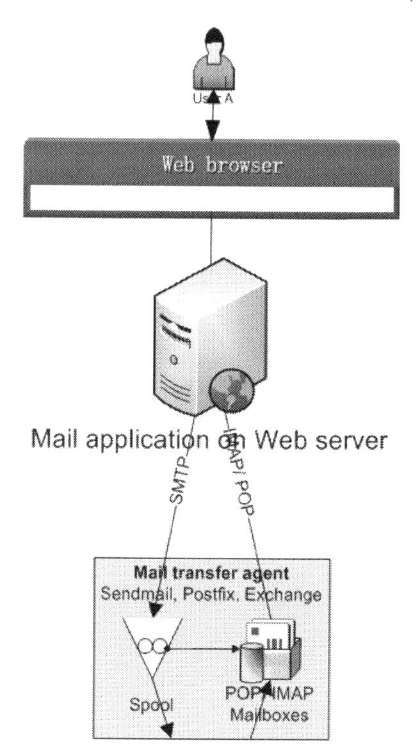

Figure 6.12: Web mail

FTP (File Transfer Protocol)

FTP was developed along with SMTP to enable end users to exchange information using the Internet. *FTP is an Internet protocol for transferring files from one computer to another, regardless of the hardware and software configurations of the two computers.* While e-mail was intended to send small messages, FTP was intended for large messages. Whereas e-mail continues to be popular, FTP has now largely been replaced by HTTP for downloads. Among young users, FTP also competes with Bit Torrent for sharing extremely large volumes of data.

For users who need FTP, the Filezilla open-source project has created a very popular FTP client.

Though many end users are not aware of FTP, FTP is routinely used by web developers who need to upload web pages to web servers. If you have worked as a web developer, chances are that you are very familiar with FTP. HTTP can be used to download information from web servers, but FTP continues to be one of the simplest ways to upload content to web servers. FTP is also more convenient than HTTP when multiple files have to be transferred. In the last few years, file transfer has reemerged in a different form.

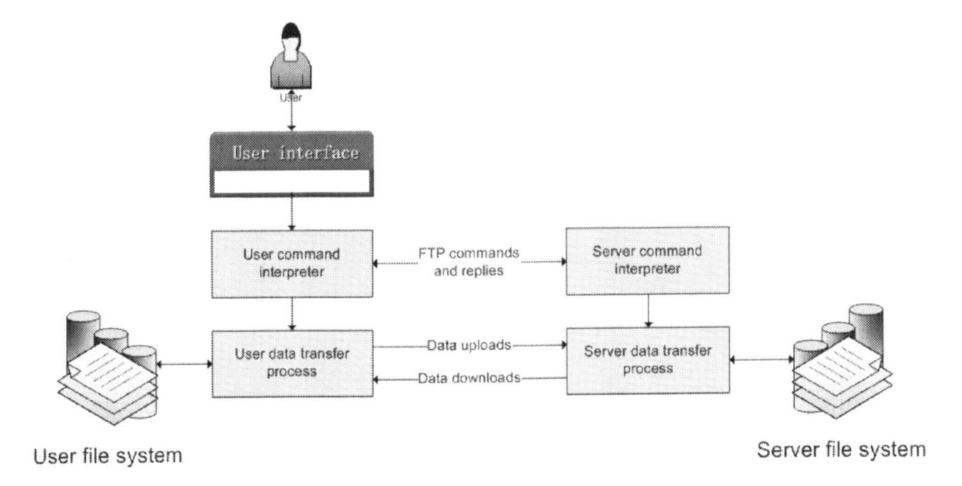

Figure 6.13: FTP operation

Photo sharing and printing sites such as Shutterfly and Flickr, and video sharing sites such as YouTube are essentially file-transfer services with a web-based user interface, which allow users to perform file-transfer operations. Finally, if you are a web-applications developer, though it is not immediately apparent, deploying web applications to remote servers is essentially the use of FTP to transfer the packaged application file to the remote application server.

Figure 6.13 shows the operation of FTP. FTP essentially allows files to be transferred between any two computers. The operation of FTP is quite similar to the operation of HTTP in the sense that the client uses FTP commands to specify the files to be transferred and these files are transferred between the client and server.

> The author of RFC 114, the earliest FTP specification, was Abhay Bhushan, the first graduate (Roll no. 600001) of IIT Kanpur, then a fledgling institution in India, and the alma mater of the first author of this book.

However, there are two major distinctions between HTTP and FTP. The first difference is the bi-directionality of FTP. Whereas the natural direction of file movement in HTTP is unidirectional (from the web server to the web browser), files can be moved in either direction in FTP—from the client to the server (upload) or from the server to the client (download). Since HTTP is inconvenient for file uploads, FTP is needed for file uploads.

The second major difference between HTTP and FTP is the presence of a control channel in FTP. HTTP uses a single channel to transfer HTTP commands and HTTP data. FTP uses one channel for commands and another channel for data transfer. The additional channel does not necessarily make FTP faster than HTTP because most of the delay in data transfer is caused by network latency. However, the control channel ensures that commands do not get delayed by getting trapped behind large data transfers.

FTP and SMB

End users commonly obtain file transfer functionality through vendor-driven protocols such as SMB (Server Message Block) or NFS (Network File System). SMB was developed by Microsoft and NFS by Sun Microsystems. These protocols allow end users to share access to files on a network. When you mount a remote folder to your computer, you are using a protocol such as SMB.

For students interested in reading about the evolution of Internet technologies, RFC 959 describes the evolution of the FTP protocol. Though FTP is a relatively simple protocol, it passed though many steps before arriving at its final form. The earliest FTP protocol (RFC 114) accomplished the goal of specifying a common set of end-user commands independent of end-user technologies (such as differences in directory structures and control characters). The protocol accomplished these goals on two specific computers at MIT and used the proprietary transport protocols of these two computers for data transfer. By RFC 354, the idea of separating FTP commands (from the client to server) and FTP data (from server to client) had been clarified. New commands such as "Make Directory" and "Remove Directory" were added to FTP in RFC 959. RFC 1123 described FTP as a core Internet application layer protocol to interface with TCP and IP.

FTP example from the financial sector

FTP continues to be important where large file volumes are processed. For example, all publicly-traded companies upload their mandatory statements to the ftp site of the SEC (Securities and Exchanges Commission).[31] Unfortunately, companies are identified not by names but by an assigned identifier called a Central Index Key (CIK). Apple Computer's CIK is 320193, and you can access recent filings from Apple by following these simple steps:

```
> ftp ftp.sec.gov
<Type anonymous for user>
<Enter your e-mail address for the password>
> cd edgar/data/320193
> dir
```

IM

The last application layer protocol we discuss in this chapter is instant messaging or IM. *IM is an application layer protocol that allows users to send short, quick messages to each other.* The primary difference between IM and e-mail is the absence of a mailbox in IM. Since there is no mailbox in IM, messages cannot be sent to offline clients. (Services may allow you to send a short "hello" but no more than that.) If you wish to keep a record of an IM conversation, you are required to do that by saving the communication after it is completed.

---

31 https://www.sec.gov/edgar/searchedgar/ftpusers.htm.

Instant messaging was defined in RFC 2778 in 2000. In addition to message exchange, instant messaging protocols also support *presence. Presence is the ability of users to subscribe to each other and be notified of changes in state such as being online or busy or away.* When a user changes state, all users who choose to be notified of that user's state are informed of the change.

1997 was a red-letter year for instant messaging. That year was the first time that AOL (the dominant ISP of the time) sent more instant messages than the total volume of mail handled by the US Postal Service.

EXAMPLE CASE—Google Ads

The print media has traditionally been the most effective check on the misuse of power of the ruling elite. In the past, efforts by the print media have even forced a US president to leave office. In recent years though, the print media seems to have met an opponent as formidable as itself—online news sources. The print media has been unable to resist the defection of its audience to online news sources. The Internet may turn out to be the most formidable adversary the print media has ever faced.

In 1972, an investigation by the *Washington Post* culminated in the resignation of President Richard Nixon two years later. Possibly out of fear of the power of expression, in 1968 in Romania, Nicolae Ceausescu made owning an unregistered typewriter a crime punishable by death. Other prize-winning examples of investigative journalism in the print media have exposed corruption in governments at all levels and have led to many social reforms. While the print media has attacked some of the most powerful sectors of modern society, the print media may itself be under attack from the Internet in the coming years.

Newspapers reached their peak daily circulation in 1984 at 63 million subscriptions. Since then, circulation has been falling at approximately 1% each year. Coinciding with the rapid rise in the online population, newspaper circulation has fallen sharply since 2004. In 2014, daily newspaper circulation in the US was almost 40 million copies per day, a drop of 4% from the previous year, and a level last seen in the 1940s.[32]

The drop in circulation has been accompanied by a drop in advertising revenues for the print media. Newspapers typically earn more than 75% of their revenues from advertising, and only about 25% from subscriptions. The Newspaper Association of America reported that newspaper advertising revenues fell 4.1% in 2014. Newspapers, which have put up with open hostility from powerful governments, have been unable to withstand this onslaught of change in reader preferences. *The Tribune*, owner of the *Los Angeles Times*, has sought bankruptcy protection. *Business Week*, the popular magazine that published the widely followed MBA rankings, was sold to Bloomberg for only an estimated $2–$5 million. In 2008, *U.S. News and World Report*, publisher of the famous college rankings, went from a weekly publication schedule to bi-monthly and, finally, to a monthly publication schedule. In early 2009, one of the nation's leading newspapers, *The New York Times*, mortgaged its headquarters building to borrow a mere $250 million at a steep interest rate of 14%.

32 http://www.naa.org/Trends-and-Numbers/Circulation-Volume/Newspaper-Circulation-Volume.aspx.

This was at a time when home mortgages could be obtained at an interest rate of approximately 5%.

What is causing this pain in the print media? Very simply, the Internet. Not only has there has increase in the amount of time people spend online. According to an estimate, people spend 8% of their waking time with newspapers, and 29% of their waking time online.

There is a logical explanation for this change in user behavior. Paraphrasing a senior Google executive, in traditional media the audience has to go where the content is—a movie theater, or Channel 298 at 8 p.m. on Thursday. On the Internet, content providers have no choice but to go where the audience is. Accordingly, TV shows are now on Hulu, and movies are on Netflix, available for viewing at any time of the day. Advertising dollars have chased this shift in audience tastes and have been diverted to online channels. Since total advertising expenditures have remained virtually constant, most of the online ad revenues have come at the cost of traditional media.

Online channels are much more advertiser-friendly than traditional media. As stated by Mel Karmazin, the CEO of Viacom, which owns channels such as CBS, "You pay $2.5 million for a spot on the Super Bowl … you pay your money. You take your chances." By contrast, Google has developed highly measurable advertising mechanisms. Whereas traditional media cannot tell advertisers what works and what doesn't, Google can state with a high degree of precision that $100 spent on advertising through its platforms will lead to $X in revenues. It can report revenues for each customer and create separate reports for advertisers from different industries. It is therefore not surprising that advertisers also prefer the Internet over other advertising channels.

It is no surprise that, with its rising popularity and better measurability, online media is growing while print media is shrinking. The number of full-time editors at newspapers fell to below 40,000 jobs in 2012, a level last seen in 1978. Google added more than 150 employees each week during 2008. While print ad revenues fell from $47 billion in 2005 to $19 billion in 2012,[33] online ad revenues rose from $12.5 billion to $36 billion during the same period. Of this, $3.4 billion in 2012 was from mobile ads. Mobile ad revenues are now the fastest-growing segment of the three, constituting $12.5 billion out of $49.5 billion total online ad revenues in 2014.[34]

Newspapers missed many chances to develop online platforms, probably because they were late to take the online medium seriously. Newspapers always had great faith in the power of their "storytelling." In 1995, Craigslist was founded. This website gave users the opportunity to post classifieds by city and by category and was clearly a threat to newspaper classifieds. But newspapers did not take this threat seriously at that time. Newspapers also do not seem to have considered the tiny text ads next to Google searches as a serious competitive threat. It was only in 2008 that CBS became the first traditional media company to open an office in Silicon Valley.

Though Google has benefitted from the shift in audience tastes, it clearly needs the newspapers to create interesting content. Google CEO Eric Schmidt has said, "There is a

33 http://www.journalism.org/2013/08/07/the-newspaper-industry-overall (accessed Dec. 4, 2015).
34 http://www.marketingcharts.com/online/4-trends-in-us-online-advertising-spending-53895/attachment/iabpwc-online-ad-revenues-2005-2014-apr2015/ (accessed Dec. 4, 2015).

systemic shift going on in how people spend their time ... we have a shared problem. We need newspapers' content. And it is critically important that they continue." At the current time, when a reader follows a news story to a newspaper's website, the newspaper gets the opportunity to display ads to the reader. Google serves newspapers by increasing traffic to their websites, but it is still not clear whether this will be sufficient for newspapers to survive in their current form. Expect a lot of changes in how the print media structures its business.

Do you recall the statistic that people spend 8% of their time with newspapers and 29% of their time online? Well, that is not the complete story. Advertisers still spend 20% of their ad budgets on newspapers and only 8% on the Internet. If advertising expenditures become proportional to the time we spend with each medium, there could be even more pain ahead for newspapers.

References

1. Ahrens, Frank. "Mexican Billionaire Gives Loan to New York Times Co." *Washington Post*, Jan. 21, 2009.
2. Auletta, Ken. *Googled: The End of the World As We Know It.* Penguin Press, 2009.
3. Crovitz, L. Gordon. "China's Web Crackdown Continues," *Wall Street Journal*, Jan. 11, 2010, A17.
4. Lowry, Tom. "Bloomberg Wins Bidding for Business Week." *Business Week*, Oct. 13, 2010.
5. Meeker, Mary, Collis Boyce, Mayuresh Masurekar, and Liang Wu. "Economy + Internet Trends." Morgan Stanley report, Mar. 20, 2009.
6. Perez-Pena, R. "U.S. Newspaper Circulation Falls 10%." *New York Times*, Oct. 26, 2009.
7. "The Watergate Story," *Washington Post*, Apr. 4, 2016.

Summary

This chapter introduced the most common application layer protocols—HTTP, e-mail (SMTP, POP, IMAP), FTP, and IM. These protocols provide convenient commands that end-user applications can use to exploit the data-transport services offered by TCP/IP.

While discussing these protocols, we learned how HTTP offers a pull form of communication whereas e-mail offers a push form of communication. We presented ways in which these differences can influence how the two protocols are used in business. Finally, it was pointed out that the key technological difference between e-mail and IM is the availability of a mailbox in e-mail and of presence in IM.

The protocols discussed in the chapter are not the only application layer protocols available. An example of an application layer protocol not discussed in this chapter is SIP (session initiation protocol) for voice. However, the protocols covered here are the ones most widely used, and they demonstrate how application layer protocols specify user commands that exploit TCP/IP to provide meaningful services to the end user. Information about the other protocols can be obtained from the RFCs for these protocols.

About the Colophon

The quote is self-explanatory, and appeared in an article in the *Wall Street Journal* by the creator of Dilbert.[35]

Additional Notes

Important tips on e-mail etiquette and a popular recommendation for managing e-mail are provided here.

E-mail etiquette

This is a text on the technology, but it is very useful to point out a few tips on e-mail etiquette. It is very important to remember that e-mail does not change the rules of effective communication. E-mail does make it easy for you to get your message to the receiver's inbox, but it is just as easy for the receiver to delete the message. Precision, courtesy, language, and background information together increase the likelihood that the receiver will read and act on the e-mail.

An excellent article by Horowitz[36] that dates back almost 15 years provides simple, timeless advice for anyone using e-mail. The advice is based on the heritage of e-mail as an alternate to business memos. Memos are business communication designed to be saved in an organized filing system.[37] Suggestions include:

Make the subject line descriptive—The subject line should help the receiver in deciding how to act on the message or file it.

Discuss one subject per message—This facilitates forwarding and replies. If someone needs to be brought into the discussion, there is less danger of unwanted information leakage if every message deals with just one subject.

Keep the message short and to the point—Where possible, fit the entire message into one screen so the user does not have to use the mouse or keyboard to scroll down.

Scrutinize the addressees every time—E-mails are notorious for going to more people than necessary. Make sure an e-mail to your boss does not include criticism of the same boss!!!

Use smileys—One major limitation of e-mail is its lack of emotion. Add the emotional touch using smileys. Smileys are keystroke combinations that indicate emotions. For example :-) indicates happy news whereas :-(indicates sad news.

AVOID ALL CAPS—This is interpreted to mean shouting. Do not jump to conclusions, however, if you receive e-mail in ALL CAPS. Many senior citizens find it easier to use ALL CAPS when composing e-mail.

---

35 Scott Adams, "What if government were more like an iPod?" *Wall Street Journal*, Nov. 5-6, 2011: C3.

36 R.B. Horowitz and M.G. Barchilon, "Stylistic Guidelines for E-mail." *IEEE Transaction on Professional Communications*, 37(4) (1994): 207–212.

37 J. Yates, "The Emergence of the Memo as a Managerial Genre," *Management Communication Quarterly*, 2(4) (1989): 485–510.

Be extra polite—Even with the best of intentions, written criticism appears harsher than verbal criticism.

Be calm—Do not "shoot off" an e-mail in anger. Most long-time users of the Internet have at least one unpleasant memory of an occasion when they sent off an angry e-mail to a friend over an issue that seems trivial in retrospect.

Avoid flaming—Sending nasty or insulting messages is called flaming. Avoid this. If something seriously bothers you, try resolving it over the phone or in person.

And finally, a summary rule: Do not send anything by e-mail that you would not want your mom to see on page 1 of the *Wall Street Journal*. It is useful to remember that almost all the e-mails exchanged by Enron executives are available in research databases online. None of these executives anticipated that their e-mails would be publicly displayed in such manner.

REVIEW QUESTIONS

1. What are the functions of the application layer? How were the earliest application layer protocols defined? What application do you spend the most time on?
2. How has the web been most useful to you?
3. What are the three most popular websites in the world today (Alexa → Top Sites is widely considered to be the best resource for this)? What primary service does each site offer?
4. Describe some changes in the patterns of Internet usage based on changes in the list of most popular websites globally.
5. Why is the web gaining popularity as a marketing tool over traditional methods such as yellow pages?
6. What is a hyperlink?
7. What is an inlink? What information about a web page can be inferred from inlinks to the page?
8. What is HTML?
9. Provide an overview of the structure of a web page.
10. What is URL? Describe the parts of a typical web URL.
11. What is a killer application?
12. In what ways do you use e-mail in your daily life? When do you prefer to use e-mail over the postal system? When do you prefer the postal system over e-mail?
13. Describe the differences between pull and push forms of communication, using the web and e-mail as examples.
14. Using examples from your own life, describe some advantages of e-mail as a communication medium compared to your other choices (such as cell phones or meetings).
15. What are some potential disadvantages of e-mail as a communication medium? Can you describe some occasions when you have run into these disadvantages?
16. What is flaming?
17. Describe the high-level structure of the e-mail system.
18. What are an MTA and an MUA in the context of e-mail?

19. What is the role of SMTP in e-mail?
20. What is the role of POP/IMAP in e-mail?
21. What are some important differences between POP and IMAP?
22. What is FTP?
23. How is FTP different from HTTP?
24. What is instant messaging? How is IM different from e-mail?
25. What is presence in the context of IM?

EXAMPLE CASE QUESTIONS

1. How do you use Google for advertising?
2. What are some emerging trends in digital marketing?
3. How can your university use digital marketing to improve student placement?
4. Take a leading national newspaper or your leading local newspaper. Briefly describe its online presence. Include information such as the following: What are the major news categories on the publication's website? What are the main products or services advertised? What customization options does the site offer? Does it charge a subscription fee for online content? What do you think is the newspaper's target audience?

HANDS-ON EXERCISE—Wireshark

In this exercise, we will use Wireshark, the leading packet-analyzing application. We will use Wireshark to explore all the protocol header fields described in Chapter 2 through Chapter 6. Wireshark is open source and is available for most of the popular operating systems.

Wireshark provides a graphical user interface to assist users in probing the various protocol header fields in packets. Wireshark recognizes a large variety of packet formats. To assist users, Wireshark also interprets the meanings of header fields and reports special packets such as TCP-handshake packets.

While the introductory content in this exercise will help you get started with Wireshark, the Wireshark website also has a number of useful resources, including short training videos. These are available from the "Learn Wireshark" link at the Wireshark website.

Installing and Using Wireshark

Wireshark is available for download from http://www.wireshark.org.

Wireshark works in conjunction with packet-capturing utilities, which go by names such as WinPcap. These utilities access the lower network layers and capture packets. The captured packets are interpreted and displayed by Wireshark.

Download Wireshark from the Wireshark website. For reference, the download page appears as shown in Figure 6.14. The application installs like any other application. Save the installer file to your desktop (for ease of locating later). After the file has downloaded, double-click the file to install the application. During the installation, Wireshark also installs the WinPcap utility and prompts you to confirm the installation of the utility. During installation, Wireshark also adds entries to the Programs menu in Windows.

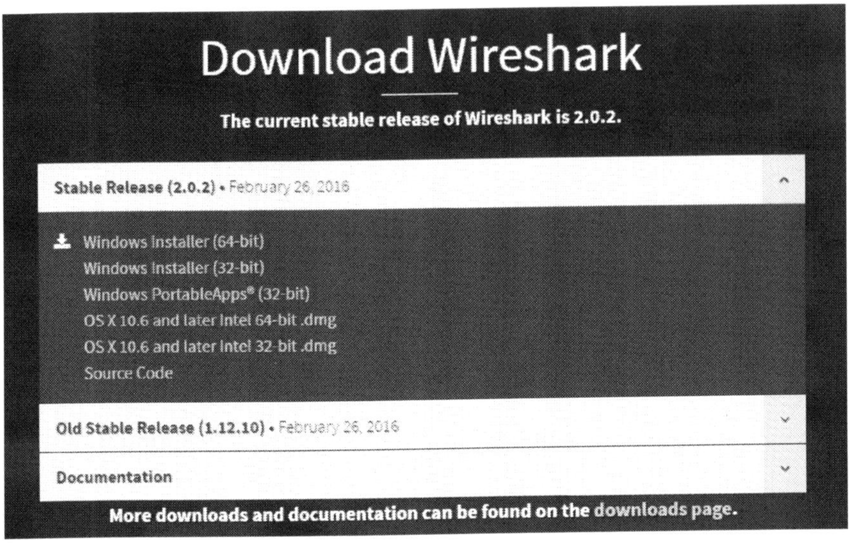

Download Wireshark

The current stable release of Wireshark is 2.0.2.

Stable Release (2.0.2) • February 26, 2016

Windows Installer (64-bit)
Windows Installer (32-bit)
Windows PortableApps® (32-bit)
OS X 10.6 and later Intel 64-bit .dmg
OS X 10.6 and later Intel 32-bit .dmg
Source Code

Old Stable Release (1.12.10) • February 26, 2016

Documentation

More downloads and documentation can be found on the downloads page.

Figure 6.14: Wireshark's download page

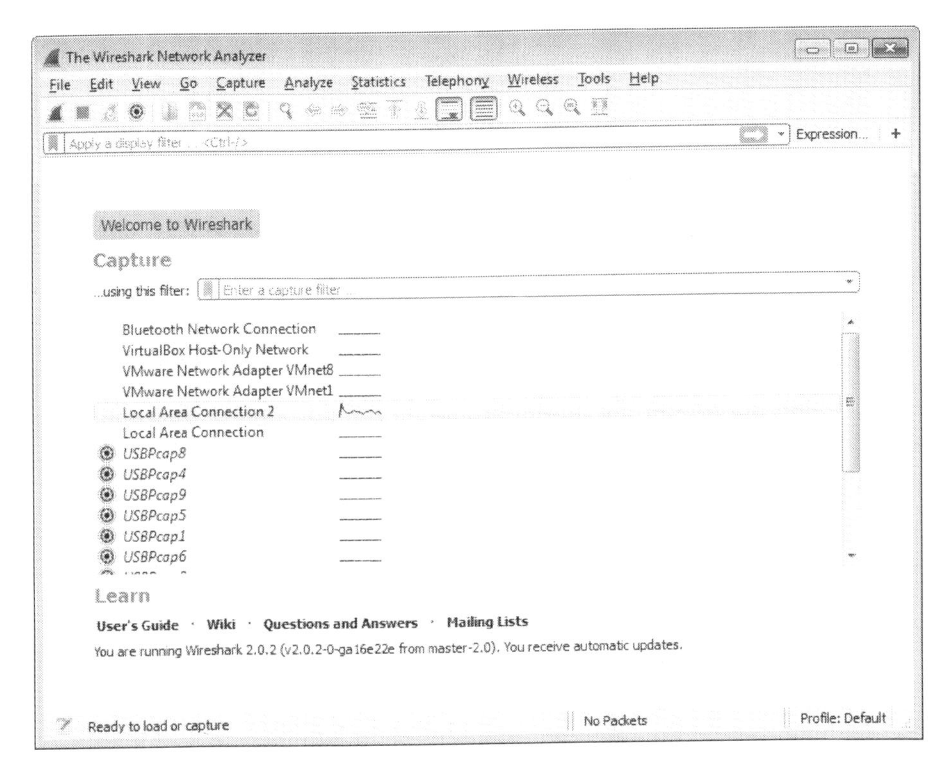

Figure 6.15: Wireshark welcome interface

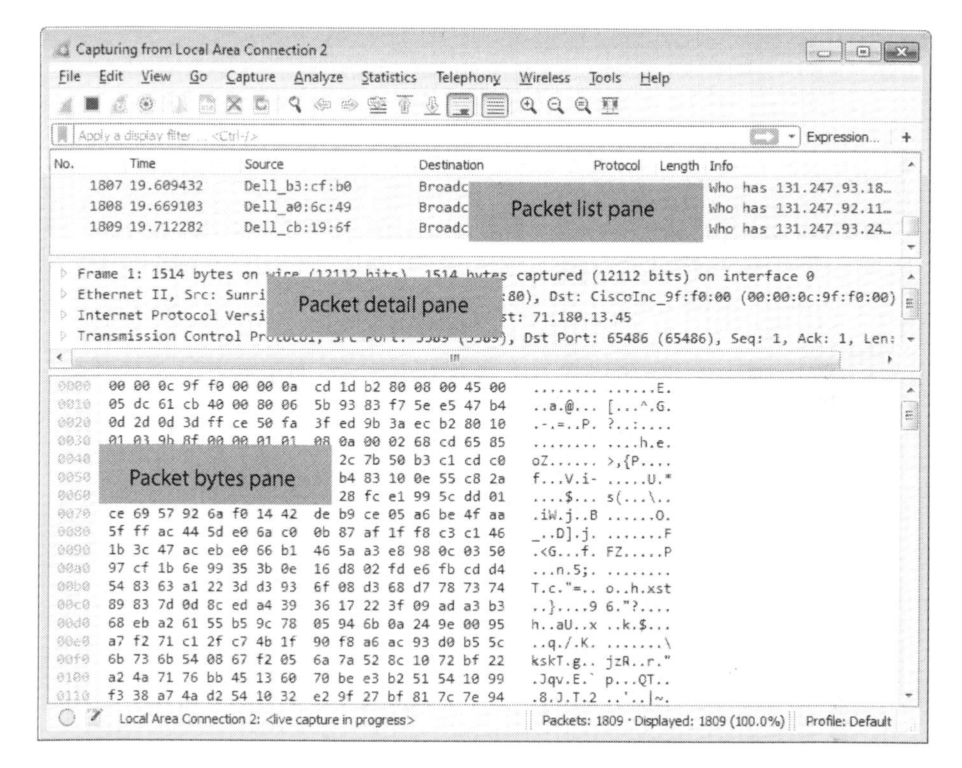

Figure 6.16: Wireshark packet-capture window

When you start Wireshark, you'll see the welcome interface shown in Figure 6.15. Interfaces currently handling traffic are indicated for convenience (LAN 2 in this example). To begin capturing packets using the default options, double-click an active LAN interface to bring up the Wireshark application window.

After starting a capture, if you use a networked application such as a web browser, Wireshark will capture all the packets generated by the application. The captured packets are displayed on the capture window as shown in Figure 6.16. This is the window in which we will spend most of our time in this exercise. To stop a capture, click on the "Stop the running live capture" button in the packet-capture window. It is a distinctly red icon, near the left in the main toolbar.

The packet-capture window (Figure 6.16) has three panes. From top to bottom, the panes show increasing levels of detail in the captured data. The packet-list pane at the top shows all the captured packets, one packet per line. The info column in the packet-list pane shows summary information about the packet, based on Wireshark's interpretation of the protocol-header fields. Clicking on a line in the packet list pane selects the packet. A very useful feature of this pane is that it highlights packets using protocol-specific colors. For example, web packets are highlighted in green. This helps quick visual scans of a capture.

When a packet is selected in the top pane (packet-list pane), the packet-detail pane shows a protocol layer-wise breakup of all the header information in the packet. For example, in

Figure 6.17, the packet-detail pane shows header information for the Ethernet, IP, TCP, and HTTP layers of the selected packet (packet 4 in the packet-list pane). By default, Wireshark shows the important summary information for each layer. Clicking on the ▷ sign next to a protocol layer expands the layer-header information, showing information about all fields at that layer for the packet. For example, in Figure 6.17, the HTTP header has been expanded displaying all HTTP header fields, such as host and GET.

The third pane, the packet-bytes pane, shows the byte-level view of the selected packet. If a field is selected in the packet-detail pane, the packet-bytes pane highlights the bytes containing that information in the packet. For example, the HTTP header has been selected in Figure 6.17, and the HTTP-header bytes have been highlighted in the packet-bytes pane. Data byes are usually shown in hexadecimal notation.

Thus, starting from the top pane, the three panes in Wireshark allow us to drill into captured packets to any desired level of detail.

Viewing and Analyzing a Capture

The companion website includes a packet capture while the htm1101.html web page used in the chapter was downloaded. Since the web page was very short, only a few packets were necessary to download the page. Predictably, the capture is mostly highlighted in green, because most packets in the capture are associated with downloading the web page.

Walking through the capture (Figure 6.17), packet numbers 1–3 are the three packets involved in the three-way handshake for TCP (connection establishment) with the web server (see Chapter 5). Once the TCP connection is established, in packet 4, the client

```
*Wi-Fi
File Edit View Go Capture Analyze Statistics Telephony Wireless Tools Help
tcp.stream eq 6                                                        Expression...
No.  Time      Source          Destination     Protocol Length Info
 59 2.894759   192.168.1.8     131.247.94.229  TCP      66 52823 → 80 [SYN] Seq=0 Win=65535 Len=0 MSS=1460 WS=256...
 67 2.917323   131.247.94.229  192.168.1.8     TCP      66 80 → 52823 [SYN, ACK] Seq=0 Ack=1 Win=8192 Len=0 MSS=1...
 68 2.917449   192.168.1.8     131.247.94.229  TCP      54 52823 → 80 [ACK] Seq=1 Ack=1 Win=262144 Len=0
 69 2.917603   192.168.1.8     131.247.94.229  HTTP    438 GET /html105.html HTTP/1.1
 72 2.941068   131.247.94.229  192.168.1.8     HTTP    596 HTTP/1.1 200 OK  (text/html)
 73 2.941168   192.168.1.8     131.247.94.229  TCP      54 52823 → 80 [ACK] Seq=385 Ack=543 Win=261376 Len=0

v Frame 69: 438 bytes on wire (3504 bits), 438 bytes captured (3504 bits) on interface 0
   Interface id: 0 (\Device\NPF_{DFB8D6B0-CC9A-491D-812C-C8E6AEC2B649})
   Encapsulation type: Ethernet (1)
   Arrival Time: Feb 27, 2016 20:44:32.497951000 Eastern Standard Time
   [Time shift for this packet: 0.000000000 seconds]
   Epoch Time: 1456623872.497951000 seconds
   [Time delta from previous captured frame: 0.000154000 seconds]

0000  18 1b eb 80 6d f3 30 59  b7 14 05 e2 08 00 45 00   ....m.0Y ......E.
0010  01 a8 7d 8d 40 00 80 06  d7 35 c0 a8 01 08 83 f7   ..}.@... .5......
0020  5e e5 ce 57 00 50 1e e0  e4 e1 e7 d9 56 94 50 18   ^..W.P.. ....V.P.
0030  04 00 9b fa 00 00 47 45  54 20 2f 68 74 6d 6c 31   ......GE T /html1
0040  30 35 2e 68 74 6d 6c 20  48 54 54 50 2f 31 2e 31   05.html  HTTP/1.1
0050  0d 0a 41 63 63 65 70 74  3a 20 74 65 78 74 2f 68   ..Accept : text/h
0060  74 6d 6c 2c 20 61 70 70  6c 69 63 61 74 69 6f 6e   tml, app lication
0070  2f 78 68 74 6d 6c 2b 78  6d 6c 2c 20 69 6d 61 67   /xhtml+x ml, imag
0080  65 2f 6a 78 72 2c 20 2a  2f 2a 0d 0a 41 63 63 65   e/jxr, * /*..Acce
0090  70 74 2d 4c 61 6e 67 75  61 67 65 3a 20 65 6e 2d   pt-Langu age: en-
00a0  55 53 2c 65 6e 3b 71 3d  30 2e 35 0d 0a 55 73 65   US,en;q= 0.5..Use
00b0  72 2d 41 67 65 6e 74 3a  20 4d 6f 7a 69 6c 6c 61   r-Agent:  Mozilla
00c0  2f 35 2e 30 20 28 57 69  6e 64 6f 77 73 20 4e 54   /5.0 (Wi ndows NT
00d0  20 31 30 2e 30 3b 20 57  69 6e 36 34 3b 20 78 36    10.0; W in64; x6
wireshark_pcapng_DFB8D6B0-CC9A_AEC2B649_20160227204429_a09168   Packets: 184 · Displayed: 6 (3.3%) · Marked: 24 (13.0%) · Dropped: 0 (0.0%) · Ignored: 1 (0.5%)   Profile: Default
```

Figure 6.17: Wireshark packet capture for htm1105.html

uses HTTP to send a GET request to the web server asking for the htm1101.html page. In packet 5, the web server sends back the web page. The response also includes HTTP code 200, indicating that the client's request was handled successfully. Once the web page is transferred completely, the web server will send a TCP FIN (final) packet to the client, indicating that the server has no more data to send and is ready to close the connection. This is analogous to the way we generally close phone conversations using phrases such as "talk to you later." If the client has no more data to send to the server, the client will also close the connection sending a FIN packet. When the connection is closed, the operating system can release resources, such as port numbers and memory that have been dedicated to the connection.

Since TCP connections are very common, Wireshark has a very convenient view to strip out lower-level protocol-header information and aggregate all the application layer information into one view. This view is accessed by right-clicking any TCP packet and selecting "Follow TCP Stream." This brings up the window shown in Figure 6.18. This window is an excellent tool for understanding the operation of application layer protocols. The HTTP transaction begins with the HTTP GET request from the browser. The server responds with the status code and the HTML page. When the browser asks for the favicon image, the server responds with error code 404 and a formatted page, displaying the code in an error page. For the convenience of users, Wireshark highlights client requests and server responses in different colors.

Wireshark has numerous features to simplify analysis. One of these is filters. For example, if you are interested in only capturing traffic to and from your university's website (say www.usf.edu), you can enter the text host www.usf.edu in the capture filter textbox at the welcome interface. After stopping the capture, the statistics tab has a number of useful options, including flow graph, which shows a visual timeline of the traffic, and protocol hierarchy, which shows a layer-by-layer breakdown of traffic statistics.

To answer the questions for this exercise, install Wireshark, start Wireshark, begin capturing packets on an active interface, and then use a web browser to visit the website of your university. Stop the capture after the page has downloaded completely. Scroll through these packets until you reach the web packets in your capture.

Answer the following:

1. Using online and any other resources, write a brief summary of how Wireshark is used to manage computer networks.
2. What are the sequence and acknowledgement numbers of the three TCP packets performing the three-way handshake? These packets were exchanged just before the GET request for the web page.
3. Right-click an HTTP packet and select "Follow TCP stream." What are the HTTP header fields in the first client request and the first server response? What are the values in these fields? As an example, in Figure 6.18, one of the client fields is "Host" and the value of the field is "www.ismlab.usf.edu."
4. How many different GET requests did your browser have to make to download the entire page? What were the arguments to these GET requests? For example, in Figure 6.18,

Figure 6.18: Follow TCP Stream window

the browser made two GET requests. The first request asked for "/htm1101.html" and the second GET request asked for "/favicon.ico." If there are more than five GET requests, only list the arguments for the first five GET requests. (A very convenient way to do this is to first select "Follow TCP stream" and then sort the results by the Info column. All the GET packets will be collected together.)

5. What is the IP address of your default gateway? This is obtained from ipconfig.
6. What is the MAC address (physical address) of the default gateway identified above? This is obtained from arp –a.
7. Select an HTTP packet. List its source and destination IP addresses as well as source and destination Ethernet addresses.
8. Referring to Figure 6.19 and the information collected in questions 5 and 6 above, what machines does each of these addresses (MAC and IP) refer to? You may be able to relate this to the idea that whereas MAC addresses are hop-by-hop addresses, IP addresses are end-to-end addresses.
9. What are the source and destination port addresses in the selected packet?
10. Look up your etc\services file (in Windows, this file is usually located in C:\Windows\System32\drivers\etc). Which of these ports is a standard port? Paste the entire line from the etc\services file that contains information about this port.

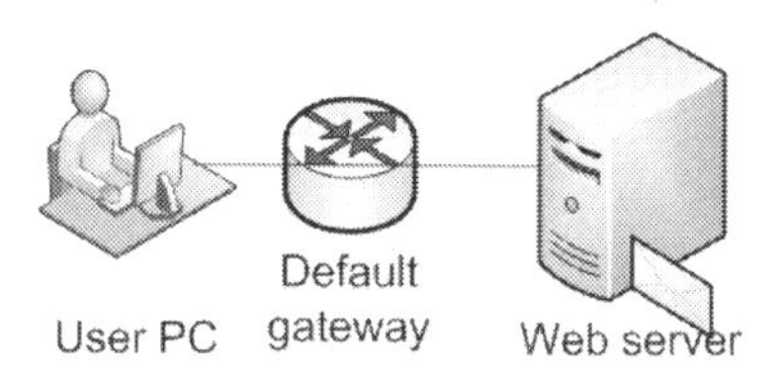

Figure 6.19: Typical network setup

CRITICAL-THINKING EXERCISE—Protesting SOPA

In January 2012, many leading websites, including Wikipedia and Google, blacked out their websites (Figure 6.20) to protest against the Stop Online Piracy Act (SOPA). These protests highlight how web content can intersect with legal frameworks. Reflect on these concerns by answering these questions.

1. What were the provisions of SOPA?
2. What was the motivation for SOPA?
3. What were the concerns of the protestors?
4. Who do you agree with—the lawmakers who drafted the act, or the protestors?
5. Under what conditions would you support laws such as SOPA?

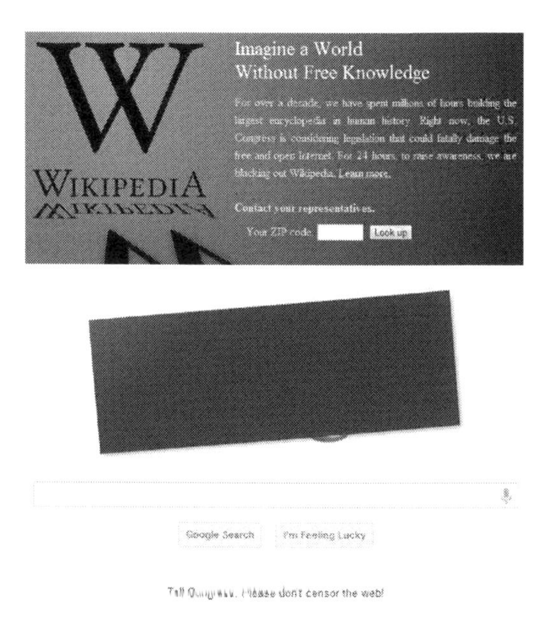

Figure 6.20: Blacked-out pages of Wikipedia and Google

CRITICAL-THINKING REFERENCE

1. https://en.wikipedia.org/wiki/Protests_against_SOPA_and_PIPA.

IT INFRASTRUCTURE DESIGN EXERCISE—Identifying Market Leaders

In Chapter 1, you identified different ways in which TrendyWidgets could use computer networks at its different offices. Answer the following question related to these uses of computer networks:

1. For at least five of these uses of computer networking, use the Internet or other sources to identify the market-leading software applications used to obtain the required functionality (for example, to serve web pages, the market-leading software is the Apache web server). In a table, provide the following information for the five uses: the application (use) category, the market-leading software application in the category, and a paragraph or two about the strengths of the selected software application. (You can use the marketing information from the application's website to describe the strengths of the application.)

CHAPTER 7

Support Services

What's in a name? That which we call a rose
By any other name would smell as sweet;

Juliet to Romeo

Overview

In the previous chapters, we have examined each layer of the TCP/IP stack. To make these technologies easy to use, the designers of the Internet also created some supporting technologies. For example, one of these services, DHCP, automatically assigns IP addresses to computers that join networks. Another service, DNS, allows users to address hosts using names such as www.uta.edu instead of IP addresses such as 129.107.56.31. At the end of this chapter you should know:

- what DHCP is and how it works
- what non-routable addresses are and how they are used
- what NAPT is and how it works
- what ARP is and why it is useful
- what DNS is and how it works
- how all the above services are put together in your home network

DHCP

DHCP, or dynamic host configuration protocol, is used to automatically provide network information to hosts on the network. We saw in Chapter 4 how every computer on the network is uniquely identified by its IP address. Comparing the computer addresses at the MAC layer (Chapter 3) and IP addresses at the network layer, we saw that the great advantage of IP addresses over MAC addresses is that network administrators can assign IP addresses according to the needs of the network

But the flexibility of IP address assignment introduces a new chore for network administrators. They must find a way to assign IP addresses for every computer that joins a network. This works fine if there are very few computers on the network and if these few computers remain connected to the network forever. Examples of such networks include public computing labs at most universities and corporate data centers. In these cases, the network administrator can permanently assign an IP address for each computer manually.

Unfortunately, most networks do not meet these criteria of limited size or relative stability. University wireless networks are an example. Students enter the network at all times of the day, stay connected for some time during class, and then return home. In these networks, manual IP address assignment can get very cumbersome and error prone.

Pause for a moment to think how you might provide IP addresses manually to these mobile wireless users on a typical college campus. Perhaps each student could be given an IP address when they register as a student at the university. Unfortunately, most schools do not have enough IP addresses to do this. If students cannot be allocated an IP address for the whole year, perhaps students could obtain an IP address for the day on entering campus, perhaps by logging into a secure kiosk. In this case, we would only need as many IP addresses as the number of students who visit campus during the day.

But then, in the rush to return home after class, many students are likely to forget to return their allocated IP addresses at the end of the day. Periodically, when all IP addresses available to campus IT are lost in this manner, some kind of reset procedure will be necessary to reclaim all addresses lost since the last reset. Hopefully, this example has convinced you that it is not easy to manually assign IP addresses in large networks.

Another problem with manual IP address assignment is that if IT is not careful while handing out IP addresses, they might occasionally assign duplicate addresses, causing network connectivity problems to the users assigned these duplicated addresses.

Manual assignment of IP addresses also makes the organization vulnerable in case the network administrator quits without notice. If the departing administrator did not keep good records, the new administrator would have no easy way to collect information about all IP addresses already allocated in the organization.

DHCP, or dynamic host configuration protocol, addresses these issues by automating the allocation of IP addresses in networks. *DHCP is a technology that enables automatic assignment and collection of IP addresses.* Almost every network has a DHCP server to allocate IP addresses. On your home network, the wireless router typically acts as the DHCP server.

Besides making it easier to use IP addresses by automating their allocation, DHCP is also designed to improve the efficiency of allocating IP addresses. We have seen in Chapter 4 how the 32-bit size of the IP address field limits the number of available IP addresses to about 4 billion. Unfortunately, however, since these addresses are allocated in blocks by the registries, the available IP addresses are not used very efficiently. According to some estimates, only about 5%–15% of the available IP pool of 4 billion addresses would actually get used.[1] This meant that when we reached about 200–600 million hosts on the Internet, we would run out of IP addresses to assign to new computers joining the network. As of July 2015, there were already more than 1.03 billion hosts online. Thus, we have already reached the point where, until recently, experts believed that new computers would not be able to join the Internet, limiting the growth of the Internet. On May 20, 2014, IANA

1 George Lawton, "Is IPv6 Finally Gaining Ground?" *IEEE Computer*, 34(8) (Aug. 2001): 11–15.

distributed the last /8 block of free IPv4 addresses to the regional registries, bringing the IPv4 generation to its closing phase.[2, 3]

The long-term solution to remove the shortage of IP addresses is to increase the number of IP addresses available. IPv6 accomplishes this by using a 128-bit IP address field. As we have seen in Chapter 4, IPv6 provides trillions of IP addresses per square foot of the earth's surface, ensuring that we will never run out of IP addresses.

However, since introducing IPv6 requires significant investments on the part of network operators, a short-term solution has been devised until such time that IPv6 becomes more prevalent. This solution is expected to allow IPv4 addresses to meet network needs for several more years. The solution is implemented using three technology components— DHCP, reusable IP addresses, and address translation. DHCP is the first of these three components of the IP-address reuse mechanism that we will examine.

To understand the role of DHCP, it may be noted that most computers that are allocated IP addresses do not use the addresses efficiently. Only a small fraction of the computers that have been allocated IP addresses are actually performing any network activity at any given time. We could greatly improve the efficiency of IP address utilization if we could allocate an IP address to a computer only when the computer needed network connectivity. DHCP is the standard mechanism used to accomplish this goal.

DHCP was introduced in 1997 in RFC 2131. When using DHCP, network administrators run one or more DHCP servers on the network. These servers are given the responsibility of managing the IP addresses available for the network. At home, your wireless router acts as the DHCP server. Figure 7.1 shows the DHCP server settings of a typical home wireless router. The DHCP server on this router has been configured to allocate 150 IP addresses in the range 192.168.1.1–192.168.1.150. This is more than adequate for the average home network.

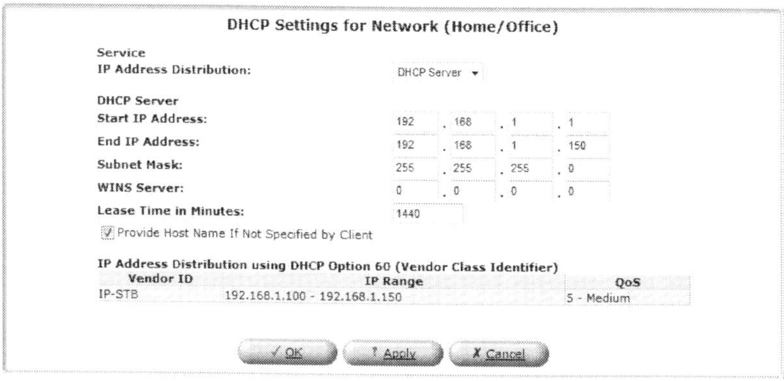

DHCP Settings for Network (Home/Office)
Service
IP Address Distribution: DHCP Server
DHCP Server
Start IP Address: 192 . 168 . 1 . 1
End IP Address: 192 . 168 . 1 . 150
Subnet Mask: 255 . 255 . 255 . 0
WINS Server: 0 . 0 . 0 . 0
Lease Time in Minutes: 1440
Provide Host Name If Not Specified by Client
IP Address Distribution using DHCP Option 60 (Vendor Class Identifier)

Vendor ID	IP Range	QoS
IP-STB	192.168.1.100 - 192.168.1.150	5 - Medium

OK Apply Cancel

Figure 7.1: DHCP settings to allocate IP addresses in specified range

2 http://www.internetsociety.org/deploy360/blog/2014/05/goodbye-ipv4-iana-starts-allocating-final-address-blocks/.

3 https://www.icann.org/news/announcement-2-2014-05-20-en.

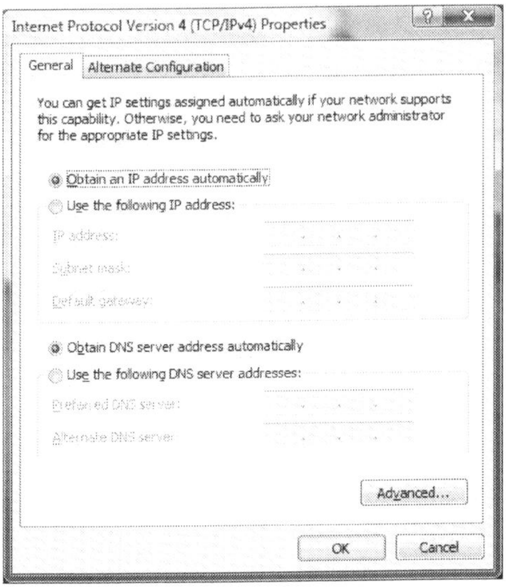

Figure 7.2: Windows PCs use DHCP by default

By default, all PCs are configured at the factory to look for a DHCP server out of the box and obtain an IP address from the server. PCs do this as part of the booting up process. Figure 7.2 shows the dialog in Windows that shows this setting (accessed from Control Panel → Network and sharing center → View Status → Properties → IPv4 → Properties). As the figure shows, you can also assign an IP address manually if you prefer, but for convenience of end users, the default setting on all PCs is to look for a DHCP server for an IP address.

DHCP—Operation

The basic DHCP operation is very simple. Every network that uses DHCP has at least one DHCP server listening for DHCP client requests on the network. During startup, DHCP clients broadcast requests asking for network parameters from DHCP servers listening on the network. Since the client DHCP request is broadcast, clients need not know the IP address of the DHCP server to locate it. When the broadcast reaches the DHCP server, the server responds by providing the client with an IP address and other essential network parameters for a fixed duration. *The duration for which an IP address is provided is called the DHCP lease-time.* Before the lease-time expires, clients can request an extension of the lease.

DHCP servers try to be courteous toward clients. If the DHCP server had provided an IP address to the client before (say the previous day) and if the address is still available, the server provides the same IP address to the client. For this reason, you may notice that the IP address provided to you by your ISP can stay the same for many days, even though the ISP uses DHCP to allocate IP addresses.

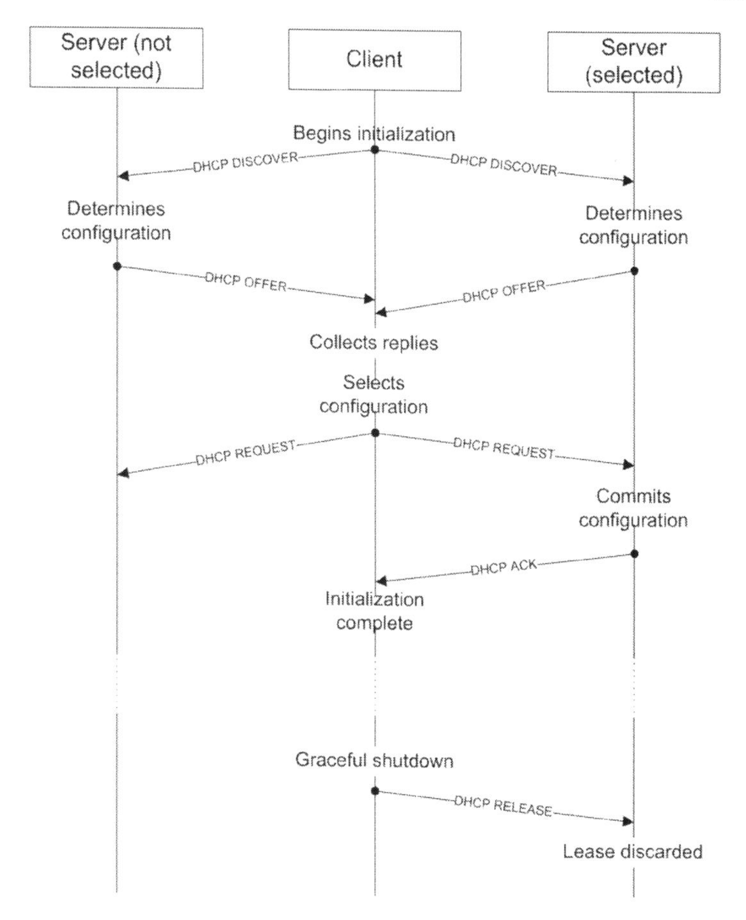

Figure 7.3: DHCP operation timeline

The sequence of messages exchanged between DHCP clients and servers is shown in Figure 7.3. At a high level, a client requests an IP address and DHCP servers provide one. When the client accepts an IP address, the server marks this address in its records as allocated.

At a greater level of detail, at boot time, clients broadcast DHCP DISCOVER messages asking for network parameters. DHCP servers intercept these messages and search their configuration files to determine the IP address and network configuration to be allocated to the client. Servers send the IP address to clients in a DHCP OFFER message. If the client receives more than one DHCP OFFER, it selects one of the offers and broadcasts a DHCP REQUEST message informing all DHCP servers of its selection. The server whose offer was selected records this selection in its own configuration files so that the offer is not repeated to other clients. The DHCP server then sends a DHCP ACK to the client to confirm that the client may now use the OFFERed IP address. All these messages are shown in the figure and are exchanged while the computer is booting up (before you get the login prompt).

When the computer is shut down, clients send a DHCP RELEASE message to the server. Upon receipt of the DHCP release message, the server discards the lease and adds the IP address into its pool of available addresses so that the address can be reused if necessary. If no DHCP RELEASE message is received, the lease automatically terminates at the end of the lease-time. This usually happens when the computer is abruptly shut down.

DHCP Allocation Schemes

To allow maximum flexibility for network administrators, DHCP can allocate IP addresses in three different ways.

Automatic allocation In automatic allocation, DHCP allocates IP addresses to hosts on the network on a first-come, first-served basis. Once allocated, this IP address remains assigned to the host forever. This can be very convenient in networks where devices are referred directly by IP address. For example, if a networked printer is assigned an IP address by automatic allocation, client PCs only need to be set up once to use the specified IP address. The network administrator can be assured that each time the printer is switched on, it will be assigned the same IP address by the DHCP server, and the software on client PCs will have no difficulty in locating the printer on the network. For this reason, some home routers use automatic allocation to distribute IP addresses.

Manual allocation In manual allocation, the network administrator manually specifies the IP addresses to be allocated to individual devices on the network. Each time a device with a manually allocated IP address is switched on, it is given the IP address manually specified by the network administrator. Thus, both automatic allocation and manual allocation permanently assign IP addresses to clients.

If both automatic and manual allocation assign IP addresses permanently, are there any cases where you would prefer manual allocation over automatic allocation? Manual allocation is useful when the network administrator wishes to use some preferred address-allocation scheme for specific devices. Automatic allocation would simply assign the first available IP address to each device when it is switched on for the first time. Manual allocation is therefore most commonly used for printers and other non-server devices. Usually this is done for security reasons, to simplify the configurations of firewalls. Firewalls are discussed in the chapter on network security.

Dynamic allocation In dynamic allocation, IP addresses are leased out for short durations by the DHCP server. Before the lease expires, clients can ask the DHCP server to extend its lease. If the DHCP server does not hear from the client before the lease expires, the DHCP server reclaims the IP address and the address becomes available for another client.

Dynamic allocation is very useful for computers such as client PCs that are only intermittently connected to the network. Since most residential customers need Internet connectivity only in the mornings and evenings, they fall into this category. Dynamic allocation is therefore popular among ISPs to allocate IP addresses to subscribers. Dynamic allocation is also popular in enterprise networks to provide IP addresses to client PCs, since these are usually connected to the Internet only during office hours. It is easy to see that of the three allocation mechanisms, dynamic allocation enables address reuse and helps administrators use IP addresses efficiently when their availability is limited.

Dynamic allocation in phone numbers[4]

Phone numbers are recycled in a manner analogous to the dynamic allocation mechanism used in computer networks. A group of about 10 people, the North American Numbering Plan Administration (Nanpa), located in Sterling, Virginia, recycles almost 37 million phone numbers each year, as users change carriers or locations.

Each DHCP server can mix and match allocation methods for different devices on the network. For example, the DHCP server may use manual allocation for printers, automatic allocation for a few selected clients, and dynamic allocation for all other client devices. You may note that DHCP is typically not used to configure servers. Servers are almost always assigned static IP addresses. This ensures that DHCP errors will not interrupt server operations.

```
option domain-name                       "datacomm.example.com";
option domain-name-servers     10.1.1.1, 10.2.1.1, 10.3.1.1;
option routers                           10.1.1.254;
option subnet-mask                       255.255.255.128;

default-lease-time                       21600;

subnet       10.1.1.128 netmask          255.255.255.128    {
             range        10.1.1.236     10.1.1.253;
}

host www {
        hardware          ethernet       00:06:5B:CE:39:05;
        fixed-address     10.1.1.2;
        host-name    "www.datacomm.example.com";
}
```

Figure 7.4: Sample DHCP server-configuration file

4 Alyssa Abkowitz, "Wrong number? Blame companies' recycling," *Wall Street Journal*, Dec. 1, 2011.

DHCP Configuration

For reference, Figure 7.4 shows a sample DHCP server configuration file for one of the most popular DHCP servers—the DHCP server maintained by the Internet Systems Consortium, ISC. (You may remember ISC from the domain survey in Chapter 4.) The configuration creates a dynamic pool of 18 IP addresses in the range 10.1.1.236–10.1.1.253 in the subnet with base address 10.1.1.128. The lease time is 6 hours (21,600 seconds). The configuration also specifies a manual allocation of 10.1.1.2 for the web server. As seen from the manual allocation entry, DHCP identifies hosts by their Ethernet MAC addresses. The other information in the configuration file allows DHCP to provide additional information to clients, including the addresses of DNS servers, the subnet mask, and default gateway. (DNS is covered later in this chapter.)

IPv6 auto-configuration

IPv6 hosts can self-configure their own IP addresses. After inheriting the network address from the network it is connected to, it uses a process called MAC-EU164 conversion to obtain its 64-bit interface ID by expanding its 48-bit Ethernet address to 64 bits. This is done by inserting the hex value FFFE between the 3rd and 4th octets of the MAC address. Also, the local/universal bit (7th bit of first octet) is forced to 1 (universal). An example is below.

Original MAC address: D0:C1:B0:OA:2D:51
Insert hex FFFE in middle: D0:C1:B0:FF:FE:OA:2D:51
Force universal bit: D2:C1:B0:FF:FE:OA:2D:51

Non-routable (RFC 1918) Addresses

DHCP was the first component of the three-part short-term solution to address the problems arising from the shortage of IP addresses. Address reuse is the second important piece of the solution.

When reusing IP addresses, more than one computer on the Internet can be assigned the same IP address. If this can be made to work, a small pool of IP addresses can be used to network a large number of computers. The main problem with reusing IP addresses is that if IP addresses are no longer globally unique, routers will not know how to route packets addressed to computers sharing an IP address. Say two computers, one in California and one in New York, have the same IP address, 192.168.2.8. If a packet with a destination address of 192.168.2.8 reaches a router located in Chicago, the router will have no way of deciding whether to send the packet east to New York, or west to California. For this reason, reused IP addresses have to be handled very carefully. Most importantly, the challenge is to ensure that the destination address of every packet reaching every router on the Internet is unique, even if IP addresses are reused in different parts of the Internet.

RFC 1918, defined in 1996, provides a very interesting solution to this challenge. RFC 1918 defines three pools of IP addresses for IP address reuse. These address pools are:

10.0.0.0/8	:	10.0.0.0–10.255.255.255 (Class-A pool)
172.16.0.0/16	:	172.16.0.0–172.16.255.255 (Class-B pool)
192.168.0.0/16	:	192.168.0.0–192.168.255.255 (Class-B pool)

IP addresses within these three address pools may be used by anybody within any network without permission from the Internet registries. Thus, these IP addresses may be reused as often as necessary. One IP address can only be used once within a network, such as the network in a home or apartment, but other networks can reuse the same IP address for addressing within their own networks.

RFC 1918 addresses are a lot like common names, such as "John" or "Jane." If you were to go to the visitors' reception center at your university and ask to see John, the staff at the reception desk would find it almost impossible to locate "John" among the many people on campus named John. However, if you reached the department where John worked, chances are high that there would be only one person named John within the department, making it likelier for you to locate him. Similarly, RFC 1918 addresses are locally unique but globally common.

To preserve the uniqueness of IP addresses on the global Internet, routers and firewalls at the edge of the enterprise limit the packets using RFC 1918 addresses to the LANs in which they originate. Packets do not leave any router with an RFC 1918 address in either the source IP address or destination IP address field. If a packet were to leave a router with an RFC 1918 address in the destination address field, other Internet routers would have no way of uniquely identifying which of the many computers on the Internet with that IP address to deliver the packet to. Similarly, if a packet left a router with an RFC 1918 address in the source address field, replies to this packet would have an RFC 1918 address in the destination address field. Again, it would be impossible to deliver the reply because many computers on the network would have the same IP address.

Figure 7.5 shows an example of how RFC 1918 IP addresses are used in practice. In the example, the home has a wireless router, a laptop, and a PC. The laptop has IP address 192.168.2.2, the PC has IP address 192.168.2.3, and the wireless router has the IP address 192.168.2.1 on its interface facing inside the home. Computers within a home can connect to each other using these IP addresses.

So, what is the point of getting an RFC 1918 IP address if it cannot be used for routing? After all, isn't routing pretty much the only thing IP addresses are used for? The answer to this question lies in NAPT (Network Address Port Translation)—the third component of the three-part solution to address the problem of shortage of IP addresses. RFC 1918 addresses are meant to be used in conjunction with NAPT. In this section, we will see how the combination of RFC 1918 and NAPT allows routing even while using RFC 1918 addresses.

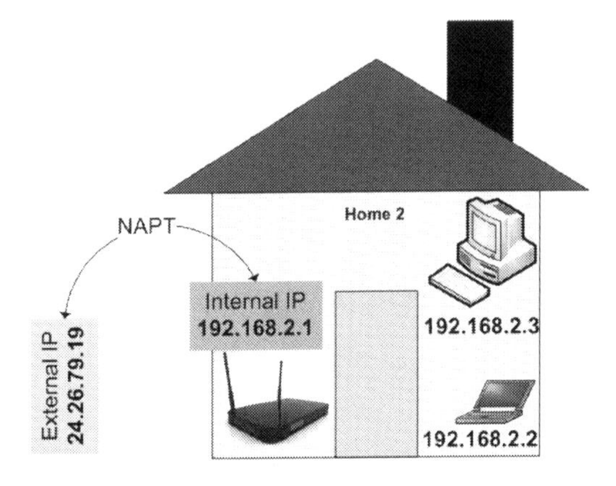

Figure 7.5: Using non-routable RFC 1918 IP addresses

Network Address Port Translation (NAPT)

NAPT, often abbreviated as NAT or PAT, is the method by which IP addresses are mapped from one address block to another address block, providing transparent routing to end hosts. The details of NAPT have been specified in RFC 3022. Network address translation (NAT) refers to changing IP addresses from one IP address block to another. Port address translation (PAT) refers to changing port addresses during network address translation if necessary. NAPT (network and port address translation) combines both, and is typically used to connect networks using RFC 1918 non-routable IP addresses to networks using globally unique addresses. As shown in Figure 7.5, home networks are almost always set up as RFC 1918 networks.

A common use of NAPT is by ISPs to serve computers within customer homes. NAPT is often also used to connect desktops in large enterprises. In conjunction with RFC 1918 non-routable IP addresses, NAPT vastly expands the availability of IP addresses.

NAPT and enterprise IP address assignment

An important design feature of NAPT is that all the complexity of address translation is located at the edge router, and no change is required in end clients inside the network. This greatly simplifies the deployment of NAPT into a network. If all clients on a network use DHCP, a simple change in the DHCP address pool will ensure that when the current IP-address lease expires, clients will be allocated addresses from the new internal IP-address pool. Clients will not even be aware of the change.

The sequence of operations when using NAT is shown in Figure 7.6. Say, a student using his home PC wishes to browse his university's website. His home wireless router has been allocated an IP address of 65.32.26.70 by his ISP, most likely using dynamic allocation by a DHCP server running within the ISP network. His home PC has the IP address 192.168.2.3, most likely assigned using automatic allocation by the DHCP server running at the wireless router. The IP address of the university's website in the example is 131.247.80.88 (USF). In the first step, the PC sends out an HTTP request with source IP address 192.168.2.3 and destination address 131.247.80.88. As the packet leaves the home, it is intercepted by the home router, which also acts as the NAPT translator. In step 2, when the packet leaves the wireless router, the source IP address of the packet is translated from 192.168.2.3 to 65.32.26.70.

RFC 1918 and NAPT are designed to support outbound connections from computers inside networks. If an incoming packet does not match an existing entry in the NAT forwarding table, the NAT translator will not know which computer in the internal network to send the packet to. For this reason, in many organizations, clients use RFC 1918 addresses, while web servers, e-mail servers, and a small number of other devices that need to accept connection requests from external hosts are given unique addresses.

In step 3, the web server sends a response to the source IP address, i.e. to the home router at 65.32.26.70. Finally, in step 4, the home wireless router translates the destination address of the incoming packet from 65.32.26.70 to 192.168.2.3 using NAPT. This packet is sent to the PC over the home network. To correctly translate incoming packets, NAT routers maintain a record of outgoing requests in a NAT forwarding table or translation table.

The combination of the source and destination IP address and port address makes each translation unique. Since there can be 65,536 ports associated with each IP address, and each port can support one communication channel, each globally unique IP address can support

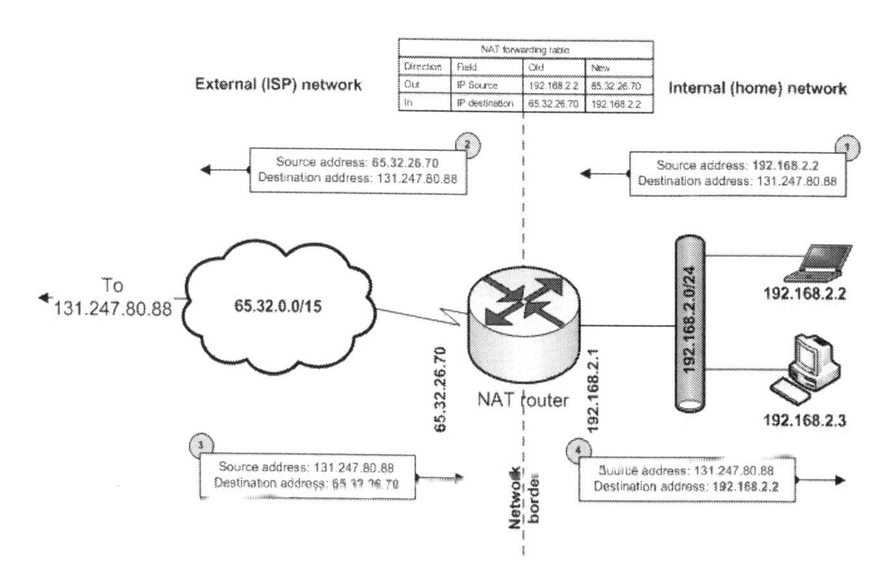

Figure 7.6: Basic NAT operation

65,536 separate connections with one specific host using NAPT. If you talk to a second host, you can have another set of 65,536 unique connections to the second destination from the same source.

The example of Figure 7.5 may be extended using NAPT, as shown in Figure 7.7. The homes in the example are served by an ISP with address pool 24.24.0.0/14. Two IP addresses from this pool, 24.26.79.18 and 24.26.79.19, are used to serve the two homes. The outward-facing ports of the routers in Home 1 and Home 2 are assigned these IP addresses. Since these IP addresses are not in the RFC 1918 pool, they are globally unique. These addresses can be used to route packets to the two customer homes. When packets leave Home 1, they leave with source IP address 24.26.79.18 and packets leaving Home 2 have source IP address 24.26.79.19. When replies are received at Home 1, the router uses NAPT to translate from the ISP's IP address 24.26.79.18 to the internal RFC 1918 IP address 192.168.2.1 used inside the home.[5]

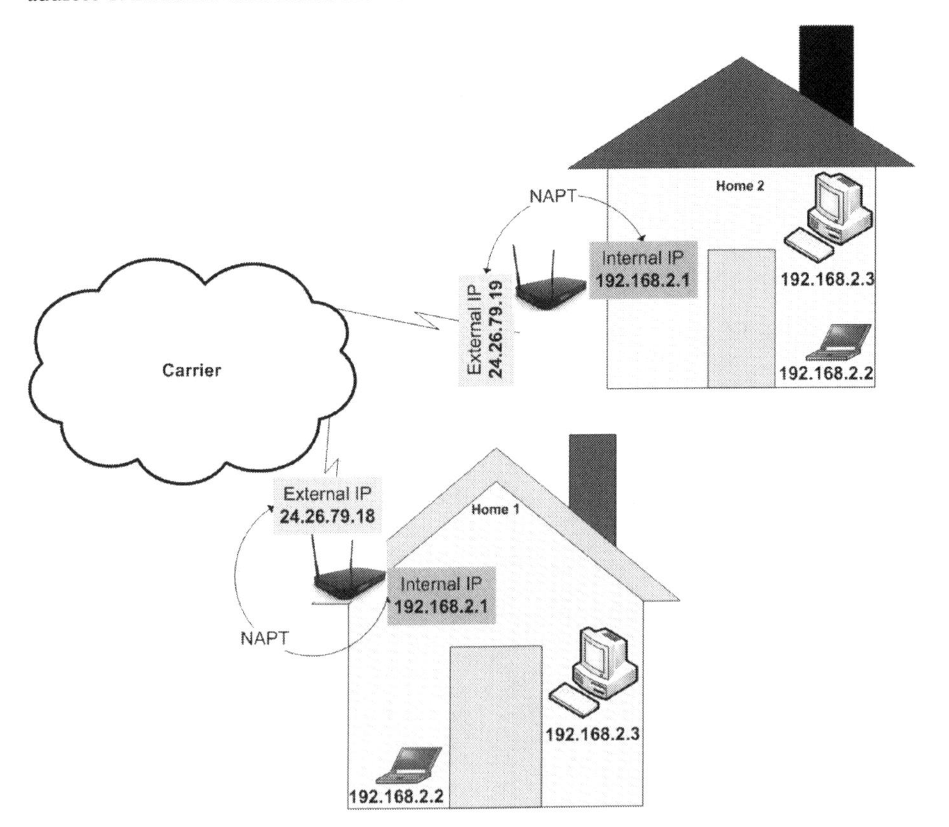

Figure 7.7: Using NAPT and non-routable RFC 1918 IP addresses in ISPs

5 At this point, you will find it useful to read the excellent white paper on the topic of IPv4 address exhaustion, which talks about some additional issues, "Next Generation Internet: IPv4 Address Exhaustion, Mitigation Strategies and Implications for the U.S.," https://www.ieeeusa.org/policy/whitepapers/IEEEUSAWP-IPv62009.pdf (accessed Dec. 2015).

Since NAPT is so useful and easy to use, it has become a huge commercial success. All wireless routers sold in retail stores implement NAPT by default. However, many experts involved in the development of networking technologies dislike NAPT because IP addresses are expected to be preserved from end to end, and NAPT disrupts this model. Many applications do not work behind NAT. Some experts also believe that the success of NAPT has delayed the deployment of IPv6.

Address Resolution Protocol (ARP)

After discussing the three-part solution used to increase the availability of IP addresses, we turn our attention to two other support services used to facilitate network operation. ARP, discussed in this section, is used by devices within networks to obtain MAC addresses corresponding to known IP addresses. DNS, discussed in the next section, is used by devices to translate hostnames to IP addresses.

From Chapter 3 and Chapter 4, we know that each computer connected to the network has at least one MAC address (used as a name), and at least one IP address. As packets traverse networks, computers and routers use the IP addresses to forward packets. Within the LAN, computers typically obtain the IP address of their immediate neighboring device—the gateway router—from DHCP. To send the packet to the gateway router or the next hop, computers and routers will use a data-link layer technology such as Ethernet. To form an Ethernet frame, these devices need to know the MAC address of the gateway or the computer whose IP address has been supplied by the application. ARP is used to obtain the MAC address for a given IP address. *ARP is a protocol that dynamically determines the data-link layer address associated with a given IP address.* ARP therefore links the addresses at the two layers—data link and network—within the same device. ARP is a very simple protocol. The sequence of operations in ARP is shown in Figure 7.8.

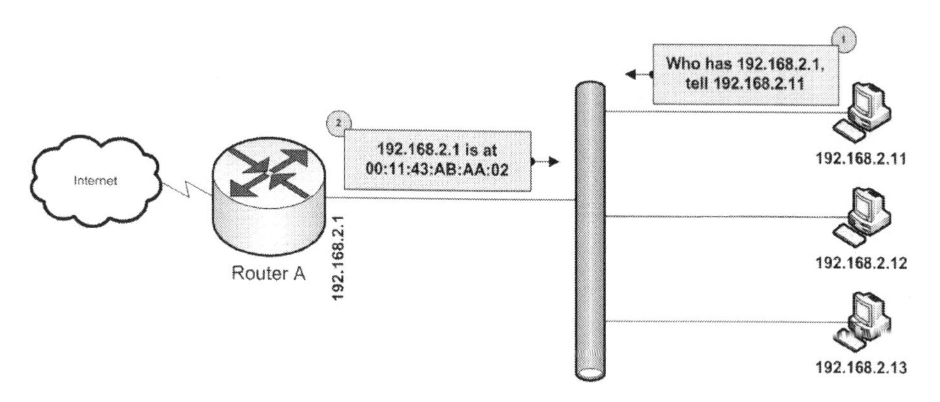

Figure 7.8: ARP sequence of operations

ARP in elementary school

If you can recall picking up someone in elementary school, you typically went through something like an ARP process. In that scenario, you had the name of the student, and the front office looked up the name of the student in the records to find the name of the home room teacher (ARP lookup), and called the room for the student to be dropped off at the front office.

```
ARP request
Sender MAC Address   : 00:11:50:3a:da:22
Sender IP address    : 192.168.2.11
Target MAC Address   : 00:00:00:00:00:00
Target IP address    : 192.168.2.1

ARP response
Sender MAC Address   : 00:18:8b:c9:24:6b
Sender IP address    : 192.168.2.1
Target MAC Address   : 00:11:50:3a:da:22
Target IP address    : 192.168.2.11
```

Figure 7.9: ARP packets exchanged in Figure 7.8

```
Administrator: Command Prompt

C:\>arp -a

Interface: 192.168.1.3 --- 0x8
  Internet Address      Physical Address      Type
  192.168.1.1           00-18-3a-c0-9a-43     dynamic
  192.168.1.153         00-d0-4b-84-8d-4e     dynamic
  192.168.1.255         ff-ff-ff-ff-ff-ff     static
  224.0.0.22            01-00-5e-00-00-16     static
  224.0.0.251           01-00-5e-00-00-fb     static
  224.0.0.252           01-00-5e-00-00-fc     static
  239.255.255.250       01-00-5e-7f-ff-fa     static
  255.255.255.255       ff-ff-ff-ff-ff-ff     static
```

Figure 7.10: ARP cache displayed using arp –a

In the figure, the desktop with IP address 192.168.2.11 has some data to send to the Internet. The gateway router is the neighboring device in the route. The desktop knows from its network configuration that the IP address of its gateway router is 192.168.2.1.

To obtain the MAC address of this router, the desktop broadcasts an ARP request on the LAN. When the gateway router sees this ARP request, it provides its MAC address in its reply. The PC can now use this MAC address to format the Ethernet frame for transmission on the LAN.

Figure 7.9 shows the contents of the fields in the ARP request and ARP response frames exchanged in Figure 7.8. Observe the destination MAC address of 00: 00: 00: 00: 00: 00 in the ARP request. This is a placeholder that is to be populated by the receiver. The ARP request is encapsulated in an Ethernet frame with destination address ff:ff:ff:ff:ff:ff, which is the broadcast address on the LAN.

Hosts typically cache the MAC addresses obtained from ARP for short durations, say 2–10 minutes. The cached arp information can be retrieved in Windows using the `arp -a` command at the DOS prompt as shown in Figure 7.10. Netstat –p can also be used to view the ARP cache on the local computer.

ARP works in the background and completely hides data-link layer addresses from end users.

DNS

Let us now look at the last important support service that simplifies the use of computer networks on a day-to-day basis. This service is called the Domain Name Service, DNS for short. This service is provided by the Domain Name System. *The Domain Name System is the set of databases that performs the correspondence between the domain name and its IP address.* The Domain Name System is also abbreviated as DNS. Therefore, the term DNS is used to refer to the system as well as the service offered by the system.

DNS translates domain names such as www.slu.edu to IP addresses, wherever in the world the domain name may be located. DNS is such an important service that it is probably the very first networked service used by every computer user. However, like all support services, DNS runs in the background and users are largely unaware of its existence.

We know from Chapter 4 that computers are uniquely identified on the network using IP addresses. We also know from personal experience that we rarely ever type IP addresses into the address bar of a browser or e-mail address field. We almost always refer to sites by user-friendly names such as www.osu.edu. DNS allows us to refer to computers by these names, instead of IP addresses. Only after DNS completes the translation from domain name to IP address does the application create a packet to send to the destination over the Internet.

From an end-user perspective, using DNS is very simple. Clients typically obtain the IP address of a designated DNS server from DHCP during boot-up. Whenever a user types in a host name in any network application, the client sends a DNS request for the host name to the designated DNS server. The designated DNS server looks up the host name in the domain name system to retrieve the IP address, which it forwards to the client. This is shown in Figure 7.11 for a client resolving the IP address corresponding to UB's website. In the example, the DNS server is at 192.68.2.1 (the local wireless router). This is common in home networks.

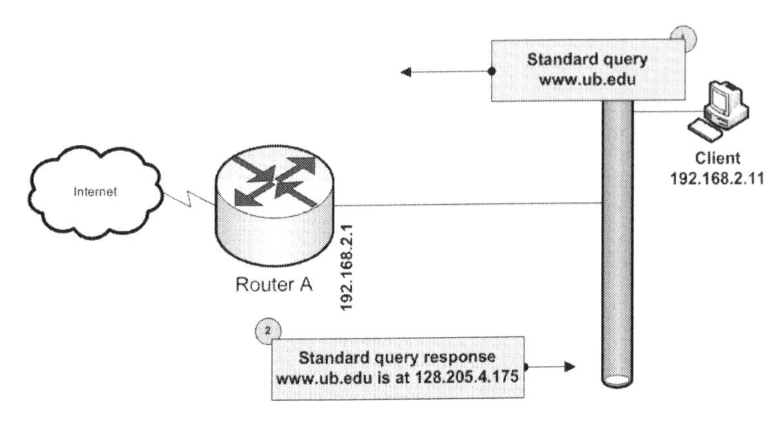

Figure 7.11: DNS use

> An important thing to note is that all DNS requests are sent to the local DNS server. This server will take care of all the complexity of translating the host name, wherever in the world it might be located.

Though the use of DNS by clients is easy, creating the DNS databases is more involved. In the early days of the Internet, long before the advent of the web, the domain-name-to-IP-address mapping was done using a simple text file called hosts. This file had entries such as the following, one line per networked host:

```
127.0.0.1              localhost
131.247.222.249        www.usf.edu
161.116.100.2          www.ub.edu
```

Every user maintained a local copy of the hosts file. Users could use the hostname www.ub.edu in any application and the computer would resolve the host name to the IP address 161.116.100.2. As users became aware of more computers on the network, they manually added entries to their hosts[6] file.

Gradually however, the Internet became too large for end users to be able to maintain their hosts files. This motivated the need for a name → IP address mapping system that was (1) accurate in mapping hostnames to IP addresses; (2) easy to maintain; and (3) easy to use.

DNS, defined in 1987 in RFC 1034 and RFC 1035, emerged as the solution to this need. DNS created the concept of domains. *A domain is the part of the naming system that is administered by an entity*. Domains are arranged hierarchically, originating from a common root. The key feature of DNS is that the responsibility for maintaining the name → IP address mappings within each domain is delegated to administrators within the domain.

---

6 The hosts file is still present in Windows at C:\Windows\System32\drivers\etc\hosts. It typically contains one entry: `127.0.0.1 localhost`, to map the hostname localhost to the IP address 127.0.01. For a visual on what the Internet directory looked like in 1982, please see Danny Hillis's 2013 TED Talk, http://blog.ted.com/what-the-internet-looked-like-in-1982-a-closer-look-at-danny-hillis-vintage-directory-of-users/.

DNS poisoning

The trust between DNS servers as they exchange domain information can be abused if DNS servers deliberately return incorrect IP addresses. This threat is called DNS spoofing or DNS cache poisoning. To see what this can do, add an entry such as the following in your hosts file:

127.0.0.1 www.microsoft.com

And try to visit www.microsoft.com on your browser. Where will the browser go?

DNS Structure

The root level of the naming universe is written as ". ." Starting from the root level within the naming universe are the top-level domains such as ".com."; ".edu."; and ".org." *Top-level domains identify the highest hierarchical level in the geographical or organizational structure of the addressing system in the Internet.* These top-level domains generally delegate mapping responsibilities to organizations and networks. These organizations get domain names such as fau.edu. or youtube.com. Domains can further delegate to subdomains. Organizations usually delegate further until endpoint domains such as www.fau.edu. are reached. Endpoint domains are the host names that users typically access on the Internet.

The first set of top-level names of .com, .org, .edu, .gov, and .mil were created by Elizabeth Feinler, who directed the Network Information Systems Center at the time.[7]

With this method of decentralized domains, the root of the domain name system is written as ". ." Domain names are read from right to left. The domain name www.slu.edu. (observe the "." at the end of the domain name) is interpreted as the domain "www," located within domain "slu," which is located within the "edu" domain. "Edu." is a top-level domain and originates from the root of the domain name system. Though every domain name ends in a "." it is customary for end users to write domain names without the "." representing the naming universe. DNS completes this detail for you in the background.

A number of top-level domains (TLDs) have been defined.[8] TLDs are maintained by the Internet Assigned Numbers Authority (IANA). You may recall from Chapter 4 that this is the same organization that also maintains the entire pool of IP addresses. The IANA delegates the responsibility of maintaining the name servers at each TLD to a sponsoring organization. End users can sign up with any accredited registrar, such as GoDaddy, to obtain a subdomain within any open TLD. Registrars inform the sponsoring organization of the TLD about the new subdomain registration and the subdomain is added to the DNS system.

7 https://en.wikipedia.org/wiki/Elizabeth_J._Feinler.

8 The list of TLDs is available at http://www.iana.org/domains/root/db/.

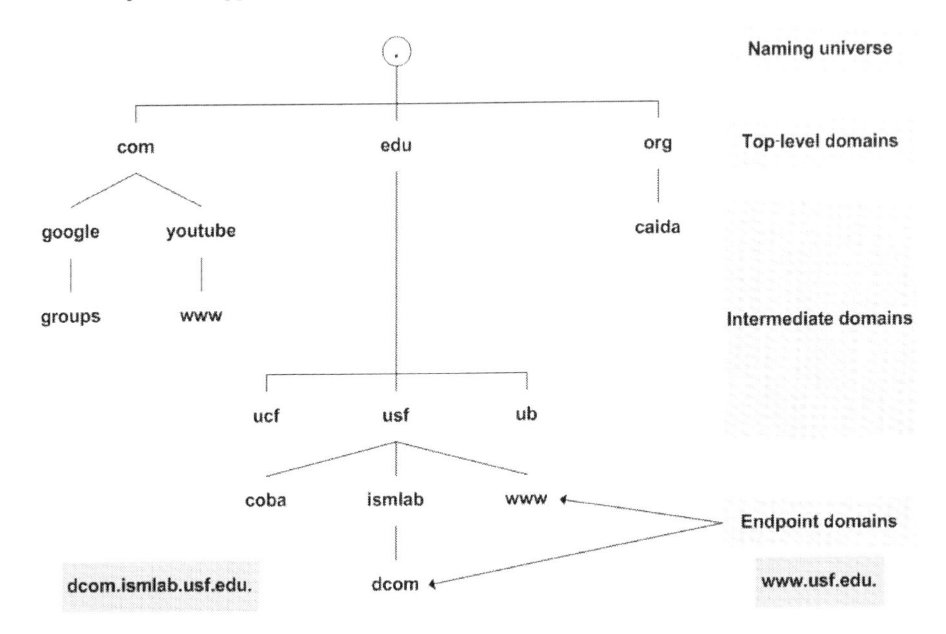

Figure 7.12: View of a section of the domain name hierarchy

There are various kinds of TLDs. The most familiar TLDs are open domains such as .com, .org, and .info. Anybody can register a domain as a subdomain of any of these open TLDs. There are also limited domains such as .gov, .edu, and .mil. The administrators of these domains require that organizations must satisfy certain conditions before they can register themselves as subdomains of these domains. For example, only US federal, state, and local governments can register themselves as subdomains of the .gov domain.

Some industries have been given their own top-level domains. These include the .aero and .travel TLDs. However, these industry-specific TLDs do not seem to have become very popular. (Even www.aaa.travel is redirected to www.aaa.com.) Finally, there are country domains such as .us and .af, one for each country. You might think that only residents of the country are allowed to register a country domain name. However, contrary to what you might think, sponsoring organizations of most country domains keep their domains open. A resident in the US can just as easily register a subdomain of .us as a subdomain of .vg (British Virgin Islands).

On June 20, 2011, IANA approved the creation of any TLD by sponsors. These TLDs don't seem to have caught much traction.[9]

The complete name of any domain is obtained by including the path to the domain from the top level. Paths are written from right to left as we go down the naming hierarchy.

9 Sam Holmes and Christopher Rhoads, "Web addresses enter new era," *Wall Street Journal*, June 21, 2011.

In Figure 7.12 for example, google.com, ub.edu, and caida.org are domains. Child domain names must be unique within the immediate parent domain. Thus, there can only be one www subdomain in google.com, but we can have www.google.com, www.yahoo.com, www.netcraft.com, etc.

DNS Lookup

Until this point we have described how domain names are interpreted. We can now examine why it is useful to organize domain names in a hierarchical manner. Recall that the Domain Name System was created to make it convenient to maintain an up-to-date mapping between computer names and IP addresses. To accomplish this, each domain maintains jurisdiction over its immediate subdomains and *only* over its immediate subdomains. For example, the .edu domain maintains the mapping between the domain name usf.edu and the IP address belonging to usf.edu. The usf.edu domain, in turn, maintains the mapping between names within usf.edu, such as www.usf.edu, and the corresponding IP addresses. The administrator of the .edu domain has no idea about the IP address of www.usf.edu. It has to defer to the administrator of the usf.edu domain to retrieve the IP address of www.usf.edu. Similarly, the administrator of usf.edu has to defer to the administrator of bsn.usf.edu to resolve www.bsn.usf.edu.

In short, if anyone asked the .edu domain for the IP address of www.usf.edu, the .edu domain would reply by saying, "I do not know what the IP address of www.usf.edu is. But I do know that .usf.edu knows the IP address of www.usf.edu. I also know that the IP address of .usf.edu is 131.247.100.1. Please contact .usf.edu at 131.247.100.1 to obtain the IP address of www.usf.edu." The user would then contact .usf.edu to obtain the IP address of www.usf.edu, which is 131.247.92.201.

The delegation of host-naming responsibilities to child domains ensures that each domain administrator only has a relatively small set of domain names to maintain. No single administrator is responsible for maintaining the name → IP address mappings for the entire Internet. As a result, name → IP address mappings are usually always current.

The use of a hierarchical system for managing domain names follows the examples of using a hierarchical system for naming in other large networks. We saw in Chapter 4 how the telephone number system, zip codes, and credit cards also use a hierarchical system for managing their address space. Most large networks find it convenient to use a hierarchical system to manage names.

Say a user enters the hostname www.yahoo.com on their browser. How does the browser get the IP address corresponding to the host name?

The DNS server is a piece of software that is used to resolve host names. The domain name server maintains the name → IP address for a domain. For clients it serves, the DNS server also retrieves IP addresses for host names maintained by other DNS servers. To use DNS, a computer needs to know the IP address of a domain name server it can use. It typically gets this at boot time from DHCP. You can see the DNS server information on your PC using the command `ipconfig/all` at the DOS prompt. All operating systems also have a DNS resolver or a DNS client.

When a user enters a URL in a browser, the browser passes the host name in the URL to the DNS resolver on the PC. The resolver contacts its DNS server to obtain the required IP

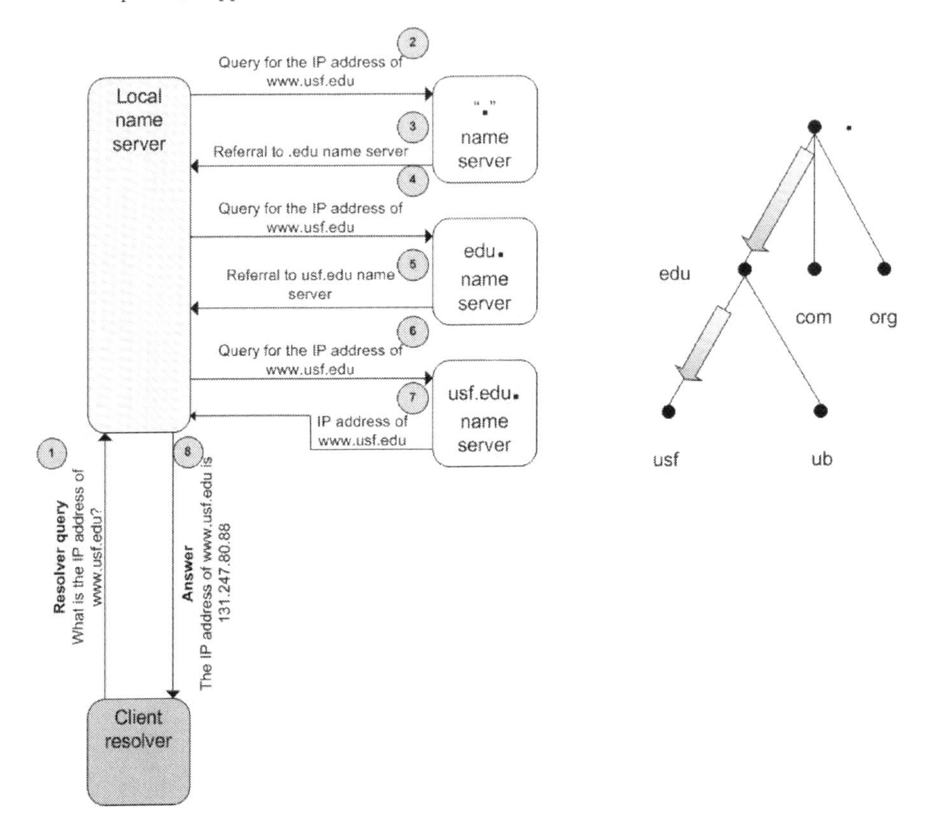

Figure 7.13: Recursive DNS query resolution

address. The DNS server performs all necessary lookups in the DNS system to obtain the IP address. When the IP address is obtained, the DNS server passes it to the DNS client, which passes the IP address to the browser. The process is shown in Figure 7.13. Now, the browser can populate the header fields in all layers and send out the request packets to the Yahoo web server.

It is easy to see how DNS works on a Linux/UNIX terminal. The DNS client is accessible using the `dig` command. Figure 7.14 shows the results of a DNS query to obtain the IP address of www.buffalo.edu. The answer section of the response provides the desired information—that the IP address is 128.205.4.175. In the example, the authority section lists the name servers that have authoritative information about the domain. These are the name servers that maintain the name → IP address mappings for buffalo.edu. It is interesting to see that not only is the university maintaining a backup name server on its own premises, it is also maintaining a third backup name server at northwestern.edu. The additional section provides the IP addresses of the authoritative name servers.

```
# dig www.buffalo.edu

;; Got answer:

;; QUESTION SECTION:
;www.buffalo.edu.   IN     A

;; ANSWER SECTION:
www.buffalo.edu.   86400  IN     A      128.205.4.175

;; AUTHORITY SECTION:
buffalo.edu. 71951  IN     NS     ns.buffalo.edu.
buffalo.edu. 71951  IN     NS     sybil.cs.buffalo.edu.
buffalo.edu. 71951  IN     NS     accuvax.northwestern.edu.

;; ADDITIONAL SECTION:
ns.buffalo.edu.              71951  IN     A      128.205.1.2
sybil.cs.buffalo.edu.        53404  IN     A      128.205.32.8
accuvax.northwestern.edu     11624  IN     A      129.105.49.1

;; Query time: 3 msec
;; SERVER: 131.247.100.1#53(mother.usf.edu)
```

Figure 7.14: Typical DNS query (to obtain the IP address of www.buffalo.edu)

DNS domains are sometimes also called zones. Thus buffalo.edu is a zone. As described earlier, each name server is responsible for its zone and only for its zone. Zones may have sub-zones, and each sub-zone is responsible for its own zone.

To reduce the number of DNS queries, name servers and clients cache resolved domain names. The administrators of each zone specify the time to live (TTL) for their zone. They may even specify TTLs for each record in their zone. Domain names within the zone can be cached till the TTL expires. In Figure 7.14, the IP address for www.buffalo.edu can be cached for 86,400 seconds (1 day) while the IP address for buffalo.edu can be cached for another 71,951 seconds. If another user requests this name server within 86,400 seconds for the IP address of www.buffalo.edu, the name server will not perform a DNS query; instead it will provide the requested IP address from its cache.

What happens if your domain name server does not have a cached entry for a TLD for a domain it is asked to search? For example, say a user types in the URL www.gm.com.cn. Let us also assume that no one using this DNS server has visited a site in China in the last few days. The DNS server therefore has no cached entry for the .cn TLD.

On Windows, the nslookup utility provides functionality similar to dig in UNIX. Please see the hands on exercise for an example run of nslookup. When the nslookup command is run without arguments, it enters into an interactive mode. In the interactive mode, the set d2 command causes nslookup to enter debug mode. At this point, nslookup shows complete output for any query. The example shows a query for www.google.com.

DNS and the phone system

You may note that the phone system has some similarities with DNS. Each entity in the phone system only knows about 1,000 other entities (areas or end offices). Yet, any of these devices can connect to any of the 10 billion phones in North America.

However, the big difference is that the phone system hierarchy is hard-wired and geographically organized. An organization in Tampa, Florida, USA, will always have country code 1, and area code 813. However, if the same organization changes ISPs, it can get a completely new address block, and DNS will ensure that it continues to be reachable by all hosts across the Internet.

In this case, the name server starts the search from the naming universe. There are 13 name servers[10] distributed around the world that maintain information about the top-level domains.[11] These are called root-name servers. Information about these root-name servers is hardwired in all name servers. When a DNS query is made, the DNS server can start with one of the root-name servers and search down the domain name hierarchy till it finds the authoritative name server for the domain being searched. This process is shown in Figure 7.13 for a client asking its local name server for the IP address of the URL www.usf.edu; this is step 1. In step 2, the local DNS server queries a root-name server, which refers the local DNS server to the "edu" name server. The process continues until, in step 7, the usf.edu name server authoritatively provides the IP address for www.usf.edu. Finally, in step 8, the local DNS server sends this IP address to the client. The client sees only the information exchanged in steps 1 and 8, as shown in Figure 7.13. The complete trace using dig +trace is shown in Figure 7.15.

```
pns:~# dig +trace www.usf.edu
; <<>> DiG 9.2.4 <<>> +trace www.usf.edu

.            77639        IN     NS     E.ROOT-SERVERS.NET.
(and other root name servers

edu.         172800       IN     NS     E.GTLD-SERVERS.NET.
(and other .edu name servers)

usf.edu.     172800       IN     NS     justincase.usf.edu.
usf.edu.     172800       IN     NS     mother.usf.edu.
(and other usf.edu name servers)

www.usf.edu. 600          IN     A      131.247.80.88
```

Figure 7.15: Tracing the DNS query for www.usf.edu

10 https://miek.nl/posts/2013/Nov/10/13%20DNS%20root%20servers/ (13 is an unusual choice, but is based on the early packet-size limit of 512 bytes defined by IP).

11 http://www.root-servers.org/ has an image of the locations of all the root-name servers.

The dig command also has a +trace option. Using the trace option shows the information retrieved as the DNS server traverses the domain name hierarchy. The trace results from a query for www.usf.edu are shown in Figure 7.15. The DNS server first locates the root name server, then the .edu name server and finally, the usf.edu name server, which provides the IP address for www.usf.edu. This re-creates the query of Figure 7.13.

An important advantage of using DNS is that it hides changes in IP addresses from end users. Network administrators can relocate web and other servers to hosts with different IP addresses when required, without affecting end-user connectivity. For example, Figure 7.16[12] shows that the IP address of the popular e-commerce website Amazon.com changed quite frequently in December 2015. However, DNS hides these changes from end users. As long as Amazon can ensure that DNS resolves its IP address correctly, it does not have to worry about informing end users about its current IP address.

ipconfig and DNS

Having done some exercises with ipconfig, you may like to check out the displaydns option of ipconfig using the command, `ipconfig /displaydns | more.` Use the space bar to scroll forward one page, and q to return to the command prompt. Displaydns will display the current contents of the local DNS cache on Windows. In particular, you may find it interesting to observe how the time to live for the records counts down to 0.

Hosting History

Netblock owner	IP address	OS	Web server	Last seen Refresh
Amazon.com, Inc. 1918 8th Ave SEATTLE WA US 98101-1244	54.239.25.200	unknown	Server	3-Dec-2015
Amazon.com, Inc. 1918 8th Ave SEATTLE WA US 98101-1244	54.239.26.128	unknown	Server	3-Dec-2015
Amazon.com, Inc. 1918 8th Ave SEATTLE WA US 98101-1244	54.239.25.192	unknown	Server	1-Dec-2015
Amazon.com, Inc. 1918 8th Ave SEATTLE WA US 98101-1244	54.239.26.128	unknown	Server	29-Nov-2015
Amazon.com, Inc. 1918 8th Ave SEATTLE WA US 98101-1244	54.239.25.200	unknown	Server	28-Nov-2015
Amazon.com, Inc. 1918 8th Ave SEATTLE WA US 98101-1244	54.239.26.128	unknown	Server	27-Nov-2015
Amazon.com, Inc. 1918 8th Ave SEATTLE WA US 98101-1244	54.239.17.6	unknown	Server	25-Nov-2015
Amazon.com, Inc. 1918 8th Ave SEATTLE WA US 98101-1244	54.239.25.192	unknown	Server	24-Nov-2015
Amazon.com, Inc. 1918 8th Ave SEATTLE WA US 98101-1244	54.239.17.6	unknown	Server	23-Nov-2015
Amazon.com, Inc. 1918 8th Ave SEATTLE WA US 98101-1244	54.239.26.128	unknown	Server	22-Nov-2015

Figure 7.16: Changes in IP address of www.amazon.com over 10 days

12 http://toolbar.netcraft.com/site_report?url=http://www.amazon.com.

DNS and router configuration error[13]

An interesting example of the impact of DNS configuration error on network connectivity is reported by CAIDA. Microsoft had placed all its externally visible name servers on the same subnet. In January 2001, a router configuration error prevented the subnet from communicating with the outside world. The DNS address record for the microsoft.com domain had a TTL of 2 hours. As a result, 2 hours after the outage, name servers around the world had cleared their cached records of Microsoft's IP address.

When the microsoft.com name servers continued to be unreachable due to the wrongly configured router, name servers around the world started querying the root servers for microsoft.com and related names (expedia.com, passport.com, msn.com, msnbc.com, etc.). Data showed that during this period, the query load for Microsoft names went from normal rates of approximately 0% of the overall load at the root-name servers to more than 25% of the total query load. Thus, local problems in DNS at popular sites can significantly disrupt the operations at root-name servers. According to the CAIDA article, Microsoft outsourced their DNS provisioning after the incident.

I-root server and The Great Firewall[14]

It is well known that many popular websites are not accessible within certain countries for various reasons. However, the indifference of the Internet to geographical boundaries can lead to interesting errors. In March 2010, many users outside China were suddenly unable to access popular sites such as Facebook and Twitter, and had the same Internet experience as if they were in China.

What had happened was that some DNS requests to the root-name servers from outside China were being forwarded to the I-root servers located in China. This can happen when an ISP depends upon upstream ISPs for network services such as DNS.

One of the most popular DNS servers is BIND, maintained by the ISC (which also maintains the most popular DHCP server). Figure 7.17 shows a sample configuration file for the BIND DNS server for a domain example.com. The configuration specifies a TTL of 1 day (86,400 seconds). The e-mail address of the administrator is hostmaster@example.com (any problems with the DNS server can be automatically e-mailed to the administrator). The name server (NS) for the domain is pns.example.com, which is located at IP address 192.168.16.129. The e-mail server (mail exchange, or MX) is at mail.example.com, with IP address 192.168.16.130. The web server, www.example.com, is at 192.168.16.129. In other words, the same host (192.168.16.129) acts as both the web server and name server in this domain. A subdomain, test.example.com, is also defined. The name server for the test.example.com domain is located at 192.168.16.143. The administrator of test.example.com can define subdomains of test.example.com.

---

13 Nevil Brownlee, K.C. Claffy, and Evi Nemeth, "DNS Measurements at a Root Server," Cooperative Association for Internet Data Analysis, San Diego, 2001, https://www.caida.org/publications/papers/2001/DNSMeasRoot/dmr.pdf.

14 "DNS: When governments lie," http://research.dyn.com/2010/11/dns-when-governments-lie-1/ (accessed Dec. 2015).

```
$TTL    86400
     @       IN   SOA  pns.example.com.  hostmaster.example.com. (
                           serial 2008072701
)
                           IN           NS     pns.example.com.
                           IN           MX     10     mail.example.com.

pns          A       192.168.16.129
www          A       192.168.16.129
mail         A       192.168.16.130

test         NS      demo
demo         A       192.168.16.143
```

Figure 7.17: Sample BIND DNS server configuration file

The caching of domain names can lead to websites becoming unreachable if their IP addresses are changed. DNS servers will pass obsolete IP addresses from their cache to clients. Eventually, as the DNS entries are timed out in the caches of name servers around the world, the new IP address is retrieved and the site becomes reachable again. Therefore, if you plan to change IP addresses of any of your Internet servers, it is a good idea to reduce the TTL of your name server to say 15 minutes. This should be done sufficiently in advance so that records with the old TTL expire before the change. This way, when you do change IP addresses, it will only be 15 minutes before the old IP address expires within the cache of all DNS servers on the Internet.

In most home networks, the wireless router also acts as the DNS server to perform DNS lookups.

Home Networking

Starting from Chapter 2 and ending in the previous section of this chapter, we have covered all the components that work together to bring Internet connectivity to your home. In this section, we will see how all the different components are put together in your home network.

The focal point of your home network is your wireless router. This router may have been provided by your ISP, or you may have purchased it off the shelf from a retail store such as Best Buy. This router typically acts as your DHCP server, your DNS server, and your NAPT router. With the addition of the router and the DNS and DHCP servers, the network looks like in Figure 7.18. The home router is a multifunctional device that does not just provide wireless access. It also provides all the services discussed in this chapter, and acts as the gateway to the Internet.

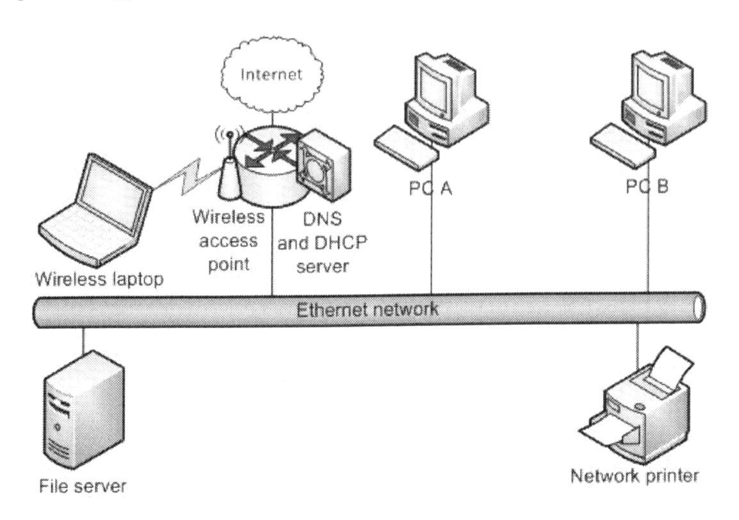

Figure 7.18: Home LAN with wireless router

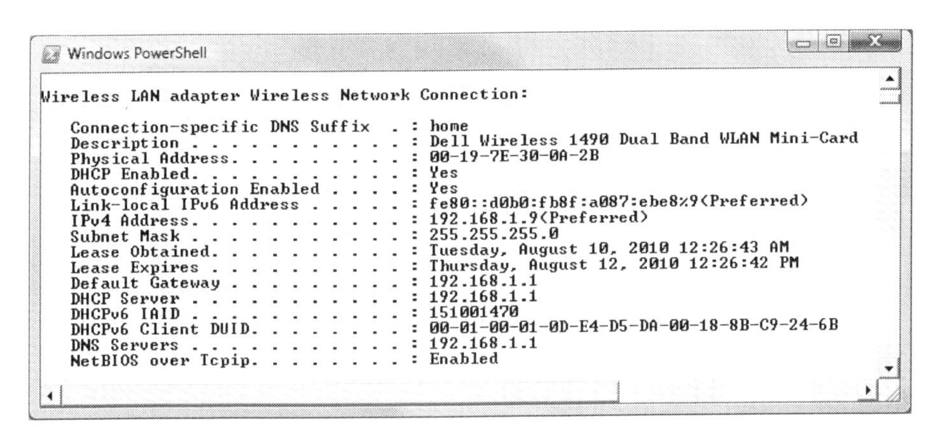

Figure 7.19: Home PC IP configuration

To understand your network, perform the following three steps. In step 1, type `ipconfig /all` at your DOS prompt. This will show you the network configuration of your PC as shown in Figure 7.19. You will generally see an RFC 1918 address ending in .1 as your default gateway, DHCP server, and the DNS server. This is the internal IP address of your wireless router. In step 2, check your wireless router's network configuration. Most routers allow you to see and modify the router configuration by pointing your browser to this IP address from a computer in the home. Manufacturers have different interfaces, but if you type in your router's RFC 1918 IP address in your browser, you should see a page similar to that in Figure 7.20. Most routers let you see information such as the different computers that have been assigned IP addresses by the DHCP server on the router.

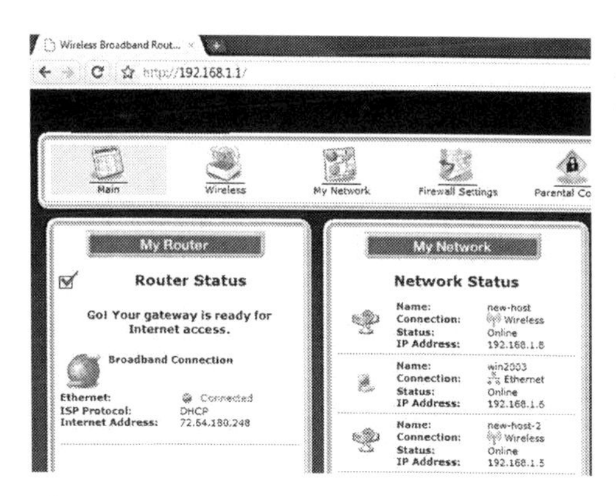

Figure 7.20: Home wireless router web interface

Finally, in step 3, you can visit the website: http://www.whatismyip.com. This website will show you the IP address assigned to the external port of your router by your ISP. This IP address is generally a globally unique IP address from the IP address pool assigned to your ISP by the network registry.[15]

Putting these pieces together, how does your home network work? All PCs in your home get an RFC 1918 IP address from your wireless router at boot time. The DHCP server in your router also informs the computers in your home that the wireless router is the next hop (default gateway) as well as the DNS server on the local network (though some ISPs use one of their own DNS servers for end users). When a packet leaves your home network, your home router translates the source IP address field in the packets from the internal RFC 1918 address to the external IP address assigned by the ISP. When reply packets reach the wireless router, the router translates IP addresses back from the ISP's address to the correct internal IP address.

EXAMPLE CASE—DNS and Virtual Hosts

Electronic commerce has reduced the barriers faced by entrepreneurs in trying to sell their products and services to a global audience. Leading technology firms and IT entrepreneurs have developed technology solutions so that anybody can create and operate e-commerce websites at reasonable costs. In this case, we see how DNS and another technology, virtual hosting, help in improving the utilization of computer hardware powering these websites. By improving hardware utilization, DNS and associated technologies significantly lower e-commerce costs.

It is quite likely that you have set up a simple HTML website in some class or have tried using one of the many free website builders to create a website. Building a simple HTML website is quite easy. However, e-commerce websites need many additional functions, such as the ability for the seller to add products, be notified about orders, print shipping labels, and notify customers when orders are shipped. Since there is a reasonably large customer base interested in these services, most major metro areas have many competing providers offering web hosting and related services to enable e-commerce websites. Yahoo is one of the national players in the market.

How do these service providers offer their services at such reasonable prices, often less than $50 per month? It turns out that service providers can host multiple websites at a single server, thus amortizing hardware and software costs for each server over all the

15 If you are adventurous, you can type in your external IP address into the "whois" search box at www.arin.net to confirm that your IP address indeed belongs to your ISP.

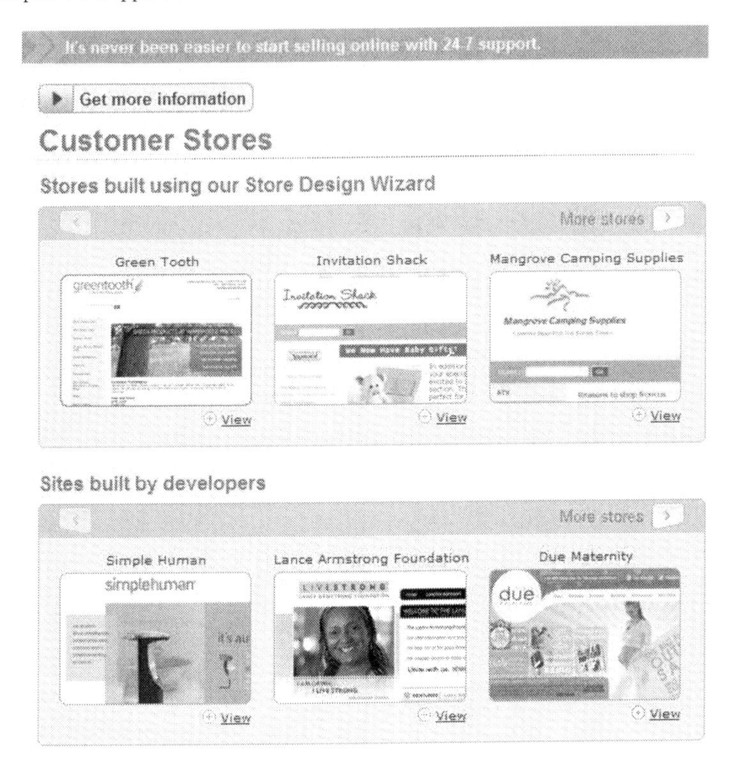

Figure 7.21: Example websites hosted by Yahoo!

websites hosted on the server. DNS plays an important enabling role in making this happen. The other technology is virtual hosts, the ability of a single web server to host multiple websites. Let us see these technologies in action using the Yahoo small business website as an example.

Figure 7.21 shows some example websites hosted by Yahoo. Let us find out the IP addresses of these websites. The utility to do that is nslookup. What do we see?

We see that the websites hosted by Yahoo are all accessible at the same IP address, 68.142.205.137. These e-commerce websites hosted on Yahoo also use the Yahoo! favicon on their websites, indicating their affiliation with Yahoo.

How is it beneficial for businesses if many websites can be hosted at the same computer? Think what it would cost businesses if each website had to be hosted on a different computer. Web-hosting companies would need to find space to locate all of these computers. Each computer would draw power. More hardware would lead to more repair needs and more personnel for maintenance and updates. Each of these would raise costs.

Hosting multiple websites on a single computer saves costs. However, if there are many websites hosted at one website, how is the web server to know which site a user is requesting? Here, DNS and virtual hosts come into play. When a user types in a URL, DNS directs the user to the computer with the IP address at which the site is hosted. The user request contains the name of the target website in the URL.

nslookup www.green-tooth.com
Non-authoritative answer:
www.green-tooth.com canonical name = stores.yahoo.net.
stores.yahoo.net canonical name = html.store.yahoodns.net. Address: 68.142.205.137
www.invitationshack.com canonical name = stores.yahoo.net.
stores.yahoo.net canonical name = html.store.yahoodns.net.
Name: html.store.yahoodns.net Address: 68.142.205.137
www.mangrovecampingsupplies.com canonical name = stores.yahoo.net.
stores.yahoo.net canonical name = html.store.yahoodns.net.
Name: html.store.yahoodns.net Address: 68.142.205.137
www.duematernity.com canonical name = stores.yahoo.net.
stores.yahoo.net canonical name = html.store.yahoodns.net.
Name: html.store.yahoodns.net Address: 68.142.205.137

Figure 7.22: Virtual hosts architecture

The web server uses the name of the target website in the URL to determine which site the client is requesting and sends the request to the site for handling, as shown in Figure 7.22. The requested resource is served from the site's folder on the web server.

The following example shows how the popular Apache web server can be configured to serve three websites: www.example.com; www.example.net; and www.example.org.

```
<VirtualHost *:80>
DocumentRoot "/export/home/websites/example_com"
ServerName www.example.com
DirectoryIndex index.php index.html
ErrorLog "/export/logs/example_com/example_com-error_log"
CustomLog /export/logs/example_com/example_com-access_log combined
</VirtualHost>
<VirtualHost *:80>
DocumentRoot "/export/home/websites/example_net"
ServerName www.example.net
DirectoryIndex index.php index.html
 . . .
</VirtualHost>
<VirtualHost *:80>
DocumentRoot "/export/home/websites/example_org"
ServerName www.example.org
DirectoryIndex index.php index.html
 . . .
</VirtualHost>
```

Each virtual host section specifies one website and maps a folder on the computer to the website. For each client request, the web server uses the information in the virtual host definitions to determine which folder on the computer to serve pages from. Client requests for all websites hosted on a server arrive at the server. Based on the URL, the web server responds with data in the corresponding folder.

References

1. Yahoo merchant solution integrations, https://developer.yahoo.com/stores/.

Summary

In this chapter we looked at a number of support services that make IP addresses easy to use even by novices who know nothing about how the Internet works. DHCP allocates IP addresses to computers during booting up and before users see a login prompt on their PCs. Most organizations allocate reusable RFC 1918 addresses to client PCs within their networks so that they can use a small number of IP addresses to provide connectivity to a large number of computers in the organization. This conserves IP addresses. ARP finds the MAC addresses corresponding to known IP addresses and is used to populate the destination MAC address-field of frames.

DNS allows end users to address computers using friendly names such as www.yahoo.com instead of the more cumbersome IP addresses like 69.147.76.15.

Finally, we saw how all these services and the TCP/IP stack help end users seamlessly obtain network connectivity when they start up a PC at home. The smooth operation of these services allows users to get online within minutes of opening up a brand-new PC out of the box, even if they have no idea about signals, MAC addresses, ARP, IP addresses, DHCP, RFC 1918, default gateway, routing, NAPT, port addresses, or DNS.

About the Colophon

What's a name worth? What do names indicate? In one of Shakespeare's most famous plays, the names indicated family affiliations and eventually led to the deaths of the two principal characters in the play. On the Internet, names are a way for companies to leverage the investments they make in their brands.

Can names influence personal and commercial outcomes? In Shakespeare's play, changing names would probably not have changed consequences for the doomed characters. On the Internet, however, names are significant. Changing names may mean lost business. Cars.com is a better name for a site that sells cars than A123.com. Ford.com is a better name for Ford Motor Company's website than www.48126.com. (48126 is the zip code of the company's headquarters.)

On the Internet, as in drama, there's a lot in a name. Fortunately, there is DNS.

REVIEW QUESTIONS

1. What is *DHCP*? Why is it useful?
2. What are some of the reasons for the inefficiencies in allocating the available IP addresses?
3. What are the three types of address allocation schemes in DHCP? Under what conditions is each of these categories of address allocation preferred?
4. What is *address leasing* in DHCP?
5. Use ipconfig/all at the command prompt of a Windows computer. What is the lease duration of the IP address?
6. Briefly describe the sequence of operations that allow a freshly booted DHCP client to obtain an IP address from a DHCP server on the network.
7. What are *non-routable IP addresses*?
8. IP addresses are used for routing. Why are non-routable IP addresses useful?
9. Why can't a computer with an RFC 1918 IP address be used as a public-facing web server?
10. What is *network address translation*? Why is it useful?
11. How do DHCP, non-routable addresses, and NAT help improve the efficiency of utilizing IP addresses and reduce the shortage of IP addresses?
12. What is *NAPT*? How can it improve the efficiency of utilizing IP addresses, compared to NAT, without port translation?
13. Describe the NAPT operation and how the IP addresses in a packet change as a request packet travels from a source with an RFC 1918 address to a destination and the reply comes back to the source.
14. What is *ARP*? What is it used for?
15. Briefly describe the operation of ARP.
16. List the entries in the ASP cache of your computer using the arp –a command.
17. What is *DNS*? What is it used for?
18. Describe the hierarchical organization of domains on the Internet.
19. Why is it useful to organize domain names such as www.usf.edu hierarchically as they are done in DNS?
20. What are the different kinds of top-level domains?
21. Describe the process used by a name server to resolve the IP address of a URL typed by a user.
22. What is a *zone* in the context of DNS?
23. What is a recursive query in DNS? When does it become necessary?
24. What are the different network services provided by the typical home wireless router provided by ISPs?
25. Use the nslookup command to obtain the IP address of www.google.com. (You may have to type in "." after com.) Which name server performed the name resolution for you—your local name server or the Google name server?

EXAMPLE CASE QUESTIONS

1. What are some of the essential features needed in an e-commerce website? A good starting point for your answer would be Yahoo's small-business site.
2. What is an important agricultural or other natural product that your state is known for? Find a business that sells this product online. Describe some important e-commerce capabilities of the company's website. If there is no such product, select one from a nearby state.
3. Find a hosting service provider in your city or the nearest metropolitan area. Visit the company's website and describe some of the services offered by the company. (Select up to three services if the company offers many services.) What are the monthly fees for each of these services? For each of these services, think of a business or nonprofit that might find the service useful.
4. From the ISP's website, what are some of the job openings at the hosting service provider you selected? What are the required skills for these openings? If the selected provider has no openings at this time, pick another provider whose website lists at least one job opening.
5. Instead of using the full suite of e-commerce services from a hosting provider, you could limit yourself to hosting services and use free software such as Zen Cart to create your online store. What are some of the capabilities of Zen Cart (www.zen-cart.com)?
6. What is a *favicon*? What is your university's favicon?

HANDS-ON EXERCISE—nslookup

The nslookup utility is included with almost every operating system to resolve domain names to IP addresses. In this exercise you will use nslookup to resolve a few URLs. Figure 7.23 shows an example of using nslookup to obtain the IP address of www.ucf.edu. The IP address of the website is 132.170.240.131. To convince ourselves that this is indeed the case, we can use the IP address in the browser to bring up the web page as shown in Figure 7.24.

Answer the following: (Note: you may find that you need to end the URL with a "." to resolve the names correctly, as shown in Figure 7.23.)

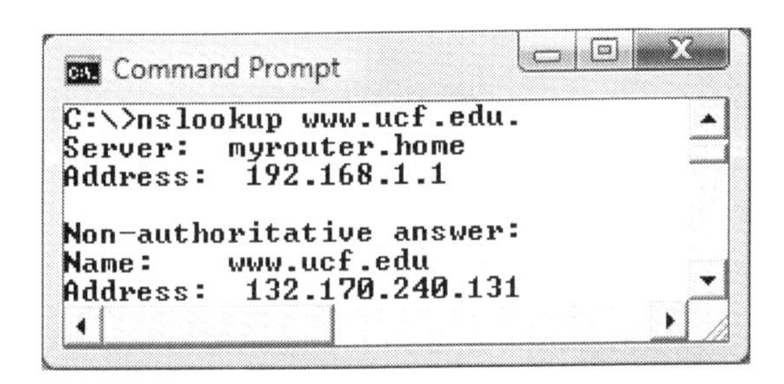

Figure 7.23: Using nslookup to resolve a URL

Figure 7.24: Using IP address to navigate to a website

1. Use nslookup to obtain the IP address of your university's website. Show the output. Which name server was used to obtain the IP address?
2. Use nslookup to obtain the IP address of the home page of one of the major employers in your area. Show the output. Which name server was used to obtain the IP address?
3. Use nslookup to obtain the IP address of the home page of one of the companies included in the S&P 500 index. The list of these companies can be found from many sources by searching online for the term "s&p 500 companies." Show the output from nslookup. Which name server was used to obtain the IP address? What pattern do you observe in the identities of the name servers for the three queries?

nslookup set d2

The nslookup utility on Windows has an interesting option that gives more information than the standard lookup in Windows. To use the debug mode, do the following. At the Windows command prompt, just type `nslookup`, without arguments. At the prompt, enter `d2`. At the next prompt, enter the URL you would like to resolve, such as www.google.com. (make sure to have the "." at the end). Figure 7.25 is an example.

CRITICAL-THINKING EXERCISE—Nissan Computer Corp.

Read about the lawsuit filed by Nissan Motors Corporation against Nissan Computer Corporation. Information can be accessed at http://www.nissan.com.

1. Which of the two parties do you think is rightfully entitled to own the Nissan.com domain name? What are your principal arguments in favor of the party?

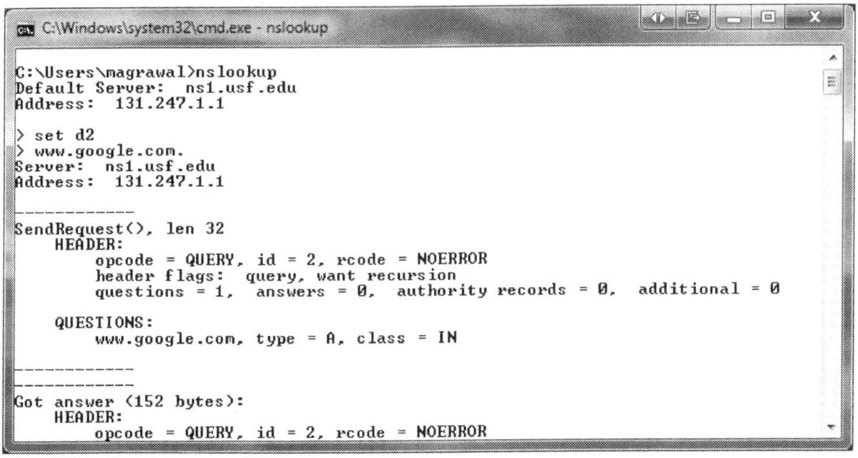

```
C:\Windows\system32\cmd.exe - nslookup

C:\Users\magrawal>nslookup
Default Server:  ns1.usf.edu
Address:  131.247.1.1

> set d2
> www.google.com.
Server:  ns1.usf.edu
Address:  131.247.1.1

------------
SendRequest(), len 32
    HEADER:
        opcode = QUERY, id = 2, rcode = NOERROR
        header flags:  query, want recursion
        questions = 1,  answers = 0,  authority records = 0,  additional = 0

    QUESTIONS:
        www.google.com, type = A, class = IN

------------
------------
Got answer (152 bytes):
    HEADER:
        opcode = QUERY, id = 2, rcode = NOERROR
```

Figure 7.25: Using set d2 option with nslookup

IT INFRASTRUCTURE DESIGN EXERCISE—Start Infrastructure Diagram

TrendyWidgets eventually decided not to obtain an address block from ARIN. Instead, it chose to obtain a few static externally-addressable (non-RFC 1918) IP addresses from its ISP for its Internet-facing applications such as web servers. The company also decided to obtain a few more IP addresses from its ISP to share internally using NAPT. It will use a /21 address block to assign IP addresses to all the other computers in the company. It also decided to have a single connection to the Internet from Tampa. Answer the following questions:

1. What services in TrendyWidgets's network will require externally addressable IP addresses?
2. The companion website has an initial infrastructure diagram in the document Ch1_NetworkDesignCase. This diagram only has the Tampa office shown. Please feel free to use the diagram and icons in the document to draw your network. Update the infrastructure diagram in the document by adding an Internet connection from the Tampa office. The Internet is typically drawn as a cloud.
3. Further update the diagram to include a NAPT device that translates between the internal /21 network and the external IP addresses obtained from the ISP.

CHAPTER 8

Routing

I think that I shall never see
A graph more lovely than a tree.
A tree whose crucial property
Is loop-free connectivity.
A tree which must be sure to span
So packets can reach every LAN.
First the root must be selected
By ID it is elected.
Least-cost paths from root are traced.
In the tree these paths are placed.
A mesh is made by folks like me,
Then bridges find a spanning tree.

Radia Perlman, *Algorhyme*

Overview

Routing is the process of moving information from the source network to the destination network through intervening LANs and WANs. Routing is used every time a computer sends packets to another computer that is not a member of its own local network. This chapter describes how routing works and describes the components, such as routing tables, that make it work. At the end of the chapter, you should know:

- how routing is different from switching
- what autonomous systems are
- what routing tables are
- how to view routes to a network
- the two categories of routing protocols
- what route aggregation is and why it is useful

Introduction

In Chapter 3, we saw how broadcasting is used to transfer data between computers on the same local network. Unfortunately, broadcasting is generally very inefficient for exchanging data across networks. For example, your home network is directly connected to your ISP network through your home router. Your ISP's network serves thousands of homes and handles very large amounts of data, most of which is not relevant to any computer within your home. Broadcasting the ISP data to computers within your home would create an unnecessary processing burden on the computers. There would also be privacy concerns if you could read data from other subscribers to the ISP's service.

Therefore, when networks are connected together to form larger networks, some suitable non-broadcast mechanism is necessary to transfer data that needs to go from one network to another. For example, a request from a PC in the home network to a web server on a remote site will need to be transferred from the home network to the ISP's network. The reply from the remote web server will need to be transferred from the ISP's network to the home network. These non-broadcast data transfers between networks are performed by routing.

Therefore, when data traffic reaches a network boundary, instead of broadcasting the data to all connected networks, it is efficient to direct the data along a suitable path toward its destination. *Routing is the process of selecting paths to move information across networks from the source network to the destination network*. Within LANs, the simple technique of broadcasting is preferable to directed transmission for its simplicity. But in large networks, routing improves efficiency by limiting broadcasts.

Routing is performed in networks by devices called routers. *Routers are devices used to interconnect two or more networks*. Routers are one of the most important components of data networks. They are made by companies like Cisco, Juniper, and Force 10. Routers can be expensive, costing tens of thousands of dollars, and need to be operated by trained administrators. The commercial importance of routers may be gauged from the fact that router manufacturers are some of the largest companies in the IT industry. You may not have heard of many of these firms because routers are located deep inside computer networks where end users are generally not aware of their existence. Routers are sold directly to corporate IT, and not through traditional retail channels. End users see only the home routers made by companies such as Linksys and Belkin, which are sold at many retail outlets.

Switching vs. Routing

Before we look at the details of routing, it is useful to contrast it with switching. Switching was covered in Chapter 3 and, at first glance, it seems to perform the same functions as routing. When frames arrive at a switch, the switch examines the MAC header and directs the frame to the device with the destination MAC address. This improves network efficiency by curtailing broadcasts. Routers seem to perform the same task. They look at the destination IP address and send packets to the network closest to the destination.

Thus, both switches and routers look at destination addresses and direct packets toward the destination. So, are there any differences between routing and switching? Indeed, there are. Figure 8.1 shows a prototype switched network and a prototype routed network.

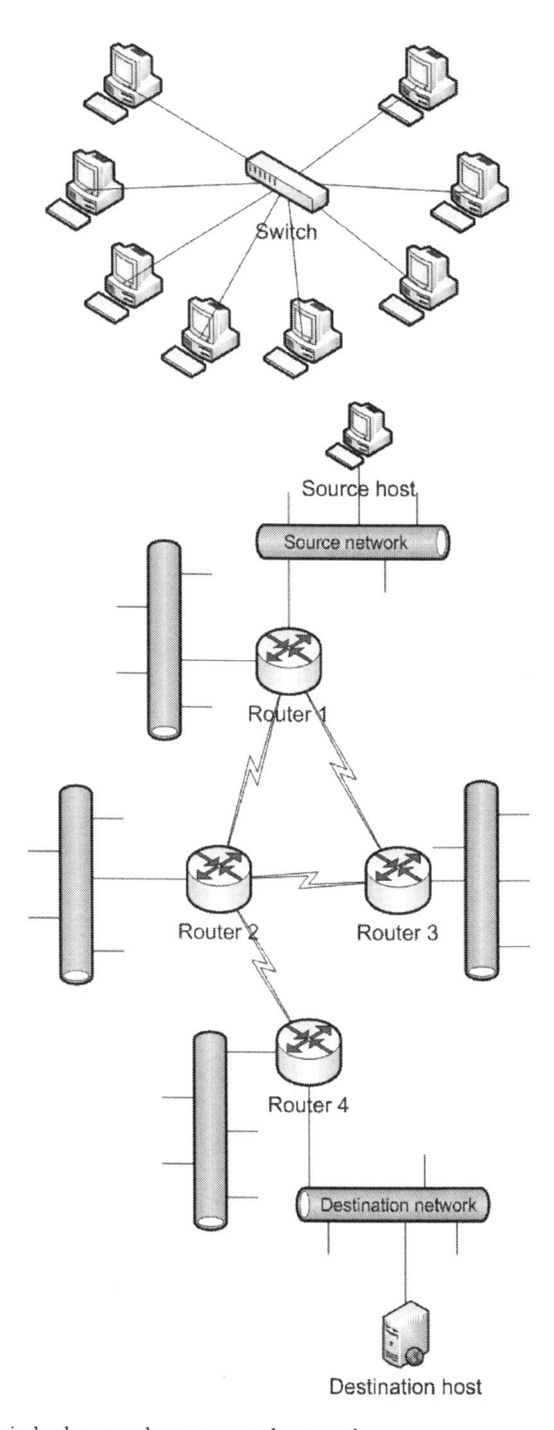

Figure 8.1: A switched network vs. a routed network

As seen in the figure, switches connect hosts within LANs and there is only one path from source to destination in a switched network.[1] Routers connect networks and there can be multiple paths from source to destination in routed networks. Hence switches have a relatively simple job—to select the correct path within a LAN. Routers, on the other hand, need to be capable of choosing the best possible path among multiple possible paths through networks across the globe.

Routers in Networks

Routing protocols help routers gather the required information to select a route. Routers save the gathered information in routing tables. In the rest of this chapter, we will see the contents of routing tables and other elements of routers. Let us begin by looking at the typical placement of a router in computer networks.

Routers are the glue that holds the Internet together. Without routers, each individual LAN and WAN would function very well internally, but we would have no way to guide packets across networks. Routers are placed at the interface between two or more networks and guide packets across networks in the right direction to their respective destinations. For example, ISPs place a router at the interface between your home network and the ISP's network. Figure 8.2 shows an example of one such router, which is located at the interface of two networks—USF and Bright House, an ISP serving the Tampa Bay area.

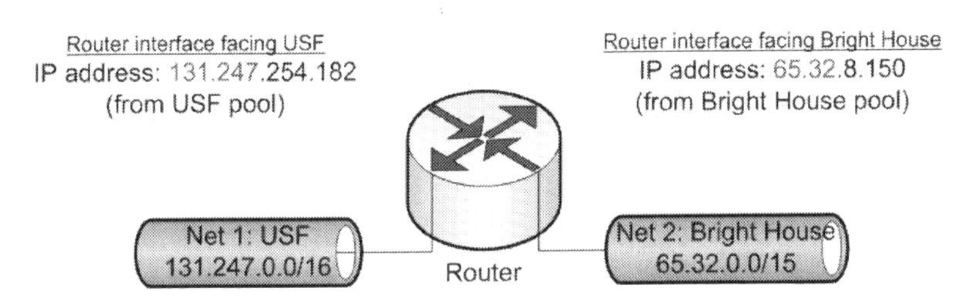

Figure 8.2: Router at the interface between USF and Bright House networks

Routers have at least two interfaces (connection points). In the example above, one of the interfaces of the router connects to the USF network and the other interface connects to the Bright House network. Each interface of the router gets an IP address from the network to which it is connected. Thus, the interface connected to Bright House has an IP address from the Bright House address pool, and the interface connected to the USF network has an IP address from the USF address pool. When packets arrive at a network interface, the router looks at the destination address of the packet, figures out the next hop for the packet, determines which of the connected networks contains the IP address of the next hop, and sends the packet out through the interface connected to that network.

1 Switches have evolved to the point where switches can also select from multiple available paths. But for this book, we limit our treatment to the basic capabilities of switches.

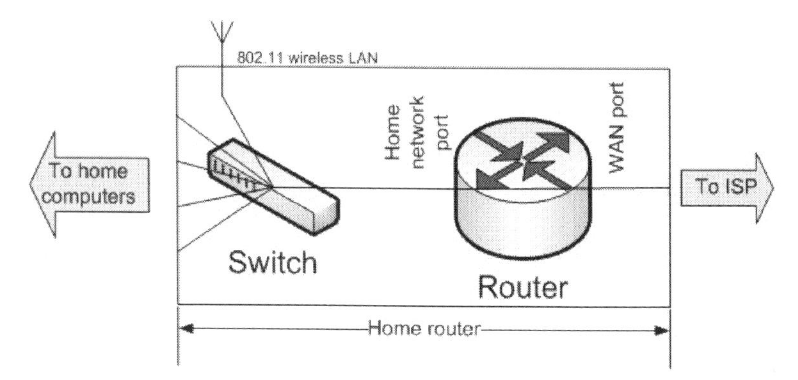

Figure 8.3: Home router

The router provided to you by your ISP also operates essentially the same way. It acts as the interface between two networks—your home network and your ISP's network. One of the ports of the ISP router acts as the interface with the ISP's network. This port is generally labeled "WAN" on most ISP routers (Figure 8.4). The other interface of the router connects to the home network. Generally, this port is internally connected to a switch as shown in Figure 8.3. The switch usually provides a wireless port and two to four ports to connect computers and printers in the home.

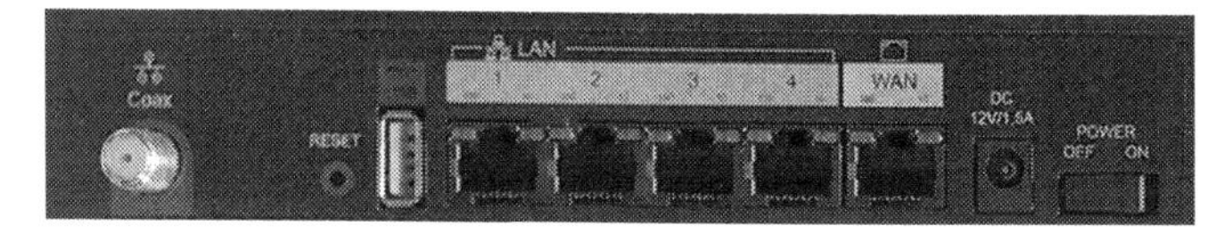

Figure 8.4: Typical wireless router ports, including WAN port

Autonomous Systems

Before we look at routing tables, there is one more concept to introduce—autonomous systems (AS). *An autonomous system is a collection of routers that fall under one administrative entity.* RFC 1930 defines an AS as a connected group of one or more IP prefixes (network IDs) that has a single and clearly defined routing policy. What this means is that once a data packet reaches an AS, the administrator of the AS guarantees the delivery of data to destination computers within the AS. Routers on the Internet only have to worry about getting data packets to the destination AS. The details of packet delivery within the AS are left to the administrator of the AS.

To use an analogy, post offices are a lot like autonomous systems. Each post office announces the zip code(s) for which it takes responsibility to deliver mail. We may treat the zip code served by the post office as its AS number. The USPS system knows that all it has to do to get a piece of mail delivered to a home address within a zip code is to pass the mail item to the post office announcing responsibility for the zip code. The USPS system in the rest of the country does not have to worry about how the post office internally manages its mail carriers, mail routes, new addresses, staff shortages, and the like. The post office will take care of the details of timely delivery within its area.[2]

Autonomous systems are the basic units of Internet routing. Having introduced the idea of autonomous systems, we can redefine routing as sending data from the source AS to the destination AS. In our example of Figure 8.2, both USF and Bright House are autonomous systems. Each AS has a globally unique AS number assigned by the IANA in a manner similar to the assignment of IP address blocks. IANA allocates AS numbers to the Internet registries, which in turn allocate AS numbers to organizations within their jurisdiction.[3] For example, the AS number for USF is AS 5661.[4] As of 2014, there are more than 47,000 autonomous systems in the global Internet.[5]

How are ASes relevant to routing? As we will see shortly, routes on the Internet are written as a chain of ASes. For example, the path from AT&T (AS 7018) to USF (AS 5661) through Cogent (AS 174) is written as 7018 174 5661.

Peering between autonomous systems

Autonomous systems connect to each other on mutually agreeable terms to exchange data traffic, usually without any payments. This is called peering. You can search for the peering relationships for any network at the peering database, located at https://peeringdb.com.

For an example of the motivations for peering, please take a look at Facebook's peering page at https://www.facebook.com/peering/. Facebook's peering policies are aimed at improving user access to its resources.

Now that the key elements of routing have been introduced, let us take a peek inside the Internet to see some routes between networks. The simplest way to do this is to use BGPlay, a utility developed at the University of Oregon and Roma Tre University in Italy to display changes in network connectivity around a specified network.[6] The network neighborhood around USF (prefix 131.247.0.0/16) looks as shown in Figure 8.5. You may note that in the figure the networks are identified by their AS numbers. Each link (dotted line) in the figure represents a connection between two ASes through a router. Packets use these connections to traverse from any AS to any AS. Routers enable this transfer by maintaining connectivity information between ASes on the Internet in a routing table.

---

2 The analogy of AS with post offices is not perfect, though. All Internet traffic moves through ASes. However, once the source post office (PO) collects all the mail to be delivered to other post offices, the USPS has a distribution system that bypasses post offices and uses aircraft, subcontractors, and other mechanisms to directly deliver the mail to the destination PO, where it gets delivered. If the analogy between post offices and ASes were perfect, mail would be handed from PO to PO until it reached the destination PO.

3 http://www.iana.org/assignments/as-numbers (accessed Nov. 30, 2015).

4 AS numbers can be found from www.cidr-report.org/as2.0/autnums.html.

5 https://en.wikipedia.org/wiki/Autonomous_system_(Internet).

6 BGPlay is accessed from https://stat.ripe.net/special/bgplay.

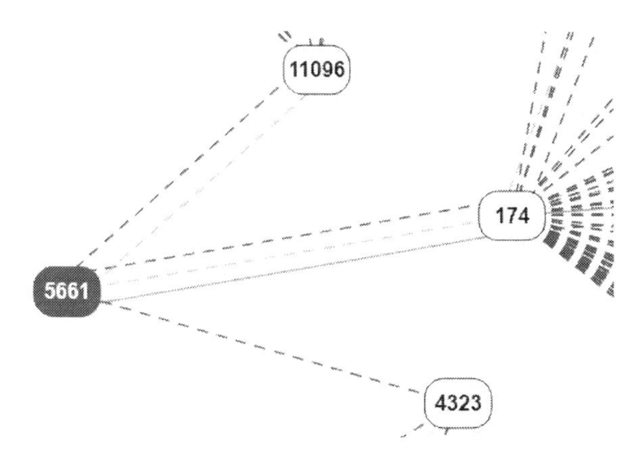

Figure 8.5: Network routes in the neighborhood of 131.247.0.0/16 (USF)

> Modern GPS systems have made it simple for drivers to navigate around obstacles. Routing made this feature available on the Internet from the very beginning.

Routing Tables

Routers help navigate data packets through networks. Routers do this by maintaining the network connectivity information available to them in the form of a routing table. *A routing table is a collection of paths that can be reached from the router along with information about the path.* As each packet reaches a router, the router looks at the destination IP address of the packet, and searches its routing table to find a suitable route to the destination. The router then forwards the packet to the next router on the selected path to the destination. The process is repeated at every router until the packet reaches the destination AS. Inside the destination AS, the local network delivers the packet to the destination computer.

The discussion above shows that each router is only responsible for passing the packet on to a suitable neighboring router. As long as the neighbors are correctly chosen, the packet will reach the destination. This process is very similar to the process of following a route map as discussed in Chapter 4. You follow the route map hop by hop until the destination is reached. As long as you get each hop right, you are assured of reaching your destination.

To enable routers to select the correct neighbor, each entry in the routing table lists a network that can be reached, the neighboring router to forward the packet to in order to reach this network, and details about the path to the destination through this neighbor. Since a router is only responsible for passing the packet to the correct neighboring router, this information is adequate for the router to make its forwarding decision. Consider the following route in the routing table (from the Route Views project):

```
131.247.0.0/16    64.71.255.61  0      812 174 5661 5661 i
```

This route shows that the network 131.247.0.0/16 (USF) can be reached from the Route Views network by sending the packet to the router at 64.71.255.61 (ARIN lookup indicates that this is owned by Rogers Cable Communications). The third column in the route indicates that a measure of the cost of sending the packet to the destination through this router is 0. This number is called the routing metric. Finally, the last column in the route is the actual path, which in this case is Rogers Cable (AS 812) → Cogent (AS 174) → USF (AS 5661).

Example routes to USF (131.247.0.0/16) from the Route Views project are shown in Table 8.1 for reference.[7]

Table 8.1: Part of routing table from Route Views project

Network	Next hop	Routing metric	Path
131.247.0.0/16	64.71.255.61	0	812 174 5661 5661 i
131.247.0.0/16	66.185.128.1	563	1668 174 5661 5661 i
131.247.0.0/16	217.75.96.60	0	16150 3549 174 5661 5661 i
131.247.0.0/16	208.51.134.246	13186	3549 174 5661 5661 i
131.247.0.0/16	12.0.1.63	0	7018 174 5661 5661 i
131.247.0.0/16	67.17.82.114	2503	3549 174 5661 5661 i
131.247.0.0/16	192.203.116.253	0	22388 11537 11096 11096 5661 i
131.247.0.0/16	203.181.248.168	0	7660 22388 11537 11096 11096 5661 i
131.247.0.0/16	64.57.28.241	1045	11537 11096 11096 5661 i
131.247.0.0/16	216.18.31.102	0	6539 11164 11096 5661 i
131.247.0.0/16	216.218.252.164	0	6939 11096 5661 i

One common feature of all the paths in Table 8.1 may be noted—they reach USF (AS 5661) either through Cogent (AS 174), or through FloridaNet (AS 11096). You can relate this to the network neighborhood around USF shown in Figure 8.5, which illustrates that USF is connected to the rest of the Internet through Cogent and FloridaNet. Hence, packets from the outside world can get to USF only through Cogent or FloridaNet. This is reflected in the routes in the routing table maintained at the Route Views project in Oregon.

7 If you have access to a UNIX/Linux system, it is easy to view the paths to any network from the Route Views router using the following steps:

a) $ wget http://archive.routeviews.org/oix-route-views/2009.08/oix-full-snapshot-latest.dat.bz2 (replace 2009.08 with the current year and month). This downloads the compressed routing table.

b) $ bunzip2 oix-full-snapshot-latest.dat.bz2. This uncompresses the table.

c) $ ggrep --after-context=10 <prefix> oix* (for example ggrep --after-context=10 131.247 oix*). This prints out the routes to the selected network.

As seen from the example paths to USF in Table 8.1, there are generally multiple paths available to get to any network on the Internet. Routers maintain records of all these available routes to networks. The routing metric in each route is a measure of the cost of sending packets through the route. Paths with smaller routing metrics are considered cheaper and are preferred to paths with higher routing metrics. Administrators can assign routing metrics to enforce any suitable policy on their network. The metric for a path is the sum of the metrics for the components of the path. An example routing metric could be hop count, since the total hop count will be the sum of the component hop counts.

We can put all these components together and see how routing works. The Internet is organized as a collection of autonomous systems. Routers connect autonomous systems to one another. Routers maintain a routing table that aggregates information about all known paths between autonomous systems. If one of these paths goes down, routers can pick one of the other available routes to deliver packets. This gives routers their resilience. Routers also periodically exchange information about these routes with each other, informing other routers about known changes to these paths so that other routers can update their routing tables.

When a packet arrives at a router, the router looks at its routing table to identify the best neighboring router to forward the packet to. The neighboring router then identifies the next best neighbor and so on, until the packet reaches its destination autonomous system. At the destination AS, the network uses whatever technology is available to direct the packet to the destination host.

Default routes

A default route is the destination to send packets to when the router has no route to the destination. For a home router, it is the first router at the ISP. Default routes offer many advantages. For one, the home router (or other routers using default routes) can get by with a minimal number of routing rules, sending all other packets to the default router. Further, default routes save the router from having to process route changes across the Internet.

Viewing Routes

All operating systems have utilities to help you see Internet routes from your computer to a specified destination. In Windows, this utility is called *tracert*. An example of the use of tracert is shown in Figure 8.6. The example shows a trace of the route from a PC at USF to the website of *The Wall Street Journal* (www.wsj.com). The trace shows that the path from USF to WSJ passes through Cogent and Sprint. The presence of Cogent on the path should not be surprising, given the information in Figure 8.5. The presence of Sprint Networks on the path (hops 7–11) indicates that the WSJ is connected to the Internet through Sprint. Hence, Cogent has to pass on the packets from USF to Sprint for onward transmission to the WSJ.

```
Windows PowerShell
PS C:\> tracert www.wsj.com

Tracing route to uslb.wsj.akadns.net [205.203.132.1]
over a maximum of 30 hops:

  1    <1 ms    <1 ms    <1 ms  vlan95.edu-msfc.net.usf.edu [131.247.95.254]
  2    <1 ms    <1 ms    <1 ms  vlan254.campus-backbone2.net.usf.edu [131.247.254.46]
  3    <1 ms    <1 ms    <1 ms  vlan256.wan-msfc.net.usf.edu [131.247.254.81]
  4     1 ms     1 ms     1 ms  gi1-5.ccr01.tpa01.atlas.cogentco.com [38.104.150.41]
  5    10 ms    10 ms    10 ms  te4-4.ccr01.mia01.atlas.cogentco.com [154.54.29.197]
  6     7 ms     7 ms     7 ms  te3-1.ccr01.mia03.atlas.cogentco.com [154.54.24.234]
  7    10 ms    10 ms    10 ms  sl-crs1-mia-.sprintlink.net [144.232.24.213]
  8    54 ms    38 ms    38 ms  sl-crs1-atl-0-0-0-0.sprintlink.net [144.232.18.216]
  9    42 ms    42 ms    41 ms  sl-crs1-nyc-0-5-3-0.sprintlink.net [144.232.20.49]
 10    40 ms    40 ms    40 ms  sl-gw35-nyc-14-0-0.sprintlink.net [144.232.13.37]
 11    44 ms    42 ms    42 ms  sl-dowjo-129545-0.sprintlink.net [144.232.234.142]
 12    42 ms    42 ms    42 ms  online.wsj.com [205.203.132.1]

Trace complete.
PS C:\> _
```

Figure 8.6: Example tracert output

Route print

You can view the routing preferences of your computer using the route print command. Figure 8.7 is the output on my machine.

At the heart of understanding routing is an understanding of how routers generate their routing tables. We have already seen how routing tables are used. Given the size of the Internet, it is clearly inefficient to expect network administrators to manually enter routes into their routing tables. Automated procedures are necessary to help routers exchange connectivity information with each other so that all routers on the Internet are informed about the connectivity of all other routers on the Internet. This information should change in a timely manner so that, as new routes are added and old routes are dropped, the changes are automatically propagated to all routers on the Internet. All these activities involved in maintaining routing tables are done using routing protocols.

Routing protocols are the mechanisms used by routers on the Internet to maintain routing tables. Details of routing protocols are beyond the scope of this book. However, it is useful to be aware of the broad outlines of how routing protocols work, which will be done in this section.

All routers are configured by their administrators with information about their immediate network neighborhood. The challenge for routing protocols is to propagate this local information about network connectivity to routers in other parts of the Internet. To consider a simple case, take the routed network shown in Figure 8.1. Here, router 4 is connected to the destination network, but is not directly connected to either router 1 or to router 3. The challenge then is, how can router 4 inform all other routers in the network that it is connected to the destination network and can deliver packets to the destination network?

Broadly speaking, there are two kinds of routing protocols to accomplish this task—exterior and interior.

```
Command Prompt
Microsoft Windows [Version 6.1.7601]
Copyright (c) 2009 Microsoft Corporation.  All rights reserved.

U:\>route print
===========================================================================
Interface List
 17...00 15 83 11 65 69 ......Bluetooth Device (Personal Area Network)
 13...00 0a cd 1d b2 80 ......Realtek PCIe GBE Family Controller
 11...f4 6d 04 56 20 a3 ......Realtek PCI GBE Family Controller
 20...00 50 56 c0 00 01 ......VMware Virtual Ethernet Adapter for VMnet1
 22...00 50 56 c0 00 08 ......VMware Virtual Ethernet Adapter for VMnet8
  1...........................Software Loopback Interface 1
 12...00 00 00 00 00 00 00 e0 Microsoft ISATAP Adapter
 14...00 00 00 00 00 00 00 e0 Microsoft ISATAP Adapter #2
 15...00 00 00 00 00 00 00 e0 Teredo Tunneling Pseudo-Interface
 19...00 00 00 00 00 00 00 e0 Microsoft ISATAP Adapter #4
 23...00 00 00 00 00 00 00 e0 Microsoft ISATAP Adapter #5
===========================================================================

IPv4 Route Table
===========================================================================
Active Routes:
Network Destination        Netmask          Gateway       Interface  Metric
          0.0.0.0          0.0.0.0   131.247.95.254  131.247.94.193     10
        127.0.0.0        255.0.0.0         On-link        127.0.0.1    306
        127.0.0.1  255.255.255.255         On-link        127.0.0.1    306
  127.255.255.255  255.255.255.255         On-link        127.0.0.1    306
     131.247.92.0    255.255.252.0         On-link   131.247.94.193    266
   131.247.94.193  255.255.255.255         On-link   131.247.94.193    266
   131.247.95.255  255.255.255.255         On-link   131.247.94.193    266
     192.168.38.0    255.255.255.0         On-link     192.168.38.1    276
     192.168.38.1  255.255.255.255         On-link     192.168.38.1    276
   192.168.38.255  255.255.255.255         On-link     192.168.38.1    276
    192.168.117.0    255.255.255.0         On-link    192.168.117.1    276
    192.168.117.1  255.255.255.255         On-link    192.168.117.1    276
  192.168.117.255  255.255.255.255         On-link    192.168.117.1    276
        224.0.0.0        240.0.0.0         On-link        127.0.0.1    306
        224.0.0.0        240.0.0.0         On-link   131.247.94.193    266
        224.0.0.0        240.0.0.0         On-link    192.168.117.1    276
        224.0.0.0        240.0.0.0         On-link     192.168.38.1    276
  255.255.255.255  255.255.255.255         On-link        127.0.0.1    306
  255.255.255.255  255.255.255.255         On-link   131.247.94.193    266
  255.255.255.255  255.255.255.255         On-link    192.168.117.1    276
  255.255.255.255  255.255.255.255         On-link     192.168.38.1    276
===========================================================================
Persistent Routes:
  None

IPv6 Route Table
===========================================================================
Active Routes:
 If Metric Network Destination      Gateway
 11    266 ::/0                     fe80::1
 11    266 ::/0                     fe80::2
  1    306 ::1/128                  On-link
 11     18 2607:fe50:0:6201::/64    On-link
 11    266 2607:fe50:0:6201:ccdb:619d:77b5:f0e3/128
                                    On-link
 11    266 2607:fe50:0:6201:f97f:9c64:7c7a:13c/128
                                    On-link
 11    266 fe80::/64                On-link
 20    276 fe80::/64                On-link
 22    276 fe80::/64                On-link
 20    276 fe80::a55b:f447:187f:7f4f/128
                                    On-link
 11    266 fe80::ccdb:619d:77b5:f0e3/128
                                    On-link
 22    276 fe80::e9f5:d9ea:cca2:58b6/128
                                    On-link
  1    306 ff00::/8                 On-link
 11    266 ff00::/8                 On-link
 20    276 ff00::/8                 On-link
 22    276 ff00::/8                 On-link
===========================================================================
Persistent Routes:
  None

U:\>
```

Figure 8.7: Route print output routing protocols

In exterior routing protocols, each router informs its immediate neighbors about the networks it can connect to. Routers collect this information from all their neighbors, aggregate it, and pass this information on to their other neighbors. For example, in Figure 8.1, once router 2 is informed by router 4 about a path to the destination network, router 2 can then pass this information on to router 1 and router 3 as a route advertisement. In routing parlance, *a route advertisement is information transmitted from a router to another router indicating that a specific network is reachable via a specified neighboring IP address.* Thus, router 4 advertises a path to the destination network to router 2. Router 2 then advertises this path to router 1 and router 3. Eventually, all routers on the network become aware of all routes on the network. The most popular implementation of an exterior routing protocol is BGP, or the Border Gateway Protocol.

BGP and China's 18-minute mystery[8]

On April 8, 2010, between 15:50 UTS and 16:08 UTC, China Telecom announced ownership of more than 50,000 IP address blocks, including networks from more than 170 countries. Since BGP essentially works on an honor system where announced routes are generally accepted as true, many ISPs added this information to their routing tables. The result was that some routes that normally took about 100 msecs took more than 400 msecs during this time. Such accidents happen routinely due to configuration errors, but this was probably the largest such incident.

In interior routing protocols, routers periodically broadcast information about their immediate neighborhood to all other routers on the network. Continuing with our example, if an interior routing protocol is used, router 4 will broadcast its connectivity to the destination network to all routers on the network. Once broadcasts from all routers are received, each router independently uses the information in all the broadcasts to compute the layout of the entire network. Thus router 1 will receive information from router 4 about a connection to the destination network. It will also receive information from router 2 about a connection to router 4. Router 1 already knows that it is directly connected to router 2. From these pieces of information, router 1 will infer that a path to the destination network exists through router 2 and router 4. The most popular implementation of an interior routing protocol is OSPF, or the Open Shortest Path First protocol.

Routing protocols and social networks

I like to point out the analogies between routing protocols and social networks. Think of Facebook as analogous to an interior routing protocol. Information posted by people is instantly visible to their entire social network. Face-to-face meetings such as coffee breaks and water-cooler conversations are analogous to exterior routing protocols.

---

8 Jim Cowie, "China's 18-minute mystery," Dyn Research, http://research.dyn.com/2010/11/chinas-18-minute-mystery/ (accessed Apr. 9, 2016).

People only meet a small number of neighbors at a time, and only get updated on the status of their larger network of contacts if one of these neighbors becomes aware of a change (did you know Alice and Bob are getting married?).

If you have ever been concerned about the amount of time you spend on Facebook, you are aware of the difficulty of processing status updates from the entire network.

Both interior and exterior routing protocols lead to the same routing table. The trade-offs are in speed of propagation and computational complexity. Exterior routing protocols can take time to propagate path information because the information has to be passed from neighboring router to neighboring router. Typically, routers send updates about once a minute, so it can be a few minutes before the entire network becomes aware of changes in remote networks. Interior routing protocols propagate information quickly because connectivity information is broadcast to all routers. However, the trade-off is that interior routing protocols can require more computation to process the information received from all the broadcasts.

BGP, DNS, and Internet censorship

In the 2000s there have been repeated reports of state-controlled authorities manipulating BGP and DNS to selectively block websites and general Internet access in parts of the affected countries. Some countries where such incidents have been reported are Syria and Libya.[9]

As the names suggest, interior routing protocols are generally used within autonomous systems and exterior routing protocols are used by ASes to exchange routes with other ASes. A summary comparison of these protocols is in Table 8.2.

Distance vector and link state

You are likely to come across these two terms in discussions of networks. They refer to the general principle followed to maintain routing tables in the two routing protocols we have discussed. Distance vector protocols such as BGP inform neighbors about the distance and path (vector) to accessible networks. Link-state protocols such as OSPF inform all routers about the state of a router's links to its own neighbors.

Distance-vector protocols are sometimes described humorously as rumor-based protocols. When a neighboring router tells you that it has a path to a destination network, you have no choice but to trust that when needed, your neighbor can indeed safely deliver packets to that destination.

9 https://en.wikipedia.org/wiki/Internet_censorship_in_the_Arab_Spring.

Table 8.2: Comparing interior and exterior routing protocols

Characteristic	Exterior routing protocols	Interior routing protocols
Update speed	Slow	Fast
Processing need	Low	High
Communication partners	Neighbors	All routers in network
Information exchanged	Global changes	Neighbors connected
Example protocols	BGP	OSPF

Routers also do not advertise routes that include RFC 1918 IP addresses to other routers outside the enterprise. All routers are aware that since these addresses can be reused, they may not be unique on the Internet and are not useful for routing. Since packets cannot be routed to IP addresses defined in the RFC 1918 address pool, these addresses are also called non-routable addresses or private addresses.

Simplifying Routing Tables—Route Aggregation

As the Internet has grown in popularity, more and more organizations have come online, resulting in an increasing number of IP address blocks being advertised. Clearly, if not managed properly, routing tables can get quite large when all address blocks are added to routing tables. For example, the routing table downloaded from the Route Views project is more than 900 MB in size.

For each packet that arrives at a router, the router has to search through this vast 900-MB database to select the most suitable route to forward the packet. The tables keep growing in size as new organizations join the Internet. As routing tables get larger, searching the routing tables for a path can take longer and longer. Alternately, to keep the search time low, routers can get more and more expensive as they add hardware to speed up the search.

There is, therefore, an effort in the Internet community to reduce the size of routing tables. This is done through route aggregation, which is specified in RFC 1338 and RFC 1518. You may recall from Chapter 7 that RFC 1518 also specified CIDR addressing. The goal of route aggregation is to reduce the number of routes advertised by networks.

Route aggregation is the combination of two or more IP address blocks into one larger address block. The idea behind route aggregation is to stop assigning IP address blocks to small organizations and, instead, assign larger blocks of IP addresses to large network service providers. Small organizations would be asked to obtain address blocks from network service providers. The key idea is that network service providers would take care of routing details to smaller organizations within their own networks. Routes to organizations within the network service provider would not be advertised to the outside world. If successful, routing tables would only record routes between large ISPs. Let us look at an example of this, in Figure 8.8 and Figure 8.9.

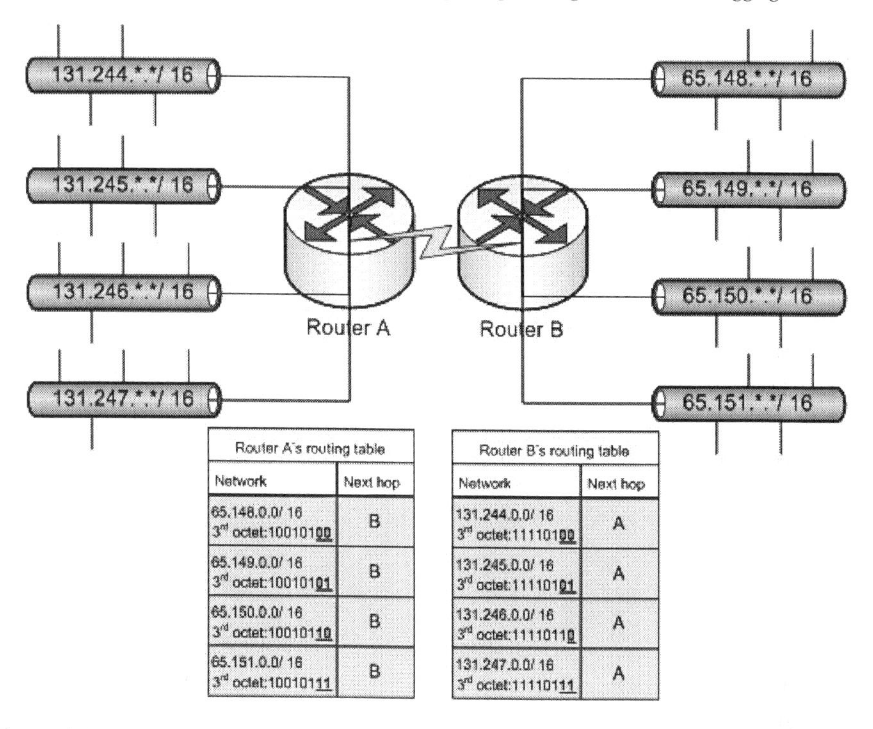

Figure 8.8: Routes without route aggregation

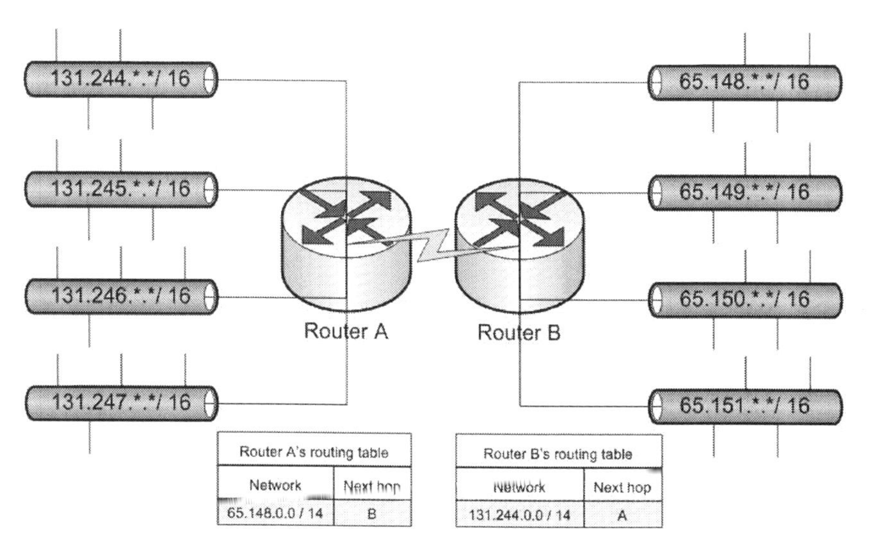

Figure 8.9: Routes with route aggregation

The example shows the interface between two networks. Both networks have four constituent /16 networks. Each of these networks has $2^{16} = 65{,}536$ hosts/network. Without route aggregation, the routing tables at each of the two routers will record four routes as shown in Figure 8.8, one for each constituent /16 network on the other side. The routing table of router A then has four routes, one to each /16 network connected to router B. Observe that since all the constituent networks are connected to the other networks through the same router, the next hop information in all the four routes in each router is the same. For example, all the four routes in router A point to router B as the next hop. Clearly, there is redundancy in the information and some optimization is possible. Can we condense the four routes to just one entry in the routing table?

Indeed, it is possible. Pay close attention to the last 2 bits in the third octet, shown in Figure 8.8. We know from the binary numbers overview that there are four possible combinations with 2 bits. From Figure 8.8, we can see that regardless of the combination of these bits, the outcome is the same—sending the packet to router B. This means that these 2 bits have no impact on the routing decision, and can therefore be treated as being in the host part of the IP address. The network part of the IP address will then only have 14 bits (16 - 2), and this network will have $2^{(32-14)} = 2^{18} = 262{,}144$ hosts. So for routing purposes, we replace four smaller networks with 65,536 hosts each, with one network that is four times larger.

By using this insight, the routes in the same network with route aggregation are shown in Figure 8.9. This time, the four routes on each side are compressed to one /14 route.

How can one aggregated route consolidate four un-aggregated routes? It may be noted that each route in the aggregated routing table points to a /14 network, whereas the routes in the table without aggregation point to /16 networks. Recall from our discussion of CIDR in Chapter 7 that in a /14 network, the network part of IP addresses is 14 bits long. The remaining 18 bits (32 - 14) are used to identify hosts within the network. This allows a /14 network to have 2^{18} computers within the network. By comparison, a /16 network can accommodate 2^{16} computers. Thus, the 2 extra bits in the host part of the /14 network allow the /14 network to be four times as large as the /16 networks ($2^{18}/2^{16} = 2^2 = 4$). The routes in Figure 8.9 point to networks that are four times as large as the networks in Figure 8.8, and the two routing tables are equivalent. In general, the routes in the aggregated routing tables point to larger networks. Since the total number of IP addresses is fixed (2^{32}), if we distribute these addresses among larger networks, we will have fewer networks. This will reduce the size of routing tables because routing tables keep track of networks, not individual IP addresses.

Figure 8.10 shows another example of how a network service provider may hide the details of its internal structure and only advertise its entire address block to the outside world. In the example, the 38.0.0.0/8 network has many smaller networks within it. However, the network hides the details about its interior structure and only advertises one route to the outside world. This spares the rest of the world from having to record paths to each network within the 38.0.0.0/8 network. These internal paths are irrelevant to routers outside the 38.0.0.0/8 AS if the 38.0.0.0/8 network acts as an AS and takes care of routing within its boundaries.[10]

---

10 The website www.cidr-report.org provides a wealth of information relating to route aggregation, including the biggest ASes that could meaningfully reduce the size of the global routing tables if they aggregated their routes.

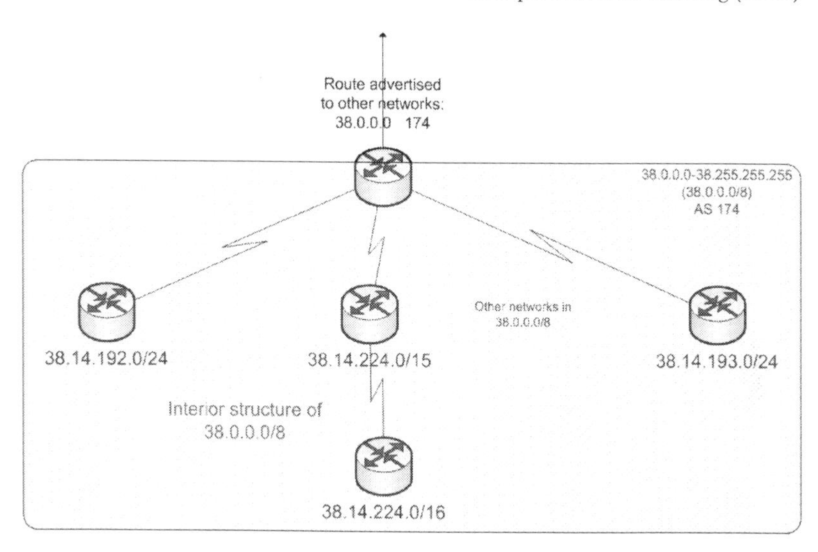

Figure 8.10: Example of advertisement of aggregated route

Multiprotocol Label Switching (MPLS)

Route aggregation, discussed in the previous section, is one of the methods by which attempts are being made to simplify routing. In recent years, a second method to simplify routing has become increasingly popular among carriers. This is called Multiprotocol Label Switching, or MPLS. *MPLS is a packet-forwarding mechanism that uses predefined labels to determine how to deliver packets.* The technology eliminates the need to examine IP header information and has been specified in RFC 3031, which was published in 2001.

The development of MPLS is motivated by the observation that whereas routers have only one job—selecting the next hop to forward the packet—the packet header that is processed by routers contains a lot more information than is needed to do the job. For reference, the IP header introduced in Chapter 4 is shown in Figure 8.11. Of all the fields

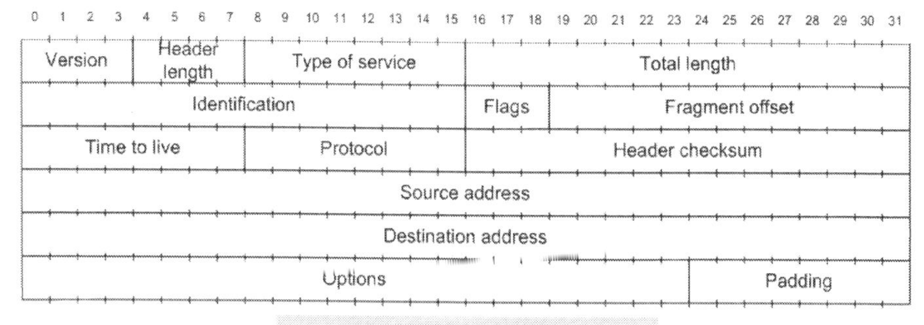

Figure 8.11: IP header

in the header, the only information field that is required to make the routing decision is the destination address field. Processing other header fields such as TTL puts an unnecessary processing load on routers. What if we could find a way so that only the destination address field was presented to routers for processing? This would simplify the processing required at each router, potentially simplifying routers.

Also, in traditional routing, as a packet makes its way to the destination, each router along the path will independently make a routing decision to forward the packet. A lot of this processing can be redundant. For example, consider two students at Michigan State University, browsing the websites of two schools in the Golden State—University of San Diego (sandiego.edu) and San Diego State University (sdsu.edu). The schools are located within 10 miles of each other. Looking up the network neighborhood of the University of San Diego (CIDR 192.55.87.0/24) at BGPlay, we see that sandiego.edu is actually connected to the Internet through the California State University network (AS 2152), which also includes sdsu.edu (Figure 8.12).

Given this network arrangement, consider how the two packets flowing from MSU to the two schools will be processed by routers along the path. Both packets will essentially follow the same path almost to the very end. The path to SDSU is (MSU → Internet 2 → California State University Network → SDSU) and the path to USD is (MSU → Internet 2 → California State University Network → USD). In traditional routing, each router on the path will independently make a routing decision on each packet. If one packet is sent to each of the two networks, assuming three routers handle the packet within each AS, a total of 24 routing decisions have to be made to deliver the two packets (three routers/network * four networks * two packets).

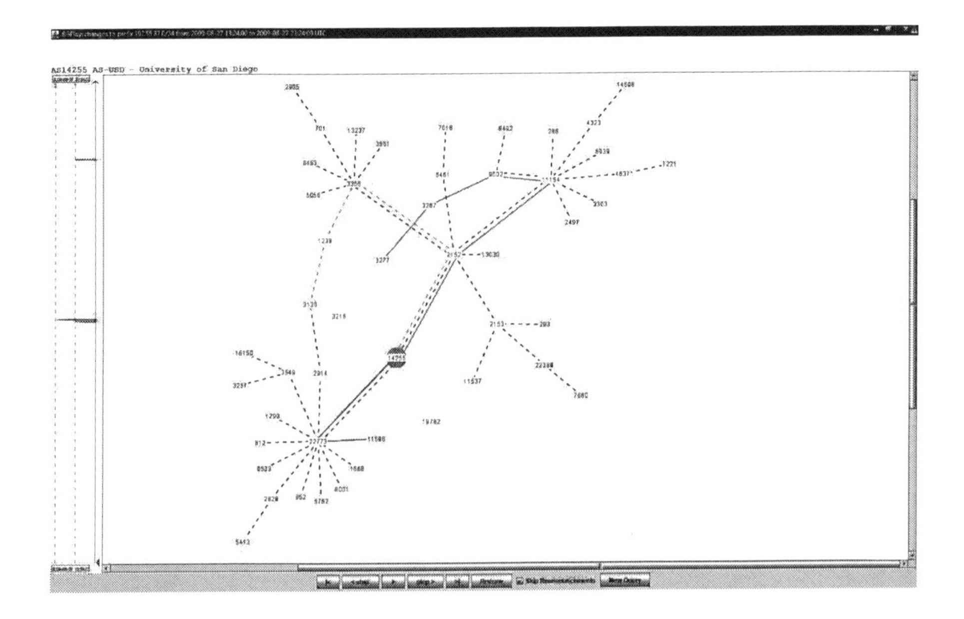

Figure 8.12: Network neighborhood around sandiego.edu (AS 2152)

This process appears inefficient and seems to require too much processing by routers. What if we could find a way to tell all the routers in each network on the path that the two packets are going to take the same route? Even better, what if we could make the routing decision just once near the entry of the packet into the network and give all the routers in the network an instruction such as "send these packets by route 5"? The routers in the rest of the network would not have to make any routing decisions at all. The 24 routing decisions in the previous paragraph could be replaced by eight routing decisions (one decision per network * four networks * two packets).

MPLS addresses these two problems—too much header information to process by routers, and too much processing by each router on the paths. To address the second problem, MPLS defines the concept of forwarding-equivalence classes. *A forwarding-equivalence class is a group of IP packets that are forwarded in the same manner, for example, over the same path.* In MPLS, the forwarding-equivalence class is called a label. All packets with the same label are handled identically. To address the first problem of too much packet overhead to process, MPLS adds the label to the packet. Routers in the network only look at the label to determine the next hop for the packet. MPLS routers maintain a forwarding table to determine the next hop for packets from the MPLS label of the packet.

The concept of a forwarding-equivalence class makes MPLS very useful for commercial carriers. It allows for the creation of private networks on top of a shared ISP infrastructure. Packets from different customers can be given different labels and be treated differently, even if they share the same source and destination. MPLS can also natively transport any number of protocols (ATM, Frame Relay, T1s, etc.) over a shared infrastructure, thereby providing connectivity between sites over these same diverse protocols.

If routers in the MPLS network still have to look up a table to forward packets, how is packet forwarding in MPLS simpler than traditional IP routing? The key difference is that in traditional routing, routers gather information from all routes advertised by all networks on the global Internet. This can lead to large routing tables. In MPLS, the labels are local to each network. Routers only know how to forward packets with a small set of labels. There are no global route advertisements, and administrators configure the forwarding table in each MPLS router. The forwarding table in the MPLS router shows the next hop for packets that arrive with allowed labels.

Thus, in MPLS, the routing decision takes the form of a labeling decision. The labeling decision is only done once on a packet, when the packet enters the network. The label is added to the packet at the point of entry into the network after the labeling decision is made. The process is shown in Figure 8.13. At every switch within the network, the MAC header is removed and the MPLS label is read to determine the next hop. The IP header has no role within the MPLS network. When the packet exits the network, the MPLS label is removed. If the packet enters another MPLS network, the label switch router of the second network attaches the MPLS label appropriate for that network.

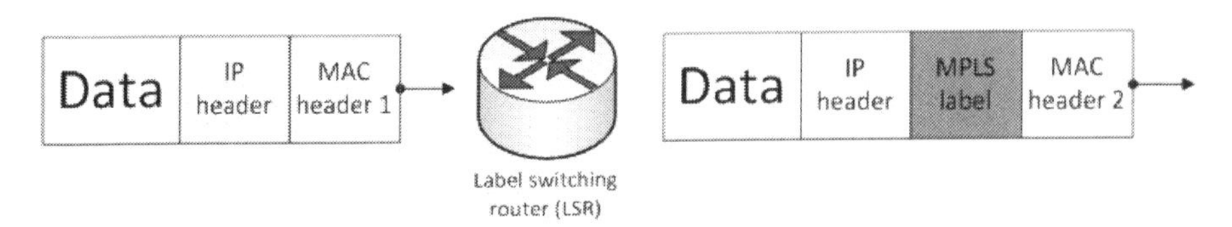

Figure 8.13: MPLS labeling

Throughout the packet's path through the network, only the MPLS label is used for packet forwarding. MPLS labels are local to a network; when the packet leaves a network, the MPLS label that was added to the packet on entry into the network is removed. If the packet now enters a second MPLS network before reaching the final destination, the second network may add a fresh MPLS label to the packet upon entry into the second network and remove this label when the packet leaves the second network.

MPLS and the daily commute

Let's look at an analogy between MPLS and the daily commute. When you drive in to work or to school, you have numerous route choices you can make. For example, when you leave your home, you can turn either left or right. You can take the interstate or a local road. Each of these is a routing decision. While any of these paths will get you to your destination, on most days you use a preferred route that you know usually works well. Within this route, you reflexively make the correct turns at each traffic light, without help from a GPS or other aid. You might describe this route to a friend as, "I get to work using I-275N." Your friend will easily infer all the turns and exits along the route from this simple route label. Using this preferred route also eliminates the need for you to make daily choices at every turn.

This is the MPLS philosophy—eliminate the need to make choices at each node. The traditional router network is designed to make explicit routing decisions at each node. The label is the network administrator's assignment of the preferred route at each node for all packets of a forwarding-equivalence class. Switches within an MPLS network do not have to compute routes.

OpenFlow—Software-defined Networking

Our discussion of routing has focused thus far on the merits and implementation of decentralized routing, where each router independently determines which of its neighboring routers to forward an incoming packet to. The discussion of routing protocols described how routers maintain information internally about network connectivity across the Internet. For the general end-user context, where users visit Internet resources at random, this is a wonderful way to manage network traffic, since this network is capable of handling any kind of traffic thrown at it.

However, these adaptive capabilities have costs. One of these costs is the addition of functionality within the router to manage routing tables. In addition to forwarding packets, routers also have to maintain their routing tables by periodically exchanging routing table information with neighbors. Another cost is that routers cannot exploit all available information about network capacity limits and traffic priorities. Routers essentially treat all Internet traffic as having the same priority, and higher-priority traffic can get stuck behind lower priority traffic, especially during busy periods. To minimize the adverse impacts of these concerns, most network providers ensure that networks operate at about 50% of their capacity, so that all traffic can get through without delay. However, this also means that most of the time, the hundreds of billions of dollars invested in network capacity are only used at about 50% efficiency.

OpenFlow is a technology that centralizes the maintenance of routing tables in a network. It was developed and is being improved at Berkeley, Stanford, and other universities and companies.[11] Centralization of the routing table allows network operators to more easily prioritize traffic on the network, which can be especially useful in important IT functions such as backups. These operations generate large volumes of data, but can be scheduled and prioritized by the network operator. Since the data sizes, end-to-end routes, and other parameters of the flow are known in advance, operators can plan for networks to operate at almost 100% capacity using simpler switches than traditional routers.

Google was one of the early adopters of OpenFlow. The company operates two different networks. The first network connects its data centers to users of its services such as search, Gmail, and YouTube. The second network is its core network that connects its data centers together to backup e-mail, transfer indexes of web content, and perform other such core operations. The traffic in this second network is large enough to qualify as one of the largest ISPs in the world. Google now almost exclusively uses OpenFlow for data transfer in its core network and reports almost 100% utilization on this network.[12]

Routing as a Metaphor for the Aerotropolis[13]

Throughout this book, we have used analogies from the road and air transportation networks to explain concepts in data networks. But things are now coming full circle in the business world, where concepts from data networks are inspiring the design of cities and traditional transportation networks. One of these developments is the Internet of Things (IoT). Another is the Aerotropolis, a city planned around its airport. More generally, an Aerotropolis is a city that is less connected to its land-bound neighbors than to its peers thousands of miles away. Examples include Dubai, Singapore, Chongqing (China), and Songdo (South Korea). Memphis, Tennessee, is an example in the US, with FedEx using that location as an operational hub.

Just as high-speed data networks have made cloud computing (Chapter 12) possible, high-capacity and dense air networks have made the Aerotropolis attractive to professionals and city planners alike. In its idealized form, the Aerotropolis offers simplified legal and tax regimes, arrays of customizable office parks, convention hotels, factories, and cargo complexes, all built around a high-volume airport at the center through which traffic flows 24/7. Using the

11 http://onrc.berkeley.edu.

12 Steven Levy, "Going with the flow," *Wired*, Apr. 17, 2012.

13 Greg Lindsay, "Cities of the sky," *Wall Street Journal*, Feb. 26, 2011.

routing analogy, the Aerotropolis is seen as a node in a global economic network that moves people and goods instead of data.

While the cities in the ancient world were built around waterways, at some point in your career you may find yourself working in an Aerotropolis, thousands of miles away from friends and family.

EXAMPLE CASE—Disasters, Katrina, and 9/11

In this book, we have focused almost exclusively on the efficiency of modern data networks. We have studied how breaking data down into small packets and interleaving data streams from multiple users allows a given network infrastructure to serve more customers. In this case, we see how the redundancy provided by routing can offer another benefit. Properly designed Internet-based networks can be more resilient to disasters than traditional circuit-switched phone networks.

Hurricane Katrina attained Category 5 status on the morning of August 28, 2005, and reached its peak strength at 1:00 p.m. CDT that day. The maximum sustained winds from the hurricane reached 175 mph. The hurricane had a devastating impact on communications networks in the Gulf Coast region. More than 3 million customer telephone lines lost service. More than 1,000 cell sites lost functionality. Thirty-eight 911 call centers went down. Approximately 100 television broadcast stations were unable to transmit. Hundreds of thousands of cable TV customers were affected.

Immediately after Hurricane Katrina struck land, the city government of New Orleans was operating out of the Hyatt hotel. Fifteen of the top officers of city government, including the mayor and the chief of police, had camped in a conference room at the hotel since it appeared to be the safest refuge from the storm.

Amidst the carnage, these people found that the Internet was the only communication service still working. The phones were dead. Cell phones were not working. The group did have satellite phones for emergencies such as this, but the phone batteries rapidly lost charge and could not be recharged. Fortunately, one of the members of the technology team, Scott Domke, had recently set up a phone account with Vonage, a nationwide voice-over-IP service provider. Just past noon on August 31, the mayor's team used this VoIP service to make its first call to the outside world in two days. A few hours later, President George Bush contacted the mayor at the same VoIP number from Air Force One.

The Katrina panel established by the FCC also noted the fragility of traditional (circuit-switched) voice networks and lauded the robustness of Internet technologies such as VoIP and text messaging.

9/11

The reliability of the Internet was tested in another major disaster, 9/11. Due to the importance of the Internet, the Association for Computing Machinery (ACM) and the Computer Science and Telecommunication Board (CSTB) established an expert committee to report on the performance of the Internet on 9/11.[14] Overall, the committee found that

---

14 National Research Council, *The Internet Under Crisis Conditions: Learning from September 11* (National Academies Press, 2003).

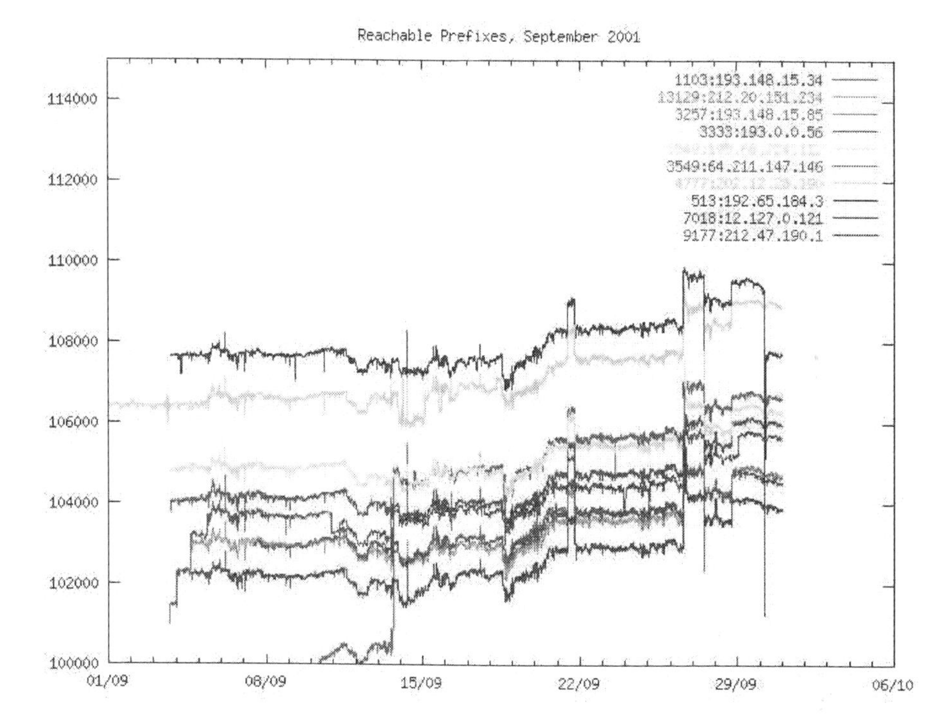

Figure 8.14: Reachable prefixes around 9/11
(Source: Renesys)

although 9/11 had only a marginal impact on the Internet, there were also some surprising effects around the world. Due to the unique structure of the Internet, the Nimda virus (released one week later, on September 18, 2001) caused more damage to the Internet than 9/11 (Figure 8.14). South Africa suffered more damage in Internet connectivity than New Jersey. The most significant issues arose from power outages in Manhattan.

The events of September 11, 2001, damaged the physical infrastructure of the Internet in one of the network's most important hubs—New York City. Verizon had a central office at 140 West Street, very near the World Trade Center complex. This office was destroyed. Electrical power in Lower Manhattan was disrupted, and backup power systems to telecommunications facilities in the area quickly ran out of fuel. At least 74 US and multinational telecommunications carriers have equipment in New York. The city is served by more than 100 international Internet carriers, and it has direct links with 71 countries. Many of these connections take place over the network of the local exchange carrier, Verizon, and the long-haul fiber networks pass through Verizon's central offices. Less than half a mile from ground zero, Telehouse operates an Internet exchange point on 25 Broadway, serving about 40 Internet providers from the NYC area, Europe, South America, and South Africa. About 70% of Internet traffic from Europe to the US passed through 25 Broadway.

In spite of this, serious impacts on communications networks were confined to New York City and some other regions that were highly dependent on NYC for their Internet connectivity. In some cases, automatic rerouting allowed Internet traffic to bypass many of the damaged areas. This was very different from the experience of other communications media such as the cellular phone services in greater New York, which suffered due to the infrastructure damage and congestion around NYC.

Internet routing, which dynamically adjusts the routes that packets follow in response to changes in the network (such as failures of communications links), can easily leverage redundancy. The redundant design of the Internet assisted the quick healing of the network on 9/11. When the New York Academy of Medicine lost its connection to the Internet, it found it relatively easy to restore Internet connectivity using a jerry-rigged wireless link outside a window to a nearby ISP. Some network operators used instant messaging and voice-over Internet Protocol (IP) to coordinate activities when telephone service was impaired through local damage to telephone circuits.

Interestingly, while the effects of September 11 on the Internet were largely limited to the immediate neighborhood of the WTC, some parties far from the physical disaster sites were affected. ISPs in parts of Europe lost connectivity because they interconnected with the rest of the Internet in New York City. Some Internet customers in Western New England could not dial in to their ISPs because the ISPs had located their modems in Manhattan. South Africa was largely cut off from the Internet because of its dependence on Domain Name System (DNS) resolution from servers in New York City. Users in South Africa had difficulty resolving domain names ending in .za, the top-level domain for South Africa, in the days following September 11. As a result, they could not access Internet services (such as web servers) within the country, even though there were no physical network disruptions in South Africa itself at that time. In fact, of all Internet users, those in South Africa were the most seriously affected by the incidents of 9/11.

The Internet did not suffer the kinds of overloads that are often associated with the telephone system in a time of crisis. The events of 9/11 do not indicate how the Internet may perform in the face of a direct attack on the Internet itself, because many ISPs had not concentrated their facilities on 140 West Street. However, the performance of the Internet on 9/11 demonstrated the benefits of the Internet's inherently flexible and robust design and its overall resilience in the face of significant infrastructural damage.

Internet Traffic Patterns on 9/11

While not directly related to routing, the National Academy report on 9/11 also noted how Internet traffic patterns were affected on 9/11. A very interesting part of the report details the steps taken by content providers to respond to the event.

Internet traffic volumes were somewhat lower on September 11 than on a typical business day. Many people who would normally have been using the Internet turned to the television for news and to phone calls for reaching loved ones. Traffic did increase in two areas—the quest for news and the use of Internet communications as a substitute for telephone calls. Low-bandwidth e-mail and instant messaging were used as substitutes for telephone service, especially where conventional telephone and cellular network congestion was high.

Internet-based communications alternatives such as text messaging and e-mail make more efficient use of limited communications capacity than do services such as telephones. By midday on September 11, the cellular phone networks in Manhattan were severely congested. Yet, there were reports that people who used their cell phones or wireless-equipped PDAs to send instant messages were able to communicate effectively.

People used the Internet very differently in the aftermath of the September 11 attacks. Less e-mail was sent overall. News sites and instant messaging were used more heavily. The overall conclusion was that individuals used the Internet to supplement the information received from television (which was the preferred source of news). Those unable to view television often substituted Internet news. The telephone, meanwhile, remained the preferred means of communicating with friends and loved ones, but chat rooms and e-mail were also used, especially where the telephone infrastructure was damaged or overloaded. Other activities on the Internet, such as e-commerce, declined. One consequence of this decrease was that in spite of larger numbers of person-to-person communications, total load on the Internet decreased rather than increased, so that the network was not at risk of congestion.

As a result of the unprecedented demand, news websites took a number of steps to enhance their ability to handle the traffic. Since the Internet technologies group at CNN was represented at the expert committee, CNN's experience, in particular, and the strategies it employed are documented in the National Academy report. On September 11, CNN's overall demand surged greatly, with page views increasing to 132 million—nearly 10 times the more typical load of 14 million on September 10. The number of page hits (pages or images requested) doubled every 7 minutes, resulting in an order-of-magnitude increase in less than 30 minutes. The demand for news continued to increase following the attack, reaching 304 million page views on September 12—more than twice that measured on September 11. CNN responded with a combination of several techniques to deal with the load.

Reduced web page complexity CNN had a strategy in place for dealing with high-demand periods that called for reducing web page complexity. Accordingly, on 9/11, the CNN.com main web page was significantly reduced in size by eliminating elements such as headline pictures and graphical menu bars. In fact, the main page was stripped down to the bare bones—even further than the usual minimum—to increase its ability to serve pages. At its minimum complexity, the CNN.com home page could fit into a single IP packet.

Adding more servers CNN's other servers normally used for other CNN and Turner Broadcasting content were experiencing significantly reduced volume that day. A number of them were reconfigured and added to the CNN.com server pool. Interestingly, CNN did retain server capacity for the Cartoon Network, which saw an increase in volume—likely reflecting parents' desire to provide children with an alternative to the disturbing news.

Temporarily employing a third-party content-distribution network CNN increased its use of the Akamai content-delivery network to reduce the load on the CNN servers themselves.

References

1. Federal Communications Commission. "Independent Panel Reviewing the Impact of Hurricane Katrina on Communications Networks: Report and Recommendations to the Federal Communications Commission." June 12, 2006.
2. National Research Council, *The Internet Under Crisis Conditions: Learning from September 11*. National Academies Press, 2003.
3. Ogielski, A., and J. Cowie. "Internet Routing Behavior on 9/11." Renesys Corp. Mar. 5-6, 2002.
4. Rhoads, C. "Cut Off: At Center of Crisis, City Officials Faced Struggle to Keep In Touch." *Wall Street Journal*, Sept. 9, 2005, A.1.
5. Von.org. "Benefits of VoIP," http://www.von.org/secpgs/02_benefits/benefits_01_ovrvw.html (accessed Apr. 9, 2016).

Summary

In this chapter we introduced routing and saw how routing is different from switching. We saw how the global Internet is organized as a mesh of autonomous systems. Routers are responsible for transferring packets from the source autonomous system to the destination autonomous system through intermediate ASes. Routers advertise known routes to all other routers using routing protocols. Routers save these advertised routes as routing tables. When a packet reaches a router, the router selects the best route from its routing table and forwards the packet to the neighboring router specified in the route. Passed from router to neighboring router, packets eventually reach their destination networks. At the destination network, Ethernet and other data-link layer technologies are used to deliver packets to the destination host.

As network traffic has grown, procedures and technologies have been developed to simplify routing. Route aggregation attempts to reduce the number of routes advertised on the Internet to reduce the size of routing tables. MPLS is a technology that attempts to reduce the number of routing decisions that have to be taken to deliver packets to their destinations. MPLS also reduces the amount of processing that has to be done at each router.

About the Colophon

Radia Perlman, the inventor of the spanning tree algorithm to find the shortest paths between nodes on a network, wrote this poem to describe the algorithm. In her book,[15] Dr. Perlman states that it took her more effort to compose the poem than to create the spanning tree algorithm.

REVIEW QUESTIONS

1. What is *routing*?

15 Radia Perlman, *Interconnections: Bridges, Routers, Switches, and Internetworking Protocols*, 2nd ed. (Addison-Wesley Professional, 1999).

2. LANs use broadcasting to ensure that data reaches its destination. Why is it not advisable to use broadcasting between LANs across the entire Internet?
3. What are *routers*? Find three carrier-grade router models made by the major vendors. What is the range of list prices on these routers? (Tip: routers are generally sold through value-added resellers. The simplest way to answer this question is to find three models from the website of a vendor and Google for the price of the model.)
4. What are the important similarities between switching and routing?
5. What are the important differences between switching and routing?
6. Consider a router at the interface of two networks, such as your university and its ISP. Draw a figure showing the IP addresses of the two interfaces of the router and the CIDR address blocks of the two networks. (You may need to use tracert to obtain the IP addresses at the two interfaces and arin.net to obtain the address blocks of the two networks. Tip: first tracert to your school to get the IP address on the ISP side. Second, tracert to this IP address from within your school to get the IP address on the school side. Finally, use ARIN to get the address blocks.)
7. What is an *autonomous system*?
8. How are post offices like autonomous systems?
9. Briefly describe how routing works.
10. What is a *routing table*? What information is stored in a routing table?
11. One route from the routing table at the Route Views project is shown below. What does each term in the route indicate?
    128.210.0.0/16 194.85.4.55 0 3277 3267 9002 11537 19782 17 i
12. Use the AS numbers website (www.cidr-report.org/as2.0/autnums.html) to find the names of all the autonomous systems referred to in Question 11.
13. What is a *routing metric*? How is the routing metric used to select the path when multiple paths are available?
14. Use the tracert utility to record the route from your home computer to your university's home page. (Many universities are locking up their routers as an information security measure. If so, report what you find or try a neighboring school.)
15. What are *routing protocols*?
16. What are the two kinds of routing protocols? Give an example of each.
17. Provide an overview of how exterior routing protocols work.
18. What is *route aggregation*?
19. Why is route aggregation useful?
20. Give an example of routes before and after route aggregation.
21. What is *software-defined networking*? Where is it most useful?
22. What is *MPLS*?
23. What is the motivation for the development of MPLS?
24. What is a *label* in MPLS? Where is the label attached to a packet? Where is it removed?
25. What is a forwarding-equivalence class (FEC) in MPLS? Give an example of two packets that may be assigned the same FEC even though they are addressed to two destinations.

EXAMPLE CASE QUESTIONS

1. Describe some of the damage caused to communications networks due to Hurricane Katrina.
2. Describe some of the damage caused to communications networks on 9/11.
3. Why is text messaging and VoIP more reliable than circuit-switched voice networks?
4. Name some leading Voice-over-IP service providers in the country. Compare their services—pricing, equipment required, etc.
5. What are content-delivery networks? How are they useful during disasters?
6. View the statistics on the nyiix website. What is the busiest time of day for Internet traffic? Why do you think traffic peaks at this time?
7. What were some changes in Internet traffic patterns on 9/11?
8. Why is it useful to reduce the size of a web page so it fits into one IP packet?
9. What is a possible extreme disaster that could strike your city? Summarize the key features of a disaster continuity plan for communication that will allow you to maintain communication with the two most important people in your life if this disaster strikes. What information will you provide each person in your plan? If you wish to maintain the privacy of these individuals, just call them A and B, or use other such anonymous names. If possible, test the plan and describe the challenges in implementation (e.g. availability of the selected technology and challenges in training people in using the technologies involved in your plan).

HANDS-ON EXERCISE—BGPlay

In this exercise, you will view the autonomous systems in the vicinity of your university using BGPlay. The Route Views project at the University of Oregon has direct connections to many major ISPs and collects routing information from around the world. This information is archived daily. When an IP address is provided as input to the BGPlay applet, the applet searches through this database of routes to provide a graphical view of the network connections around that IP address. Upon completing the exercise, you will get a visual picture of network routes and their volatility.

The BGPlay applet is available at https://stat.ripe.net/special/bgplay. This brings up the BGPlay query form. Upon entering the IP address in the form, BGPlay searches through its database and extracts all routes to the selected network and displays its results, as shown in Figure 8.15. You can click on the "play" button to see an animated view of path changes between the selected dates. You can also click on any AS icon to see the name of the network at the top left of the window.

Answer the following:

1. Write a brief summary of the goals of the Route Views project from the project home page at www.routeviews.org.
2. What is the network ID of the network to which your university belongs? (In some states, university networks are part of larger state networks.)
3. Show a screenshot (similar to Figure 8.15) of the BGPlay query results for your university's network.

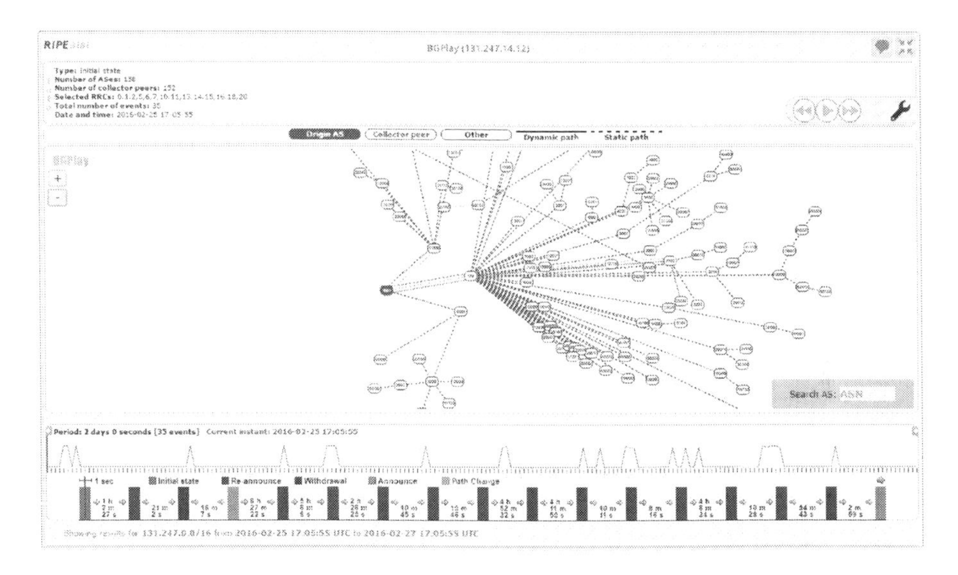

Figure 8.15: BGPlay query results

4. Pick a route originating from your university that passes through at least three autonomous systems. Write the route as a sequence of ASes (e.g. 2905 701 1239 19151 3851).
5. What are the network names of the ASes selected above?
6. Pause BGPlay during a route announcement. Show a screenshot as shown in Figure 8.15. What is the new route added?
7. Pause BGPlay during a route withdrawal. Show a screenshot. What is the route withdrawn?

CRITICAL-THINKING EXERCISE—Internet Censorship

Internet censorship is the control of what users can access on the Internet. Parental controls are a benign form of Internet censorship. But governments have used Internet censorship to block the flow of ideas within their country.

1. What would be some motivations for governments (or other institutions in positions of authority) to block Internet access?

CRITICAL-THINKING REFERENCES

1. https://en.wikipedia.org/wiki/Internet_censorship_in_Syria.
2. https://en.wikipedia.org/wiki/Internet_censorship_in_the_Arab_Spring.
3. Robertson, Jordan. "The Day Part of the Internet Died: Egypt Goes Dark." *USA Today*, Jan. 28, 2011.

IT INFRASTRUCTURE DESIGN EXERCISE—Failover

Failover is an information security control in which standby equipment automatically starts providing services when the main system fails. Failover is useful in keeping services operational. Although TrendyWidgets is not an AS, since it has chosen to maintain only one connection to the Internet, TrendyWidgets does use OSPF internally. It does this so that its routers can respond to network connectivity issues and direct traffic through the slower ISDN links in case the primary WAN links fail. This process is called failover. Answer the following questions:

1. To connect to the other locations, TrendyWidgets needs border routers at each location. Update the infrastructure diagram from Chapter 7 to include a border router at each of the four locations.
2. If you haven't already done so, update your infrastructure diagram so that the Internet connection from Tampa goes through the border router at Tampa.
3. Each location uses Ethernet to connect the various devices at the location to each other and to the Internet through the border router. Update the diagram to show an Ethernet at each location. Connect the DHCP server at each location to the Ethernet. Also, connect the Ethernet to the border router.
4. A consultant suggested that you run the commands below. Look up the documentation to the route command and describe what these commands do, and how they could be useful to you (make reasonable assumptions about the identities of router-net1 and router-net2).

```
route add host 192.168.1.2 router-net2 1
route add host 192.168.1.3 router-net1 1
```

Multi-home logistics

We will gloss over this in this text, but multi-homed sites such as TrendyWidgets typically need to have their own registry-assigned address space (not from their ISP), obtain an AS number, and run a routing protocol like BGP to communicate with their multiple ISPs.

CHAPTER 9

Subnetting

Nothing is particularly hard if you divide it into small jobs.

Henry Ford

Overview

Subnetting is the method by which network administrators divide the large number of IP addresses allocated to an organization into smaller address blocks called subnets. Each department within the organization can be assigned one or more of these smaller address blocks. Each department's network administrator can use these smaller address blocks to manage their networks. Subnetting is one of the core skills for all network administrators. At the end of the chapter, you should know:

- what subnets are and why they are useful
- what factors determine the size of subnets
- what subnet masks are
- what the relationship is between subnet masks and subnet size
- how to compute subnet masks

Subnets were discussed briefly in Chapter 4. However, given the importance of the topic, subnetting warrants its own chapter.

Why Subnetting

Subnetting helps organize IP addresses. *Subnetting is a way of breaking down large blocks of IP addresses into smaller address blocks.* We know from Chapter 4 that Internet registries prefer to allocate large blocks of IP addresses. For example, a large ISP may be allocated a /12 address block, giving the ISP about 1 million IP addresses ($2^{(32-12)} = 2^{20} = 1,048,576$ IP addresses).

This is a very large number of IP addresses to manage. DHCP servers managing so many IP addresses are likely to become bottlenecks for the ISP. Instead, if the ISP serves 20 markets nationwide, wouldn't it be convenient to divide the pool of 1 million IP addresses into 20 smaller blocks of about 50,000 IP addresses each (50,000 * 20 = 1 million)? Each block of

50,000 addresses could be used to serve one market and the network administrators in each market could be given the responsibility of managing their own pool of 50,000 IP addresses.

Subnetting is very flexible and allows organizations to delegate IP addresses as appropriate for the organization. As another example, say an organization has a /16 IP address block ($2^{(32 - 16)}$ = 65,536 IP addresses). Using subnetting, the organization may distribute these addresses as 256 subnets with 256 addresses/subnet (256 * 256 = 65,536). Alternately, the organization may choose to have 128 subnets with 512 hosts/subnet (128 * 512 = 65,536). Note that in each case, the total number of available IP addresses in the organizations is 65,536 (256 * 256 or 128 * 512). The difference is in the number of smaller address blocks that the available addresses are divided into.

Business Motivation for Subnetting

We will illustrate our discussion of subnetting with an example set in a university campus. Technically, this setting is very similar to that of a large business campus, but it has the advantage of being familiar to students who have limited real-world networking experience. Therefore, consider a typical state university. It is likely to have a /16 IP address pool such as 131.247.0.0/16. This means that it has $2^{(32 - 16)}$ = 65,536 IP addresses available. How can the network administrator at this university go about allocating IP addresses to individual computers from this large pool of available IP addresses?

A simple mechanism would be to use a first-come,-first-served scheme. If the university network uses DHCP, the first computer to come online on a given day could get the IP address 131.247.0.1. Let's say this is a student in the business school using a wireless laptop while working on a project past midnight.

The second computer that comes online could get the IP address 131.247.0.2. Let's say this is a computer being used by a student browsing on a kiosk in the library. Allocating a few more IP addresses on a given day, the university network might appear as in Figure 9.1.

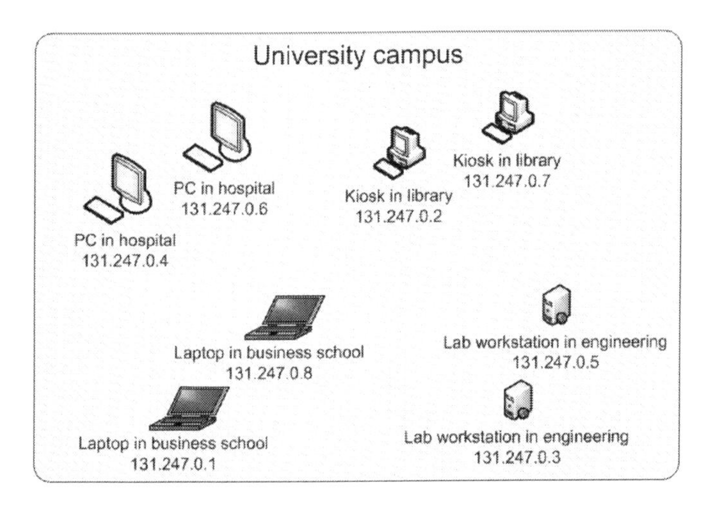

Figure 9.1: IP address allocation without subnetting

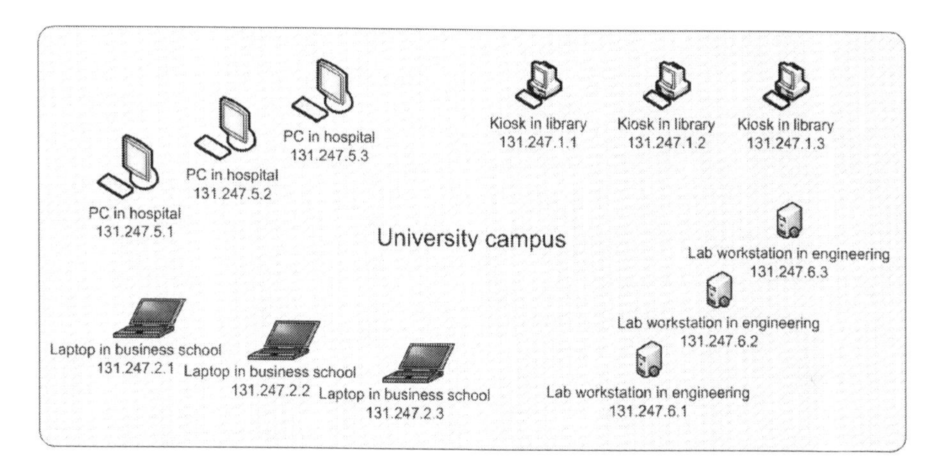

Figure 9.2: IP address allocation with subnetting

This IP address-allocation scheme is functional as far as network connectivity is concerned. Every host on the network will get an IP address and will be able to communicate with all other hosts within the university network and outside.

But think about the implications of the addressing scheme from the perspective of the network administrator. He will not have any idea where a host with a given IP address is located on campus. Therefore, if a user with a network connectivity problem calls the network administrator for help, the network administrator would have a very difficult time trying to troubleshoot the problem.

Also, most organizations break up large networks into smaller departmental networks with administrators responsible for the departmental networks. It is very convenient if these departmental IT managers can be given address blocks of their own to manage their departmental networks.

Now consider an alternate method to allocate IP addresses. In this method, each college is allocated a contiguous set of IP addresses. For example, the library is given IP addresses in the range 131.247.1.0–131.247.1.255; the college of business is given IP addresses in the range 131.247.2.0–131.247.4.255; the hospital is given IP addresses in the range 131.247.5.0–131.247.5.255; and the college of engineering is given IP addresses in the range 131.247.6.0–131.247.7.255. Figure 9.1 may be updated to reflect this departmental addressing scheme as shown in Figure 9.2.

Observe the pattern of IP addresses allocated to laptops in the college of business in Figure 9.2. Not only do they share the network part of the IP address (131.247), even the third octet in the IP addresses, 2, is common to all the laptops in the college of business. Similarly, the IP addresses of all the PCs in the hospital have three octets in common—131.247.5. The same pattern is followed in the college of engineering (131.247.6) and the library (131.247.1).

It is clear that a network administrator would much prefer the neatly organized allocation of IP addresses shown in Figure 9.2. If a user calls the administrator with a connectivity problem, the administrator can ask the user to provide their IP address (Start → `cmd` → `ipconfig/all`) and the administrator can immediately locate the computer on the network. If a network component such as a switch fails, all hosts within a department are likely to experience connectivity problems. Identifying patterns in the IP addresses of the disconnected hosts is likely to suggest the cause of the problem.

Therefore, subnetting is useful because it helps organize IP addresses within a network or organization. As the name suggests, a subnet is a network within another network.

Subnets—The Technical Motivation for Subnetting

The primary motivation for subnetting is to organize IP addresses within large organizations. This motivation is facilitated by the technical organization of large networks. We know from Chapter 3 and Chapter 4 that large networks such as campus-wide networks are composed of smaller local area networks, usually Ethernets. We have also seen in Chapter 3 that since Ethernet is a broadcast network, it has a practical limit on the number of computers per network. A university with more than 10,000 computers clearly needs many Ethernets. Its network is likely to be composed of many Ethernets, as shown in Figure 9.3.

When a large network is built up of smaller networks, each smaller network is called a subnet. Subnets are described in RFC 950. As defined in RFC 950, *subnets are logically visible subsections of a single Internet network*. In Figure 9.3, the university network may be said to be composed of the library subnet, the engineering subnet, the business school subnet, and the hospital subnet. Each subnet generally has all the support services such as DNS and DHCP required for network operation.

Subnetting allows network administrators to carve out small blocks of IP addresses from the organization's large address pool and assign these small blocks of IP addresses to the different subnets. Mapping the IP addressing scheme of Figure 9.2 to the network architecture of Figure 9.3 gives the subnet addressing scheme shown in Figure 9.4.

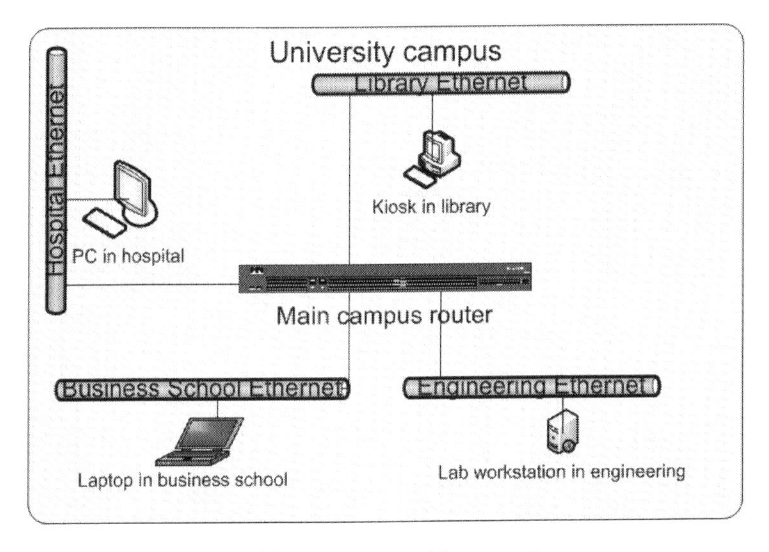

Figure 9.3: Internal structure of large campus-wide network

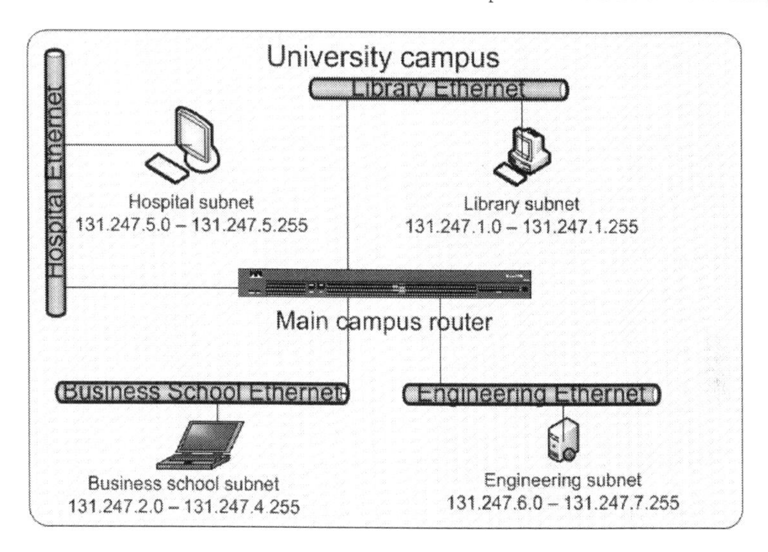

Figure 9.4: Subnet structure of a large network

Three-part IP Addresses with Subnetting

We saw in Chapter 4 that IP addresses have two parts: the network part and the host part. The network part identifies the network to which the IP address belongs, and the host part identifies the host within the network. This is shown in Figure 9.5. In this example, the first 16 bits of the IP address identify the network and the remaining 16 bits identify the host within the network.

With subnetting, the IP address shown in Figure 9.5 may be interpreted as shown in Figure 9.6. This time, instead of treating the host part of the IP address as one unit, we interpret it as being made up of two IDs—a subnet ID and a host ID. The first few bits of the host part identify the subnet to which the host belongs (e.g. library, business, hospital, etc.), and the remaining bits of the host part identify the computer within the subnet. Since IP addresses in subnetted networks are interpreted as having three parts—network ID, subnet ID, and host ID—they may be called three-part IP addresses.

Recall from the discussion of multi-part addressing in Chapter 4 that subnetting organizes IP addresses a lot like telephone numbers. As seen in Figure 9.7, telephone numbers have three parts. The first part denotes the metro area, the second part identifies the exchange within the metro area, and the last four digits identify the phone receiver within the exchange. Similarly, as seen in Figure 9.6, in an IP address, the network ID identifies the organization or ISP, the subnet ID identifies the department within the organization, and the host ID identifies the host within the department.

There is one major difference between phone numbers and subnetting. The difference is the varying lengths of the three IDs in subnetting. The phone system has fixed lengths for each part: the area code always has three digits, the exchange code is always three digits long, and the number is always four digits. But IP addresses can be broken at any bit boundary. Organizations can have /12, /13, or /16 network IDs. Subnet IDs can have

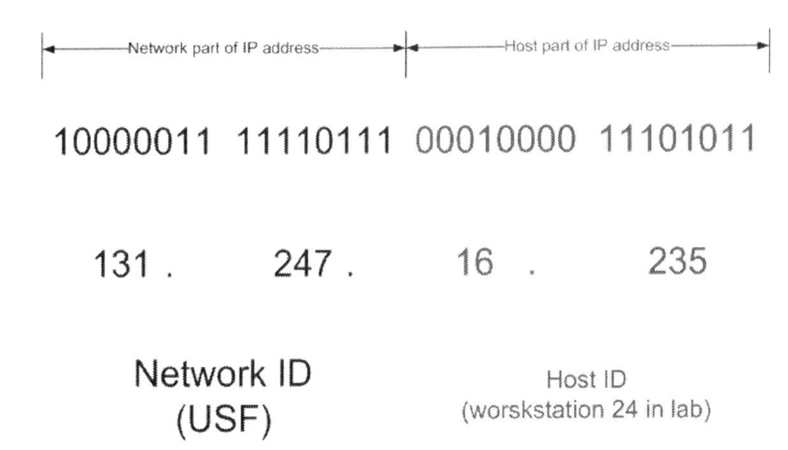

Figure 9.5: Two-part interpretation of IP address

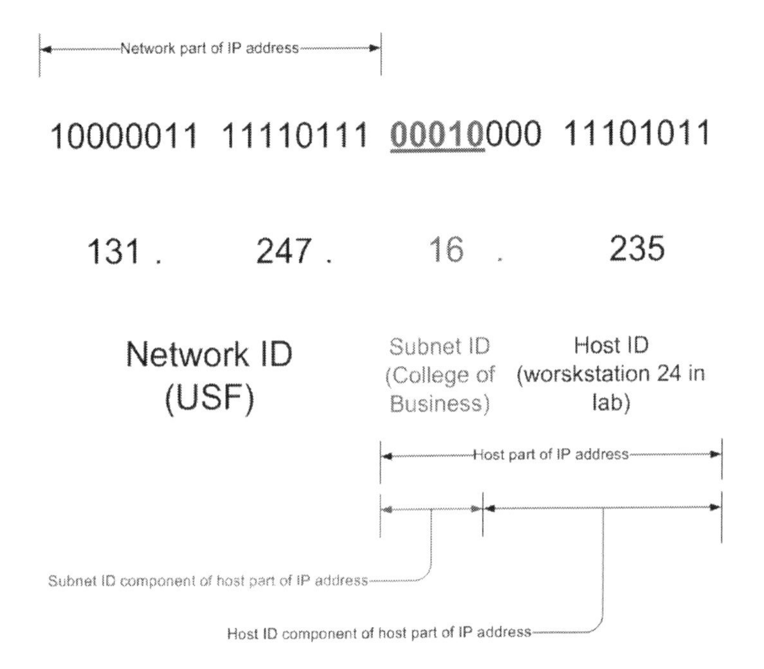

Figure 9.6: Three-part interpretation of IP address

5 bits, 8 bits, 9 bits, or 10 bits. These numbers are only illustrative. The 32-bit IP addresses can be partitioned as required between network ID, subnet ID, and host ID.

Subnetting the Network Address Block

Now that we know why subnetting is useful, we can take on the challenge of subnetting a network's address block. This is the part that appears mathematically challenging to students when they deal with it for the first time. We will proceed one step at a time. While the math can seem intimidating, subnetting essentially boils down to calculating the number of bits required to create the desired number of subnets.

An analysis of the business determines the number of subnets the organization needs. You then calculate how many bits are needed to label each of these subnets. The rest of this section describes the details of this process using the example of a university network.

In the standard subnetting scenario, the network administrator examines the technical and network structure of the organization to determine how many subnets the network should have. The network of a typical university is composed of subnets for each college. There are also some additional subnetworks, such as the dorm network, a subnet for the administration, subnets for branch campuses, and a subnet for campus IT. Add these up, and our example university may have, say, 16 subnets.

How many bits do we need to be able to assign a unique subnet ID to each of these subnets? We know from the binary numbers overview in Chapter 4 that with 4 bits, we

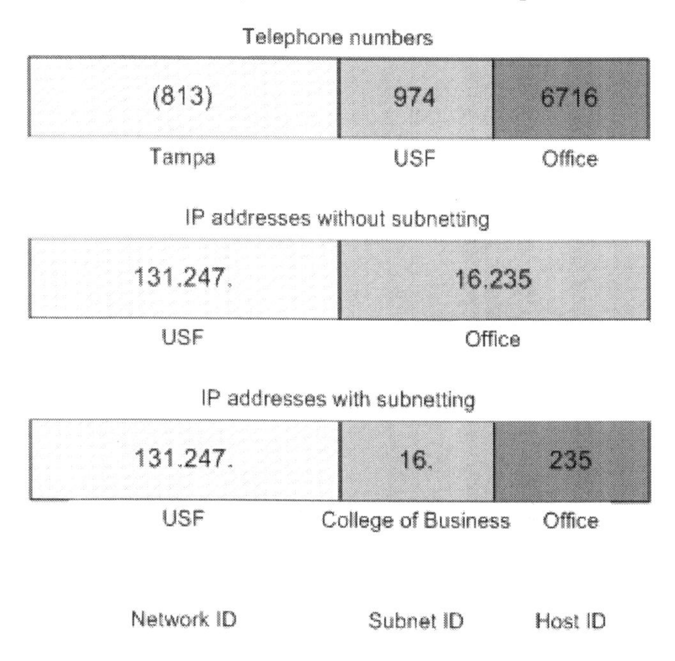

Figure 9.7: Similarities between subnetting and phone numbers

can have 16 (2^4) subnets. While this may meet the immediate needs of the university, what happens if a few new units are created on campus? If we provide for 16 subnets when our current needs are for 16 subnets, we limit our room for growth in the future. Therefore, in this situation, most network administrators will prefer to plan for 32, or even 64 subnets. Thus, using 5-bit ($2^5 = 32$) or 6-bit ($2^6 = 64$) subnet IDs seems like a good choice for this university.

Say we choose 5-bit subnet IDs. If we use 5-bit subnet IDs, one possible numbering scheme for the 32 subnets of the university is shown in Table 9.1. Subnet IDs are shown as 5-bit binary numbers. The numbers in parentheses are the decimal representations of the subnet IDs.

Table 9.1: Possible subnet ID assignment using 5-bit subnet IDs in an example university

Subnet ID	Campus unit	Subnet ID	Campus unit
00001 (1)	College 1	10000 (16)	<future dorm>
00010 (2)	College 2	10001 (17)	Branch campus 1
00011 (3)	College 3	10010 (18)	Branch campus 2
00100 (4)	College 4	10011 (19)	<future branch campus >
00101 (5)	College 5	10100 (20)	Administration
00110 (6)	College 6	10101 (21)	Campus IT
00111 (7)	College 7	10110 (22)	<future expansion>
01000 (8)	College 8	10111 (23)	<future expansion>
01001 (9)	College 9	11000 (24)	<future expansion>
01010 (10)	College 10	11001 (25)	<future expansion>
01011 (11)	<future college>	11010 (26)	<future expansion>
01100 (12)	<future college>	11011 (27)	<future expansion>
01101 (13)	Dorm 1	11100 (28)	<future expansion>
01110 (14)	Dorm 2	11101 (29)	<future expansion>
01111 (15)	<future dorm>	11110 (30)	<future expansion>

Since we have 32 possible subnets with 5-bit subnet IDs but only 16 usable subnets at this time, some subnets are as yet unused. These are highlighted in the table. The unused subnet IDs are available for future units that may get added to the university. These unused subnet IDs may also be used to accommodate growth within existing units on the campus. For example, if the college of business experiences growth and a new building is added to the college, one of the available subnet IDs may be allocated to the college of business to accommodate the new computers that would get added to the new building. Similarly, if the university decides to deploy wireless networks across the campus, the wireless network could be assigned one of the unutilized subnet IDs.

Once we know the subnet IDs for the different subnets, we know the first two parts of the IP address of any host on any subnet on campus. For example, if the network ID of the organization is 131.247.210.0.0/16, the first 16 bits of the IP address of any host in the university will be 10000011.11110111 ($10000011_2 = 131_{10}$) and ($11110111_2 = 247_{10}$). We can now use the information about subnet IDs in the table to get the second part of the IP addresses of hosts within each of the 10 colleges. These are shown in Table 9.2. Subnet IDs have been highlighted.

Table 9.2: Network IDs and subnet IDs for hosts within each of the 10 colleges in our example university

Campus unit	Subnet ID (from Table 9.1)	First two parts of IP addresses by college Network ID Subnet ID
College 1	00001	10000011.11110111.**00001** ___.________
College 2	00010	10000011.11110111.**00010** ___.________
College 3	00011	10000011.11110111.**00011** ___.________
College 4	00100	10000011.11110111.**00100** ___.________
College 5	00101	10000011.11110111.**00101** ___.________
College 6	00110	10000011.11110111.**00110** ___.________
College 7	00111	10000011.11110111.**00111** ___.________
College 8	01000	10000011.11110111.**01000** ___.________
College 9	01001	10000011.11110111.**01001** ___.________
College 10	01010	10000011.11110111.**01010** ___.________

Consider College 1 in Table 9.2. We know that the network part of the IP address of all hosts within the university is 131.247. So, the first 16 bits of the IP address of every host in the college will be 10000011.11110111. We also know from Table 9.1 that the 5-bit subnet ID assigned to the college by the university's network administrator is 00001. Therefore, we know that the IP address bits in positions 17–21 for all hosts in College 1 will be 00001. Therefore, the first 21 bits of all IP addresses in the college are 10000011.11110111.00001. Now, the remaining 11 bits (32 - 16 - 5) are available to assign to individual hosts within College 1. These 11 bits are shown as "_" in Table 9.2.

With 11 bits available for the host ID, College 1 can accommodate 2^{11} = 2,048 hosts. The network administrator of College 1 may run a DHCP server within the college to dynamically assign these 2,048 IP addresses to hosts within the college.

You may have observed that subnet ID 00000 (0) and 11111 (31) have not been shown in Figure 9.1. This is because according to RFC 943 on assigned numbers, all zeros and all ones have a special meaning in IP addressing. All zeros within any part of the IP address are interpreted to mean "this," and all ones are interpreted to mean "all." Therefore, all zeros and all ones are not used in any of the three parts of an IP address. As a result, even though a 5-bit subnet ID allows 32 subnets, only 30 subnets were traditionally possible.

RFC 1878 has since eliminated the special interpretation of all 0s and all 1s in the host part. We have, however, continued with the traditional addressing scheme. You may use all 0s and all 1s as subnet addresses in your designs.

Let us quickly recapitulate what we have learned in this section:

1. Given the required number of subnets, we can calculate the length of the subnet ID part.
2. Given the length of the network ID part, we can calculate the length of the host part. Length of host part = 32 minus length of network ID part minus length of subnet ID part.
3. Given the length of the host ID part, we can calculate the number of hosts/subnet.

Addressing a Subnet

Once the subnets shown in Table 9.1 have been created and assigned to the individual campus subnetworks, individual subnets can be addressed by setting the host ID field to all zeros following RFC 943. For example, the subnet address of College 1 is 10000011.1111 0111.00001000.00000000, which is 131.247.8.0. The subnet address of College 2 is 10000 011.11110111.00010000.00000000, which is 131.247.16.0. Similarly, we can get the subnet addresses of all other colleges in the university. Each of these subnets uses 21 bits for the network ID and subnet ID parts of the IP address. In CIDR notation, therefore, each of these college subnets is said to have a /21 address. This leaves 11 bits for the host ID part, which allows each subnet to have up to 2^{11} = 2,048 hosts. We can write these subnet addresses as shown in Table 9.3. Subnet IDs are in bold and are underlined.

With the subnet addressing scheme shown in Table 9.3, a possible configuration for the university network is shown in Figure 9.7. In the example, the university has three internal routers that serve the different colleges. Colleges 1, 2, and 3 are served by the router on the west side of the campus; Colleges 4, 5, and 6 by the router on the south side of the campus; and Colleges 7, 8, 9, and 10 by the router on the east side of the campus. The other units (branch campuses, etc.) are also connected to these routers but have not been shown for simplicity.

Table 9.3: Subnet addresses of colleges in our example university shown in Table 9.1

Campus unit	Subnet addresses (binary)	Subnet addresses (decimal)
College 1	10000011.11110111.00001000.00000000	131.247.8.0/ 21
College 2	10000011.11110111.00010000.00000000	131.247.16.0/ 21
College 3	10000011.11110111.00011000.00000000	131.247.24.0/ 21
College 4	10000011.11110111.00100000.00000000	131.247.32.0/ 21
College 5	10000011.11110111.00101000.00000000	131.247.40.0/ 21
College 6	10000011.11110111.00110000.00000000	131.247.48.0/ 21
College 7	10000011.11110111.00111000.00000000	131.247.56.0/ 21
College 8	10000011.11110111.01000000.00000000	131.247.64.0/ 21
College 9	10000011.11110111.01001000.00000000	131.247.72.0/ 21
College 10	10000011.11110111.01010000.00000000	131.247.80.0/ 21

Once we have the arrangement of Figure 9.8, how does the main campus router handle incoming packets from the Internet? If the destination address of a packet belongs to a host in College 1, the router needs to send the packet to the Campus West router. If the destination is in College 10, the packet needs to be sent to the Campus East router. Note that the main campus router is not responsible for sending the packets to the destination hosts. The main campus router is only responsible for sending the incoming packet to the correct router on campus. This means that in our example university, the main campus router only needs to figure out which of the three internal routers to direct a packet to.

How can we tell the main campus router which college a specific destination address belongs to? One possibility is to give the router an exhaustive list of all IP addresses on campus and their affiliations. For our university with a /16 address, this means that the router's lookup table will have $2^{(32-16)} = 2^{16} = 65{,}536$ entries. Given this list, when a packet comes into the university, the router can look up the 32-bit destination address against this list, identify the college that the address belongs to, and direct the packet to the appropriate router serving that college.

This process will work, but it is easy to see that it is rather inefficient. If the main campus router only needs to know which of the 16 campus units (or 30 units when the campus is fully built out) the incoming packet needs to go to, the router only needs to look at the subnet part of the destination address of incoming packets. For every incoming packet, we just need a way to tell the main campus router which of the 5 bits of the destination IP address identifies the subnet to which the address belongs. If we can do this, the router's lookup table can be reduced to just $2^5 = 32$ entries, instead of 65,536 entries—a huge improvement. This speeds up router performance and reduces the hardware requirements for routers, thereby lowering costs.

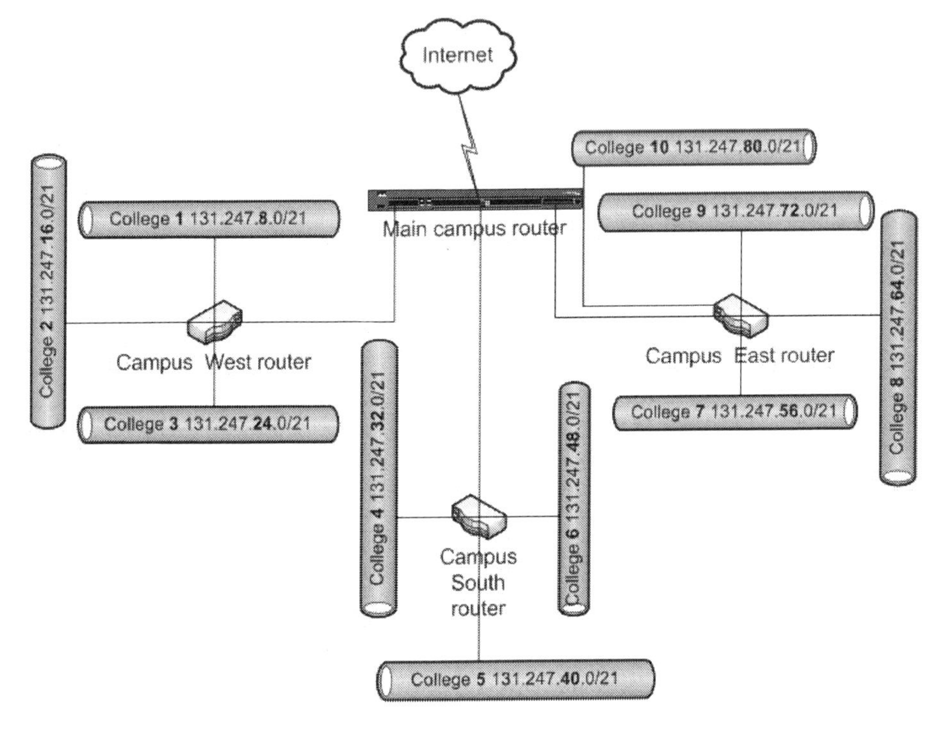

Figure 9.8: Example university college subnets

The standard technique used to help network devices identify the subnet ID of an IP address is the use of a subnet mask. This topic is covered in the next section.

Subnet Masks

We are now left with the last challenge of making subnetting work—how do hosts and routers on the network know which bits of an IP address constitute the subnet part of an IP address? For example, in our example university, how can we convey to routers and hosts that bits 17–21 identify the subnets on campus?

We start by making an observation. We really do not need to know where the subnet ID begins in an IP address. We only need to know where it ends. For example, in our university, we only need to know that the subnet ID ends at the 21st bit.

Why is this? Recall that all hosts in our university have the same network ID. If we ignore where the subnet ID begins, we add the same network ID bits to every subnet ID. In other words, instead of defining the subnet ID of College 1 as 00001 (Table 9.1), we can define its subnet ID as 10000011.11110111.00001 (Table 9.3). Similarly, the subnet ID of College 3 becomes 10000011.11110111.00011 instead of 00011. The number of subnet IDs does not change. Our example university will still have 32 subnets.

Subnetting and route aggregation

When assigning subnet addresses, it is a good idea to be watchful of opportunities for route aggregation. If contiguous addresses are kept in close network proximity, these routes could potentially be aggregated upstream, so that upstream routers will have a minimal number of routes to track.

What is the advantage of doing this? After all, isn't it easier to work with smaller numbers? Isn't it easier for routers to deal with 5-bit subnet IDs than 21-bit subnet IDs?

The big advantage of adding the network ID to define the subnet ID is that all subnet IDs begin at the first bit. We then have a uniform representation of subnet IDs for all organizations. Consider two organizations, one with a Class-A address block, and another with a Class-B address block. If both organizations decide that they need about 2,000 IP addresses per subnet, they will both have /21 subnets. In the Class-A organization, bits 9–21 will define the subnet ID, and in the Class-B organization, bits 17–21 will define the subnet ID. In both cases, the subnet ID will end at bit position 21. The only difference is that the Class-A organization will have $2^{13} = 8,192$ subnets and the Class-B organization will have $2^5 = 32$ subnets.

If we try to specify where the subnet ID part begins, we will need to come up with one subnet mask procedure for Class-A networks, another for /9 networks, and so on. Not only will subnetting become more complex to implement, the extra complexity will serve no useful purpose.

Instead, if we add the network ID to the subnet ID, we are assured that within any organization, we are prepending the organization's network ID to all subnet IDs. A Class-A organization will prepend its 8-bit network ID to all its subnets. A Class-B organization will prepend its 16-bit network ID to all its subnets. There will be no change to the number of subnets within the organization. At the cost of a slightly larger subnet ID field, we have gained a uniform representation for subnet IDs across all networks. We therefore identify subnets by the network ID + subnet ID bits within the organization.

With this simplification, we only need a mechanism to convey where the subnet ID ends. Observe that this is equal to the length of the network ID + length of subnet ID. The standard method for conveying the length of the network ID and subnet ID parts of IP addresses in a network is to use a subnet mask. *A subnet mask is a number that tells the host what bits in an IP address constitute the network ID and subnet ID of the network.* Every network interface on every host on the network is assigned a subnet mask. The interface gets the subnet mask information at the same time that it gets an IP address either from an administrator or from DHCP. You can see your subnet mask using the command ipconfig (Windows)/ifconfig (Mac/Linux). Figure 9.9 shows an example.

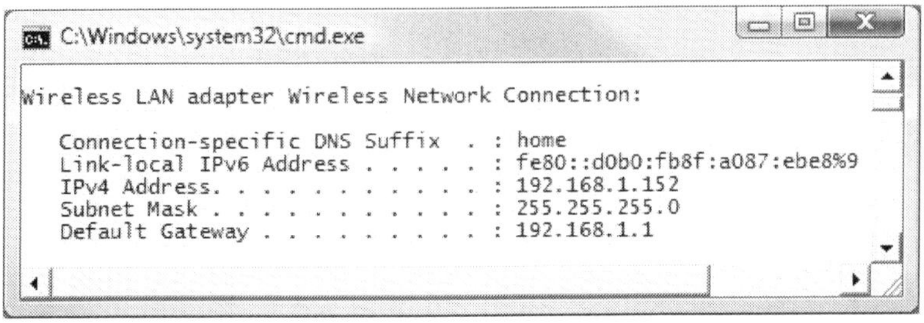

```
C:\Windows\system32\cmd.exe

Wireless LAN adapter Wireless Network Connection:

   Connection-specific DNS Suffix  . : home
   Link-local IPv6 Address . . . . . : fe80::d0b0:fb8f:a087:ebe8%9
   IPv4 Address. . . . . . . . . . . : 192.168.1.152
   Subnet Mask . . . . . . . . . . . : 255.255.255.0
   Default Gateway . . . . . . . . . : 192.168.1.1
```

Figure 9.9: Using ipconfig to find subnet mask

Subnet masks look like IP addresses, but have a very special structure. Written as 32-bit binary numbers, subnet masks are a sequence of 1s followed by a sequence of 0s. The 1s indicate the bits in IP addresses that constitute the network ID + subnet ID of the subnet. For example, since the subnets in our example university have a total of 21 bits in their network ID and subnet ID, the subnet mask used in the university will have a sequence of 21 ones, followed by 11 zeros. This gives us the subnet mask for the university, as shown in Table 9.4.

Table 9.4: Subnet mask structure

11111111.11111111.11111	000.00000000
21 ones (16-bit network ID + 5-bit subnet ID)	11 zeros (32 - 21)

We can express this subnet mask in the familiar dotted-decimal notation as 255.255.248.0. You may note that this mask has some similarity to the subnet mask shown in Figure 9.9. Both start with 255.255.

How do subnet masks help hosts and routers? It is easier to show this by example. Let us start with two hosts in two different colleges in our university. For our example, let us consider one host in College 1 and another host in College 3. Say the host in College 1 is 131.247.8.45 and the host in College 3 is 131.247.27.231. In binary notation, these IP addresses are:

Host 1: 10000011.11110111.00001000.00101101
Host 2: 10000011.11110111.00011011.11100111

What happens when we apply the subnet mask to these addresses? Recall that the 1s in the subnet mask indicate the network part + subnet ID and 0s indicate the host ID. To implement this, when we apply the subnet mask to an IP address, the bits in the IP address corresponding to the 1s in the subnet mask are passed through. The 0s in the subnet mask block the bits in the corresponding positions in the IP address.[1] The result for our example is shown in Table 9.5.

1 You may observe that this is a bit-wise logical AND operation.

Table 9.5: Example of masking using a subnet mask

	Host 1	Host 2
Host IP address	10000011.11110111.00001000.00101101	10000011.11110111.00011011.11100111
Subnet mask	11111111.11111111.11111000.00000000	11111111.11111111.11111000.00000000
Masked host IP address	10000011.11110111.00001000.00000000	10000011.11110111.00011000.00000000
College matching masked IP address	College 1	College 3

In the example of Table 9.5, consider host 1. Its IP address is: 10000011.11110111.00001000.00101101
When we mask this IP address with the subnet mask:

11111111.11111111.11111000.00000000,

since this mask has 21 ones, the first 21 bits of the IP address pass through unchanged. The remaining 11 bits are masked and replaced with 0s. We see the result as the masked host IP address.

The concept of masking is quite general in software. For example, in Photoshop, you use a mask to limit the area over which an editing effect applies. A simple web search for Photoshop masking tutorial will bring up many results.

What is interesting about the masking process is that the masked IP addresses will match the subnet IDs of the corresponding hosts. The mask essentially removes the host ID part from the destination IP address, leaving behind just the network ID + subnet ID parts of the IP address.

Put another way, the masked IP address of any host is the network ID of the host + the subnet ID of the host.

Once the router has extracted the subnet ID using the subnet mask, it can look up the subnet ID in its routing tables and forward the packet to the appropriate router within the campus. This process is shown for a few representative packets in Figure 9.10. To keep things simple, the network ID is shown in decimal.

In Figure 9.10, five packets are shown arriving at the router. The router views the destination IP address of each packet through the subnet mask. The 0s in the mask remove the host ID part of the destination address. The remaining bits in the destination address give the router the subnet ID of the destination address of each packet. The router can look up this subnet ID in its routing table to identify the router to which the packet should be forwarded. For example, packet 2 is addressed to a host in College 3, and is sent to the Campus West router.

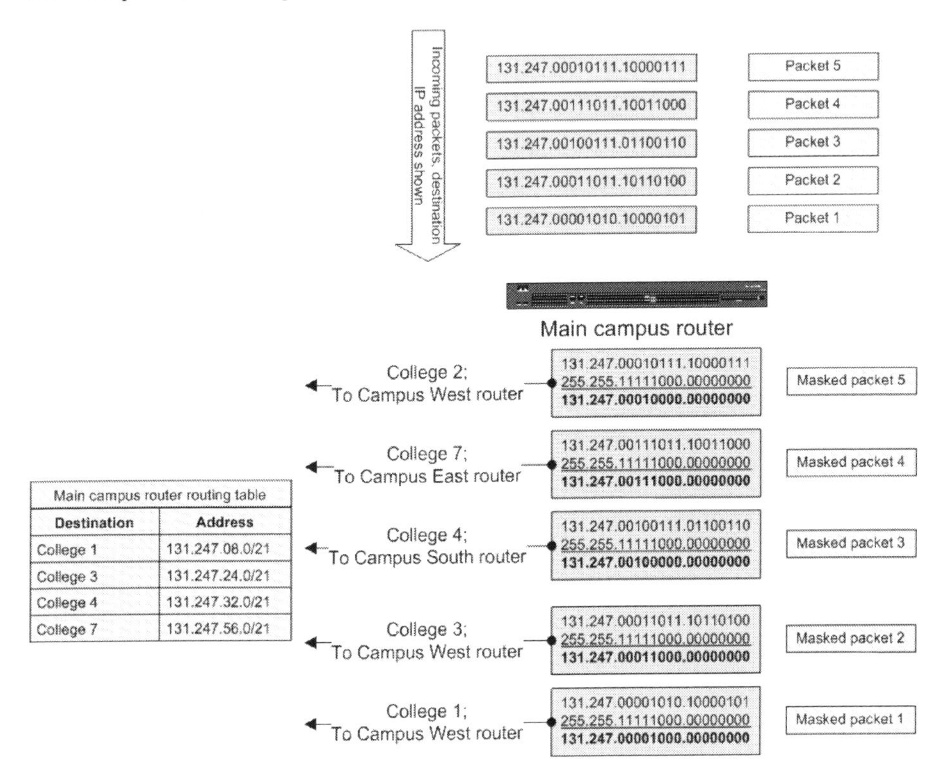

Figure 9.10: Subnet masking of packets at router to determine subnet ID

When creating subnets, the longer the subnet ID, the shorter the host ID part, leading to fewer hosts/subnet. Further, within each subnet, we lose at least one IP address for the gateway router. Therefore, it is not advisable to create very small subnets except when specifically necessary or for point-to-point links.

Benefits of Subnetting within Subnets

The previous section describes how subnetting is useful for routers to direct incoming packets. However, Figure 9.9 shows that hosts on subnetted networks also have the subnet mask information. These hosts do not perform routing functions. What is the benefit of providing the subnet mask information to hosts that do not perform routing?

It turns out that letting hosts know the subnet mask of the network can reduce the amount of work that routers need to do in processing outgoing packets. Recall from Chapter 7 that if the destination is outside the local network, hosts send outgoing packets to their gateway router for onward transmission to the destination. However, what if the destination is in the host's own subnet?

If the destination is in the host's own subnet, the router is going to send the packet back into the same subnet from which the packet originated. But, if hosts can detect that the destination of a packet is in its own subnet, the packet does not need to be routed. The sender can simply broadcast the packet over the local Ethernet. This can improve network efficiency by reducing the workload of routers.

Subnet masks help hosts detect when a destination host lies in their own subnet. Before sending packets out, hosts apply their subnet mask to the destination address. This gives the subnet ID of the destination as seen by the host. If the subnet ID of the destination matches the sender's own subnet ID, the sender sends out an ARP request to obtain the MAC address of the destination and broadcasts the packet directly into the subnet. Ethernet ensures that this packet reaches the destination.

If, on the other hand, the host detects that the subnet ID of the destination of a packet is not the same as its own subnet ID, it knows that the destination is not in its own subnet. The host then sends the packet to the gateway router. The gateway router then routes the packet to its destination.

Subnetting can therefore help in limiting network traffic at routers. If network administrators can identify hosts within networks that exchange a lot of data with each other, they should place these hosts within the same subnet. This can reduce networking costs by reducing the packet processing requirements at routers. An example is an application server and the linked database server.

We can see this mechanism in place in the simple packet capture shown in Figure 9.11. The IP address of the source host is 192.168.1.3 and its subnet mask is 255.255.255.0. The capture shows two transactions. The first transaction is a ping to host 192.168.1.153 (same subnet as sender), and the second transaction is a ping to 131.247.100.1 (different subnet from sender). The IP address of the gateway router is 192.168.1.1. Observe the difference in the two ARP queries.

In the first transaction, a user on the host pings host 192.168.1.153. The subnet ID of both hosts is 192.168.1.0. The host recognizes that the destination is on its own subnet and therefore makes an ARP request for IP address 192.168.1.153 (packet 1). It receives the reply in packet 2. Packets 3–6 are two ping requests and replies between the two hosts. The critical thing to note about packet 1 is that the ARP request is for the actual destination (192.168.1.153) and not the gateway router (192.168.1.1).

No.	Time	Source	Destination	Protocol	Info
1	0.000000	00:21:70:bf:ff:40	Broadcast	ARP	Who has 192.168.1.153? Tell 192.168.1.3
2	0.000227	LaCieGro_84:8d:4e	00:21:70:bf:ff:40	ARP	192.168.1.153 is at 00:d0:4b:84:8d:4e
3	0.000242	192.168.1.3	192.168.1.153	ICMP	Echo (ping) request
4	0.000444	192.168.1.153	192.168.1.3	ICMP	Echo (ping) reply
5	1.009324	192.168.1.3	192.168.1.153	ICMP	Echo (ping) request
6	1.009612	192.168.1.153	192.168.1.3	ICMP	Echo (ping) reply
7	[illegible]	00:21:70:bf:ff:40	WestellT_c0:9a:43	ARP	Who has 192.168.1.1? Tell 192.168.1.3
8	10.853030	WestellT_c0:9a:43	00:21:70:bf:ff:40	ARP	192.168.1.1 is at 00:18:3a:c0:9a:43
9	11.444676	192.168.1.3	131.247.100.1	ICMP	Echo (ping) request
10	11.472657	131.247.100.1	192.168.1.3	ICMP	Echo (ping) reply
11	12.444083	192.168.1.3	131.247.100.1	ICMP	Echo (ping) request

Figure 9.11: Packet transmission to hosts within and outside subnets

In the second transaction, the same host pings 131.247.100.1. The host sees that the subnet ID of the destination (131.247.100.0) is not the same as its own subnet ID (192.168.1.0). Therefore, packet 7 in the capture is an ARP request for the gateway (192.168.1.1) and not for the final destination (131.247.100.1). When the host gets the ARP response, it sends the ping request to the gateway for onward transmission to the destination. This is accomplished by setting the destination IP address of the ping requests to the IP address of the destination and the MAC address to the MAC address of the gateway.

Subnetting, therefore, improves network efficiency by limiting the volume of traffic that needs to be routed. If the destination is on the same subnet as the source, subnetting helps the source determine this fact and bypass routing so that it can reach the host directly over Ethernet.

Representative Subnetting Computations

Now that we have seen how subnetting is done and why subnetting is useful, we can wrap up the chapter by looking at representative subnetting computations to determine the subnet mask. Typically, the business requirement is to create subnets that accommodate a certain number of hosts per subnet. For example, you may need to create subnets with about 500 hosts per subnet.

To accomplish this, we calculate the number of bits necessary in the host ID part of the network address. To accommodate 500 addresses, we know that we need 9 bits ($2^9 = 512$). Now, the bits that remain after taking away the network ID and the host ID parts in IP addresses can be used for subnet IDs.

For example, if the organization has a /15 network address, the first 15 bits in IP addresses will be used for the network ID, the last 9 bits will be used for the host ID, and the remaining 8 bits (32 - 15 - 9 = 8) will be used for subnet IDs. This will give us $2^8 = 256$ subnets in the organization. This is shown in Figure 9.12.

The organization will express this information in the subnet mask assigned to routers and hosts. How do we express the subnet mask for the organization in dotted decimal notation? Since the subnet mask has 1s in the positions of network ID and subnet ID, the subnet mask in the organization will have 23 ones (15 for the network ID and 8 for the subnet ID). This leaves 32 - 23 = 9 zeros in the mask. This gives a subnet mask of 11111111.11111111.11111110.00000000. In the dotted decimal notation, this mask is written as 255.255.254.0.

Alternately, if the organization requires a certain number of subnets, we can calculate the number of bits required to label all the subnets. For example, if the organization requires 20 subnets, we will need 5 bits to label all 20 subnets. Four bits will be inadequate because we can only label $2^4 = 16$ subnets with 4 bits. Given the organization's network ID, we can calculate the number of hosts that can be accommodated in each subnet. If the organization has a /17 network address, the first 17 bits will be taken up for the network ID, the next 5 bits will be used for subnet IDs, and the remaining 10 bits (32 - 17 - 5 = 10) will be available for the host ID part within each subnet. This will allow each subnet to have 1,024 hosts ($2^{10} = 1{,}024$). The subnet mask will have 22 ones (17 for the network ID and 5 for the subnet ID). The remaining 10 positions will be zeros. The subnet mask is therefore 11111111.11111111.11111100.00000000. In the dotted decimal notation, this is 255.255.252.0.

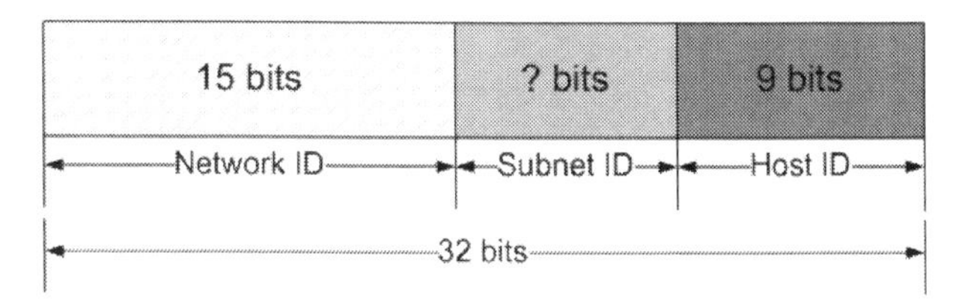

Figure 9.12: Calculating the length of the subnet ID

We can summarize the above in the following algorithm to determine the subnet mask for an organization.

- Step 1: If the required number of subnets is known, calculate the number of bits needed to uniquely label each subnet. This is the length of the subnet ID part of network addresses. Alternately, if the required number of hosts is known, calculate the number of bits needed to uniquely label each host. This is the length of the host ID part of network addresses.
- Step 2: Find the number of bits in the network part of the IP address. This is the length of the network ID part of network addresses.
- Step 3: Using the information from step 1 and step 2, calculate the lengths of each of the three parts of IP addresses within the organization (network ID, subnet ID, and host ID). The total length of all three parts is 32 bits.
- Step 4: Write the subnet mask in binary notation. The subnet mask is 32 bits long. It has (length of network ID + length of subnet ID) 1s followed by (length of host ID) 0s.
- Step 5: Write the subnet mask in dotted decimal notation.

As an example of the algorithm, consider a /18 organization requiring 200 hosts per subnet. What is the subnet mask?

- Step 1: We know the number of hosts. So we start by calculating the length of the host ID part. We need 8 bits for the host ID part because $2^8 = 256$. Thus, the length of the host ID part is 8 bits.
- Step 2: The length of the network ID part is given as 18 bits.
- Step 3: The missing information is the length of the subnet ID. Since network ID + subnet ID + host ID = 32 bits, we can calculate the length of the subnet ID as 32 - 18 - 8 = 6 bits.
- Step 4: The subnet mask has 18 (length of network ID) + 6 (length of subnet ID) = 24 ones. The remaining 8 bits (32 - 24) are zeros. In binary notation, the subnet mask is: 11111111.11111111.11111111.00000000.
- Step 5: The subnet mask in dotted decimal notation is 255.255.255.0.

Subnetting in IPv6

There is an important difference between IPv4 and IPv6 related to subnetting. Recall from the discussion on IP address classes that the length of the network part of IPv4 addresses varies widely, from as little as 8 bits to as much as 24 bits. We also saw earlier in this chapter that subnet ID bits are taken from the host part of the IPv4 address. The challenge with varying lengths of network and subnet parts of IP addresses is that it complicates routing. For every incoming packet, the router has to first determine how many bits constitute the subnet ID and then search for the ID in its routing table to locate the path to the destination.

Learning from this experience, RFC 3587 standardized the lengths of the network and subnet parts of IPv6 addresses. The field sizes of IPv6 addresses are as shown in Figure 9.13. The Interface ID (often called the host ID in IPv4) is 64 bits long, to coincide with the length of MAC addresses. The subnet ID is 16 bits long, and the global routing prefix (network ID) is 48 bits long. While these field sizes might appear wasteful for many networks, it is deemed useful for the simplicity afforded by fixed field sizes.

Global routing prefix (48 bits)	Subnet ID (16 bits)	Interface ID (64 bits)

Figure 9.13: Standard field sizes of unicast IPv6 addresses

EXAMPLE CASE—An ISP in Texas

Internet service providers (ISPs) provide Internet service to homes and businesses. In this case, we see an example of how one such ISP, Texlink Communications, which provided Internet and phone services to customers in four major metros in Texas—San Antonio, Austin, Dallas, and Houston—subnetted its IP addresses (Figure 9.14).

As can be verified from ARIN, the ISP obtained a /18 address block, 66.118.0.0/18, which gave it $2^{(32 - 18)} = 2^{14} = 16{,}384$ IP addresses. How did it organize these IP addresses?

More than half the ISP's customers were in Houston. The fewest customers were in Austin and Dallas, where the ISP estimated it needed about 2,000 IP addresses each. The ISP therefore decided to subnet its 16,384 IP addresses into 8 subnets with about 2,000 IP addresses per subnet (16,384/ 8). Since $2^{11} = 2{,}048$, the ISP needed 11 bits in the host ID part, leaving 21 bits (32 - 11) in the network and subnet ID parts. Since 18 of these 21 bits were already taken by the network ID part, 3 bits were available for the subnet IDs part in this network.

If we label the address bits as N if they represent the network ID, S if they represent the subnet ID, and 0 if they represent the host ID, the IP addresses in the organization will have the following structure:

NNNNNNNN.NNNNNNNN.NNSSS000.00000000

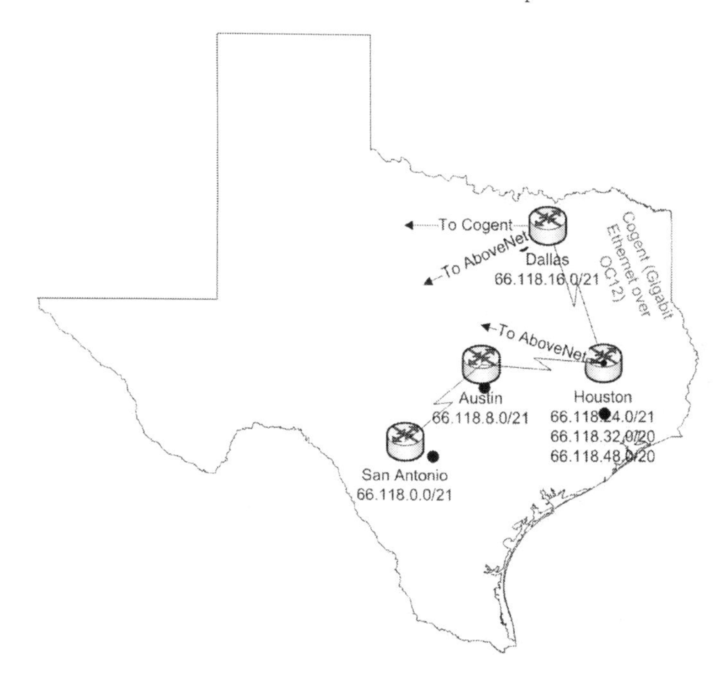

Figure 9.14: Coverage map of Texlink Communications

Table 9.6: Subnets in Texlink communications

SSS	Subnet	Aggregate	City
000	66.118.0.0 /21		San Antonio
001	66.118.8.0 /21		Austin
010	66.118.16.0/21		Dallas
011	66.118.24.0/21		Houston
100	66.118.32.0/21	66.118.32.0/20	Houston
101	66.118.40.0/21		Houston
110	66.118.48.0/21	66.118.48.0/20	Houston
111	66.118.56.0/21		Houston

We see that all the subnets are identified by the three bits labeled SSS in the third octet, giving us $2^3 = 8$ octets. Writing out these 8 subnets, we get Table 9.6.

To see how the SSS bits lead to the corresponding subnets, let us look at the last Houston subnet, 66.118.56.0/21, and see how we arrive at the subnet address. The first two octets, 66 and 118, come from the network address. Since the third octet in the network address block (66.118.0.0) is 0, the 2 NN bits in the third octet are 0. The three SSS subnet ID bits are

111 from the first column in the table. Finally, the host ID bits for any network address are 00. Combining this information, the third octet in the subnet address is 00111000_2, which translates to 56_{10}.[2] Finally, the fourth octet is 0. Combining all four, the subnet address is 66.118.56.0, of which 21 bits identify the subnet part of the address. We therefore write the subnet address as 66.118.56.0/21.

The third column shows how some subnets have been merged. To understand how this is done, let us look at the subnets 66.118.32.0/21 and 66.118.40.0/21 and see how they aggregate to 66.118.32.0/20. To keep things simple, let us focus only on the third octet in both subnets. These two octets are written below, with the subnet IDs in bold:

00**100**000 (third octet in 66.118.32.0/21)
00**101**000 (third octet in 66.118.40.0/21)

If both subnet IDs are used for the same subnetwork (a subnet in Houston), the last bit in the subnet ID (the underlined bit above), does not play any role in identifying a subnet. The subnet is then identified by just the first 2 bits of the subnet ID: **10**. If that is the case, we can include this bit in the host ID part and use it to identify hosts within the subnet. To complete the subnet address, we add the 18 bits of the network ID to the 2-bit subnet ID, giving us a subnet ID that is 20 bits long. The result is the /20 subnet, 66.118.32.0/20.

The ISP obtains upstream connectivity to the Internet through two national service providers: Cogent and AboveNet. Texlink peers with both these ISPs at Dallas. AboveNet also connects at Houston. A dedicated OC-12 line brings Cogent to Houston. Austin and San Antonio are connected to the rest of the network as shown in the figure.

Acknowledgments

Special thanks to Pete Templin of Texlink Communications (now Pac-West Telecomm) for providing the information used in this case.

Summary

Subnetting allows network administrators to divide an organization's IP addresses into smaller address blocks. This simplifies delegation of network responsibilities within the organization. In cases where a lot of network traffic is generated by a small group of computers exchanging data with each other, subnetting can be used to limit network traffic by isolating these computers within a subnet.

Subnetting is implemented using subnet masks. Subnet masks look like IP addresses and are composed of a sequence of 1s followed by a sequence of 0s. The 1s in the subnet mask indicate the positions of the network ID and subnet ID bits in IP addresses.

The chapter implicitly assumes that organizations use the same subnet mask everywhere within the network. This has been done for simplicity because this is the first exposure to subnetting for most students. In reality, subnet sizes are allocated based on the actual needs of each subnet.

---

2 The subscripts 2 and 10 used here indicate binary (base 2) and decimal (base 10) numbers.

About the Colophon

Henry Ford founded the Ford Motor Company and was one of the first entrepreneurs to use assembly-line manufacturing to mass produce affordable automobiles. His success with the technique revolutionized industrial production in the United States and the world. In a quote in *Reader's Digest* in 1936, he suggested that even extraordinary challenges such as setting up the assembly line are manageable if broken down into simple tasks.

Subnetting brings to computer networks the idea of breaking complex tasks into small jobs. The structure and traffic patterns on the global Internet are extremely complex. However, the Internet is organized as a network of networks. It is simpler to first identify closely interacting computers and use appropriate data-link layer technologies to organize these computers into subnets. These smaller networks, taken together, result in ever-larger networks and eventually create the global Internet. Also, once the Internet is broken down into subnets, even relative novices can manage some of the individual subnets. As an example, most homes are now subnets, but they are managed relatively easily by end users.

REVIEW QUESTIONS

1. What is *subnetting*?
2. What is the organizational motivation for subnetting?
3. In large organizations, what are some disadvantages with allocating IP addresses on a first-come, first-served basis?
4. How do you determine the number of subnets needed in an organization?
5. Using an example, describe the three-part interpretation of IP addresses when subnetting is used.
6. Describe how subnetting is similar to the three-part organization of telephone numbers.
7. Describe how subnetting is similar to the multi-part organization of zip codes.
8. How does the three-part numbering system used in telephones facilitate the switching of long distance calls?
9. How does the multi-part numbering scheme used in zip codes simplify the mail handling tasks at a typical post office?
10. What are the three IP addresses on any network that are not available for allocation to hosts?
11. What factors determine the subnet structure of an organization? For example, if you have a Class-B address, how will you determine if you should have 512 subnets, 256 subnets, 128 subnets, 64 subnets, or some other number of subnets?
12. What is a *subnet mask*?
13. Why are subnet masks needed?
14. What do the 1s and 0s in a subnet mask represent?
15. Can 255.255.253.0 be a subnet mask? Why or why not?
16. What information about a network can be gathered by looking at its subnet mask?
17. Say you have a /14 network address. You are asked to create subnets with at least 1,000 hosts/subnet. What subnet mask should you use?
18. How many subnets can you have on the network in the previous question?

19. The broadcast address on a subnet is obtained by replacing the bits in the host part with 1s instead of 0s. What is the broadcast address of the subnet 192.168.1.192/28?
20. How many hosts can a /18 network support?
21. Say you have a /15 network address. You are asked to create subnets with at least 1,000 hosts/subnet. What is the maximum number of such subnets you can create?
22. What is the subnet mask you will use for the question above?
23. Consider two IP addresses: 192.168.35.56 and 192.168.36.135. If the subnet mask used is 255.255.252.0, what are the masked IP addresses (subnet IDs) for the two IP addresses?
24. How is subnetting in IPv6 different from subnetting in IPv4?
25. How can subnetting help limit network traffic at routers?

EXAMPLE CASE QUESTIONS

1. How many customers can the ISP serve in San Antonio?
2. How many customers can the ISP serve in Houston, combining the capacity of all the subnets?
3. If the two /20 Houston subnets are merged together, what is the subnet address of the resulting network?

After deploying the network as shown in the case, the ISP experienced slow customer growth in Dallas and Austin, but high customer growth in San Antonio and Houston. To accommodate these growth patterns, the ISP pulled some /24 subnets from Dallas and Austin and allocated them to San Antonio and Houston. The ISP had to do this to demonstrate IP address utilization before it could apply to ARIN for more IP address blocks.

4. What are the /24 subnets that comprise the Dallas /21 subnet? (Hint: there are $2^{24-21} = 2^3 = 8$ such subnets.)
5. What are the /24 subnets that comprise the Austin /21 subnet?
6. Use BGPlay to show the network connectivity diagram for the 66.118.0.0/18 network.
7. What is the data rate of an OC-12 connection?

(Optional) Now, assume that the last /24 subnet in each case (Dallas and Austin) is pulled out and allocated to San Antonio and Houston.

8. What are the remaining subnets in Dallas and Austin? Aggregate the subnets where possible.

HANDS-ON EXERCISE—Subnet Mask

For this exercise, we will revisit the utility we have used in many prior exercises—ipconfig. On my computer, the output of ipconfig /all | more appears as shown in Figure 9.15. We see that the subnet mask on the network is 255.255.252.0.

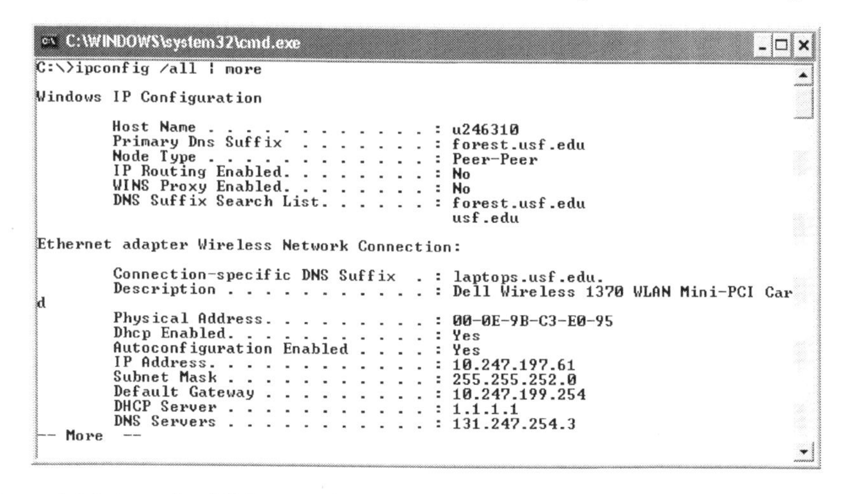

Figure 9.15: ipconfig /all | more showing subnet mask information

Answer the following questions:

1. Write the subnet mask in Figure 9.15 in 32-bit binary notation.
2. How many bits in IP addresses on the network are used for the *host part* of the IP address?
3. How many hosts can the subnet of Figure 9.15 accommodate?
4. How many bits in IP addresses on the network are used for the *network part* of the IP address?
5. The IP address of the computer is 10.247.197.61 (Figure 9.15). Write this IP address as a 32-bit binary IP address.
6. What is the masked address of the host in binary notation? (This is the network ID + subnet ID part of the IP address.)
7. Express this masked address in dotted decimal notation. Express the masked address in CIDR notation.
8. Show the ipconfig /all output on your computer. What is the subnet mask on the network?
9. Based on the subnet mask, how many computers can be addressed on the subnet on which your computer is located?

CRITICAL-THINKING EXERCISE—Subnet Design

Say you are using the 192.168.0.0/16 address block to create subnets for a typical medium-sized business with about 2,000 employees.

1. You have been told to provide a subnet for each major department in the business. Assuming that the business has a typical organization structure for a business of this size, list the different subnets you would like to set up for the business.

2. How many bits would you need in the subnet ID to provide the required number of subnets?
3. How many host IDs would be available in each subnet in your design?
4. Present your results in the form of a table like Table 9.3 in the chapter.

IT INFRASTRUCTURE DESIGN EXERCISE—Subnet Design

In this exercise, you will design the subnets for TrendyWidgets, based on the information about TrendyWidgets provided in Chapter 1. In Chapter 7, we saw that the company has decided to use a /21 address block to provide addresses to hosts within the company. For quick response, the company has decided to use a DHCP server at each of its four locations—Tampa, Amsterdam, Mumbai, and Singapore. Using this information, answer the following questions:

1. What subnets would be suitable for use in the company's four locations?
2. Update the infrastructure diagram from Chapter 8 to show the subnets you created in the previous question.
3. Further update the infrastructure diagram to show the four DHCP servers, one server at each location.

CHAPTER 10

Wide-area Networks

It's a small world.

L.A. Times, December 27, 1896

Overview

Wide-area networks (WANs) are networks that span large areas, such as states, countries, and even continents. All ISP networks are WANs. WANs serve as the backbone for the global Internet. You use WANs almost every time you connect to a computer that is located outside your immediate neighborhood. LANs (Chapter 3) connect to WANs to get connectivity to other networks. This chapter describes WANs and highlights the primary differences between LANs and WANs. It also introduces the most popular wide-area networking technologies. At the end of the chapter, you should know:

- the need for wide-area networking technologies as distinct from local-area networking technologies
- the three standard approaches to creating WANs—SDM, TDM, and FDM
- common WAN technologies, including dial-up, T links, SONET, and DWDM
- virtual circuits
- location of WANs on the TCP/IP stack

Introduction

WANs are networks that provide data communications to a large number of independent users. These users are usually spread over a larger geographic area than a LAN. As the definition suggests, WANs are distinguished from LANs in their ability to scale up to networks serving hundreds of millions of users located anywhere around the world.

As we will see in this chapter, WAN technologies are more complex than the simple broadcast technology used in LANs. Before we look at specific WAN technologies, it is useful to first understand why CSMA/CD will not work in WANs and why other, more complex approaches are required for WANs. With this understanding in place, the WAN solutions will be seen as specific approaches to addressing the particular challenges of WANs.

LANs typically connect computers in the immediate neighborhood. Computers within a classroom lab are a very good example. In such networks, the cables that connect computers to the central switch are short, usually less than 100 meters long. Since the cost of each link is much less than the cost of the computer it connects, it is not very important to optimize the utilization of these links. Thus, in Ethernet LANs, link utilization is very low, as the links are active only when network traffic is specifically directed to the connected computers. This inefficiency in link utilization is acceptable in LANs because it results in low costs for the technology, which can be built using low-cost network cards and switches.

Cables that connect computers across cities, states, or even countries are extremely expensive to install and maintain. For example, the submarine cables that connect continents can cost billions of dollars to install and hundreds of millions of dollars in annual maintenance costs. We cannot afford low rates of utilization for these expensive assets. It therefore becomes absolutely necessary to ensure that long-distance data links are utilized to the greatest extent possible.

Another issue to address is that though broadcasting is very effective for communication in small networks, it becomes very slow as the number of users on the network increases. To avoid interfering with other users, each station in a broadcast network has to wait for the network to go silent before it can transmit. It is easy to see that as more computers join the network, the wait times for transmission will get longer and longer.

WANs overcome these two challenges—the need to increase link utilization and the need to reduce wait times. WANs do this by sharing each link among multiple users and merging the traffic from all these users onto the same link.

WANs and home prices[1]

Homeowners are increasingly refusing to purchase homes in areas without broadband connectivity. A fiber-optic connection can add more than $5,000 to the value of a $175,000 home, about half the value of a bathroom.

The Road Network as an Analogue for LANs and WANs

It may be easier to visualize the distinction between LANs and WANs by relating them to the road network. There are many similarities between the architecture of the Internet and the architecture of the road network. Seen this way, LANs are the Internet analogue for neighborhood roads and WANs are the Internet analogue for interstate highways.

To see the analogy, observe how LAN links are short, like neighborhood roads. Also, similar to the way traffic entering the neighborhood road waits for the road to clear before entering it, LAN traffic waits for the link to be clear before inserting traffic into the network. Collisions are prevented by drivers looking out for other traffic on the road. In contrast, WAN links are long, like interstate highways. Similar to the way traffic entering the interstate merges seamlessly into existing traffic, WAN traffic merges with existing data traffic on the WAN without waiting for silence on the WAN network. Merge ramps facilitate integration

1 Ryan Knutson, "How fast Internet affects home prices," *Wall Street Journal*, June 30, 2015.

Figure 10.1: Neighborhood intersection as CSMA example
(Source: Google Maps)

Figure 10.2: Traffic merging onto interstate
(Source: Google Maps)

of local traffic with preexisting traffic on the interstate, just like routers merge multiple streams of traffic on WANs.

Figure 10.1 and Figure 10.2 show these features of the two kinds of roads. In the neighborhood road (Figure 10.1), the STOP sign encourages carrier-sense behavior by asking vehicles to stop and look before joining the road. For example, the white car will wait for the black car to pass the intersection before entering the road. On the interstate (Figure 10.2), there are no stop signs on the ramp or the interstate to slow down the merging traffic. Both streams of traffic (existing traffic and entering traffic) continue at their regular speeds, and the merging traffic finds gaps between vehicles and blends into the existing traffic. WANs operate in a similar manner and allow multiple streams of traffic to blend together into one seamless stream.

What are the advantages in the road network of creating two different kinds of roads—one for the neighborhood and one for interstates? Why can't all roads be built like the interstate? The problem is that while interstates can handle large traffic volumes moving at very high speeds, they are very expensive to build. Entry and exit ramps and dedicated lanes have to be created to facilitate entry and exit from the interstate. They also can be very inconvenient in the neighborhood since traffic can only enter and exit the highway at designated exits. By contrast, neighborhood roads are less expensive to build. They are adequate for carrying lower volumes of traffic at relatively slow speeds around the neighborhood. They are also convenient since homes can be built anywhere along the road and traffic can enter the road from anywhere. They are safe as long as all drivers watch out for each other and avoid collisions.

There are good reasons to have two different kinds of roads, and for similar reasons it is good to have two different kinds of networking technologies—LANs and WANs.

Categories of WANs

To summarize the above, the key capabilities of WANs are: (1) their ability to transport data over long distances, and (2) their ability to merge traffic from multiple sources into one seamless stream. We will examine some of the popular WAN technologies in this chapter. Approximately in chronological order of development, these are:

1. point-to-point technologies, including dial-up, T/DS, and ISDN links
2. statistical multiplexing, including X.25, Frame Relay, and ATM
3. TDM, such as SONET optical carriers
4. FDM/WDM/DWDM (dense wavelength division multiplexing)

Point-to-point WANs

Point-to-point connections refer to communication connections between two nodes or endpoints. Point-to-point WANs generally use the phone system infrastructure.

Dial-up Networking

Each phone line is capable of providing a data rate of 56 Kbps. The earliest WANs used the phone network to create wide-area links. A phone connection could act as a point-to-point line connecting any two networking nodes located anywhere in the country. For example, the first Internet, shown in Figure 10.3, used phone lines to connect the four Internet nodes. (You have seen this figure before, as part of Figure 1.4 on technology milestones in Chapter 1.) Each node was connected to a phone modem that was connected to a modem attached to another node on the network. The modem performed all transformations necessary to convert data into a format suitable for transmission over a phone line. These early networks were simple and did not perform any traffic aggregation or routing. Routing was performed by the nodes themselves. Thus, if node #1 (UCLA) wanted to send data to node #4 (Utah), it would send the data to node #2 (SRI), which would, in turn, route the data to node #4.

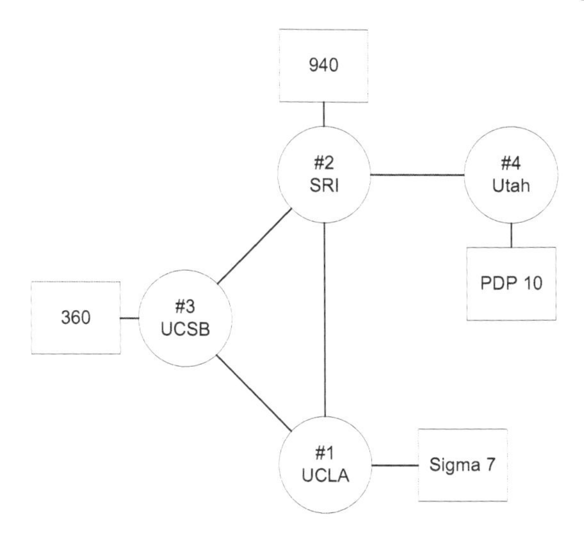

Figure 10.3: Early Internet-used phone lines

Pigeon networks

Carrier pigeons can return home from as far as 15 miles away, which has made them useful throughout history as one-way, point-to-point, wide-area networks. As late as WW1, France used 15,000 carrier pigeons as a point-to-point network between its capital and outposts. Today, France maintains a military dovecot with 150 birds. Supporters believe the birds can be useful in a disaster that takes down the cell phone network.[2] Pigeons were widespread in the United States until the late 19th century.[3]

Though dial-up networks were very rudimentary, and responsibility for traffic aggregation fell to the nodes themselves, these networks were important because they proved the viability of wide-area networking.

T-carriers/DS-signals

As wide-area networking became popular in corporations, the speed limitations of dial-up networking became apparent. Simultaneously, telecom firms saw the business opportunity in providing higher-speed point-to-point connections by aggregating the data-carrying capabilities of multiple phone lines. This led to the development of T-carriers. T-carriers, or telecom carriers, combine the data-carrying capacity of multiple phone lines to provide higher data rates. The simplest T-carrier is called T-1 and it combines the data-carrying capacity of 24 phone lines. *T-1 is a WAN service that offers 1 1.544 Mbit/s data rate.*

2 Gabriele Parussini, "In France, a mission to return the military's carrier pigeons to active duty," *Wall Street Journal*, Nov. 11, 2012.

3 Joel Greenberg, *A Feathered River Across the Sky* (Bloomsbury, 2014).

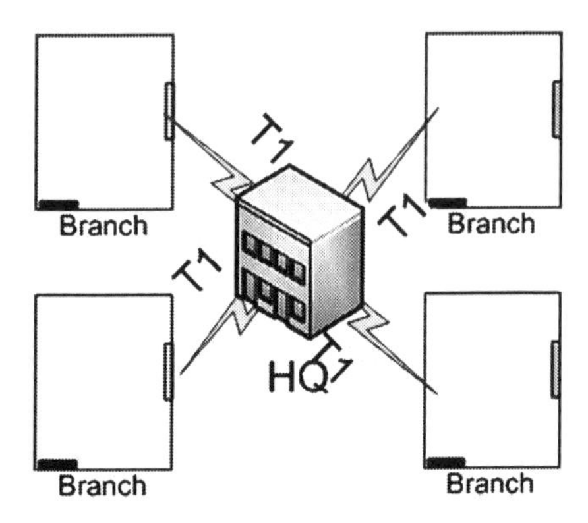

Figure 10.4: WAN built using T-1 lines

This number was presumably chosen because it resulted in a data rate of slightly more than 1.5 Mbps, which is suitable for most small- to medium-sized businesses.[4]

The data-carrying signals transmitted over the T-carriers are called digital signals, or DS for short. Formally, therefore, the T-carrier is the physical line carrying signals and the DS-signal is the signal or data transported over the T-carrier. In practice, the two terms are used interchangeably in the industry.

Higher data rates than 1.5 Mbps are possible by aggregating T-carriers. Table 10.1 shows the most common T-carriers/DS-signals popular in the industry and the number of phone lines aggregated to provide the data rates.

Table 10.1: T-carrier/DS-signal hierarchy

No. of phone lines aggregated	T-carrier name	DS-signal name	Data rate
1		DS-0	64 Kbps
24	T-1	DS-1	1.544 Mbps
96	T-2	DS-2	6.312 Mbps
672	T-3	DS-3	44.736 Mbps

T-carriers have been extremely useful in helping organizations connect offices spread across the world. An organization with one headquarters and four branch offices can create a WAN as shown in Figure 10.4. Each branch office connects to the headquarters using a T-1 line. Traffic between branches is routed through the headquarters.

4 56 Kbps * 24 = 1.344 Mbps, which is less than the T-1 data rate of 1.544 Mbps. The higher data rate is achieved because the T-1 carrier attains a slightly higher data rate than the 56 Kbps available to end users on each phone line. Phone carriers achieve 64 Kbps per phone line.

Integrated services digital network (ISDN)

Another data service developed by the phone companies was ISDN. Phone companies developed ISDN as a defensive measure to retain customers who were likely to defect to cable service providers when the latter became capable of offering data services. *ISDN is a service offered by telecom companies with the goal of delivering voice, video, and data services over the standard telephone line.* ISDN data rates are offered in multiples of 64 Kbps, with 128 Kbps being quite common. A 64-Kbps connection is called a basic rate interface (BRI). Higher data rates are obtained by combining the data capacities of multiple BRI channels.

Statistically Multiplexed WANs

In spite of their many advantages, point-to-point networks have a major limitation—there is very limited traffic aggregation on point-to-point networks. A separate link must be set up for each pair of nodes that need connectivity. While this is not a major issue in small networks, as more nodes get added to the network, the number of required links grows very fast.[5] Also, since each link only connects a pair of nodes, link utilization can be low on point-to-point networks.

To address these issues, low-link utilization, and rapid growth in the required number of links, more recent WANs use techniques to aggregate (multiplex) traffic on WAN links. We now look at some of the popular multiplexing techniques used in WANs and the technologies that use these techniques.

In statistically multiplexed networks, switches collect data packets from multiple input sources and send them out over a shared long-distance link to the next node. *Statistically multiplexed WANs allocate network resources according to need.* An example is shown in Figure 10.5, which compares how two organizations with offices across the country would be served using point-to-point T-1 links and with statistically multiplexed links. Statistical multiplexing is closest to the principles used in aggregating traffic on the interstate highway system.

The upper figure shows connectivity using point-to-point links. In the figure, organizations A and B each have their own dedicated T-1 links to connect the offices on the two coasts. It may be noted that in this configuration there is no sharing of resources among organizations A and B. The lower figure shows statistical multiplexing for WAN connectivity. When using statistical multiplexing in the WAN, offices from both organizations use dedicated short-distance T-1 links to connect to a carrier access point in their city. A switch at the access point in each city aggregates traffic from both organizations and sends it over the multiplexed link to the other city. The access point also de-multiplexes (separates) incoming traffic and directs incoming traffic from the WAN link to the appropriate office, such as B_1 or B_2. Note that the long-distance link in this configuration is shared between organizations A and B.

Comparing the two configurations in Figure 10.5, it is easy to see that statistical multiplexing is likely to give higher link utilization, albeit at the cost of slightly higher system complexity. The multiplexed system requires the carrier to create access points in each city.

5 In general, full connectivity in a network with n nodes requires n(n-1)/2 links. A network with four nodes requires six links, and a network with eight nodes requires 28 links.

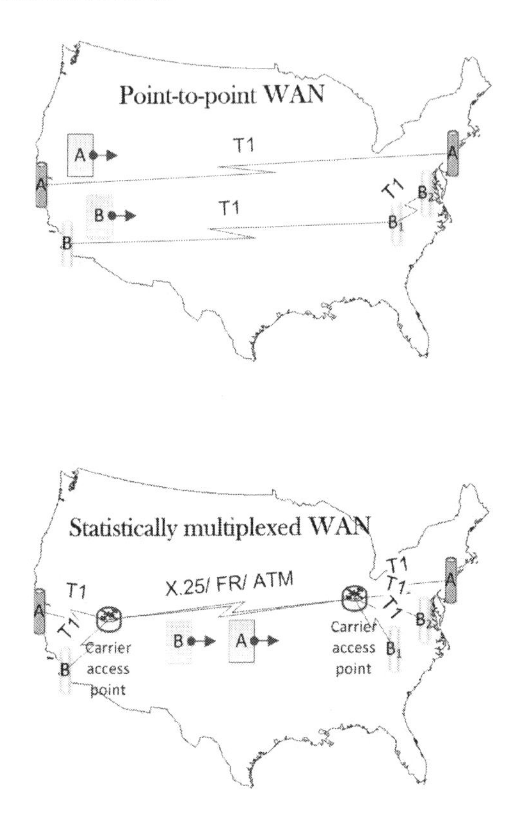

Figure 10.5: Comparing point-to-point and statistically multiplexed WANs

It also requires equipment at each access point to aggregate outgoing traffic from A and B for transmission on the WAN link. Access points must also be capable of disaggregating incoming traffic on the WAN link, to the different offices—B_1, B_2, and A. The advantage of this complexity is that the data-carrying capacity of the long-distance link—the most expensive part of the network—can now be shared among organizations A and B. This greatly improves the utilization of the capacity of the WAN link.

Depending upon the traffic patterns of customers, aggregating traffic can improve link utilization by reducing the *burstiness* of the traffic on the link. As an example, consider Figure 10.6 where most of the traffic from organization A originates in the morning and most of the traffic from organization B originates in the evening. When the traffic from the two organizations is aggregated, the overall traffic is smoother than the traffic from either organization A or B. As a result, the investment in setting up the aggregate link will be utilized at high levels throughout the day. If each of the two organizations had dedicated point-to-point links (upper figure in Figure 10.5), the link serving organization A would mainly be used during the day and the link serving organization B would only be used in the evening. In general, the higher link utilization of the statistically multiplexed WANs throughout the day allows the fixed costs of operating the WAN to be shared by A and B, leading to lower communication costs for both A and B.

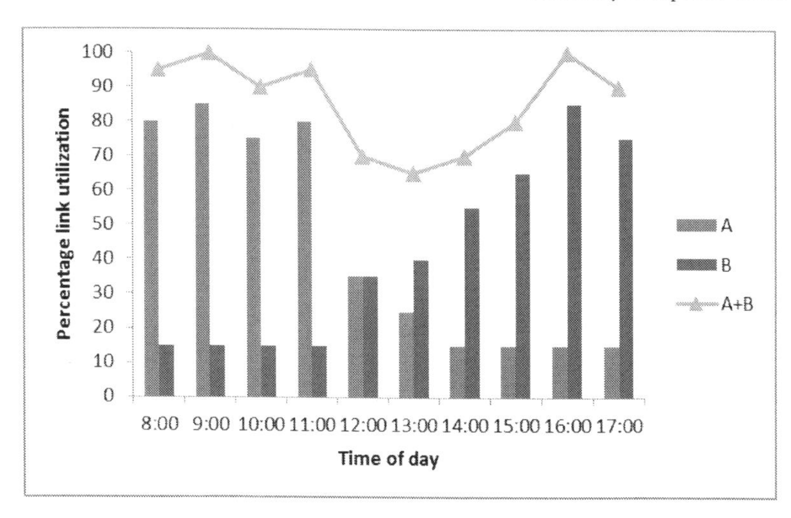

Figure 10.6: Reduced burstiness of aggregate traffic

Burstiness in computer networks results primarily from the large differences in the speed at which the network can transmit data and the speed at which human beings can process the data. The only constraint to a transmission is the TCP window size. Most computers can process torrents of data, but humans cannot. As a result, computers download a human-readable chunk, wait for the human to process it, and then download another chunk. Where data transfers are processed by machines, for example in backups and other data center operations, data transmission can be minimally bursty.

We now look at some specific statistically multiplexed WAN technologies.

Virtual Circuits

Before looking at the specific technologies, it is necessary to introduce one concept associated with statistically multiplexed WANs: virtual circuits. In the previous section, we saw that it was necessary for the access point at the receiving end to disaggregate and direct traffic to organizations A and B as appropriate. The question is, if all the data arrives over one link, how does the receiving access point know which data belongs to organization A and which data to organization B?

Table 10.2: Virtual-circuit IDs for our example

Connection	Virtual circuit ID
A-A	1
B-B_1	2
B-B_2	3

Statistically multiplexed WANs accomplish this using the concept of virtual circuits. *A virtual circuit is a communications arrangement in which data from a source may be passed to a destination over various real circuits.* Though the definition sounds complex, in practice it is quite simple. Virtual circuits are implemented by assigning a unique virtual-circuit ID for each source-destination pair that is sending data through the network. Packets entering the network from the specified source that are directed to the specified destination are labeled with the virtual-circuit ID associated with the pair. The network uses the virtual-circuit ID to direct the packet to the correct access point. At the access point, the virtual-circuit ID helps the access point pass the data to the correct customer. You may notice that the virtual-circuit ID (VCID) shares many of the properties of MPLS labels.

To see this by example, let us assign virtual-circuit IDs to the connections in our example of Figure 10.5, as shown in Table 10.2. Considering data flowing from the West Coast to the East Coast, we can see that there are three virtual circuits. As data from the sources reaches the West Coast access point of the carrier, the data packets are labeled with the appropriate virtual-circuit IDs. For example, data from B going to B_1 gets virtual-circuit ID 2. Figure 10.7 shows some packets with virtual-circuit labels flowing through the shared network link. Switches within the network use these virtual-circuit IDs to deliver the packets to the access point on the East Coast. At the East Coast, the access point reads the virtual-circuit ID and passes the packets on to the correct target. For example, the virtual-circuit ID 2 tells the access point that the data should be sent to B_1.

The term *virtual* may be interpreted as "almost but not really." For example, virtual reality refers to technology that makes you feel like you are in a place, though you are clearly not there. Similarly, a virtual circuit shares many properties of circuits, without being a real

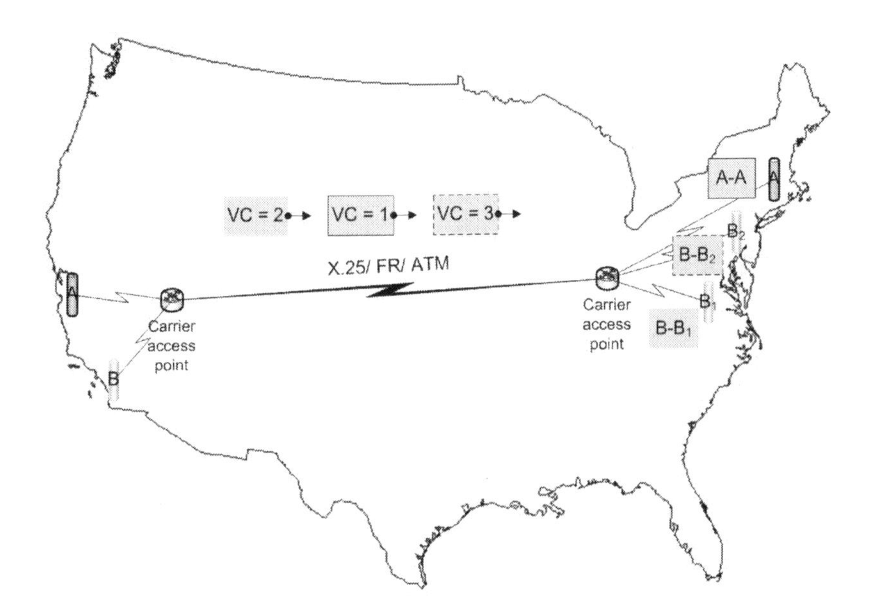

Figure 10.7: Virtual circuits

circuit. A virtual circuit is an arrangement where data can be exchanged between a pair of users using various technologies. It is different from a real circuit because a real circuit has a defined physical bidirectional path between the source and destination.

You might be wondering, why do we need yet another address? Don't we have MAC addresses and IP addresses? Are they not adequate? Why can't IP addresses indicate the source and destination? What benefit do virtual-circuit IDs provide that IP addresses do not? The primary advantage of virtual-circuit IDs is that, whereas IP addresses are global, virtual-circuit IDs are local to the telecom carrier. Virtual-circuit IDs are data-link layer addresses. This makes virtual-circuit IDs much easier to use than IP addresses. Since IP addresses are global, to make IP addressing work, routers need to advertise the accessibility of every IP network to all other networks around the Internet.[6] However, since virtual circuits are local to a carrier, virtual-circuit IDs are local to the carrier and only need to be communicated to other devices within the carrier's own network.

Whereas IP addresses and MAC addresses are assigned to the device, virtual-circuit IDs are assigned to the source-destination pair. A retailer with one head office and 50 branches would pay for 100 virtual circuits—one virtual circuit from each store to the headquarters and one virtual circuit from the headquarters to each store.

Virtual circuits in daily experience

Recall a recent flight. What was the flight number? Chances are it was something like AA1339. This is an example of a virtual circuit. Looking up AA1339 will tell you that this is a flight from Tampa (source) to Dallas/Fort Worth (destination). You can also find other details such as the fact that it is a three-hour flight that departs early in the morning and uses a Boeing MD-80 aircraft.

Using virtual-circuit numbers like this is convenient for the airline's internal operations (only one VC number is needed instead of two separate addresses), and also offers more flexibility. For example, it can assign other virtual-circuit numbers (flight numbers) to other flights between the same source and destination (e.g. AA2456), leaving at different times.

Bus routes in most cities follow a similar pattern. People know that a bus with a specific number (e.g. 51X) will take them to their destination. In the US, it is the same with interstate numbers. Users in Tampa, for example, know that I-75 will take them to Atlanta, GA. The interstate numbers have many properties of VC numbers.

Virtual-circuit numbers may be seen as a provider-centric view of the network, whereas source and destination addresses may be seen as a user-centric view of the same network.

6 There's also the issue of private addressing. If organizations A and B used the RFC 1918 address space, using IP addressing on the shared link would be impossible because of potential duplication of addresses.

X.25/Frame Relay/ATM

Three popular, statistically multiplexed WAN technologies, in the order in which they were developed, are X.25, Frame Relay, and ATM. Each technology defines a frame format suitable for the technology. As shown in Figure 10.7, customers use short-distance, dedicated links such as T-1 to connect to the most convenient access point from the carrier. Between access points, the carrier uses statistical multiplexing techniques to aggregate and switch traffic from multiple virtual circuits to the respective destinations. The primary advantage of these technologies compared to point-to-point networks is that they allow end users to pay only for the contracted data rates, much like contemporary data plans offered by mobile phone operators.

X.25 is the oldest of the three technologies. It was standardized in 1976 and is capable of data-transfer rates from 56 Kbps to 2 Mbps. It is, therefore, a relatively slow technology. However, the unique feature of X.25 is that it is capable of reliably transferring data over unreliable links. X.25 accomplishes this by using link-by-link acknowledgments as the data is transferred. This is done because data networks in the 1970s were extremely unreliable and the acknowledgments enabled X.25 to repeat as many attempts as necessary to account for link failures and ensure reliable data transfer.

X.25 was developed by telephone standards bodies to connect ATM machines. The application of the protocol to machines handing out money necessitated meticulous attention to error correction from source to destination, regardless of the reliability of the underlying networks.

As telecom networks became more reliable, the link-by-link acknowledgments of X.25 became unnecessary. The additional processing required to handle the acknowledgments made the technology relatively complex. When the underlying networks became highly reliable, the additional complexity to ensure reliability served no useful purpose. Frame Relay was developed to streamline the protocol by eliminating link-by-link acknowledgments. It was standardized in 1992 and specified data-transfer speeds in the range of 56 Kbps to 45 Mbps. Frame relay is a WAN technology that transfers data in variable-size units called "frames," while leaving error-correction responsibilities to the end-points.

How do end-points achieve reliability? Recall from Chapter 1 that the transport layer processes the data before the application receives it. When reliability is deemed useful, the sender and receiver use TCP, which ensures that any dropped data is retransmitted by the source.

The third statistically multiplexed technology is ATM (Asynchronous Transfer Mode). It was standardized in 1992 and, like Frame Relay, it eliminates link-by-link acknowledgments. *ATM is a WAN technology that can transport all forms of data traffic, including voice, data, and video.* ATM is an improvement over Frame Relay in two respects. First, it can support higher data rates than Frame Relay. ATM data rates are in the range 1.544 Mbps–622.08 Mbps.

Second, ATM is designed to meet the needs of networks that carry different kinds of traffic. Frame Relay networks treat all data packets with the same priority. But ATM networks can transfer different kinds of data with different levels of priority. For example, voice packets can be delivered at the highest priority to ensure jitter-free voice reception, and e-mail packets can be delivered at the lowest priority because users have no expectation of instant delivery of e-mail. With this capability, ATM networks have the opportunity to charge higher prices for superior classes of service.

TDM WANs

The third category of WANs is time-division multiplexed WANs. In this technology, the available data rate on the physical medium is divided into multiple time slots and each customer is allotted some specified time slots. *Time-division multiplexing (TDM) is a method of simultaneously transferring multiple data streams over a common path using synchronized switches at each end of the path so that each data stream only takes a fraction of time.* The most popular implementation of TDM WAN technology is called SONET, for Synchronous Optical NETwork. Commercially, the data service on SONET is called OC, for optical carrier.

Table 10.3: Optical carrier hierarchy

SONET service name	Data rate	Data + overhead
OC-1	50.112 Mbps	51.84 Mbps
OC-3	150.336 Mbps	155.52 Mbps
OC-12	601.344 Mbps	622.08 Mbps
OC-48	2.405376 Gbps	2.488320 Gbps
OC-192	9.621504 Gbps	9.953280 Gbps
OC-768	38.486016 Gbps	39.813120 Gbps

Unlike X.25, Frame Relay, and ATM, which are considered data-link layer technologies, SONET is generally considered a physical layer technology because SONET does not define any packet or frame formats. It is simply a method of efficiently using the data-carrying capacity of the physical medium. Each slot on a SONET network can carry bits from any technology. It is common for X.25/FR/ATM frames to be transported over SONET links. SONET data rates were first standardized in 1988 and higher data rates have been added over time as technology has evolved. The current SONET data rates are shown in Table 10.3. As seen from the table, SONET defines extremely high data rates. The slowest OC carrier can transport a T-3 link.

FDM WANs

With the improvements in data-transfer capabilities of optical fibers, it is now possible to send extremely high data rates on each strand of fiber. With current technologies, each fiber can support data rates of up to 40,000 Gbps or 40 Tbps (terabits per second). No user needs such high data rates, and current electronic technology capable of handling such high data rates can be very expensive. It is therefore much more efficient to split up the data-handling capacity of an optical fiber into multiple channels, where each channel supports a SONET channel at data rates such as OC-192 or OC-768.

To accomplish this, the data-carrying capacity of a strand of optical fiber is separated into multiple smaller channels by transmitting the signals from each channel at a different frequency. *Frequency-division multiplexing (FDM) is a signal-transmission technique in which the total bandwidth of a communication channel is divided into non-overlapping sub-bands, each of which is used to carry a separate signal.*[7] The available bandwidth of optical fiber is split into three bands—L, C, and S. Each band has 50 channels and each channel can support data rates up to 10 Gbps. Comparing these rates with Table 10.3, we see that each optical FDM channel can support one OC-192 channel. The set of specified frequencies on optical fiber are called the ITU grid and were defined in 2001.

In the early generation optical fibers, it was convenient to send data as one channel. But no end user can use the entire data-carrying capacity of a fiber-optic channel. TDM was introduced to enable the high capacity of the optical channel to be split among multiple end users. Later generations of optical fiber technologies enabled DWDM, where multiple optical channels can be sent at different frequencies. In these systems, TDM is used within each channel and the different channels are multiplexed using FDM.

DWDM is a lot like dividing a wide interstate highway into multiple, narrower lanes. Each lane is suitable for vehicles that typically use the highway. There is no point in making the entire highway available to a single vehicle because no vehicle is capable of utilizing the entire width of the road.

WANs and the TCP/IP Stack

Before closing this chapter, let us take a moment to see how WANs are related to other technologies in the TCP/IP stack. Typically, WAN technologies are used to transport IP packets. In the TCP/IP stack, WAN technologies lie between IP and the signals delivered by the physical layer, just as Ethernet does. In modern networks, WANs are therefore typically treated as long-distance equivalents of Ethernet. In general, packets traverse multiple Ethernets and WAN links on their way to the destination. As an example, a traceroute from USF to google.com is shown in Figure 10.8. We see that the first three hops in the path are within the USF LAN, and the remaining hops are WAN hops that take the packet to the destination. This traceroute example is intended to demonstrate that LAN links and WAN links are treated similarly within the network architecture. Since we have already discussed

7 In a wave, frequency * wavelength = constant. Therefore, there is a unique wavelength associated with each unique frequency. For this reason, FDM in optical fiber is also called wavelength division multiplexing. Modern fibers pack the multiplexed wavelengths very tightly. Hence the technology is also called dense wavelength division multiplexing.

```
# traceroute www.google.com
traceroute to www.google.com (74.125.159.103), 30 hops max
1 vlan272.edu-msfc.net.usf.edu (131.247.16.254) 0.997 ms
2 vlan254.campus-backbone2.net.usf.edu (131.247.254.46) 0.567 ms
3 vlan256.wan-msfc.net.usf.edu (131.247.254.81) 4.856 ms
4 tpa-flrcore-7609-1-te31-v1602-1.net.flrnet.org (198.32.166.177) 1.086 ms
5 mia-flrcore-7609-1-te24-1.net.flrnet.org (198.32.173.125) 31.991 ms
6 peer-google-flrnetcp-nota-1.net.flrnet.org (198.32.173.126) 6.269 ms
7 72.14.236.178 (72.14.236.178) 6.281 ms
8 209.85.254.14 (209.85.254.14) 20.159 ms
9 yi-in-f106.google.com (74.125.159.106) 20.255 ms
```

Figure 10.8: Traceroute showing LAN and WAN links in path

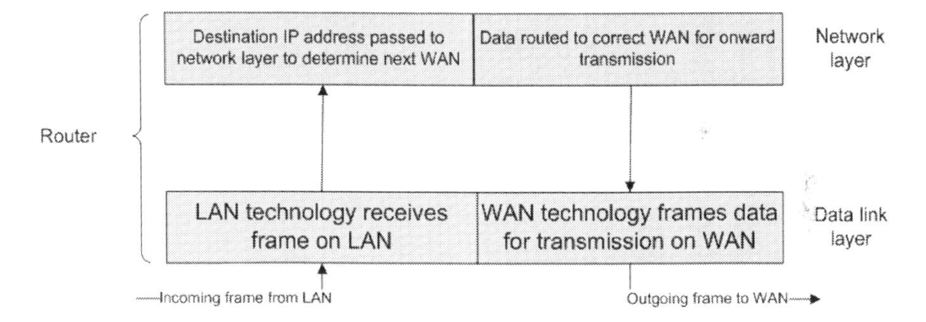

Figure 10.9: WANs in relation to IP and LANs

in detail in Chapter 3 that Ethernets are placed in the data-link layer of the TCP/IP stack, the example shows that WANs are also treated as a data-link layer technology in networks.

Figure 10.9 shows what happens at a router that interfaces between a LAN and a WAN. When a data frame arrives at the LAN interface of the router, the LAN header is stripped away to retrieve the IP header. The IP destination address is then used by the network layer to determine the appropriate WAN interface to forward the packet to. The packet is passed to this interface, which formats the packet into the appropriate frame format, adding suitable virtual-circuit IDs for correct delivery.

EXAMPLE CASE—Unmanned Aerial Vehicles

During 2000–2015, the US Air Force went through one of its most radical transformations ever. In 2001, all fighter aircraft deployed by the Air Force were conventional piloted aircraft. Fast forward to 2009, and the Air Force trained more "pilots" for unmanned aerial vehicles (UAVs) than for manned fighters. The Air Force believes that UAVs have been the most effective weapon systems in Iraq and Afghanistan for saving the lives of American soldiers. UAVs are operated entirely using global satellite-based WAN data networks.

In the war of 1973, Israeli fighter jets took a heavy beating from enemy missiles. In response, Israel developed the world's first UAV. When these UAVs were deployed in the next Arab war in 1981, the real-time images provided by the UAVs helped Israel destroy enemy air defenses, without losing a single pilot. This proof-of-concept convinced many militaries in the world of the viability of UAVs.

The US Air Force has experimented with UAVs since the 1980s. In the 1990s, UAVs, also called drones, were used mainly as remote video cameras, sending pictures of enemy movements over satellite-based data networks to intelligence analysts. Each UAV (Figure 10.10) is capable of sending a live video feed for up to 22 hours with enough resolution to detect a license plate.

In the 2000s, UAVs shot up in popularity as complete weapons systems. The Air Force barely used UAVs in the Iraq war of 2003, but in 2010 the Air Force had an extensive array of UAVs, which now account for about a third of the aircraft in the US Air Force. The Air Force also has unmanned vehicles on the ground for bomb disposal, etc. These unmanned ground vehicles help the forces remove an advantage held by enemy guerillas—their willingness to give up their lives as human bombs in order to inflict casualties upon American forces.

Figure 10.10: The Reaper UAV (Drone)
(Source: Google Maps)

UAVs would not be possible without global satellite-based WAN data networks. UAVs in war zones anywhere in the world are piloted from Air Force bases in the United States. UAVs use satellite-based WAN data networks to send detailed real-time video footage of enemies to intelligence analysts to determine the identity of enemy targets. On confirmation from intelligence analysts, sensor operators use data networks to launch and guide on-board missiles from the UAVs to targets.

The Predator UAV includes a satellite link that uses a 20-foot satellite dish. The satellite link is used to operate the aircraft when it is beyond the line of sight. The satellite link is also used to transmit video feeds. UAVs transmit thousands of hours of video in war zones each month. The military's networks allow this video to even be transmitted directly to troops on the ground.

Adoption of UAVs has forced the Air Force to rethink its training and operational strategies. While it is currently transferring combat pilots to fly UAVs, the Air Force has decided to create a separate career track for UAV pilots. The training effort required to learn to fly a UAV is comparable to the effort required to earn a master's degree. UAVs crash due to lack of proper training and device malfunctions. Some crashes are the result of old habits. Combat aircraft are capable of executing sharp turns at high speeds, but the engines used in Predators have about as much power as a snowmobile engine, and are incapable of executing these maneuvers. When former pilots execute these moves, the current generation of Predators can go out of control. Other crashes are due to the imperfect operator interface. The button to launch a missile is adjacent to the button to switch the engine off, and operators have occasionally turned the vehicles off in mid-flight.

The relative immaturity of the technology used in UAVs has caused some embarrassment to the Air Force. In December 2009, it was reported that Iraqi militants were able to intercept the video feeds being transmitted by the UAVs to ground troops. The software, SkyGrabber, used to intercept the videos was quite inexpensive, costing only $26. SkyGrabber was effective because the transmission from the UAVs was not encrypted until April 2009. Earlier, the leadership of the Air Force had dismissed the possibility that insurgents in Iraq and Afghanistan might possess the technical competence to intercept unencrypted video. (Encryption is covered in the supplement to this book.)

The improved capabilities of UAVs are enabling the Air Force to deploy new counterinsurgency techniques. UAVs fly numerous daily surveillance patrols in war zones. UAVs may also be lowering the cost of war. Where each F-22 Raptor combat aircraft costs approximately $150 million, the current generation of Predators costs $4.5 million each, and Reapers will each cost about $12 million. While combat aircraft are used to fire heavy weapons, UAVs have shot missiles on targets in Iraq and Afghanistan. Some of these missiles killed "high-value" targets, considerably reducing the security threats facing the country.

Interface improvements may further improve efficiency. Simplifying the number of steps it takes to fire a missile, one pilot could even operate a fleet of UAVs.

For all their advantages, the use of UAVs is not without controversy. Human rights advocates argue that UAVs kill innocent civilians. Militaries argue that the detailed video footage from UAVs improves the likelihood that targets are correctly identified and the lighter bombs used by UAVs reduce collateral damage compared to the heavy artillery used by combat aircraft.

The use of UAVs also creates work-life balance issues. Combat pilots at war are close to enemy territory, away from family, and engaged in the action round the clock. Piloting a UAV is almost a desk job. Pilots return home at the end of their shifts, where they may attend a birthday party just hours after destroying an enemy building or killing members of an enemy patrol.

UAVs may make war safer for the world's militaries. However, one of the biggest constraints on the willingness of political leaders to wage war is the public outcry that follows the death of soldiers in enemy territory. If UAVs eliminate that fear, will powerful countries be more willing to use force to resolve disputes?

References

1. Dreazen, Y.J., S. Gorman, and A. Cole. "Officers Warned of Flaw in US Drones in 2004." *Wall Street Journal*, Dec. 18, 2009.
2. Drew, C. "Drones Are Weapons of Choice in Fighting Qaeda." *New York Times*, Mar. 16, 2009.
3. Hagerman, E. "Point. Click. Kill: Inside the Air Force's Frantic Unmanned Reinvention." *Popular Science*, Aug. 18, 2009.
4. Levinson, C. "Israeli robots remake battlefield." *Wall Street Journal*, Jan. 13, 2010, A10.

5. "Predators and Civilians," Editorial, *Wall Street Journal*, July 14, 2009.
6. Siobhan, G., Y.J. Dreazen, and A. Cole. "Insurgents hack US drones." *Wall Street Journal*, Dec. 17, 2009.
7. Wikipedia article on MQ-1 Predator, https://en.wikipedia.org/wiki/General_Atomics_MQ-1_Predator (accessed Apr. 9, 2016).

Summary

This chapter introduced wide-area networks (WANs). WANs are used to transfer data in networks with a large number of users, typically over a wide area. We started by showing why simpler broadcast technologies as used in Ethernet are unsuitable for long-range transmission in networks with a large number of nodes. The road network was used to help visualize the differences between LANs and WANs. We saw how WAN technologies help lower costs by improving link utilization even though WANs are more complex than LANs. We then saw the four major approaches to creating WANs—point-to-point, SDM, TDM, and FDM. Virtual circuits, which are used by SDM networks, were also introduced.

About the Colophon

The first known reference to human ability to reduce barriers arising solely from geographic distance appeared in the 19th century (1896) in an article in the *Los Angeles Times*. Since then, there have been vast improvements in air transport and commercial shipping, which have further reduced geographical barriers. It is now a routine matter for customers to buy goods at the neighborhood grocery store that have been manufactured around the world. Executives also routinely fly to meetings on other continents and return to their offices within the week.

Wide-area networking is another recent technology that is bridging distances even further. People socialize with friends using websites like Facebook; collaborate on projects with partners worldwide using instant messaging; and use video conferencing to replace travel except when face-to-face communication is absolutely necessary. All of this is made possible by the extremely reliable WANs deployed today.

REVIEW QUESTIONS

1. What are *WANs*?
2. Why can't we use broadcast on WANs as is done on LANs?
3. Why is Ethernet unsuitable as a WAN technology?
4. What are some similarities between the interstate system and WANs?
5. What are some similarities between neighborhood roads and LANs?
6. What are the main categories of WANs?
7. How is the phone network used as a wide-area computer-networking technology?
8. What are *T-carriers*?
9. What are the common data rates of T-carriers?

10. What are *DS-signals*?
11. T-carriers are used to create a full mesh network with five nodes. How many links will be required?
12. What are the limitations of point-to-point WANs such as T-carriers?
13. What is *statistical multiplexing*?
14. How is statistical multiplexing useful in WANs?
15. How does statistical multiplexing reduce burstiness of traffic in the physical medium?
16. What are *virtual circuits*?
17. What is a *circuit*? How is a virtual circuit like a circuit? How is a virtual circuit different from a circuit?
18. Why are IP addresses not used for addressing within virtual circuits?
19. What is *X.25*? What are some salient features of the technology?
20. What is *Frame Relay*? What are some salient features of the technology?
21. What is *ATM*? What are some salient features of the technology?
22. What is *time-division multiplexing*? How is TDM useful in WANs?
23. What are some standard data rates of SONET, the popular TDM WAN technology?
24. What is *frequency-division multiplexing* (FDM)? How is FDM used in WANs?
25. Describe how WANs may be considered a data-link layer technology.

EXAMPLE CASE QUESTIONS

1. Identify three advantages and three disadvantages of UAVs compared to combat aircraft.
2. Read the Wikipedia article on the MQ-1 Predator and write a one-paragraph description of the evolution of the aircraft.
3. View the YouTube video on Predators and SkyGrabber at www.youtube.com/watch?v=O4I13Cnlpkk[8] (about 11 minutes). Identify as many satellite-based data-communication applications as possible that are used by Predators and their operators.
4. How did the military come to know that Predator feeds were being captured by militants? (Use online sources if necessary.)
5. Information about SkyGrabber is available at www.skygrabber.com/en/skygrabber.php. Briefly, in one paragraph, describe how SkyGrabber works. What is the intended use of SkyGrabber?
6. Technology developments can have unintended consequences. Some analysts have speculated that UAVs and satellite-based global data networks could make war more likely because the risks to soldiers' lives are reduced. What is your opinion about this assessment?

---

8 If the link does not work, the video was titled, "Skygrabber Intercepts Predator Drone Intelligence." You should be able to find it online. Even if the incident appears dated, the concept of overlooking design concerns is universal.

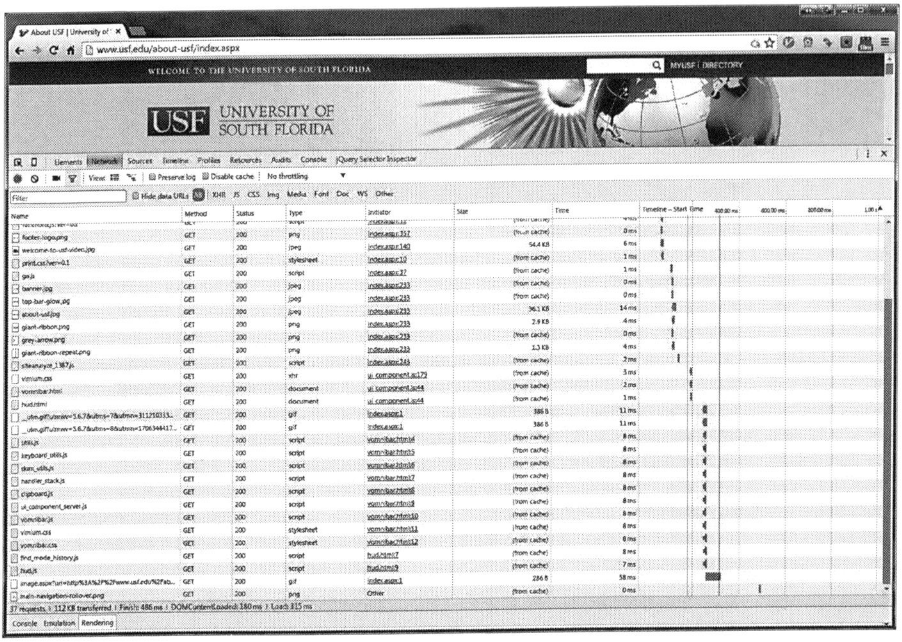

Figure 10.11: Network tab in Inspect element

HANDS-ON EXERCISE—Web Page Debugging

In the absence of an easy way to intercept WAN traffic from home machines, in this exercise we will improve our skills for debugging a common category of applications that generate web traffic—web applications. Inappropriate use of multimedia elements can generate a lot of unnecessary network traffic. But, fortunately, all contemporary browsers have a feature called "Inspect element" (or similar), which allows developers and users to debug web page performance. It is quite simple to use. Right-click on a web page and select Inspect element, which opens up a dialog similar to Figure 10.11. Every tab in the dialog is very useful. For example, elements lets you see how each element on the page is styled using CSS rules, and sources shows JavaScript source files and lets you step through JavaScript code.

For this exercise, let us use the network tab. This tab breaks down the time taken to load each component of the page. Pick one of your favorite websites and respond to the following:

1. Add a screenshot similar to Figure 10.11, showing the page and the network tab of Inspect element.
2. What are some of the slowest elements on the website?
3. What, in your opinion, is causing these elements to take time to load?
4. You will see that many elements are loaded "from cache." What is the cache on the browser? How is it useful?

CRITICAL-THINKING EXERCISE—Professional WAN

The networks we saw in this chapter connected computers located far away from each other. You may find it useful to extend this idea to develop a far-flung network of people you can count on for professional development. Weak ties, people you only meet occasionally, are a very important source of information for job opportunities.[9] Since they move in different circles, weak ties become aware of opportunities that would not typically be known to you. For example, weak ties can be maintained by creating a LinkedIn profile and nurturing it carefully. The wider your network of weak ties, the more useful it can be for you.

1. If you do not have a LinkedIn profile, please create one. What is its URL? (It is assumed that you are not averse to creating a LinkedIn profile.)
2. Distance in social networks can be measured in various ways—how often you meet someone, how different is someone's social circle compared to yours, etc. In keeping with the theme of this chapter of large distances, what are some of your farthest connections? (It is a good idea to maintain privacy, so you may report these connections in abstract terms, such as "an expert at a client firm I worked with on a project," or "a childhood friend who has relocated to a different continent," etc.)
3. In your estimate, what fraction of your LinkedIn network is a "wide-area network," i.e. composed of weak ties?
4. What are some ways your LinkedIn profile and network can help you? (A web search should be very useful.)

IT INFRASTRUCTURE DESIGN EXERCISE—WAN Design

In this chapter, we will select an appropriate WAN technology to connect all the four locations. Assuming that TrendyWidgets will use the same WAN technology for all long-haul links, answer the following questions:

1. WAN links are typically shown through a cloud using dotted lines connecting the two endpoints. Different thicknesses or colors are used to show different data rates. Assume that the company uses the following three WAN links to connect the four locations—US–Singapore, US–Amsterdam, and Amsterdam–Mumbai. The data rate on the US–Amsterdam link is 100 Mbps to accommodate the data backup traffic. All other links are 10-Mbps links. Update the infrastructure diagram from Chapter 9 to include these WAN links.
2. Of the WAN technologies covered in this chapter—dial-up, T-carriers, X.25, Frame Relay, ATM, and SONET—which technology is best suited to meeting TrendyWidgets's WAN needs?
3. After the network outage following the Middle East cable ruptures, TrendyWidgets has decided to also maintain a backup WAN capability using ISDN. Include this backup capability in your infrastructure diagram. (These backup links also terminate at the routers at each location, where the routers can use simple failover techniques to divert traffic to the backup network when needed.)

9 Mark Granovetter, "The strength of weak ties," *American Journal of Sociology*, 78(6) (1973): 1360–1380.

CHAPTER 11

Network Security

To err is human, but to really foul things up requires a computer.

"capsules of wisdom," *Farmer's Almanac*, 1978

Overview

Thus far in this book we have considered the positive aspects of computer networks, such as their ability to speed up organizational decision making and lower the costs of doing business. However, computer networks also expose the information in the organization to dangers from the network. In this chapter, we will look at the common methods used to defend organizational information against dangers originating from computer networks. At the end of this chapter you should know:

- what information security is
- what network security is and how it is related to information security
- the common elements of a risk-management model—vulnerabilities, threats, and controls
- what network security controls are commonly used to protect outgoing information
- what network security controls are commonly used to protect incoming information

Introduction

Network security is a component of information security. Considerable management attention is paid to information security. We will therefore start with a high-level introduction on information security to see why information security is important and how network security fits into an organization's overall information security setup.

Information security is the provision of confidentiality, integrity, and availability to information. Confidentiality is to preserve authorized restrictions on information to protect personal privacy and proprietary information. Integrity is to guard against improper modification or destruction of information and ensure authenticity of information. Availability is to ensure timely and reliable use of information. Confidentiality is violated when people are able to read other peoples' data, such as credit card numbers or health care records that they should not have access to. Integrity is violated when users are able to modify data they should not be able to modify. For example, if students are able to modify their own transcripts, it would constitute a violation of integrity. Availability is violated when data is not available when it should be available. For example, if the university website is down on the first day of class, or an e-commerce website is down on the last Saturday before Christmas,[1] it constitutes a violation of availability. Together, the three dimensions of information security provide assurance to end users that they can depend upon an information system to store data without fear of theft or loss, and also depend upon the output of an information system to make decisions, without fear of the data being tampered with.

There is general agreement in legal and technical circles about confidentiality, integrity, and availability (CIA) being the three dimensions of information security. Title 44, Chapter 35, section 3542 of the US Code defines information security in terms of the above three dimensions. RFC 2196, the site security handbook, also defines information security in terms of these three dimensions.

Why do organizations care so much about information security? Information security matters because we are increasingly dependent upon information for our livelihoods and way of life. As manufacturing moves offshore, professionals in the United States are increasingly involved in product design and software development. The IT sector constitutes almost 20% of the value of the S&P 500 index of large cap companies in the US. If we use the size of the IT sector in the S&P 500 as a proxy for the importance of IT in our economy, we might say that IT constitutes almost 20% of the US economy. Accounting and financial data in most companies is now stored almost exclusively on computer systems with no paper trail available for verification. Increasingly, our own personal workflows are becoming computerized. For example, most of our photographs are now stored on computers as image files. A single hard disk crash can cause permanent loss of this information. Information security, then, is critical for the smooth functioning of our economy, our businesses, and our personal and professional lives. Imagine what would happen to your career if it were trivially easy for hackers to steal designs and computer codes you created? Or tamper with financial records such as your checking account balance?

If information security is so important, how do you go about providing it? At a very high level, it helps to start with developing a risk-management model that shows how an organization's information security might be compromised. Risk-management models have three components—vulnerabilities, threats, and controls. This is shown in Figure 11.1 as a general risk-management model for information security. The figure shows the primary drivers of information risk.

> *Tom: Who do cyber sharks attack?*
> *Tim: Tell me.*
> *Tom: Web surfers!*
>
> Source: *Boys' Life* magazine, August 2013

1 In recent years, this has been the busiest shopping day of the year (not Black Friday as is popularly believed).

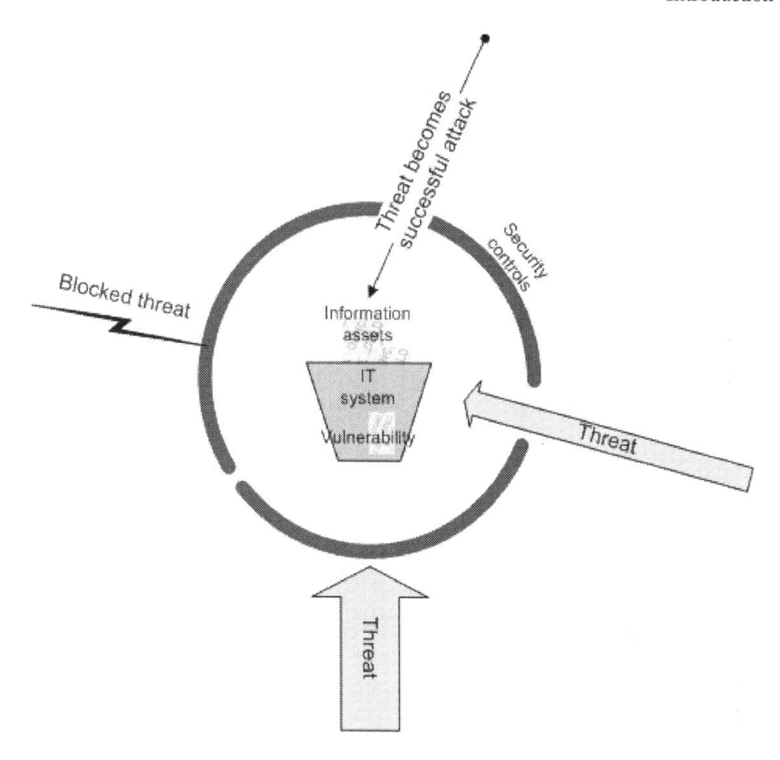

Figure 11.1: General information risk-management model

Information security begins with an identification of information assets in an organization that are worth protecting. These are shown as gold bars in Figure 11.1. Business needs generally determine what information systems are used and what information is considered worthy of protection. At a university, these assets include student academic records, staff HR records, and intellectual property developed by university faculty and students that has the potential to be commercialized. Typically, this information is stored in an information system as files and databases. In the figure, these information systems are shown as the cup holding the gold bars (information assets). It is possible for the information systems to have vulnerabilities. *Vulnerabilities are weaknesses in an information system that could be exploited by interested hackers to compromise the security of the information held in the information system.* Insecure services such as telnet, if running, are an example of vulnerabilities. Other examples are unpatched software, user accounts without passwords, or users operating their computer with administrative privileges and untrained or careless employees.

Organizations connected to networks are constantly under threat from a variety of sources that are interested in exploiting the organization's vulnerabilities. *Threats are capabilities, intentions, and attack methods of adversaries to cause harm to information.* In the figure, these threats are shown as directed arrows coming into the organization. Examples of threats include disgruntled employees, malware, viruses, motivated hackers, botnets, and even foreign governments trying to disrupt large-scale systems in the country. For example, if the organization is a retailer, hackers might be interested in reading credit card numbers as they are swiped at checkout counters. To counter these threats, professionals like you deploy information security controls in the organization. *Controls are measures taken to mitigate the dangers arising from information security threats.* Examples of controls include antivirus software, employee training, regular software patching, user account controls, and password policies. The figure shows these controls as a circle around the information system. Controls block threats to the greatest extent possible. However, due to limitations of the technology or due to limitations in the skills of the administrators responsible for these controls, it is possible that controls leave room for attackers to reach the information systems holding the information assets. If that happens, it is possible that some such attacks will successfully compromise information security.

Figure 11.1 provides a model for the responsibilities of all IT administrators and for security professionals in particular. All IT systems are constantly under threat from various sources to compromise the information security of the system. Of the three components of the risk-management model (vulnerabilities, threats, and controls), IT administrators are usually powerless to determine either the threats or the information assets that create vulnerabilities. What IT administrators can do, however, is deploy the most effective controls possible. Therefore, most discussions of information security are focused on information security controls. To facilitate understanding of these controls, they are classified in various ways. A common classification system divides information security controls into procedural, physical, and technical controls.

Figure 11.2: A typical office worker

For example, consider the typical office worker shown in Figure 11.2. What are the information security threats faced by this worker? One category of threats comes from the user's incorrect practices. For example, she might download some tempting software that also installs malware on the system. To counter these kinds of threats, she may undergo awareness training to learn best practices. This would be an example of procedural controls. *Procedural controls are controls based on formal orders and procedures.* Another category of threats comes from the office space. For example, by mistake the user might leave office doors open after work, which may lead to theft. To counter these threats, security guards may periodically verify that office doors are locked off-hours. This is an example of physical controls. *Physical controls are security measures used to deter or prevent unauthorized access.* Finally, skilled intruders may use the network connection to hack the computer. To prevent this, IT might install a firewall to block unauthorized network intrusions. This would be an example of a technical control. *Technical controls are controls that use technology to secure information.*

An important source of threats is the network. As seen in Figure 11.2, the network connection gives intruders the opportunity to compromise information security from anywhere in the world. The rest of this chapter will focus on network security.

The classification of security controls into procedural, physical, and technical controls is quite common in the industry.

Network Security

Network security is the provision of information security in the presence of dangers created by computer networks. Thus, protecting network devices (routers, computers, and other devices connected to the network) is not the primary goal of network security. Rather, the goal of network security is to provide information security from threats originating in the network. Protecting network devices is important to the extent that it secures valuable information in the organization. But at the end of the day, the information connected to the network is more valuable than the hardware connecting to the network. Viruses, phishing attacks, and intrusion attempts are all examples of information security threats inherent in being connected to the Internet. All these threats attempt to compromise valuable information in the organization. Intruders are generally less interested in causing damage to equipment. Network security controls are our defense against these dangers arising from the network.

Why does network security matter? What makes network security a particularly interesting area of an organization's overall security plan? Network security matters because increasing parts of the nation's and the world's infrastructure are connected to the network. Financial systems, payment systems, hospital records, control systems for the nation's electricity grid, and vital installations such as NASA are now connected to the network. Weakness in network security can allow hackers to obtain unauthorized access to vital information on these resources. In fact, more information security threats originate from the network than from any other source. Weaknesses in network security can get expensive. In a highly publicized incident, during 2006–2007 hackers gained access to the network of retailer T.J. Maxx by exploiting weaknesses in wireless network security, and they stole information on more than 40 million credit cards.[2] The incident is reported to have cost the retailer more than $250 million to settle the issue with banks and state regulators. In 2009, the leader of the group that was charged with the intrusion into T.J. Maxx was also charged with intruding into the databases of Heartland Payment Systems, a credit card processor, and stealing information on about 130 million credit cards. The common intruder, Albert Gonzalez, was just 28 years old and was a former government informant. There were also reports in 2009 that the US electrical grid appeared to have been penetrated by hostile interests.[3] Network security is therefore a persistent and significant challenge. In March 2010, Albert Gonzalez was sentenced to 20 years imprisonment for the offenses.

---

2 While this incident may be relatively old, it has been extensively documented, hence it is very useful for students.

3 Si Gorman, "Electricity Grid in U.S. Penetrated by Spies," *Wall Street Journal*, Apr. 8, 2009.

Table 11.1: Network security controls by category

	Incoming information	Outgoing information
Confidentiality	Patching, authentication and authorization	Encryption
Integrity	Firewalls	Digital signatures
Availability	Virus protection, end-user training	Redundancy

There are many technical controls for providing network security. Since the network is used to receive and to send information, network security controls are utilized to protect both incoming and outgoing information. In this chapter, network security controls are classified according to whether they primarily defend against incoming or outgoing data. In each case, we will look at the important controls to provide confidentiality, integrity, and availability. A summary is provided in Table 11.1. Though most of these controls defend against many threats, the table tries to identify the primary goal of each control. For example, one of the primary goals of firewalls is to maintain integrity by limiting network access to friendly computers. But firewalls also help maintain availability by blocking out intruders who send large volumes of data to networks to try to keep devices busy without reason.

We will now look at each of the six cells in Table 11.1 in more detail.

Network Security Controls for Incoming Information

What are the threats from incoming information? Most of us have encountered at least one situation where we received a virus as an e-mail attachment. We also often get e-mails trying to get our bank account credentials. Without our knowledge, intruders constantly try to send information to our computers to take control of them and install malicious software. The controls discussed in this section defend against these threats from incoming information.

Controls for Confidentiality

Confidentiality is to preserve authorized restrictions on information to protect personal privacy and proprietary information.

During 2006–2009, two of the most publicized incidents of information-security compromise related to the ability of a group of hackers to read credit card data without authorization. At Heartland Payment Systems, in 2008, the group exploited an improperly coded website to insert an application on the system that allowed the intruders to read credit card data from Heartland's databases. At T.J. Maxx, the group was able to use passwords of store managers to read credit card records.

In October 2015, Gery Shalon, Josh Aaron, and Ziv Orenstein were indicted in the US District Court, Southern District of New York, for stealing more than 100 million customer records from large financial institutions such as JPMC. These e-mail addresses were used to operate a stock "pump and dump" scheme, resulting in tens of millions of dollars in illegal profits.[4, 5]

Patching

Modern software is extremely complex. In spite of the best efforts of developers, software products, including operating systems and applications, have weaknesses that can be exploited by intruders. A common example is web applications that do not check user input. Knowledgeable attackers can exploit this weakness and compromise information security. This is what happened in the Heartland case. A web application at the company was not adequately verifying user input in a form field. The attackers were able to use this weakness to send carefully crafted SQL commands as form inputs and insert a malicious application on Heartland's systems. This application allowed the intruders to read credit card data from the databases behind the website. This attack is called an SQL injection attack. Many other software vulnerabilities to incoming data are possible.

When software weaknesses become known, developers quickly issue updates to fix problems. These updates are called patches, and the process of applying updates is called patching. A suitable patch at Heartland would have added input checks to block the hackers' ability to send improper form inputs and compromise the system.

Regular patching ensures that applications get updated as soon as updates become available. Most applications and operating systems are capable of automatic updates. Large organizations prefer to test patches before applying them to make sure that updates do not adversely affect other applications. However, for homes and small organizations, automatic updates are highly recommended.

As we will see shortly, patching is also an extremely important measure for reducing the organization's vulnerability to viruses and worms. Many viruses and worms exploit vulnerabilities in unpatched software. A specific example of the Slammer worm is discussed in a later section.

Authentication and Authorization

What if your computer did not ask you to provide a user name and password before allowing itself to be used? Anybody who succeeded in getting physical access to your computer would be able to read any file on your computer. The same thing can happen in networks. At T.J. Maxx, the attackers found a retail store where the attackers were able to read the user names and passwords of store managers. Due to poor authorization policies at the company, the user name and password of store managers gave the attackers access into the central databases at the organization where credit card records were stored.

What could T.J. Maxx have done to guard its information? Later in this chapter we will see what the company could have done to prevent the intruders from obtaining the user names and passwords of store managers. But even if the user names and passwords of

---

4 http://www.justice.gov/usao-sdny/pr/attorney-general-and-manhattan-us-attorney-announce-charges-stemming-massive-network.

5 http://www.justice.gov/usao-sdny/file/792506/download.

some users at retail stores had been compromised, if T.J. Maxx had limited database access to a small set of system administrators, the unfortunate incident that cost more than $250 million could possibly have been avoided.

Authentication is the verification of a claimed identity. Authentication is what users do when they provide a user name and password to access a secure website. No person other than the rightful user is expected to know the correct password. Therefore, providing the correct password is used as a confirmation of identity. Authentication should be required before allowing users to perform sensitive operations. For example, banks ask you to prove your identity before allowing you to open an account. Various forms of proof are acceptable, such as a driver's license or a passport. The important thing is that satisfactory proof of identity is required before the bank will allow you to open an account. Similarly, computers connected to protected data should be configured to require users to provide proof of identity.

To give access to new users, IT administrators typically assign new users a username and the system immediately generates a random password that the user changes at the earliest opportunity. The new password is known only to the user. Therefore, the username and password authentication scheme is reasonably secure. When higher levels of security are required, users may be asked to provide thumbprints or other forms of identity for authentication.

Typically, authentication is performed in three ways—something you know (e.g. a password); something you have (e.g. an ATM card); and something you are (e.g. a thumbprint). Each of these ways is called a factor of authentication. Increasing the required number of factors for authentication typically improves security.

Authentication is useful, but it is only half of the process for maintaining information confidentiality from outsiders. The other half is authorization. *Authorization is the granting of rights to a user to access, read, modify, insert, or delete certain data, or to execute certain programs.* Authorization limits opportunities for accidents and abuse. After a user is authenticated, they are granted specific permissions to a defined set of resources they are authorized to access. This way, even if a user account is compromised, the damage from the compromise is limited to the resources the user had access to. Authorization is what happens in normal security procedures. Employee ID cards allow employees access to the offices where they should have access, but not to all offices in the organization. Now, consider what happened at T.J. Maxx. The hackers were able to obtain the credentials of store managers due to a weakness in the wireless setup at a store. There was no reason for store managers to have permissions to access corporate databases. Accordingly, the company should have given store managers access only to the terminals in the stores. Had this been done, the problem may not have escalated. However, since store managers had the required authorization on corporate databases, when their accounts were compromised, the attackers were able to read credit card data stored in these databases. T.J. Maxx did perform authentication, but it does not seem to have performed proper authorization. Limiting authorization on the database to a small set of IT administrators could have prevented the incident even if the passwords of store managers were compromised.

Therefore, authentication should be followed by authorization. A good initial set of permissions might be the ability to read and write files in the user's own home directory and the permissions to execute the company's default set of computer applications.

This would allow employees to do their jobs, while at the same time maintaining confidentiality and integrity.

Patching and authentication and authorization are commonly used to thwart attacks on confidentiality from incoming data.

Password best practices

Since passwords are so important, best practices for passwords have been developed to guide administrators. The goal of a good password policy is to prevent intruders from being able to guess passwords. To that end, recommendations from Microsoft include the following:[6] (1) good passwords should include characters other than just the alphabets; (2) actual names or words should be avoided; (3) passwords should have at least eight characters (in fact, 12–14 characters are increasingly becoming the recommended standard); and (4) passwords should be changed regularly.

Controls for Integrity

Integrity is to guard against improper modification or destruction of information and ensure authenticity of information.

During the battle over South Ossetia between Russia and Georgia in August 2008, a number of Georgian government websites were defaced. These included the websites of the President of Georgia, the Ministry of Foreign Affairs, and the National Bank (the country's central bank). In most cases, the home pages of the sites were replaced with images of well-known dictators.

Firewalls

A firewall is a computer that lies between two networks and regulates traffic between the networks in order to protect the internal network from electronic attacks originating from the external network. Typically, firewalls peek into incoming packets to obtain the source and destination IP addresses and destination port address of the packet. This information helps the firewall determine where the packets are coming from and what applications the packets are trying to reach. Firewalls allow administrators to specify rules that limit which packets may be allowed to enter the network. The rules are specified as an access list. *An access list is a list of permissions associated with specified objects.* In the context of network security, access lists define reachable hosts and networks for packets, based on information in the IP and UDP/TCP headers.

With a good set of rules, administrators can ensure that only friendly packets reach computers within the network. The typical use of a firewall is shown in Figure 11.3. The firewall examines every packet reaching the local network from the Internet and verifies that the packet passes the specified rules before allowing the packet into the local network.

---

6 Microsoft, "Tips for creating a strong password," http://windows.microsoft.com/en-us/windows-vista/tips-for-creating-a-strong-password (accessed Apr. 9, 2016).

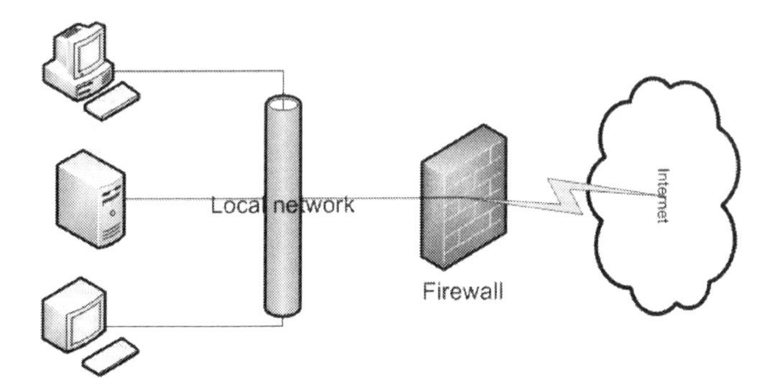

Figure 11.3: Typical firewall setup

In general, firewalls can do two things to packets—allow them, or block them. Typically, the first rule that matches the packet is applied. For example, consider a firewall which has the following set of rules:[7]

```
block in from 165.228.0.0/16 to any
pass in proto tcp from any to 131.247.95.68 port = 80
block in all
```

The first rule in this rule set blocks all incoming packets from the network 165.228.0.0/16. The network 165.228.0.0/16 has been blocked because the administrator identified it as a blacklisted network.[8] The second rule allows all remaining TCP packets to reach the web server (port 80) located at IP address 131.247.95.68. The third rule blocks all other packets. Effectively, this rule set allows packets from non-blacklisted networks to come into the network, but only if they want to access the specified web server. All other incoming packets are blocked. Administrators can build more elaborate rule sets based on experience.

As a starting point, firewalls should block any access to insecure services inside the organization. An example of an insecure service is telnet. The service allows remote users to control computers. However, telnet data is unencrypted and most implementations of the telnet service have not been patched for a long time. As a result, any computer in the organization that runs telnet by mistake is a valuable target for intruders. Firewalls can protect such computers by blocking all attempts to access telnet.

Once insecure services are shielded, the firewall should only allow remote computers from trusted networks to access safe and secure services inside the network. Examples of safe or secure services include remote desktop, DNS, e-mail, and http.

Large organizations typically implement firewalls in two stages as shown in Figure 11.4. Facing the Internet is a network that hosts public services such as e-mail, web, and DNS. This network is popularly called the demilitarized zone or the DMZ. *The DMZ is a network that contains the organization's external services and connects them to the Internet.* The first firewall (public firewall in the figure) faces the Internet and provides basic protection to the DMZ and the organization by keeping out hostile networks.

7 These rules use the syntax for ipfilter, a popular software firewall.

8 Wikipedia, "Blacklist (computing)."

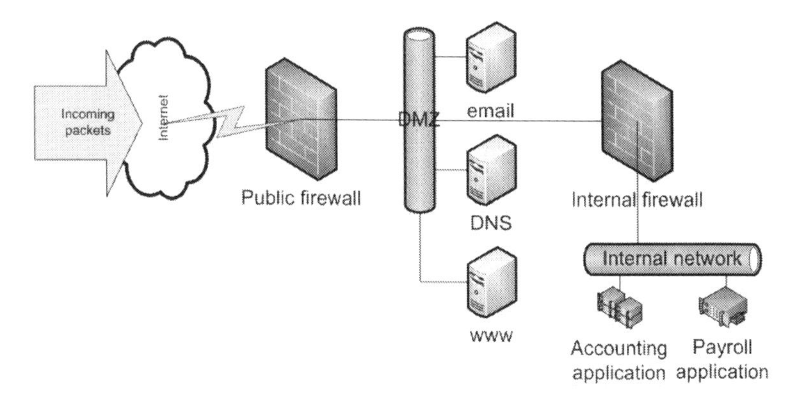

Figure 11.4: Typical enterprise firewall configuration

The second firewall (internal firewall in the figure) connects the DMZ to the organization's internal network. This internal firewall blocks out all incoming requests from outside the organization. Sensitive organizational data is placed behind the internal firewall. Here, sensitive data is only accessible to authenticated and authorized users and services. For example, internal users may access proprietary information inside the organization. Or, a web application in the DMZ may be allowed access to a database hosted on a host in the internal network.

The two-tier firewall architecture for network security in the figure parallels the physical security architecture of most businesses. Typically a business such as a store or a bank has an exterior door, which has a function analogous to the functions of the public firewall. This door allows all legitimate customers to enter the establishment, while offering some rudimentary protections. For example, the door prevents people from bringing large vehicles into the store. Inside, customers are free to browse products and use the checkout counters or tellers for transactions. This area may be considered the DMZ of the store. In a corner of the store, there is usually an area marked for employees only. This is the area where the store management keeps its records such as sales, salaries, etc. This area is typically out of bounds for customers. This area is analogous to the internal network of the company as shown in the figure.

At a typical pharmacy such as Walgreens or CVS, users are allowed to walk through all corners of the store except inside the pharmacy. The retail store selling groceries and supplies is like the demilitarized zone in the store, offering services for end users. The pharmacy is blocked off because it contains controlled drugs as well as sensitive health information about users. The pharmacy usually has a counter that opens up to the store (demilitarized zone) where customers can go and drop off prescriptions or pick up medications. When the pharmacist receives the request, he goes into the pharmacy section—the militarized zone—and returns with the items requested by the end user.

Firewalls have limitations. First, if an internal computer is compromised, it can successfully attack other computers in the local network. Second, firewalls are unable to protect services that Internet users are allowed to access. For example, if you run a web application inside your organization, you probably would want the world to visit it. If the web application has any vulnerability, the firewall cannot prevent it from being exploited because the firewall rules will allow access to the web application. Finally, a firewall is only as good as the rules specified by its administrator. If the firewall rules do not block packets from attackers, the firewall will not stop the attacker.

In the case of the Georgian attacks, clearly, the fact that the contents on the sites were modified indicates that the attackers were able to overcome any authentication and authorization controls the Georgian sites may have implemented. In this case, in addition to better patching, authentication, and authorization, suitable implementation of firewalls could have prevented the attackers from gaining access to the computers in the first place. Firewall rules could have been created to block suspicious traffic and all incoming packets from suspicious networks, particularly those in Russia.

Controls for Availability

Availability is to ensure timely and reliable use of information.

On January 25, 2003, at around 12:30 pm EST, the Slammer worm scanned the Internet for computers running an unpatched version of MS SQL server. Within 10 minutes of its release, it had infected more than 50,000 hosts, more than 90% of all vulnerable computers on the Internet, causing outages and consequences such as ATM failures and flight cancellations.

Viruses and Worms

Have you encountered through personal experience or heard from a friend the experience of being attacked by a computer virus or worm?[9] Typically, when this happens, the computer operates very slowly without an obvious reason. Occasionally, some data files get corrupted and have to be deleted. If a computer file is damaged or has to be deleted, it is no longer available. Therefore, the end result of most virus attacks is a loss of availability.

According to the ATIS Telecom glossary, *a computer virus is an unwanted program that places itself into other programs, which are shared among computer systems, and replicates itself. A worm is a self-contained program that causes harm and can propagate itself through systems or networks.* You may have read in many sources that the distinction between worms and viruses is that worms replicate themselves, whereas viruses do not. However, the above definition is more formal. Both viruses and worms can replicate themselves. The major difference between the two is

9 Early worm programs were quite an intellectual curiosity. For example, see John F. Shoch and Jon A. Hupp, "The 'Worm' programs—Early experience with a distributed computation," *Communications of the ACM*, 25(3) (1982): 172–180.

that worms are self-contained programs whereas viruses use e-mail clients or other software to cause damage and to replicate themselves. Viruses and worms are often designed to use up available resources such as storage or processing time, thereby compromising availability.

Most surveys of CIOs indicate that of all information security threats, virus attacks cause the greatest financial losses to organizations.[10] Therefore, IT professionals should pay particular attention to protecting computers within their jurisdiction from computer viruses and worms. A major challenge in defending against viruses and worms is that new viruses and worms are actively being developed and they are now designed to spread very fast. The Code Red virus of 2001 is considered the first global-scale worm attack. It affected 359,000 computers in less than 14 hours of being launched. The Slammer worm of 2002 reached 90% of its susceptible 75,000 targets in less than 10 minutes.[11]

How can organizations defend themselves against viruses and worms—particularly if new viruses and worms are emerging all the time and can cause extensive damage within minutes of release? Defense against viruses and worms typically involves two controls. The first is patching. Most viruses and worms are designed to exploit some vulnerability in software. Usually these vulnerabilities are known well in advance and patches for these vulnerabilities are almost always available. For example, the Slammer worm was released on January 25, 2003, and a patch to fix the vulnerability exploited by the worm had been issued by the vendor (Microsoft) six months earlier in July 2002. Computers that had applied the patch were immune to the worm.

The second control used to defend against viruses and worms is the use of antivirus programs. These programs are aware of a large number of viruses and worms. Antivirus programs constantly scan incoming network traffic (e-mails, web pages, etc.) to check whether the bytes in the traffic look like viruses or worms. When they do, the programs alert the user and delete the offending incoming viruses or worms before they can cause any damage.

Since viruses and worms are constantly being developed, and viruses can cause most of their damage within minutes of being released, before the mainstream media becomes aware of their existence, the virus knowledge base of antivirus programs needs to be constantly updated with information about the newest viruses and worms. Hence, it is good practice to update the antivirus definitions regularly. If the antivirus program has the option to automatically apply updates, the use of this feature is highly recommended.

Patching and the use of antivirus software would have protected computers against the Slammer worm.

Not all available controls are technical. It is also important to educate users so that they are careful when they are on the network. For example, suspicious e-mail should be deleted, and suspicious links in e-mails should not be followed through. Only reliable websites should be visited. End-user awareness has many benefits and affects all aspects of organizational security, not just network security or information security.

---

10 A.G. Lawrence, M.P. Loeb, W. Lucyshyn, and R. Richardson, "CSI/FBI Computer Crime and Security Survey," *CSI/FBI Computer Crime and Security Survey*, GoCSI.com, editor, 2006.

11 David Moore et al., "The spread of the Sapphire/Slammer worm," CAIDA, https://www.caida.org/publications/papers/2003/sapphire/sapphire.html.

Denial-of-service Attacks

A special kind of threat called a denial-of-service attack creates availability problems. *A denial-of-service (DOS) attack is when an attacker consumes the resources on a computer or network for things it was not intended to be doing, thus preventing normal use of the resources for legitimate purposes.* Home users are generally not much affected by DOS attacks, but, occasionally, well-known websites have become unavailable because hackers or commercial rivals have sent a large volume of irrelevant traffic to the website. The computer becomes busy trying to process these unnecessary requests, and in doing so, is unable to respond to legitimate requests. These attacks are popularly called DOS attacks.

An excellent and highly readable analysis of a DOS attack was written up by a system administrator.[12] Essentially, the site grc.com was brought down by a DOS attack initiated by a 13-year-old kid who bore a grudge against the administrator of grc.com. The analysis also shows how easy it can be to launch a DOS attack.

As described in the GRC case, a firewall is a good initial defense against a DOS attack. Once the source of the attacking packets is identified, a firewall rule can be added to block all incoming packets that match the rule. This prevents the offending packets from being able to reach their target and cause loss of availability. However, if the attack is consuming all of your incoming bandwidth, you may have to contact your upstream ISP to block it from reaching your network.

Network Security Controls for Outgoing Information

What information security threats affect outgoing information? The major concern is that strangers may be able to read passwords and sensitive credit card information as it passes through the network. Another important concern is that an occasional hardware or software failure may cause a website to become unavailable. The controls we will see in this section can defend against these threats.

Controls for Confidentiality

How do you prevent strangers from reading your passwords when you log into a remote website? After all, it is relatively easy to use network sniffers such as Wireshark to read packets flowing through a network. Is there anything you can do so that even if someone captures data packets leaving your computer, they could still not read the information contained in the packet? If strangers cannot read the information in the packets, how can you ensure that the target destination can read the information in the packets?

---

12 S. Gibson, "The Strange Tale of the Denial of Service Attacks Against grc.com," Gibson Research Corp., 2001, http://www.crime-research.org/library/grcdos.pdf.

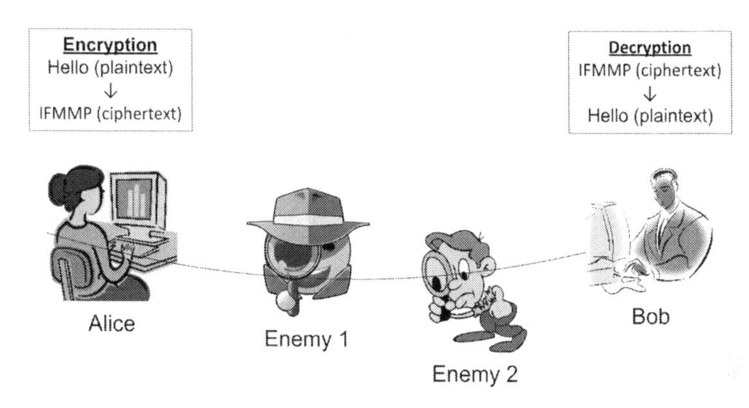

Figure 11.5: Encryption prevents enemies from reading data

Thanks to the power of mathematics and to the efforts of researchers who have discovered the rules of mathematics that make it possible, we can indeed send information to a destination over wires in such a way that enemies cannot read the information in the wire even though the destination can easily read it. This is done through encryption. As shown in Figure 11.5, using encryption, Alice and Bob can carry on a private conversation even in the presence of enemies trying to snoop on the wire.

Encryption is the process of rendering plain information unintelligible in such a manner that it may later be restored to intelligible form. The output from encryption is called ciphertext. Decryption is the process of converting ciphertext to plaintext. Figure 11.5 shows the role of each of these activities in encryption. When using encryption to send data confidentially, the raw data for transmission is called plaintext. An encryption algorithm uses an encryption key to create ciphertext. The ciphertext is transmitted. The receiver uses a decryption algorithm and a decryption key to convert the ciphertext back to plaintext.

There are two general kinds of encryption techniques: symmetric key encryption and asymmetric key encryption.

Code talkers as encryption[13]

Historically, encryption has been used in wartime. In addition to all the formal encryption techniques described below, another popular encryption technique used historically is the use of local languages.

Before technology made encryption militarily feasible, the US military used speakers of obscure native languages for secret communication in wartime. These people were called code talkers. The US Marine Corps had approximately 400–500 Native Americans for secret communications. During World War II, the US Army and Marines used Native American code talkers including Basque, Lakota, Meskwaki, and Comanche soldiers.

13 https://en.wikipedia.org/wiki/Code_talker.

Symmetric Key Encryption

Recall questions on general intelligence tests that look like this—If cat is dbu, what is dog? This is a simple example of symmetric encryption. In this example, we observe that for each character of the plaintext (cat), the encrypted text (dbu) is created by picking the next letter of the alphabet. In semiformal terms, we may write this encryption scheme as:

Character in encrypted text = character in plaintext + 1

More generally, our encryption scheme can generate encrypted text by replacing each character in the plaintext with a character n letters away in the alphabet. In our example, n = 1. In general, encryption involves an algorithm and a key. *An encryption algorithm is a mathematically expressed process to create ciphertext.* In our example, we may write the encryption algorithm as: add n. *An encryption key is a sequence of symbols that control the operation of the encryption algorithm.* In our example, we may write the encryption key as +1. This way, encryption algorithms use encryption keys to convert plaintext into ciphertext. Once we understand this scheme, it is easy to encrypt dog as eph.

Now, how would the receiver decrypt the encrypted text in this example? In other words, how would the receiver know that to decrypt the message, it has to subtract 1 from each character in the encrypted text? In our simple example, Bob could guess the key by simply looking at the ciphertext. In the more general case, before they begin communicating over the wire, Alice and Bob would have to agree upon both the encryption algorithm and the encryption key. This could happen in a face-to-face meeting at work. Or, Bob may send the algorithm and key through a courier or by mail. Whatever mechanism is used to exchange the encryption key, the wire cannot be used to exchange the encryption key because the enemies could read the key as it is being transmitted over the wire and successfully decrypt any information flowing through the wire.

Symmetric key encryption is so called because the same key is used for both encryption and decryption. In our example, we added 1 to encrypt and subtracted 1 to decrypt.

Symmetric key encryption has many advantages. Its biggest advantage is that it is computationally simple. However, its limitation is the key-exchange problem. Symmetric key encryption cannot be used to securely transmit the encryption key in such a way that the receiver can decrypt it but intruders cannot. Symmetric key encryption would be useful to secure network communication if a technique could be developed to solve the key-exchange problem. If we could somehow safely exchange the encryption key, no other known technique provides the same level of security for a given level of computation effort as symmetric key encryption.

Fortunately, safe key exchange over an unsafe medium is possible using asymmetric key encryption.

Symmetric key encryption is very commonly used to store information. A number of symmetric encryption techniques have been developed for commercial use including the Data Encryption Standard (DES), International Data Encryption Algorithm (IDEA), and Triple DES. After evaluation, in 2001 the National Institute for Standards and Technology standardized a symmetric encryption technology called the Advanced Encryption Standard

(AES) to securely store sensitive non-classified US Government information. AES uses an encryption algorithm developed by two Belgian cryptographers.

Asymmetric Key Encryption

Table 11.2: Simple asymmetric key encryption example (mod 10 table)

Number to encrypt *n* → (plaintext)	*m* ↓	**0**	**1**	**2**	**3**	**4**	**5**	**6**	**7**	**8**	**9**
	0	0	0	0	0	0	0	0	0	0	0
n * 1 mod 10 →	1	0	1	2	3	4	5	6	7	8	9
n * 2 mod 10 →	2	0	2	4	6	8	0	2	4	6	8
n * 3 mod 10 (ciphertext) →	3	0	3	6	9	2	5	8	1	4	7
	4	0	4	8	2	6	0	4	8	2	6
	5	0	5	0	5	0	5	0	5	0	5
	6	0	6	2	8	4	0	6	2	8	4
	7	0	7	4	1	8	5	2	9	6	3
	8	0	8	6	4	2	0	8	6	4	2
	9	0	9	8	7	6	5	4	3	2	1

Table 11.3: Decryption in the example above

Ciphertext (x) →		0	3	6	9	2	5	8	1	4	7
		Decrypt using x * 7 mod 10							↓		
Decrypted plaintext →		**0**	**1**	**2**	**3**	**4**	**5**	**6**	**7**	**8**	**9**

We saw that the symmetric key cannot be exchanged over the network because our enemies would be able to read the key during the exchange. What if we could encrypt information in such a way that even if our enemies knew how the information was encrypted, they could still not decrypt it, even though the receiver could decrypt it easily? Miraculously, this is possible using asymmetric key encryption.

Instead of dealing with asymmetric key encryption in abstract terms, let us look at it by example.[14] Most asymmetric key encryption techniques involve the *modulo* operation. Modulus is the remainder when a number is divided by another number. For example, when we divide 21 by 10, the remainder is 1. We write this as 21 mod 10 = 1. How can we use this operation for asymmetric key encryption?

---

14 This example is based on an example in *Network Security: Private Communication in a Public World*, 2nd Edition, by Charlie Kaufman, Radia Perlman, and Mike Speciner (Prentice Hall, 2002).

An example of using the modulo operation for encryption is shown in Table 11.2. Asymmetric encryption uses mathematical operations and it is convenient to describe asymmetric encryption using numbers. This is not a limitation because, as we have seen in Chapter 2, text can be converted to numbers using ASCII or Unicode.

The example encrypts the 10 digits 0–9 using modulo 10. The plaintext is in the first row of the table as n. To encrypt a digit, we multiply it by another digit m, and take modulo 10 of the product. Each row in the table shows the result for the value of m in the row. For example, in row 2, m = 0 and all cells in row 2 are 0 because n * 0 = 0, and 0 mod 10 = 0. Let us look at the highlighted row in the table, for m = 3. The first number, 0, is obtained as 0 * 3 mod 10, which is 0. To encrypt 4, we calculate 4 * 3 mod 10 = 12 mod 10 = 2. Similarly, we encrypt 9 using 9 * 3 mod 10 = 27 mod 3 = 7. The highlighted row shows the encrypted values for n using m = 3.

How do we decrypt these numbers? That is, given encrypted digit 7, how do we get back the unencrypted digit 9? Table 11.3 shows how the information encrypted in Table 11.2 can be decrypted. The first row in Table 11.3 is the ciphertext and is identical to the highlighted row in Table 11.2. To decrypt the numbers in this row, we multiply the numbers by 7 and take modulo 10 of the result. For example, to decrypt 3, we calculate 3 * 7 mod 10 = 21 mod 10 = 1. We can confirm that the decryption is correct, in other words, 1 was indeed encrypted as 3. As another example, to decrypt 1, we calculate 1 * 7 mod 10 = 7 mod 10 = 7. Similarly, the decryption of all other numbers may be verified.

How can Table 11.2 and Table 11.3 be used for encryption? The tables used mod 10 and multiplied plaintext by 3. If Alice wanted Bob to be able to send messages to her, she would send Bob the key (3, 10) over the wire where even her enemies might see the key as it was being transmitted over the wire. To encrypt messages, Bob would perform the operations in the highlighted row in the encryption table. For example, Bob would send 2 as 6, 4 as 2, and so on. The result would be sent as encrypted messages over the wire. Even enemies would be free to read this encrypted message as it was transmitted over the wire. Since they would not know the decryption key (7, 10), they would not be able to decrypt the message, even though they would know the key (3, 10) that was used to encrypt the message. However, when the encrypted message reached Alice, Alice would decrypt the messages using the decryption key (7, 10).

Why do 3 and 7 work as an encryption-decryption pair when using modulo 10? It turns out that if we use modulo *n*, we can use any pair of numbers (*a*, *b*) such that a * b mod n = 1. Since 3 * 7 mod 10 = 1, the pair can be used for encryption and decryption with modulo 10.

In our example, it would be easy for enemies to figure out the decrypting key (7, 10) from the encrypting key (3, 10) because the numbers are very small. If enemies know that Alice and Bob are using modulo 10 operations, it will not take them too long to figure out that the encryption-decryption pair is (3, 7). To prevent this, in real life, very large numbers are used. Current practice is to use modulo n where n is 1,024 or 2,048 bits long (300–600 decimal digits). It is believed that with current technologies, enemies would not be able to figure out how to decode such large numbers.

In our example, the encryption key (3, 10) would be publicly known and could be used by anyone who wanted to send an encrypted message to Alice. For this reason, asymmetric key encryption is popularly called public key encryption. The key used for encryption is shared publicly and is called the public key. The key used for decryption, (7, 10) in our example, is kept private and is called the private key.

The most popular implementation of asymmetric key encryption is RSA. RSA is named after its three creators—Rivest, Shamir, and Adleman.

Asymmetric key encryption can be used to encrypt data for transmission. However, as we can see from our example, a major limitation of asymmetric key encryption is that the security of asymmetric key encryption depends critically upon the use of very large numbers for security. As a result, it is extremely computation-intensive and could easily overwhelm even the most powerful computers if popular websites used asymmetric key encryption to encrypt all data between the websites and users. Therefore, the most common use of asymmetric key encryption is to exchange a symmetric key. The great strength of asymmetric key encryption is that it makes key exchange very easy. Once the symmetric key is exchanged using asymmetric key encryption, the rest of the communication can be securely transmitted using secret key encryption. Most commercial encryption technologies, such as SSL (Secure Sockets Layer), which is used by e-commerce websites, use this procedure.

In the T.J. Maxx case, had the company used WPA, the current implementation of wireless security, it is unlikely that the intruders would have been able to read the passwords, even if they had obtained the encrypted versions of the passwords.

Encryption in Practice

Encryption is used in data communications in two common ways: virtual private networks (VPN) and Transport Layer Security (TLS). VPN was defined in RFC 2764 in 2000 and SSL was created by Netscape communications in 1995.

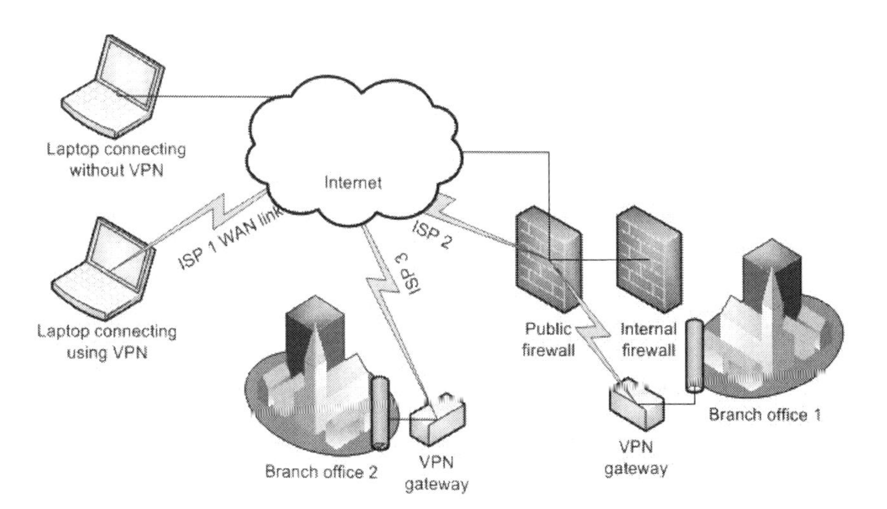

Figure 11.6: VPN example

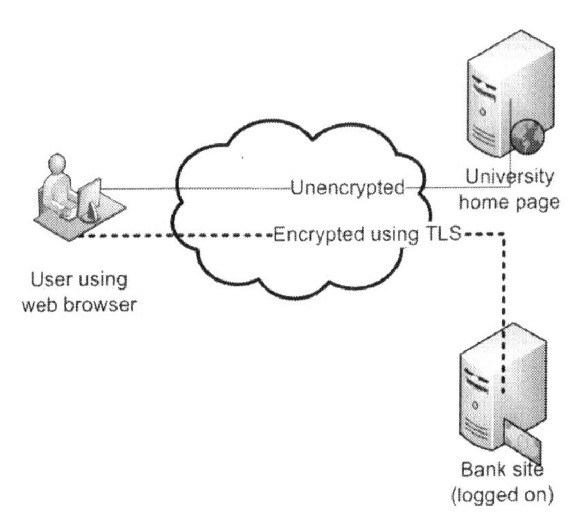

Figure 11.7: TLS example

In VPNs, all the network traffic over a communication channel is encrypted. To implement VPN, a router at the edge of a network negotiates encryption schemes and keys with devices it communicates with. *IPSec is the technology used to authenticate and encrypt each IP packet in a communication channel.* As a result, all the communication over the link is encrypted. Mobile computers in the organization can also negotiate encryption parameters with the edge router implementing VPN.

Figure 11.6 shows an example of using VPN. The VPN gateway at branch office 1 negotiates encryption parameters with all devices that wish to communicate with the branch office using VPN. The example shows how a laptop or another branch office would connect to branch 1 using VPN. These devices may use WAN links from different ISPs, but the data travelling across the WAN links would not be readable by anybody on the link. Connections from the outside world that do not use the VPN gateway would be treated as insecure and could be blocked by the inner firewall.

Transport Layer Security or TLS is the standardized version of SSL, which was created by Netscape communications. In TLS, a client application running on a computer negotiates encryption parameters with a server using TLS to secure the server application. TLS uses encryption in a similar way as VPN, but the biggest difference is that TLS is used to encrypt traffic from specific applications. For example, TLS may be used to encrypt web e-mail traffic and traffic for banking transactions, while traffic from other browser tabs may be sent unencrypted. All TLS encryption is done at the end devices without any need for specialized gateway devices required in VPN.

Figure 11.7 shows an example of TLS. In the example, one tab in the browser connects to an informational website without encryption, and another tab connects to the bank's account details page using TLS encryption. If the user used VPN instead of TLS, all the traffic leaving his laptop would be encrypted until it reached the VPN gateway.

SSH (Secure Shell)

Whereas VPN encrypts IP packets (network layer) and TLS encrypts all datagrams associated with a connection (transport layer), SSH is an application layer protocol (application layer) for secure remote login and other secure network services over an insecure network. SSH is a relatively recent protocol. It was defined in 2006 in RFC 4250–RFC 4254. Of these five documents, I find RFC 4253 to be the most useful in understanding the protocol.

Wireless security

A major limitation of wireless networks is that the information broadcast over the air is exposed to everybody within range of the transmitting station. Anybody with a wireless laptop can listen in on messages being sent over the air. Wired local area networks built using Ethernet also use broadcast. However, this broadcast stays confined to the wires. Only users who plug into one of the network jacks in the building can access these broadcasts. Office buildings have some level of access control that prevents unauthorized users from entering office spaces and plugging into one of the network jacks. This provides some level of confidentiality in a wired network. However, wireless signals can bleed outside to parking lots and lobbies where unauthorized users can read them.

To prevent such unauthorized access to organizational data, wireless networks use encryption. *Encryption is the transformation of data into a form that makes it unreadable by anyone except authorized users.* Wireless LAN standards define three encryption techniques for securely transmitting information over the air: WEP (wired equivalent privacy), TKIP (temporal key integrity protocol), and CCMP (Counter-mode with Cipher-block chaining message authentication code protocol). Network administrators can configure access points to use any one of these encryption techniques. Of these three techniques, WEP is extremely insecure and is now considered obsolete. TKIP and CCMP are generically called WPA (Wi-Fi protected access). TKIP was created to provide an easy way to improve the security of WEP devices. TKIP is popularly called WPA 1. The current recommendation is to use CCMP, also called WPA 2. The access point informs all stations on the basic service set to use the specified encryption technique. As a result, all communication on the network uses the specified encryption technique.

It is highly advisable to use encryption on wireless networks. For end-user convenience, though, most access points do not use encryption by default. Figure 11.8 shows a dialog allowing the user to specify the encryption technique on a home router.

SSH is a very interesting protocol. Unlike other application layer protocols (Chapter 6), SSH is a protocol for direct usage. As described in RFC 114, direct usage implies that users are "logged into" a remote computer and execute commands on it just as local users. The Windows remote desktop is another example of direct usage. Protocols such as HTTP, e-mail, and FTP allow indirect usage since users do not log into the remote system and do not need to know how to use the remote system. Users of indirect protocols may not even be aware of the specific hardware and software running on the remote computer. (Do you know what operating system your university web server runs on? Even if you knew, would that knowledge affect your usage of the university website?)

SSH uses encryption to secure data using procedures similar to VPN and TLS. SSH operation may be seen as advancing through five stages: (1) connection establishment, (2) protocol identification, (3) key exchange, (4) shift from public to private keys, and (5) secure data exchange. A brief overview of these five stages is below.

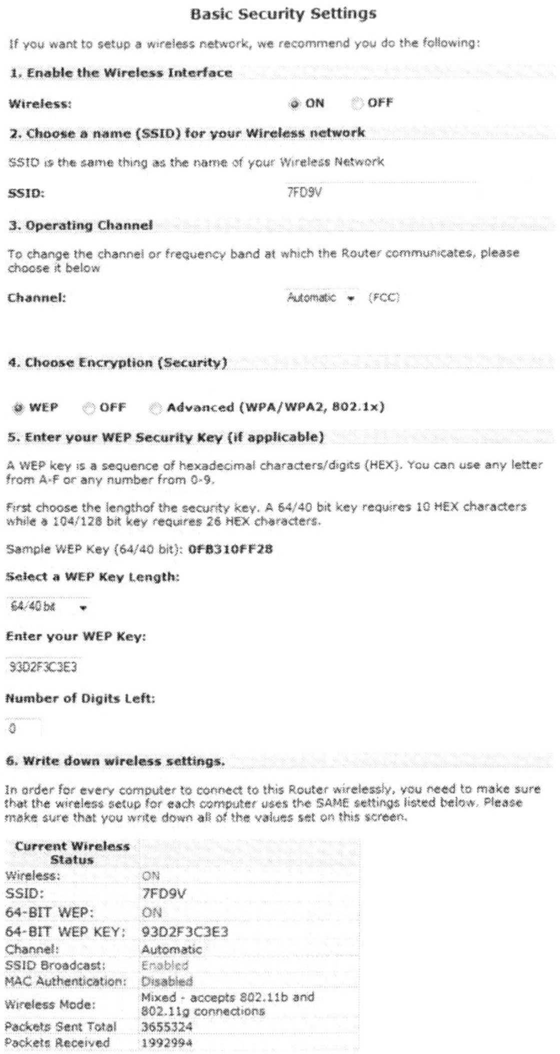

Basic Security Settings

If you want to setup a wireless network, we recommend you do the following:

1. Enable the Wireless Interface

Wireless: ON OFF

2. Choose a name (SSID) for your Wireless network

SSID is the same thing as the name of your Wireless Network

SSID: 7FD9V

3. Operating Channel

To change the channel or frequency band at which the Router communicates, please choose it below

Channel: Automatic (FCC)

4. Choose Encryption (Security)

WEP **OFF** **Advanced (WPA/WPA2, 802.1x)**

5. Enter your WEP Security Key (if applicable)

A WEP key is a sequence of hexadecimal characters/digits (HEX). You can use any letter from A-F or any number from 0-9.

First choose the lengthof the security key. A 64/40 bit key requires 10 HEX characters while a 104/128 bit key requires 26 HEX characters.

Sample WEP Key (64/40 bit): **0FB310FF28**

Select a WEP Key Length:

64/40 bit

Enter your WEP Key:

93D2F3C3E3

Number of Digits Left:

0

6. Write down wireless settings.

In order for every computer to connect to this Router wirelessly, you need to make sure that the wireless setup for each computer uses the SAME settings listed below. Please make sure that you write down all of the values set on this screen.

Current Wireless Status	
Wireless:	ON
SSID:	7FD9V
64-BIT WEP:	ON
64-BIT WEP KEY:	93D2F3C3E3
Channel:	Automatic
SSID Broadcast:	Enabled
MAC Authentication:	Disabled
Wireless Mode:	Mixed - accepts 802.11b and 802.11g connections
Packets Sent Total	3655324
Packets Received	1992994

Figure 11.8: Home router security settings

Like all network communication, SSH remote login begins with a TCP connection to a computer waiting to receive SSH connections. SSH typically listens on port 22. After connection establishment, the two computers negotiate the method used for encryption.

To account for differences in computational capabilities between local and remote computers, many secure technologies are available for use in SSH, each offering different levels of security, but with higher security coming at the cost of increased computational complexity. Therefore, before data communication can begin, the two computers decide on which of the available security technologies to use for the current connection.

At this point, public key encryption is used by the two computers to openly exchange a secret key that will be used in private key encryption for the rest of the communication. Now the two computers simultaneously switch to private key encryption in the fourth stage.

Finally, in the fifth stage, secure data communication can occur using private key encryption. At this point, the remote user has a direct "logged on" connection to the remote system through an account on the remote system. The user can execute applications, write documents, and perform any actions on the remote computer that he could have performed if he had direct access to the computer. All information is hidden (encrypted) during transmission over the network.

To enhance security, SSH also provides for a mechanism to authenticate computers. The first time a computer is used to log into a remote computer, the local computer remembers the address and some secret information associated with the remote computer. Later, if the user tries to use the same IP address and its secret information does not match the stored secret information about the remote computer, the client computer will detect the mismatch and report it to the user for necessary action.

Controls for Integrity

Encryption prevents an enemy from being able to read a message. Even if an enemy cannot read a message, it could cause considerable harm by simply damaging the message as it is transmitted. How is a receiver to know that the message it has received is indeed the message that the sender has sent? Error-detecting algorithms such as cyclic redundancy check (CRC) may not work because the enemy can re-calculate any unencrypted CRC.

It turns out that asymmetric encryption can be used to also verify integrity. In our example, it may be verified from Table 11.2 and Table 11.3 that if Alice encrypts a message using her private key (7, 10) to send a message, Bob could decrypt it using Alice's public key (3, 10). If Alice wanted to assure Bob of the integrity of a message she was transmitting to him, in addition to the message, Alice could also send some additional information. Specifically, she could encrypt the message with her own private key and send the result as an integrity verifier. At the receiving end, Bob would use Alice's public key (which he has) to decrypt the integrity verifier. If the result is the same as the received message, Bob would be convinced that the message he has received could not have been modified along the way. This works because the enemy cannot know Alice's private key (assuming of course that Alice has kept it safely):

$\text{Decrypt}_{\text{By Bob using Alice's public key}}(\text{Encrypt}_{\text{By Alice using Alice's private key}}(\text{Plaintext})) = \text{plaintext}$ (done for integrity verification)

For reference, this can be contrasted with the operation used by Alice to send a secure message to Bob:

$\text{Decrypt}_{\text{By Bob using Bob's private key}}(\text{Encrypt}_{\text{By Alice using Bob's public key}}(\text{Plaintext})) = \text{plaintext}$ (confidential transmission)

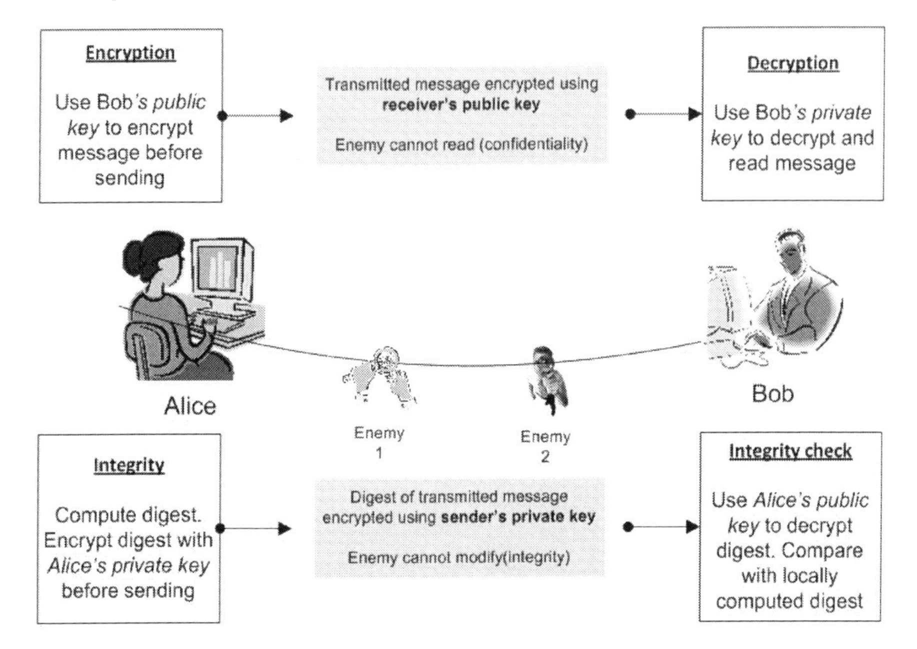

Figure 11.9: Comparing asymmetric keys for confidentiality and integrity

In practice, to reduce the size of the integrity check, Alice could first create a summary of the message. The summary is called a digest and can be calculated in many ways. For example, a CRC may be used as a digest. More commonly, algorithms called MD5 or SHA are used to calculate the digest. As an integrity check, Alice can encrypt the digest with her private key and send the encrypted digest to Bob.

Upon receiving the message and the digest, Bob would decrypt the encrypted digest with Alice's public key. He can do this because he has Alice's public key. This would give him the unencrypted digest. Then, analogous to what is done in CRC verification (Chapter 3); Bob would compute the digest from the data sent by Alice and compare the results to the digest sent by Alice. If the two numbers match, Bob would be assured that the message was received without loss of integrity.

The reason this integrity check works is the same as the reason why encryption works. Even though the enemy knows the public key, it does not know the private key. Therefore, the enemy would not be able to figure out how to manipulate the digest in such a way as to successfully mislead Bob. If the enemy manipulates either the message or the digest, Bob will detect the change. Figure 11.9 shows how asymmetric key encryption can be used both as a control to protect outgoing data for confidentiality and also to protect against integrity threats. If Alice sends a message to Bob, for confidentiality she uses Bob's public key, and for integrity she uses her own private key. The main thing to note about the figure is that the private key is only known to the owner of the key.

Controls for Availability

On May 14, 2009, about 14% of Google's users experienced an outage that lasted about an hour, beginning at 10:48 a.m. EST.[15] These users were unable to access e-mail and other Google applications such as Google analytics. Google maintains a highly redundant system. This outage resulted from a system error.

How could outgoing data become unavailable? The most common reason for outgoing data to become unavailable is a hardware or software failure that shuts some service down.

The most common control to mitigate availability threats in outgoing data is redundancy. *Redundancy is surplus capability provided to improve the availability and quality of service.* We saw an example of redundancy in Chapter 7 where we learned how most organizations maintain multiple DNS servers so that if one DNS server goes down, another DNS server can take its place. Similarly, most organizations maintain multiple web servers so that if one web server goes down, another can process user requests.

Redundancy is also common with network connections. Most organizations maintain multiple network connections to the Internet so that if one connection goes down, another connection can be used to handle network traffic. Figure 11.10 shows network redundancy at Google. The figure shows that Google maintains direct connections to many networks.[16]

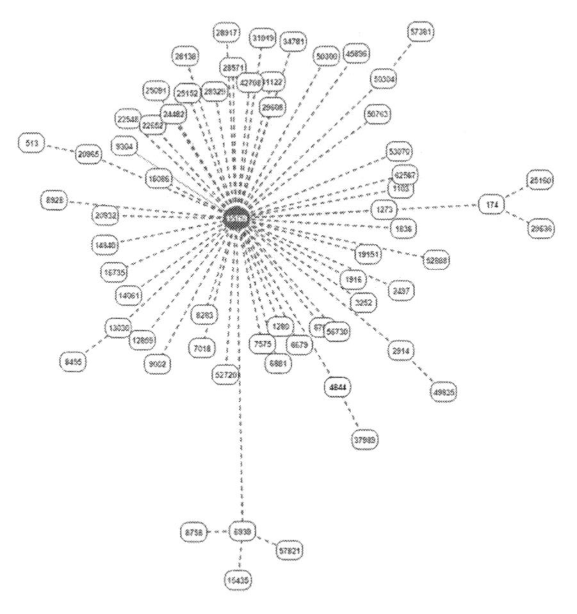

Figure 11.10: Network redundancy at Google

Redundancy is also a very difficult control to implement correctly. Unanticipated issues often merge when large catastrophes strike, resulting in situations such as that seen in Figure 11.11. ConstantContact is a publicly traded software company that provides customer relationship management services for organizations. The University of South Florida is a user of ConstantContact, but on one occasion users saw this message when they tried accessing the service.

15 http://googleblog.blogspot.com/2009/05/this-is-your-pilot-speaking-now-about.html.

16 Readers may find it interesting to note that in fall 2008, when Twitter was a critical medium of political messaging, it had no backups of its database. A database crash would have destroyed all of Twitter's data. At the time, the company only had 22 employees. In part, this discovery was responsible for the dismissal of Jack Dorsey, one of the co-founders of Twitter, from the position of CEO. Source: Nick Bilton, *Hatching Twitter* (Portfolio, 2014).

Most home users should consider using redundant data storage. The last few years have brought about a substantial change in the way we manage a very important part of our personal lives—those photographs and videos that record our memories. Most of this information is now stored as computer files instead of photo prints. Consider using a RAID storage device to save these files. Computer hard disks can crash without warning and with them can go all the memories saved on the disks, never to be recovered again.

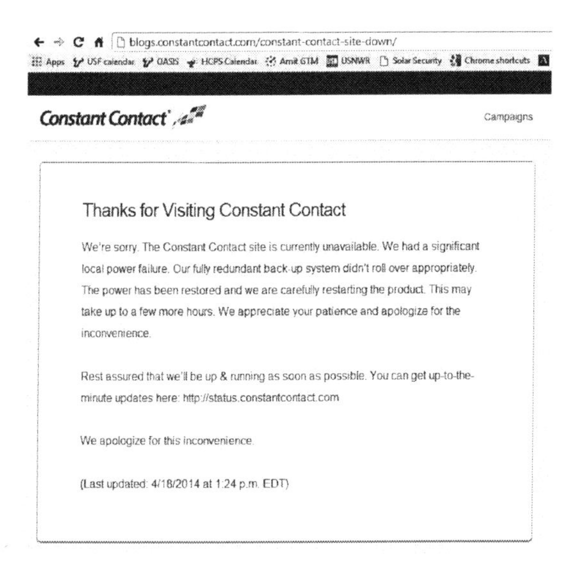

Figure 11.11: Outage example

EXAMPLE CASE—T.J. Maxx

One of the best reported cybersecurity incidents is the T.J. Maxx incident of 2007. It led to numerous changes in cybersecurity procedures, including the addition of wireless security requirements for credit card processing. Thus, while the incident may appear dated, it is very useful from a learning perspective since most related information is now in the public domain.

The year 2007 was marked by some startling revelations of hackers getting access to credit card databases at many of the leading retailers in the country. In addition to T.J. Maxx, Barnes & Noble and OfficeMax were victims of cyber attack. It was initially believed that the attacks were being led by hackers outside the country. However, the hacking spree culminated in the prosecution of 11 men in 5 countries, including the United States. The ringleader had been an informer for the US Secret Service.

On August 5, 2008, the US government charged 11 individuals with wire fraud, damage to computer systems, conspiracy, criminal forfeiture, and other related charges for stealing credit card information from prominent retailers such as T.J. Maxx, BJ's Wholesale Club, OfficeMax, and Barnes & Noble.

In August 2009, many members of the same gang were again charged with compromising Heartland Payment Systems, a credit card processing company, and stealing approximately 130 million credit card numbers. With approximately 100 million families in the United States, this translates to about one credit card per family being stolen.

Figure 11.12: Albert Gonzalez, at the time of his indictment in August 2009

The gang had been in operation since 2003. Between 2003 and 2007, they exploited weaknesses in wireless security at retail stores. This was the method used in the attacks that formed the basis of the 2008 indictment. Beginning in August 2007, the gang refined its skill set and began to use SQL injection attacks to place malware on web applications and gain access to corporate databases. This was the method used in the attacks for which the gang was indicted in 2009.

Albert Gonzalez (Figure 11.12), the ringleader, was a resident of Miami, Florida. Beginning around 2003, he is believed to have driven around Miami using his laptop computer to locate insecure wireless access points at retail stores. Stores typically use these networks to transfer credit card information from cash registers to store servers. When an open network was located, the gang would use a custom-written "sniffer" program to collect credit card account numbers, which were then sold in the gray market. The biggest victim was T.J. Maxx, which lost information on more than 40 million credit cards. The information was stored on servers in the United States, Latvia, and Ukraine. Later, when the gang graduated to SQL injection attacks, it would visit stores to identify the transaction processing systems these companies used. The gang used this information to determine suitable attack strategies for targeting these companies. The gang also studied the companies' websites to identify their web applications and to develop appropriate attack strategies for these websites.

Ringleader Gonzalez earned more than $1 million in profits by selling this credit card information. Apparently, at one time his counting machine broke and he had to manually count $340,000 in $20 bills. In August 2009, Gonzalez agreed to plead guilty to charges in the T.J. Maxx case. On March 26, 2010, he was sentenced to 20 years and 1 day in prison.[17]

Gonzalez became an informant for the Secret Service in 2003 after being arrested for various crimes. In October 2004, he helped the Secret Service indict 28 members of the website Shadowcrew.com. Shadowcrew stole credit card information and sold it for profit. While in operation, Shadowcrew members stole tens of thousands of credit card numbers. After the Shadowcrew operation was completed, however, Albert began his own exploits. Beginning with the wireless hacks into T.J. Maxx and other retailers, his attacks culminated

17 Siobhan Gorman, "Hacker Sentenced to 20 Years in Massive Data Theft," *Wall Street Journal*, Mar. 27, 2010.

in the SQL injection attacks against Heartland Payment Systems. In each case, he succeeded in obtaining tens of millions of credit card numbers.

The direct damage from the attacks in terms of fraudulent charges on customer credit cards was limited. In March 2007, one gang in Florida was caught using cards stolen from T.J. Maxx (TJX) to buy approximately $8 million in goods at various Wal-Mart and Sam's Club stores in Florida. However, the collateral damage from the incident has been colossal. TJX Companies, Inc. (T.J. Maxx Stores is one of the companies owned by the group, Marshalls is another), settled with Visa for $40 million in November 2007 and with MasterCard in April for $24 million.

The impact was nationwide. Tens of millions of customers had to be reissued credit cards. Customers (including the author of this book) who had set up automated payments on these cards received collection notices from service providers when charges did not go through because the cards had been cancelled and new ones had been issued in their place.

Surprisingly, sales at T.J. Maxx were not significantly affected by the intrusion. Customers who noticed fraudulent charges had their accounts made good by the automatic protection programs offered by credit card companies.

References

1. "Albert Gonzalez." Wikipedia, http://en.wikipedia.org/wiki/Albert_Gonzalez.
2. Gorman, S. "Arrest in Epic Cyber Swindle." *Wall Street Journal*, Aug. 18, 2009.
3. Gorman, S. "Hacker Sentenced to 20 years in Massive Data Theft." *Wall Street Journal*, Mar. 27, 2010: A1.
4. Pereira, J. "How Credit-Card Data Went Out Wireless Door." *Wall Street Journal*, May 4, 2007.
5. Pereira, J., J. Levitz, and J. Singer-Vine. "U.S. Indicts 11 in Global Credit-Card Scheme." *Wall Street Journal*, Aug. 6, 2008, A1.
6. T.J. Maxx, 10-K report, Mar. 28, 2007.
7. T.J. Maxx, 8-K filing, Jan. 18, 2007; Apr. 2, 2008; Nov. 30, 2007.
8. United States of America vs. Albert Gonzalez, criminal indictment in US District Court, Massachusetts, Aug. 5, 2008 (the T.J. Maxx case).
9. United States of America vs. Albert Gonzalez, criminal indictment in US District Court, New Jersey, Aug. 17, 2009 (the Heartland case).
10. Zetter, K. "TJX Hacker Was Awash in Cash; His Penniless Coder Faces Prison." *Wired*, June 18, 2009.

Summary

This chapter described how network security is an important component of information security. Information security controls mitigate threats to confidentiality, integrity, and availability of information. Information security and network security are becoming increasingly important as more components of our personal and professional lives revolve around computer networks and information.

In this chapter, we looked at controls that provide information security to data entering and exiting computer networks. With incoming data, we are primarily concerned that malicious attackers may be actively trying to harm our computers and information. With outgoing data, we are primarily concerned about preventing theft and manipulation of data as it flows through the network.

About the Colophon

A radical with a computer can cause more harm today than a radical with a gun. In this environment, it is difficult to put a humorous spin on computer and network insecurity. However, the pithy remark by the *Farmer's Almanac* succeeds in doing just that.

REVIEW QUESTIONS

1. What is *information security*? Why is it important?
2. What are *vulnerabilities*? Give some examples.
3. What are *threats*? Give some examples.
4. What are *controls*? Give some examples.
5. What is *confidentiality*? Give some examples showing a violation of confidentiality.
6. What is *integrity*? Give some examples showing a violation of integrity.
7. What is *availability*? Give some examples showing a violation of availability.
8. What is *network security*? Why is it important?
9. What controls can be used to ensure confidentiality from incoming threats?
10. How does authorization offer additional protection after authentication?
11. What controls can be used to ensure integrity in the presence of incoming threats?
12. What is a *demilitarized zone*? What are some of the network services offered in the demilitarized zone? What network services are not recommended to be offered in the demilitarized zone?
13. Recall your visit to a store that also includes a pharmacy. Describe how the organization of the store is similar to the two-tier firewall architecture used in network security.
14. What is a *firewall*? What are the capabilities of firewalls?
15. What are the limitations of firewalls?
16. What controls can be used to ensure availability in the presence of incoming threats?
17. How are viruses different from worms? Give an example of a well-known virus and a well-known worm.
18. What are *denial-of-service* attacks? What can you do to reduce the losses from denial-of-service attacks on your network?
19. What controls can be used to ensure confidentiality of outgoing information?
20. What is *encryption*? What is an encryption algorithm? What is an encryption key?
21. What are the advantages and limitations of symmetric key encryption?
22. What are the advantages and limitations of asymmetric key encryption? What is the most popular asymmetric key encryption technology?
23. Briefly describe how you can use asymmetric key encryption to ensure the confidentiality of outgoing information. Clearly show the sender, receiver, and how the two keys are used to provide confidentiality.

24. Briefly describe how you can use asymmetric key encryption to ensure the integrity of outgoing information. Clearly show the sender, receiver, and how the two keys are used to provide integrity.
25. What controls can be used to ensure availability of outgoing information?

EXAMPLE CASE QUESTIONS

1. Read the 10-K statement filed by T.J. Maxx on March 28, 2007, with the SEC (Symbol: TJX).[18] Use the section on "Computer intrusion" to list the major events related to the security breach at the company and the dates on which they occurred. (SEC filings are usually most easily obtained from Yahoo finance. When you search for a company and follow the link to all filings on EDGAR, all filings made by the company are listed in reverse chronological order. Filings from the last four years are available here. Filings from earlier years are obtained directly from the SEC's EDGAR database using FTP.)
2. Read the 8-K statement filed by TJX in connection with the intrusion (January 18, 2007). What information did the company report in the filing? What is an 8-K statement?
3. What legal actions were initiated against TJX as a result of the computer intrusion? (The 10-K statement for 2007 filed by TJX will be useful.)
4. What are some best practices for securing wireless networks?
5. What is a SQL injection attack?
6. Compare the sales at T.J. Maxx in its latest financial year and in 2007. Discuss any trends.
7. What does Heartland Payment Systems (NYSE: HPY) do? How does its financial performance compare to its competitors? (Yahoo Finance and Wolfram Alpha are good sites for such comparisons.)
8. What offense is defined by 18 USC, section 371? (Search online.)
9. Read the indictment against Albert Gonzalez filed in the District Court of New Jersey (available at many places online). What evidence is provided in the indictment to support the charge of conspiracy?

HANDS-ON EXERCISE—https

One of the most commonly used techniques to maintain network security is encryption. Encryption converts information into a form such that it cannot be read by intruders. This is very useful to prevent information leakage on the network. In this exercise, you will use Wireshark to capture and view encrypted traffic. An easy way to do this is to visit an e-commerce website, such as a bank, a retail store, or your university's student portal, and view the traffic after you authenticate yourself using your username and password. At that point, the website typically uses SSL to encrypt network data. Since you are familiar with Wireshark from Chapter 6, Wireshark will not be introduced here. Figure 11.13 shows a Wireshark capture after logging into a secure site. In the figure, if you observe packet number 25 (the selected packet), you will notice that this packet starts a key exchange to

18 The "SEC filings" link on TJX's investor site provides a simple search function. You can search for "10-K" or "8-K" as needed in this exercise.

```
*Wi-Fi
File Edit View Go Capture Analyze Statistics Telephony Wireless Tools Help
tcp.stream eq 6                                                          Expression...
No.  Time       Source        Destination   Protocol  Length Info
175  4.142033   192.168.1.8   23.14.84.48   TCP       66  54383 → 443 [SYN] Seq=0 Win=65535 Len=0 MSS=1460 WS...
181  4.159173   23.14.84.48   192.168.1.8   TCP       66  443 → 54383 [SYN, ACK] Seq=0 Ack=1 Win=29200 Len=0 ...
183  4.159454   192.168.1.8   23.14.84.48   TCP       54  54383 → 443 [ACK] Seq=1 Ack=1 Win=262144 Len=0
184  4.159674   192.168.1.8   23.14.84.48   TLSv1.2   474 Client Hello
199  4.175553   23.14.84.48   192.168.1.8   TCP       54  443 → 54383 [ACK] Seq=1 Ack=421 Win=30272 Len=0
201  4.179133   23.14.84.48   192.168.1.8   TLSv1.2   169 Server Hello, Change Cipher Spec, Encrypted Handsha...
203  4.179378   192.168.1.8   23.14.84.48   TCP       54  54383 → 443 [ACK] Seq=421 Ack=116 Win=261888 Len=0
204  4.179933   192.168.1.8   23.14.84.48   TLSv1.2   105 Change Cipher Spec, Hello Request, Hello Request
205  4.181306   192.168.1.8   23.14.84.48   TLSv1.2   575 Application Data
v Frame 184: 474 bytes on wire (3792 bits), 474 bytes captured (3792 bits) on interface 0
  Interface id: 0 (\Device\NPF_{DFB8D6B0-CC9A-491D-812C-C8E6AEC2B649})
  Encapsulation type: Ethernet (1)
  Arrival Time: Feb 27, 2016 21:32:57.038744000 Eastern Standard Time
0000  18 1b eb 80 6d f3 30 59  b7 14 05 e2 08 00 45 00   ....m.0Y ......E.
0010  01 cc 45 17 40 00 80 06  87 26 c0 a8 01 08 17 0e   ..E.@... .&......
0020  54 30 d4 6f 01 bb 78 10  fd 7f 0f f5 7a 93 50 18   T0.o..x. ....z.P.
0030  04 00 a7 13 00 00 16 03  03 01 9f 01 00 01 9b 03   ........ ........
0040  03 56 d2 5c 59 48 81 39  d8 c5 8a 2d 0c cc 7b 6c   .V.\YH.9 ...-..{l
0050  30 08 eb 52 1f 87 3e 8e  90 fa f8 cd 78 b2 f3 27   0..R..>. ....x..'
0060  a0 00 00 34 c0 2c c0 2b  c0 30 c0 2f 00 9f 00 9e   ...4.,.+ .0./....
0070  c0 24 c0 23 c0 28 c0 27  c0 0a c0 09 c0 14 c0 13   .$.#.(.' ........
0080  00 9d 00 9c 00 3d 00 3c  00 35 00 2f 00 0a 00 6a   .....=.< .5./...j
0090  00 40 00 38 00 32 00 13  01 00 01 3e 00 00 00 20   .@.8.2.. ...>...
00a0  00 1e 00 00 1b 69 6d 67  2d 73 2d 6d 73 6e 2d 63   .....img -s-msn-c
00b0  6f 6d 2e 61 6b 61 6d 61  69 7a 65 64 2e 6e 65 74   om.akama ized.net
00c0  00 05 00 05 01 00 00 00  00 00 0a 00 06 00 04 00   ........ ........
00d0  17 00 18 00 0b 00 02 01  00 00 0d 00 14 00 12 06   ........ ........
00e0  01 06 03 04 01 05 01 02  01 04 03 05 03 02 03 02   ........ ........
wireshark_pcapng_DFB8D6B0-CC9A-491D-812C-C8E6AEC2B649_20160227213252_a09204   Packets: 1209 · Displayed: 37 (3.1%)   Profile: Default
```

Figure 11.13: Transition from HTTP to HTTPS at secure website

switch from unencrypted transfer to encrypted data transfer. Packets following the key exchange are encrypted.

Figure 11.14 shows the TCP stream of the encrypted packets. The most noticeable feature in the figure is that the text is completely unreadable.

Answer the following questions:

1. Why is SSL useful?
2. Use Wikipedia or other information resources to write a brief summary of the evolution of SSL and TLS.
3. Log in to a secure website and examine packets exchanged after the secure connection was established. What is the port number used by the remote web server for SSL connections? (These packets are easily located as using the TLS protocol. The question asks for the port associated with "https.")
4. Right-click on a packet sent using SSL and select "Follow TCP Stream." Show the screenshot of the TCP stream.

CRITICAL-THINKING EXERCISE—Identifying Threats

Consider the typical office worker shown in Figure 11.2. List as many information security threats as possible involving this worker. These include threats faced by the worker from working in the organization, as well as threats faced by the organization as a result of the actions of this worker. Specify any assumptions you make about the worker's qualifications and personal situation, and the employer.

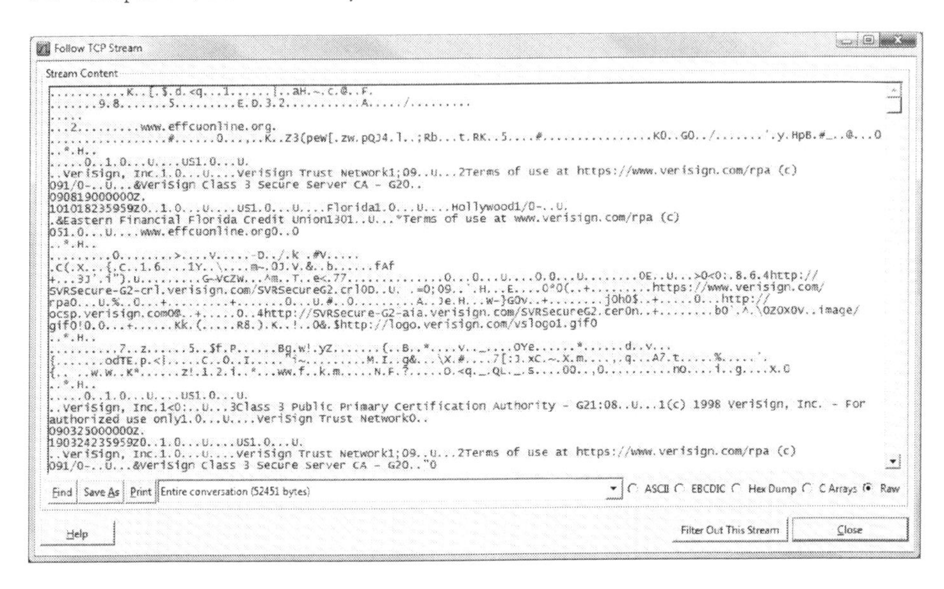

Figure 11.14: TCP stream showing encryption

IT INFRASTRUCTURE DESIGN EXERCISE—Adding Security

Our network thus far is functional but provides no protection for confidentiality and integrity of information transmitted over the WAN links. Therefore, TrendyWidgets has decided to invest in network security to protect its intellectual property. Answer the following questions:

1. How can a firewall help TrendyWidgets? Update your infrastructure diagram from Chapter 10 to include a firewall at TrendyWidgets's gateway to the Internet.
2. How can VPNs help TrendyWidgets in securing traffic that flows through its WAN? The router located in Tampa has hardware capabilities to perform the required encryption for VPN service over the Internet connection. Update your infrastructure diagram to reflect the VPN capabilities of the Tampa router (it is enough to update the label for the Tampa router, reflecting its VPN capability).
3. What encryption technology will you use to secure the wireless traffic in Amsterdam? Update your infrastructure diagram to reflect the encryption technology used in the wireless network.

CHAPTER 12

Computing Infrastructures

"Oh, yeah, it typed something?"

Ed Roberts, March 1975[1]

Overview

In earlier chapters, we have examined the workings of the communication system in modern IT infrastructures. In this chapter, we will take a look at the workings of the computing system. We will look at the structure of an individual computer, and see how these individual computers work together to handle the large-scale computing tasks thrown at computers today. At the end of this chapter you should have a broad understanding of:

- the components of a computer—memory, ALU, and operating system
- the operating system—processes, threads, and file systems
- security concerns in operating systems—privileged states, attack protection, and security principles
- virtualization, cloud architectures, and big-data architectures

Origins—von Neumann Architecture[2]

We have come a long way. Figure 12.1 is an ad for a computer from the 19th century. Modern computing traces its history to the von Neumann report[3] published by the University of Pennsylvania in 1945.[4] The report was based on the ENIAC (Electronic Numerical Integrator And Calculator) computer being developed by J. Presper Eckert and

1 Stephen Manes and Paul Andrews, *Gates: How Microsoft's Mogul Reinvented an Industry—and Made Himself the Richest Man in America* (Doubleday, 1993), 75.

2 This section, in particular the parallel between the von Neumann architecture and the programming constructs in modern programming languages, is inspired by Lecture 1.1 in the Coursera course, "Functional programming," June 2014, taught by Prof. Martin Odersky, the creator of the Scala programming language.

3 John von Neumann, First draft of a report on the EDVAC, University of Pennsylvania, June 30, 1945.

4 Appendix L.2 in the classic text on computer architecture by Hennessy and Patterson is an excellent resource on the origins of modern computers. http://booksite.elsevier.com/9780123838728/historical.php (accessed June 5, 2014).

John Mauchly at UPenn. In section 2 of the report, the required parts of a computer were identified:

1. A central arithmetic part (CA): A part that is capable of natively performing the required computational operations (such as +, -, ×, ÷).
2. A central control part (CC): A part that sequences operations properly so that the computer can perform its tasks.
3. Memory (M): This part stores instructions (programs), intermediate results from the CA's operations, any data that the computer is working on, as well as inputs and outputs from and to the outside world.
4. Input (I): This part allows the computer to receive information from the outside world.
5. Output (O): This part allows the computer to present its output to the outside world.

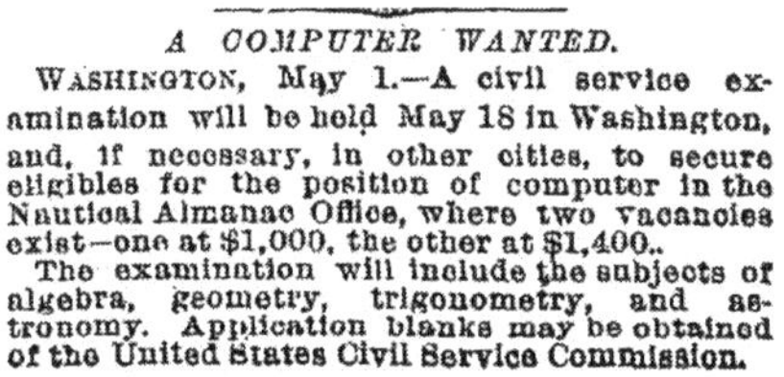

A COMPUTER WANTED.

WASHINGTON, May 1.—A civil service examination will be held May 18 in Washington, and, if necessary, in other cities, to secure eligibles for the position of computer in the Nautical Almanac Office, where two vacancies exist—one at $1,000, the other at $1,400..

The examination will include the subjects of algebra, geometry, trigonometry, and astronomy. Application blanks may be obtained of the United States Civil Service Commission.

The New York Times
Published: May 2, 1892
Copyright © The New York Times

Figure 12.1: Computer, circa 1892

Modern computers follow this design, which is called the von Neumann architecture. The von Neumann report suggested that the CA, CC, and M were analogous to the human brain. The input and output were collectively called the outside recording medium (R) of the device, and were analogous to the human sensory organs. The report also called CA and CC together as C (perhaps as an abbreviation for Computer). In the von Neumann model, C and M were to be able to natively quickly move data and instructions between each other. The von Neumann architecture is shown in Figure 12.2. This architecture has obviously been improved over time, but it continues to be the basis of all modern computers. In particular, the Harvard architecture added one additional information path between the memory and CPU. So, there was one path dedicated for data and one path dedicated to moving instructions between memory and CPU. This dramatically speeded up computer operations.

What is interesting about the von Neumann architecture is that not only is it the basis for all modern computers, it also has greatly influenced the development of modern computer programming languages. Most modern computer languages (e.g. C, C++, and Java) are built around constructs that correspond directly to the von Neumann architecture. For example, variables (e.g. int x) in programming languages correspond directly to memory cells in the architecture, and variable assignments (e.g. x = 3) correspond directly to store instructions in the architecture.

In fact, one of the challenges of modern computer programming is that most students find it very challenging to conceptualize relevant problems (e.g. predict tomorrow's weather) in terms of a vocabulary that is limited to the constructs directly supported by the von Neumann architecture. John Backus, the creator of Fortran, called it the "*primitive word-at-a-time style of programming inherited from [a] common ancestor—the von Neumann computer.*"[5]

5 John Backus, "Can programming be liberated from the von Neumann style?" Turing award lecture, *Communications of the ACM*, 21(8) (1978).

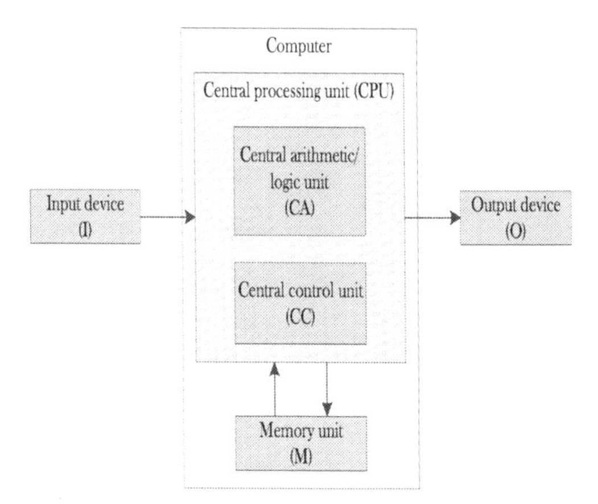

Figure 12.2: The von Neumann architecture

Efforts such as functional programming are aimed at improving the vocabulary of modern computers so that it better corresponds to the way people think about business problems, instead of having to first translate business problems to the vocabulary supported by the von Neumann architecture.[6] If you have always wanted to excel at computer programming but found it incredibly tedious, you now at least know the source of the issue.

The Operating System

Origins

The ENIAC machine described by von Neumann did not have an operating system. To get the machine to do useful work (for example, the ENIAC was created to compute the correct elevation angle at which to position a gun to reach a target), it was necessary to configure the machine appropriately so that it performed the entire sequence of operations correctly, including reading data and instructions, moving data and instructions in and out of the CPU, and gathering the output. It could take several days for a team of five experienced operators to perform one set of computations.

As the use of computers proliferated, even though computers became faster, users still had to take responsibility for specifying how to access input and output devices, and specifying how the controller was to process the data and instructions in order to provide the desired results. Over time, people began to notice that they were taking much longer to write programs to do common tasks (e.g. counting CPU cycles, counting pages printed, counting volume of data read and memory used, requesting operator intervention, recording audit trails of file access, etc.) than on the actual business they were interested in. In fact, these mundane tasks became the bottleneck in computer operations.

6 For those interested, an excellent resource for learning about functional programming and its motivations is the Functional Programming course offered by Coursera, taught by Prof. Martin Odersky of EPFL, Switzerland.

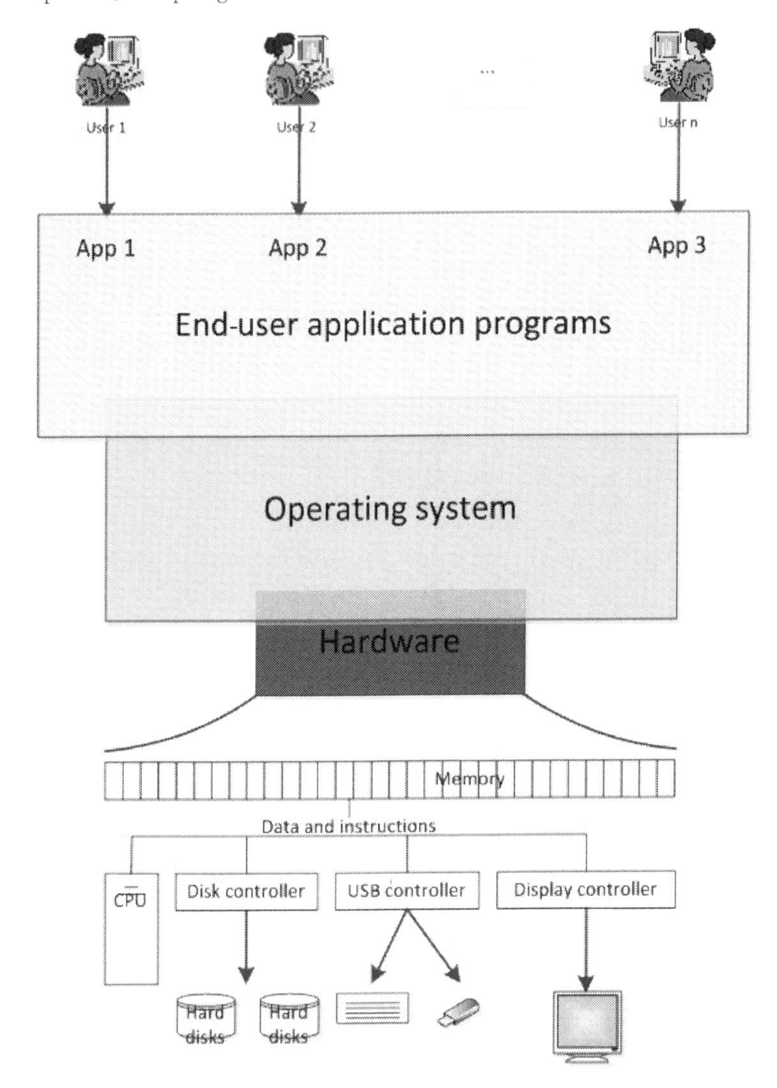

Figure 12.3: Modern computer architecture with operating system

Eventually, computer user groups and vendors began to write programs to perform these common functions. For example, IBM promoted the development of SHARE (Society to Help Alleviate Redundant Effort), to maintain a repository of common programs to perform tasks such as advancing the input to a certain position or outputting a block of data. These shared routines eventually developed into the "kernel" of the modern operating system. End-user operating systems provide additional services such as graphical interfaces, text editors, system administration utilities, etc.

With the inclusion of the operating system as the underlying software performing all operations common to applications, the architecture of modern computers looks as shown in Figure 12.3. We see how end-user applications interface with the hardware through the operating system.

System Calls

Applications frequently need to perform tasks that computers only allow the operating system to perform. Examples include reading and writing files to disk, or sending and receiving data over the network. When a user application (such as a browser or word processor) needs to perform such a task, the application invokes the corresponding operating system function.[7] This is called a system call. *A system call is an invocation by a user program of a function provided by the operating system.* System calls are requests to the operating system to perform tasks reserved exclusively for the operating system on behalf of a user program. These tasks include network communication, process control, and device or file manipulation.

Why are some functions reserved exclusively for the operating system? Typically, these functions are invasive and can cause harm to the computer if used improperly. For example, if a user wants to delete a file, it is necessary to confirm that the user has the privileges to delete that file. Reserving such invasive tasks to the operating system ensures that end-user applications cannot cause avoidable harm, either by accident or by choice.

Computer designers recognize that instructions received from the operating system are inherently more trustworthy than instructions received from user programs. Accordingly, most computer systems have a mechanism to distinguish between operating system code and user (application) code. When the hardware is executing operating system code, it runs in kernel mode. While executing user code, the hardware shifts to user mode. The kernel mode is also called the privileged mode or the system mode. Typically, this is implemented in hardware as a bit called the mode bit. The convention is to use 0 to indicate that the mode bit is in kernel mode, and 1 to indicate that the mode bit is in user mode. Hardware in modern computers offers appropriate mechanisms to ensure that end-user programs cannot bypass this protection. On Windows, you can use the Task Manager to see a graphical display of the CPU utilization in user mode and kernel mode (Figure 12.4).[8]

This mechanism of operating modern computers in one of two modes—privileged or user—is called the dual mode of operation of computers. Older computer architectures, such as the Intel 8086 and 8088 (which fueled the original PC boom along with the MS-DOS operating system), did not have a mode bit. In this environment, end-user applications had the same privileges as the operating system and an offending (or sloppy) end-user application could even wipe out the operating system since the hardware had no way of knowing which instructions came from the operating system and which instructions originated from an end-user application. Modern PC hardware such as Intel processors supports dual-mode

---

7 For example, the function to write a file to disk in Windows is WriteFile (http://msdn.microsoft.com/en-us/library/windows/desktop/aa365747(v=vs.85).aspx), accessed May 29, 2014. Most business programmers will not use the WriteFile system call, instead preferring the C# System.IO.FileStream.Write() method (http://msdn.microsoft.com/en-us/library/vstudio/system.io.filestream.write), accessed May 29, 2014. Using C# lets developers benefit from error handling built into the method by Microsoft's developers. Interested students may see the code for the C# implementation of the Write method, which uses system calls in the background on line 1775 at (http://referencesource.microsoft.com/#mscorlib/system/io/filestream.cs).

8 To get this view, start Task Manager (Ctrl+Shift+Esc), go to the Performance tab and in the View menu, select "show kernel times."

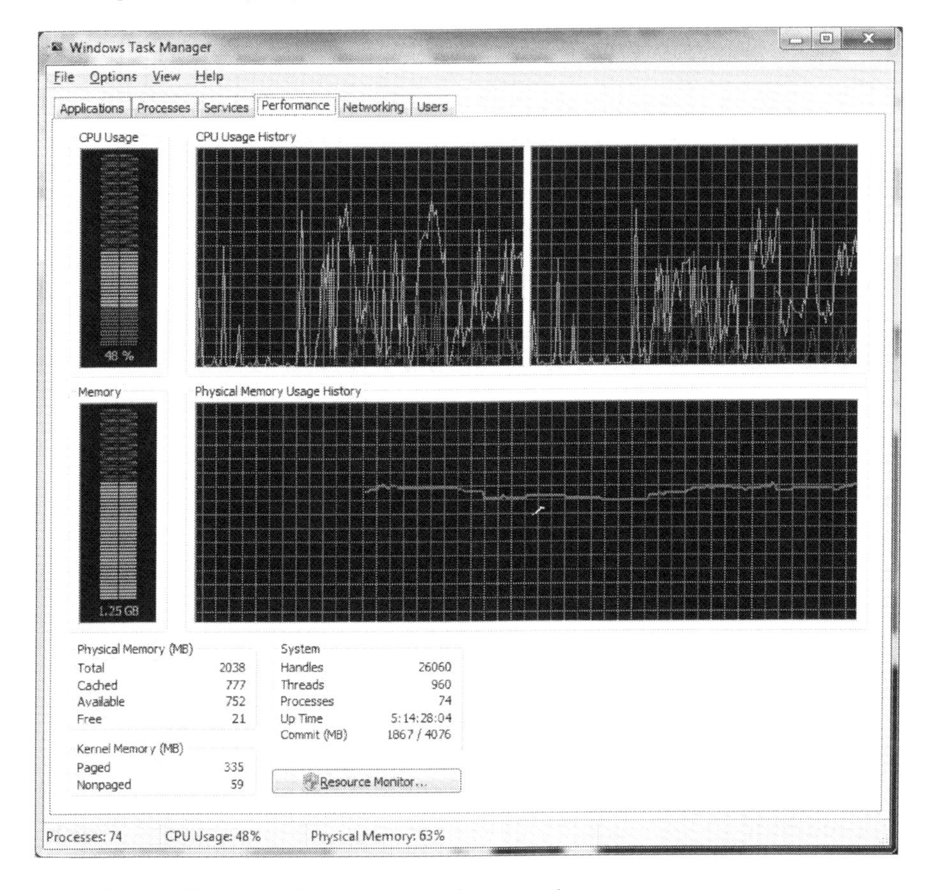

Figure 12.4: CPU usage in kernel mode and user mode

operation. When an end-user application has control of the CPU, the hardware is operating in user mode. Now, if an end-user application attempts to directly execute a CPU instruction that is reserved for the privileged mode, the hardware will refuse to execute it and instead transfer control to the operating system, alerting it to the offense. The standard procedure in this case is for the operating system to terminate the program, alert the user, and optionally save the current state of the offending application so the developer of the application can fix the problem.

Operating System Components

One of the most important architectural features of modern operating systems is their support for multitasking. As CPUs have become extremely fast, their capabilities now far exceed the requirements of most individual users or applications. But most user activities require multiple independent tasks to be performed simultaneously. For example, when a user is editing a spreadsheet stored on a network, the computer is reading and writing data on the network, and displaying content on the monitor. Simultaneously, it could also be gathering e-mail and updating the calendar in the background. To efficiently perform all these tasks, modern operating systems support multitasking. Multitasking organizes computing jobs in such a way that the CPU can be instantly provided with everything it needs to complete one job, while it is waiting for another job to complete.

For example, while the CPU waits for an acknowledgment from the remote server that the spreadsheet file has been successfully transferred over the network and saved to disk, the operating system can load the CPU with the e-mail program. If the e-mail program just received an e-mail, it can request the CPU to display a notification to the user's display. Once this notification has been displayed, the e-mail program can be unloaded from the CPU, and the operating system can reload the spreadsheet program back to allow the user to continue working. All this can happen so quickly that the user would never realize that the CPU had momentarily stopped working on the spreadsheet and was focusing on a different task.

To support multitasking, the operating system keeps all pending jobs readily available in memory. In the computing world, a job is formally known as a process. In addition to managing processes, operating systems also manage threads, memory, and data stores. In the next sections, we will look at each of these components and the important concerns while managing them. We will also look at the important ways in which operating systems support security in computer operations.

Processes

A process is a computer program that has been loaded into memory and is executing. A process is the unit of work in computers and a computer is really a collection of processes. Some of these are operating system processes, and others are user processes. By rapidly switching processes in and out of the CPU, all of these processes can appear to be executing concurrently on the computer. These concurrent processes may be independent of each other, in which case they do not impact other processes. Processes may also be cooperating, in which case they can affect each other, and may even share data with each other. One of the roles of an operating system is to provide an environment where processes can effectively cooperate with each other. To perform this role, the operating system is responsible for starting and stopping processes, handling deadlocks (where two processes are waiting for results from each other), and handling exchange of information between processes.

A process contains the program instructions as well as all the information needed for the process to run. This includes: (1) the process stack, which contains the functions currently in execution and the local variables within these functions; (2) the process heap, which is the memory currently allocated to the process; as well as (3) other related information such as the current values of the different registers in the CPU.

Process State

How does the computer execute all the different waiting processes? The way a computer handles all waiting processes is quite similar to the way a dentist handles all patients waiting for her attention. At a dentist's office, a patient's visit begins by checking in and being admitted into the patient queue. Nurses prepare the admitted patients on multiple stations, with all the required tools, devices, and patient information so that they are ready for the dentist to process. When the dentist arrives at the station, she looks at the computer record to determine the current status of the patient in the treatment procedure, and the next operation to be performed. She also reviews any reports received from other doctors. The dentist then proceeds to perform the procedure. Some procedures require a waiting period while the applied chemicals do their job. During this time, the dentist does not wait. Rather, she moves on to the next patient and performs the required procedures there. Once the waiting period is over, the nurse calls for the dentist's attention and the procedure continues when the dentist becomes available.

Computers operate much the same way, the primary difference being that the processes move to the CPU whereas dentists move to patients (Table 12.1). The difference arises because it is much easier to move information in and out of the processor than the other way around, and it is much quicker for the dentist to move between patient stations than the other way around. Table 12.1 summarizes the analogy.

Table 12.1: Analogy between a computer and a dentist office

Computer	Dentist office analogy
Process	Patient
Program instructions	Treatment plan
Program counter	Current status
Function call	x-ray request
Function call return	x-ray results
Device I/O request	Brace bracket bonding dry
Device interrupt	Bonding complete

Just as patients and the dentist move between various states in the dental office, the process moves between various states while it is executed on the computer. *A state is the present condition of an entity*. The states and the transitions between them are shown in Figure 12.5.

Threads

Threads are the smallest sequence of instructions that an operating system can manage independently. One process can be composed of multiple threads, and all these threads associated with a process share resources such as global variables and open files.[9]

So, how are threads different from processes, and what is the advantage of breaking up a process into threads?

The main advantage of breaking up a process into threads is efficiency in resource utilization within the computer. Creation of a new process is very resource intensive for the computer. It requires the operating system to load the program instructions from the hard drive (extremely slow operation), and the identification and allocation of spare memory for the new process to use. If the computer is running short of memory, the operating system

9 http://stackoverflow.com/questions/5201852/what-is-a-thread-really (accessed June 20, 2014).

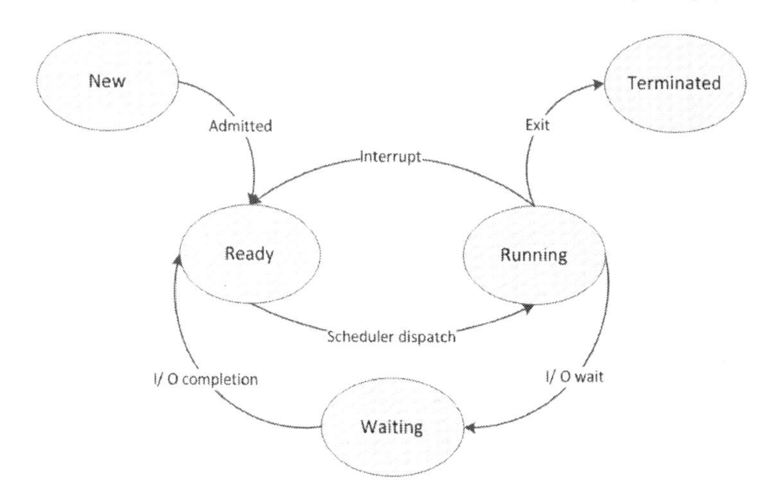

Figure 12.5: Process state transitions

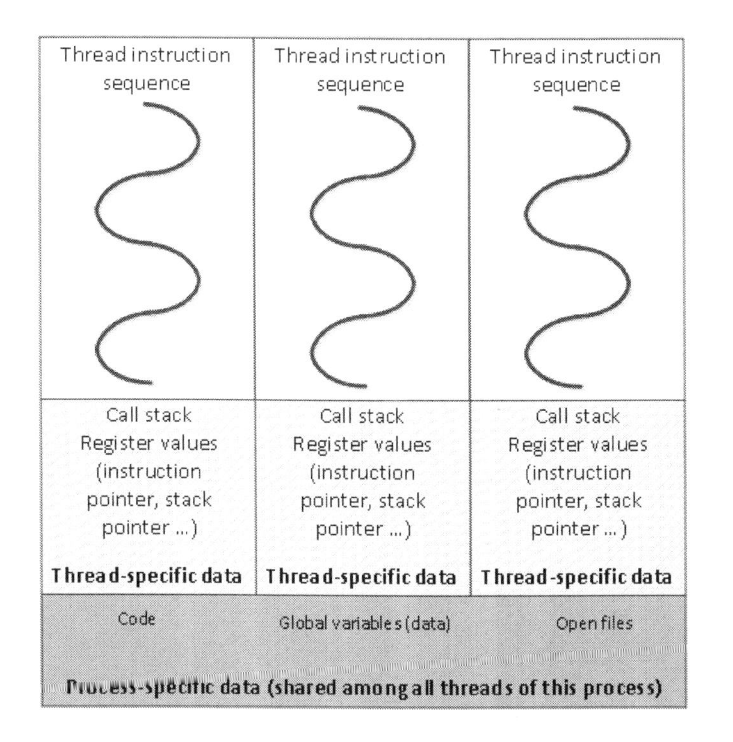

Figure 12.6: Threads within a process

will identify a process that has not been active for a while and save the state of this idle process on the hard disk (again an extremely slow process). As can be seen from Figure 12.6, all threads within a process share memory and program instructions with the process that creates the threads. Thus, creation of a new thread is very inexpensive because it does not require any fresh loading of program instructions or new memory allocation. In essence, a new thread just requires memory space to save the values of any variables unique to the thread, and some pointers to the location of the next instruction for the thread to execute.

Threads are very useful on both the server and client ends. On the server end, most servers perform the same set of actions repeatedly. For example, a web server gets requests for files and transfers the files back to the client. Prior to the use of threads, web servers would fork new processes for each new web page request. However, all modern web servers simply create a new thread to serve a new request. This can dramatically speed up the response time of the web server, while simultaneously reducing the resources required to operate a web server. In fact, modern web servers go a step further and create a set of threads in anticipation of receiving web requests. This set of threads is called a thread pool. When a fresh web request is received, the server does not even have to pause to create a new thread. It simply picks up an available thread from the thread pool to serve the request. Figure 12.7 shows an example of this configuration in Internet Information Server (IIS), a popular web server.

On the client side, threads significantly improve the user experience of most applications. For example, consider the tasks a browser has to perform when loading a web page: (1) it has to open a network connection to the server and request the page; (2) if the page involves images or other files located on other servers, it has to open connections to all these different servers and request files from all these locations; (3) it has to collect all the page contents from these different locations and render the page on the display; (4) if the page contains multimedia elements, it has to open the associated player and play the

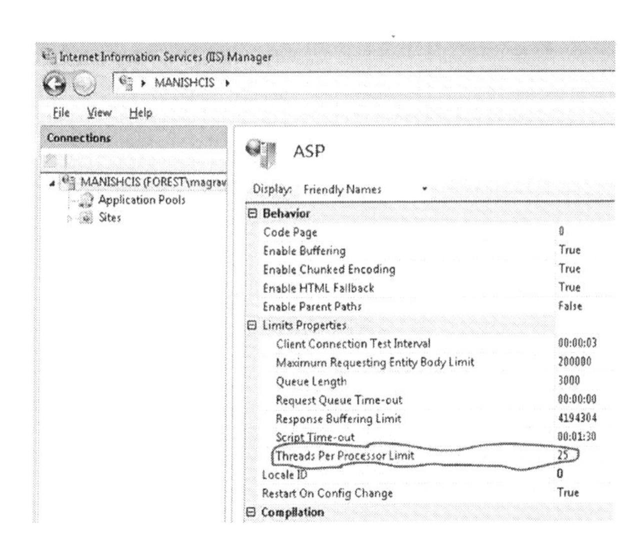

Figure 12.7: Specifying web server threads in IIS

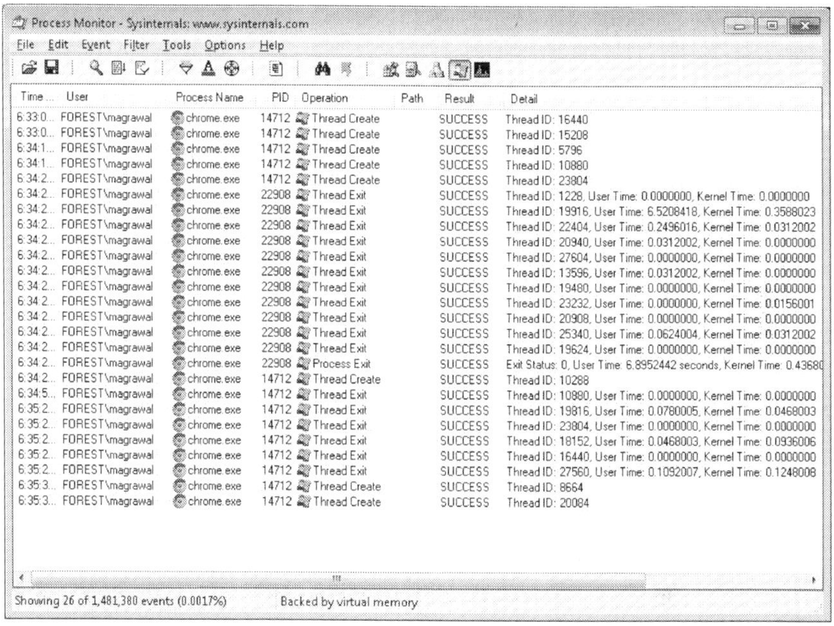

Process Monitor - Sysinternals: www.sysinternals.com

File Edit Event Filter Tools Options Help

Time ...	User	Process Name	PID	Operation	Path	Result	Detail
6:33:0...	FOREST\magrawal	chrome.exe	14712	Thread Create		SUCCESS	Thread ID: 16440
6:33:0...	FOREST\magrawal	chrome.exe	14712	Thread Create		SUCCESS	Thread ID: 15208
6:34:1...	FOREST\magrawal	chrome.exe	14712	Thread Create		SUCCESS	Thread ID: 5796
6:34:1...	FOREST\magrawal	chrome.exe	14712	Thread Create		SUCCESS	Thread ID: 10880
6:34:2...	FOREST\magrawal	chrome.exe	14712	Thread Create		SUCCESS	Thread ID: 23804
6:34:2...	FOREST\magrawal	chrome.exe	22908	Thread Exit		SUCCESS	Thread ID: 1228, User Time: 0.0000000, Kernel Time: 0.0000000
6:34:2...	FOREST\magrawal	chrome.exe	22908	Thread Exit		SUCCESS	Thread ID: 19916, User Time: 6.5208418, Kernel Time: 0.3588023
6:34:2...	FOREST\magrawal	chrome.exe	22908	Thread Exit		SUCCESS	Thread ID: 22404, User Time: 0.2496016, Kernel Time: 0.0312002
6:34:2...	FOREST\magrawal	chrome.exe	22908	Thread Exit		SUCCESS	Thread ID: 20940, User Time: 0.0312002, Kernel Time: 0.0000000
6:34:2...	FOREST\magrawal	chrome.exe	22908	Thread Exit		SUCCESS	Thread ID: 27604, User Time: 0.0000000, Kernel Time: 0.0000000
6:34:2...	FOREST\magrawal	chrome.exe	22908	Thread Exit		SUCCESS	Thread ID: 13596, User Time: 0.0312002, Kernel Time: 0.0000000
6:34:2...	FOREST\magrawal	chrome.exe	22908	Thread Exit		SUCCESS	Thread ID: 19480, User Time: 0.0000000, Kernel Time: 0.0000000
6:34:2...	FOREST\magrawal	chrome.exe	22908	Thread Exit		SUCCESS	Thread ID: 23232, User Time: 0.0000000, Kernel Time: 0.0156001
6:34:2...	FOREST\magrawal	chrome.exe	22908	Thread Exit		SUCCESS	Thread ID: 20908, User Time: 0.0000000, Kernel Time: 0.0000000
6:34:2...	FOREST\magrawal	chrome.exe	22908	Thread Exit		SUCCESS	Thread ID: 25340, User Time: 0.0624004, Kernel Time: 0.0312002
6:34:2...	FOREST\magrawal	chrome.exe	22908	Thread Exit		SUCCESS	Thread ID: 19624, User Time: 0.0000000, Kernel Time: 0.0000000
6:34:2...	FOREST\magrawal	chrome.exe	22908	Process Exit		SUCCESS	Exit Status: 0, User Time: 6.8952442 seconds, Kernel Time: 0.43680
6:34:2...	FOREST\magrawal	chrome.exe	14712	Thread Create		SUCCESS	Thread ID: 10288
6:34:5...	FOREST\magrawal	chrome.exe	14712	Thread Exit		SUCCESS	Thread ID: 10880, User Time: 0.0000000, Kernel Time: 0.0000000
6:35:2...	FOREST\magrawal	chrome.exe	14712	Thread Exit		SUCCESS	Thread ID: 19816, User Time: 0.0780005, Kernel Time: 0.0468003
6:35:2...	FOREST\magrawal	chrome.exe	14712	Thread Exit		SUCCESS	Thread ID: 23804, User Time: 0.0000000, Kernel Time: 0.0000000
6:35:2...	FOREST\magrawal	chrome.exe	14712	Thread Exit		SUCCESS	Thread ID: 18152, User Time: 0.0468003, Kernel Time: 0.0936006
6:35:2...	FOREST\magrawal	chrome.exe	14712	Thread Exit		SUCCESS	Thread ID: 16440, User Time: 0.0000000, Kernel Time: 0.0000000
6:35:2...	FOREST\magrawal	chrome.exe	14712	Thread Exit		SUCCESS	Thread ID: 27560, User Time: 0.1092007, Kernel Time: 0.1248008
6:35:3...	FOREST\magrawal	chrome.exe	14712	Thread Create		SUCCESS	Thread ID: 8664
6:35:3...	FOREST\magrawal	chrome.exe	14712	Thread Create		SUCCESS	Thread ID: 20084

Showing 26 of 1,481,380 events (0.0017%) Backed by virtual memory

Figure 12.8: Threads associated with a browser tab in Chrome

multimedia content. If the browser performed all these tasks in sequence, it would lead to long delays before the page contents could be displayed. Instead, all these operations run on parallel threads and the browser can render the page fragments in appropriate locations as they arrive so users can start viewing the page long before all the page contents are received. An example is shown in Figure 12.8. The five thread create operations at the beginning of the figure are associated with two separate downloads as may be inferred from the time stamps. The next 12 thread exit and process exit operations that follow are associated with closing the browser tab.[10]

Threads and processes

A good example of the difference between processes and threads can be seen at any retail bank. The tellers are analogous to threads of the retail transaction process. All tellers share the same procedures (code). They also share account balances in a common backend database (data and open files). The mortgage banker may be considered a separate process because the expertise required (code) is different.

The analogy is not perfect, though, because both the mortgage banker and tellers share the same common backend database.

---

10 When Chrome was introduced in 2008, Google released a comic strip describing their philosophy behind the development of the browser. It is an interesting read and introduces the idea of using a separate process for each tab to isolate failures to the single tab: http://books.google.com/books?id=8UsqHohwwVYC (accessed June 28, 2014).

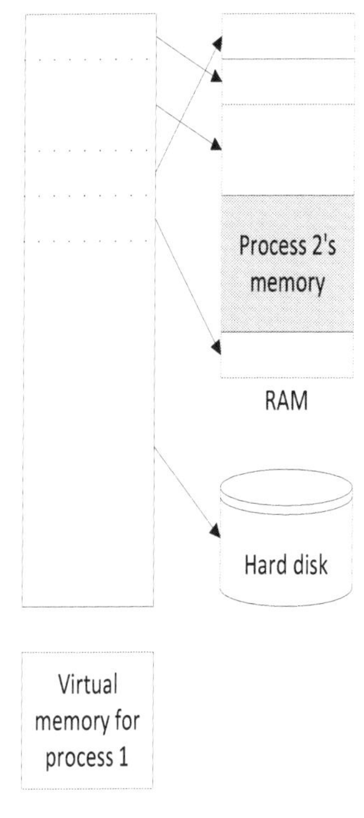

Figure 12.9: Virtual memory

Memory

Memory is the storehouse of data on the computer that can be quickly and directly accessed by both the CPU and the I/O (input/output) devices on a computer. This is seen in the architecture shown in Figure 12.3. Since both the CPU and I/O devices can access memory, the memory becomes the meeting place on the computer where the CPU and devices can share data. For example, when an application wants to print data, the CPU saves the data to be printed in a specific location in memory and tells the operating system the location of this saved data. The operating system then instructs the printer to collect the data from the specified memory location and print it.

As we saw in Figure 12.2, the main memory of a computer (popularly known as RAM) is a special entity in a computer. It is the only data-storage device that the CPU can access directly. Program instructions (for both the operating system and applications) as well as data reside in main memory. When the CPU is performing its job, it reads instructions and manipulates data stored in memory. When the operating system finds that the CPU is looking for data stored on the hard disk, it first fetches the data from disk into memory (by quickly loading the appropriate file-fetching instructions into the CPU) where the CPU can read it. Similarly, when the user clicks on an icon to start a program, the operating system first loads the program instructions into memory where the CPU can read them.

When the user closes a program, the operating system notes that the memory locations used by the instructions and data associated with the program processes are available for other programs to use.

Memory is a critical component of computers, yet most users have very little of it. Why? Customers are dazzled by CPU speeds and hard disk spaces, but, not realizing the importance of memory, try to economize on it. Operating systems have developed a method to deal with situations where the computer does not have enough memory to handle all open applications—virtual memory (Figure 12.9). *Virtual memory is a technique to take up space on the hard drive and use it as an extension of the main memory.* Processes address virtual memory and the operating system takes responsibility for mapping calls to virtual memory addresses to the correct locations on the main memory or disk. The area of the disk used as an extension of the main memory is called the page file. If you have ever encountered a situation where you opened up a long-dormant application and it took some time to load, and perhaps you heard the hard disk churning while the dormant application was loading, you have experienced virtual memory in action. The state of the dormant process had been pushed to disk to make space for foreground applications in main memory to improve end-user performance.

Memory management involves many interesting challenges. When users open multiple applications simultaneously, the operating system may find that there isn't enough room in memory to accommodate all running processes. In this situation, the operating system uses various approaches to optimize the utilization of memory. For example, the operating system may save processes that have not been invoked for a long time on the hard disk. This is why some open applications appear to reload quickly, while other open applications may appear to take time to reload. The memory management techniques used by operating systems try to minimize the instances where the user has to wait for a process to load up.

Data Stores

Available memory technologies have had two main limitations: size and volatility. By size we mean the fact that user needs vastly exceed the memory capacity they can afford. For example, most users have almost a terabyte of digital data in the form of images, videos, and documents. At current prices of approximately $10/GB, 1TB of RAM would cost almost $10,000, which would make computers prohibitively expensive for most users. By volatility we mean that the contents of memory are lost when power is turned off. There are emerging memory technologies that retain data even when power is turned off (flash memory). However, they aren't yet usable as a computer's main memory because after about 100,000 writes, their ability to retain data begins to deteriorate. Since the CPU may rewrite memory locations many times a minute as processes are moved in and out of the CPU, flash memory and other non-volatile memory technologies with limited life cannot yet be used as a computer's main memory.

Computer manufacturers have developed many innovative technologies to accommodate the needs of users to economically save data even when power is turned off. These mass storage media include magnetic disks (hard drives), optical disks (CDs, Blu-Ray), and magnetic tapes (similar to vintage audio tapes), each with widely differing characteristics. For example, magnetic tapes are slower than magnetic disks but are more economical than disks when storing large volumes of data.

The operating system simplifies the use of mass storage media by taking responsibility for the management of these media and their controllers. The primary mechanism for accomplishing this is the file system. *The file system is the component of the operating system that provides efficient and convenient access to storage devices by allowing data to be stored, located, and retrieved easily.* Apart from managing access to file contents, the file system also maintains meta-data about the files it manages.

The file system of the operating system allows end users and applications to work with files representing both application programs and data. The operating system's file management role includes enabling operations to create and delete files, read and write files, and move files between media. To support these operations, the operating system keeps track of free space on storage media and allocates available space when requested by applications.

Operating System Security Protections

The operating system is obviously a prime target for attackers to try and gain access to. The operating system provides security protections that allow multiple users and applications to safely share the same computer without intruding on each other. This includes restricting access to computer resources, such as files and memory locations, to authorized processes. When the operating system takes responsibility for these core security functions, application developers can focus exclusively on providing end users with desired functionality, without worrying about writing defensive code. As another example, modern operating systems prevent any single process from monopolizing the CPU, thereby preventing many classes of denial-of-service attacks.

Operating systems use some fundamental security design principles to defend themselves, including domain separation, process isolation, resource encapsulation, and least privilege. We now take a brief look at each of these principles and their role in protecting the operating system.

Domain separation Domain separation is the close scrutiny of communications between two domains. A domain is an area controlled by an identified entity. Domain separation typically involves restricting communications to a few identified points, as happens, for example, through the main door of a house. The interior of the house is the domain of the homeowner and the exterior is the domain of the community. By restricting entry and egress through one door, homeowners are able to simplify home security. Similarly, the operating system with support from the hardware, enforces domain separation between the user domain and kernel domain. This way, applications are not able to perform sensitive operations, which are restricted to the operating system. This also prevents application bugs from damaging the operating system.

Process isolation Process isolation is the act of limiting the effects of one process upon other processes in the system. The operating system does this using numerous methods including by limiting memory access to the memory space allocated to the process, limiting file access to one process at a time, etc. This ensures that a compromised process cannot compromise other processes or the operating system.

Resource encapsulation Resource encapsulation hides implementation details of a resource and allows users to access the resource by invoking simple, easy-to-understand API calls. Purchasing an airline ticket is an example of resource encapsulation. Users do not need to know the details of flying a plane or operating an airport. Simply by purchasing a ticket, they are given a simple mechanism by which to access the transportation service offered by the airline. Similarly, system calls and file system calls provided by an operating system allow end-user applications to safely and easily access resources on the computer. *An application programming interface (API) is a formalized set of software function calls that can be referenced by user programs in order to access supporting services.* APIs are the standard mechanism by which different software programs interact with each other. The operating system API ensures that all required security properties are addressed by the implementation of the system call. Not only is this approach easier for application developers, it is also more secure.

Least privilege Peter J. Denning articulated the fundamental principle of preventing unauthorized actions as "[creating] an essentially closed architecture within which every process has no more capabilities than needed for its task and can interact with no other process in an unexpected way."[11] At almost the same time, Jerome Saltzer wrote that "every program and every privileged user of the system should operate using the least amount of privilege necessary to complete the job."[12] This is perhaps one of the most elementary principles of security in general. When deployed by an operating system, users and processes are only given those privileges that are essential to performing their task. For example, a user performing their day-to-day tasks does not need to install a program, which can require writing into privileged areas of the computer. Therefore, an end user should not have installation privileges while performing day-to-day work. This principle protects users and processes from creating unintended harm since a compromised process will not be able to perform sensitive tasks on the machine. And when the compromised process ends, all resources associated with the process are cleaned up, removing any potential sources of harm from the computer.

Operating System Installation and Startup

Operating systems are typically distributed as files with an .iso file for download over the network. Once downloaded, the .iso file can be unzipped using any standard utility and the resulting files can be saved on a disk. When the disk is inserted into a computer's disk drive, the operating system's installation program usually prompts the user to walk through the steps of installing the operating system. When a modern operating system (e.g. Windows, Mac OS) is installed on a computer, it typically installs all the components of the operating system, support for multiple languages, support for 32-bit as well as 64-bit hardware, as well as available device drivers onto the computer.[13] Typically this takes up about 6GB–8GB of space.

On older systems such as the IBM 360, the operating system installation was more customized to the hardware configuration of the specific system using a utility known as SYSGEN. Administrators could conserve disk space by configuring SYSGEN so that unnecessary device drivers and supporting software were not installed. However, SYSGEN required specialized users who understood how to use SYSGEN to install the OS. Modern operating systems are typically designed for end users with a diverse range of skills, and, fortunately, modern hard disks are quite spacious. So, modern operating systems focus more on convenience of installation and less on optimization of disk utilization.

Boot-up

Once the operating system is installed on the hard disk, how does it become available to the computer when the computer starts up? The process of making the operating system available for use by the computer hardware is called system boot.[14] Computers are hardwired to read the first sector of the hard disk when they are powered on. This first sector has a

11 Peter J. Denning, "Fault tolerant operating systems," *ACM Computing Surveys*, 8(4) (1976).
12 Jerome H. Saltzer, "Protection and the control of information sharing in multics," *Communications of the ACM*, 17(7) (1974).
13 http://apple.stackexchange.com/questions/1160/is-it-possible-to-reduce-the-disk-space-required-to-install-mac-os-x-10-6 (accessed June 19, 2014).
14 The term comes from the word bootstrapping, which is generally understood to mean improving oneself by one's own efforts without external assistance (http://en.wikipedia.org/wiki/Bootstrapping).

special name, the Master Boot Record. The master boot record has just enough information to ask the computer to load the operating system kernel. Once control passes to the kernel, the kernel then takes responsibility for loading up the rest of the operating system in the right sequence.

Scaling Up—VMs and WSCs

The previous section described the structure of a single computer. Though computers are now very powerful, modern computing needs have expanded even faster to the point that for many industrial tasks, it is necessary to combine the capabilities of multiple computers. This has led to a lot of experimentation and the emergence of some standard designs to handle large workloads. We start this section by presenting some facts that contribute to the problem. This is followed by a discussion of two concepts that improve efficiencies in utilization of computing resources for industrial tasks: virtual machines (VMs) and warehouse scale computers (WSCs). The discussion of WSC is largely motivated by US Patent 8218322, issued to Google for modular computing environments.

Big Data

What is the memory and data-storage capacity of the largest computer you have encountered? Say, memory of 128 GB and storage of perhaps 10 TB? Compare this to the amount of data that is processed by some intense users today. Facebook received 500 TB of data/day in 2016.[15] The Large Hadron Collider rejects data on 199,999 observations out of every 200,000. Yet it collected 13,000 TB of data in 2010.[16] Even small-business websites can collect large volumes of data if they try to track buyer traffic or offer social media services.

Big data is the general term used to describe data that is difficult to process using traditional database management systems. IBM has popularized a 5-V vocabulary to define data sets that are typically considered big data—volume, velocity, variety, veracity, value.[17] Volume refers to the basic size considerations for big data. Big datasets are usually large, on the order of tens of terabytes or more. Velocity refers to the speed at which data arrives and needs to be processed. Variety refers to the formats of data. Traditional transactional datasets map directly into relational database structures. Big data can include images, tweets, blog posts, and other types of information that do not map naturally to a relational structure and which require a diverse set of techniques to process. Veracity refers to the correctness of data. If the data is not verified before it is processed, it can lead to faulty inferences that can be very difficult to debug. The final consideration is value. Big data requires capital-scale investments. These can only be justified if the investments pay off in measurable business terms. Hence, the focus on big data in the C-suite.

Business leaders read about interesting insights that leading companies such as Google and Amazon are able to draw by exploiting the information in their databases and are keen

15 https://zephoria.com/top-15-valuable-facebook-statistics/ (accessed Mar. 2016).
16 Geoff Brumfiel, "High Energy Physics: Down the Petabyte Highway," *Nature*, 469 (2011): 282–283.
17 http://davebeulke.com/big-data-impacts-data-management-the-five-vs-of-big-data/ (accessed June 29, 2014).

to replicate these successes in their own organizations. Technology vendors are happy to sell hardware and software to help these leaders pursue their dreams. But in the end, each organization has its own unique big-data requirements (variety). The analysis that works for Google may only be partially applicable to your organization. That is where analysts like you come in—to develop statistical models unique to your organization to address the problems unique to your organization using data unique to your organization.

One especially relevant and challenging area of development in big-data methods is social media information. Organizations are interested in knowing what their customers are saying about them on Twitter, Facebook, and other social media channels. However, there is wide variety in this data and posts have less structure than typical point-of-sale data. Standard SQL-based data-analysis methods are therefore less useful in this space. Big-data methods are outside the scope of this text and are an area of rapid development.

> *Bob: Why did the computer technician miss the target?*
> *Alex: I don't know.*
> *Bob: He had troubleshooting.*
>
> Source: *Boys' Life* magazine, August 2014

Virtual Machines

Virtual machines are one technology used to build the large-scale computing environments of today. For what they do, computers are very inexpensive. For this reason, companies have invested large sums of money on computers and related IT infrastructure. According to Gartner, in 2013, worldwide IT spending exceeded $3,700 billion. Of this, almost $800 billion was on devices and data-center systems. Increasingly, IT leaders began to observe an interesting phenomenon—while their computer needs were growing, most of their individual computers were idle most of the time. For example, the computer systems devoted to process payrolls may be busy for four days every two weeks processing the information. For the remaining 10 days, the computers are relatively idle. Similarly, retail systems dedicated to processing peak loads between Thanksgiving and the end of the year are relatively idle for the remaining nine months of the year. Most corporate desktops barely use 5% of their processing capacity. Even when they are not doing any useful work, the computers take up space, require administration to check for viruses and other malware, consume power, and dissipate heat, which then necessitates using even more power to cool the room. Could the spare computing capacity be used to perform other computing tasks, so that all these expenditures become productive?

Virtualization was the answer. Figure 12.10 is an overview of virtualization. Virtualization software (e.g. VMWare) runs as an application on a host computer and operating system. Guest operating systems are installed on the virtualization application. Each guest operating system can have its own set of user accounts, network address, and end-user applications.

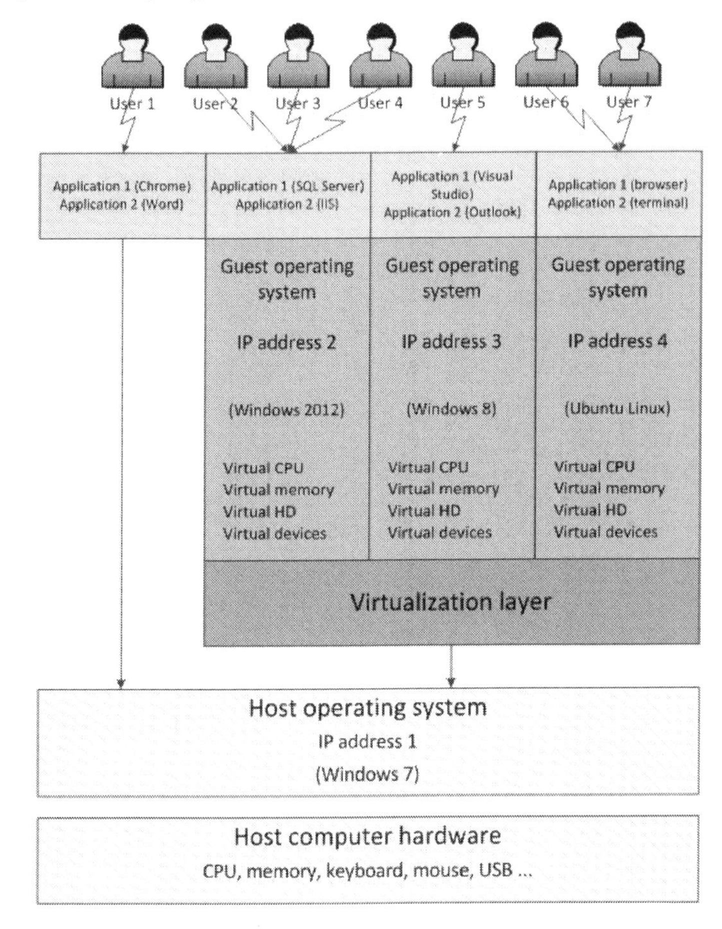

Figure 12.10: Virtualization overview

The guest operating systems can be started and stopped depending upon need. Applications running on the guest operating system have no awareness of the fact that they are accessing the host computer's resources through the virtualization layer. Each guest OS is identified by its own IP address, and users connect directly to the applications as needed.

The host operating system takes responsibility for partitioning the host system's resources among the guest operating systems. A special challenge is in implementing the dual-mode operation. Since the virtualization layer runs as an application on the host system, it is always running in user mode. When an application on a guest operating system makes a system call (for example to save a file to disk), the virtualization layer transfers the request to the host operating system, and simulates a switch to kernel mode for the guest operating system. When the system call completes, the virtualization layer receives the response from the host operating system, transfers the response to the guest operating system, and simulates a switch back to user mode on the guest operating system.

Getting the shifts between user mode and kernel mode is one of the challenges of correctly implementing computer virtualization.

Virtualization is useful not just to enterprises, but also to power users. For example, developers can install multiple operating systems on their computers and test their application on all different operating systems on their own machines.[18]

In the industry, there are two kinds of virtual machine technologies. The technology discussed above is called type-2 virtualization. Type-2 virtualization depends upon a host operating system to provide virtualization services. Type-1 virtualization software runs directly on the hardware, with no host operating system.

Type-1 virtualization software is generally the preferred approach because it eliminates the overhead of the host operating system and allows the virtualization software to directly access the capabilities of the hardware. However, type-2 virtualization is easier to use since the virtualization software can be installed over the host operating system just like any end-user application. For this reason, type-1 virtualization is more popular on servers, and type-2 virtualization is more popular on desktops. Examples of type-1 virtualization software include XenServer. Examples of type-2 virtualization include VirtualBox.

Security Concerns with VMs

Virtual machines offer many advantages but introduce a few new security concerns. VMs running on the same server may have different security requirements, which might break existing security protections. For example, network traffic to all the VMs running on the same physical machine will now reach the network interface of the machine. If the VM setup is not correct, this might open up new classes of exploits. Also, if all the VMs on a server run the same resource-intensive antivirus software, it may eliminate much of the benefit of using VMs. To address this, modern VM designs allow VMs to share services. For example, one VM may run the antivirus software and whenever a process on another VM requires an antivirus check, it can hand off the request to the VM running the antivirus.

Warehouse Scale Computers

Most big-data applications have some interesting properties that have led to the development of a unique low-cost computing model for big-data applications—warehouse scale computers (WSCs). *A WSC is a set of many (tens of thousands) hardware and software resources working in concert to appear as one large computer that efficiently delivers desired levels of Internet service performance.*[19]

In the late 1990s, the standard solution for computing-intensive applications was high-performance-computing (HPC) systems. However, the problems of that time (e.g. weather predictions, financial simulations, terrain modeling for oil-and-gas exploration) typically required direct access to large data sets, generated large intermediate result sets that were

18 VirtualBox is another popular virtualization software. It is free and can be downloaded from https://www.virtualbox.org/. The site also links to pre-built VMs for installation on VirtualBox, available at http://www.oracle.com/technetwork/community/developer-vm/index.html (accessed June 19, 2014).

19 Luiz André Barroso, Jimmy Clidaras, and Urs Hölzle, *The Datacenter as a Computer—an Introduction to the Design of Warehouse-scale Machines*, 2nd ed., Synthesis Series on Computer Architecture (Morgan & Claypool Publishers, May 2013).

interdependent on each other and kept the entire computer busy for long times (sometimes weeks) before the optimal solutions were found. HPC designs emphasize shared memory architectures to address these requirements.

By comparison, typical big-data applications have modest requirements for individual operations and the operations are usually mutually independent. For example, each e-mail is independent of all other e-mails on the server. In addition, most big-data applications do not even require the typical ACID (atomicity, consistency, isolation, and durability) and relax these requirements in favor of relaxed consistency or eventual consistency. For example, it is not necessary that the index of a search engine be correct at every instant. Or, in the case of video sharing, it is not necessary that all formats of an uploaded file be available instantly. WSCs exploit these relaxations in their designs to lower costs. The important requirements for WSCs are listed below:

- Performance/dollar of capital invested: A typical WSC costs approximately $150 million. While the efficiency of a desktop costing $750 is not material to a company, the efficiency of a WSC has material effects upon the finances of the organization.
- Performance/joule of energy consumed: About 30% of the operational costs of a WSC is the cost of electrical power. An additional 20% of the operational cost is the amortized cost of the power and cooling infrastructure, which is used to bring electrical energy in and drain the heat energy out, after the computers have done their job.
- Reliability: Users have come to expect constant availability. WSCs must be designed so that they can withstand outages to entire data centers in the event of hurricanes and earthquakes.
- Request level parallelism: Whereas desktop PCs are optimized to serve one user as effectively as possible, WSCs are optimized to serve millions of simultaneous users satisfactorily. The difference may be compared to the difference between a Ferrari and a bus. The Ferrari can transport one person very fast, but the bus will take less time to get 50 people to their destination. Modern WSCs are designed to be buses, not Ferraris.[20]
- Handling device failures: A WSC has, on average, 50,000 servers. Data center studies suggest a hard disk failure probability of 2%–10%. If we take a 4% annual probability and assume that each server in a WSC has four hard disks, we would have 8,000 hard disk failures/year at the WSC, which translates to about one hard disk failure/hour. This is only one category of failures. In addition to hard disk failures, there are also software failures, power outages, software updates, and other challenges. These must be managed efficiently. Today, the leading WSC operators use automated solutions wherever possible so that one operator can manage more than 1,000 servers.

20 I first read this analogy in connection with Sun Microsystem's introduction of the SunFire servers. A related post is at http://jonathanischwartz.wordpress.com/2006/08/24/explaining-suns-share-gains/ (accessed Mar. 2016).

Studies of deployed WSCs show that WSC designs offer the following economies of scale:[21]

- WSCs use one administrator for more than 1,000 servers, whereas the typical enterprise data center uses one administrator for every 140 servers (a 7:1 ratio). Facebook has said it wants to automate IT infrastructure enough that eventually a single person could oversee its entire network.[22]
- WSCs spend $4.60/GB/year for storage, compared to $26/GB/year for the typical enterprise datacenter (a 5+:1 ratio).
- WSCs achieve power utilization efficiencies of almost 90% compared to an approximately 50% power utilization efficiency at a typical enterprise data center.
- The processing capacity at a typical data center server is utilized to about 10%–20%, whereas the processing efficiency at a typical WSC exceeds 50%.
- Hardware vendors fulfill WSC orders within a week compared to typical waits of more than four months for enterprise customers

Figure 12.11 shows the architecture of a typical Google WSC. The system is made up of identical servers as the basic computational unit. Companies are understandably secretive about their server specifications, but it is known that they use commodity servers in WSCs. In 2009, the most recent information found, they were based on standard AMD Opteron processors (Figure 12.12). There are three interesting things to note about the server: (1) power connections at the right are attached to a spare battery; Google decided to add a battery to each server to improve power efficiency as well as to make battery costs incremental; (2) there is no cover to the server, lowering costs and speeding up access to internal components for maintenance; (3) larger components such as the hard disks and power supply are connected using Velcro instead of conventional screws to facilitate removal.

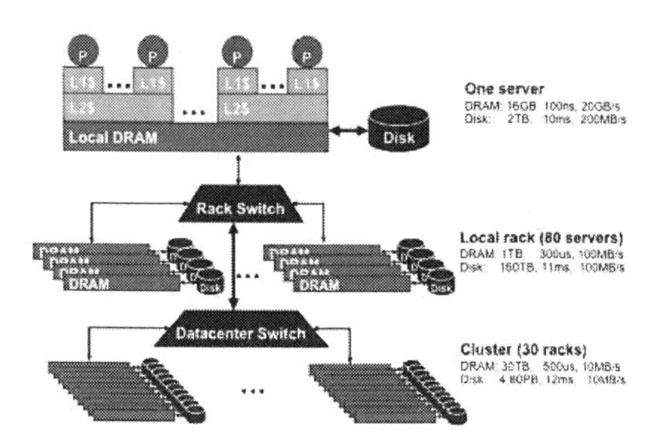

Figure 12.11: Warehouse scale computer array architecture
(Source: Barroso et al., 2013)

Each industry standard 7-foot-tall rack can hold up to 48 servers. Some servers are designated as storage servers, in which case they take up twice as much room. Figure 12.11 shows that Google connects two such racks, constituting 80 servers, to a rack switch. This rack is the basic modular unit of the WSC. Up to 30 racks constitute an array and are connected to an array switch (Hennessy and Patterson use the term *array* to distinguish from the traditional computer cluster. Barroso and Holzle use the term *cluster* for arrays).

---

21 http://awsmedia.s3.amazonaws.com/Economies_of_Scale_Hamilton_ExecSymposiumNov2009.pdf (accessed Mar. 2016).

22 Shira Ovide, "Facebook introduces own designs for networking equipment," *Wall Street Journal*, Feb. 11, 2015.

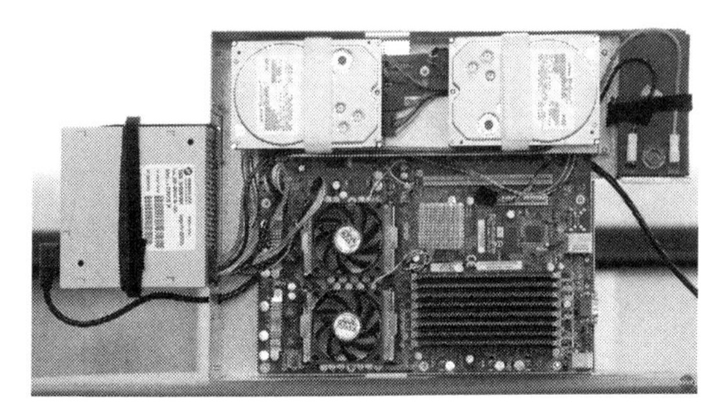

Figure 12.12: Individual server at Google WSC

An array, therefore, constitutes up to 2,400 servers. About 20 of these arrays are connected to each other and to the Internet using a router.

Increasingly, companies are using standard 1AAA containers (40' x 8' x 9'6") to build WSCs. Using recycled standard shipping containers lowers construction costs and greatly simplifies shipping and handling. Google attracted a lot of publicity when it first built containerized data centers on barges located in the ocean.[23] Figure 12.13 shows a cutout of such a container from Google's patent filing for such a design.[24] Each container contains 1,120 servers and the WSC is built out of 45 such containers. From the outside world, the container only needs power, network connections, and cooling. As currently configured (information available from 2007), each container draws 250 kW of power.

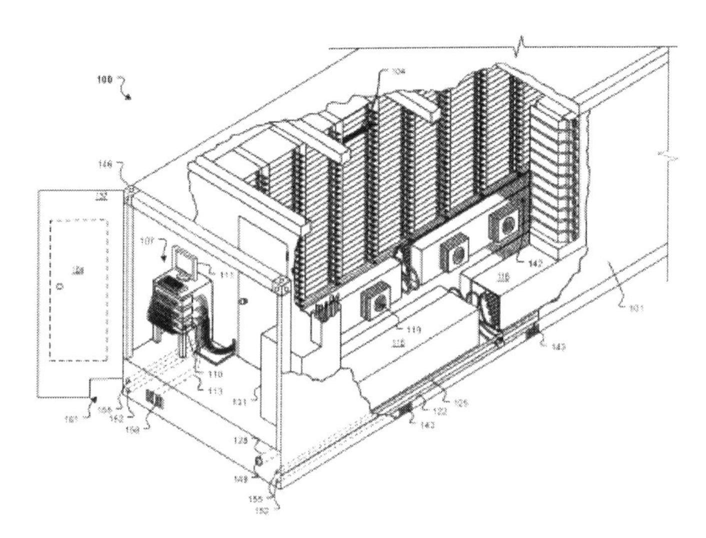

Figure 12.13: Google container WSC cutout
(Source: fig. 1 from US Patent 8,218,322)

23 http://www.kcra.com/news/local-news/news-stockton/5-things-to-know-about-googles-mystery-barge/24850762#!5Zruy (accessed June 29, 2014).

24 http://patft.uspto.gov/netahtml/PTO/srchnum.htm (enter 8218322 in search box. Accessed June 29, 2014).

In a typical warehouse the containers are stacked two high. Within the container, the racks form two long rows with the servers facing inward. Cool air is blown up from the center and hot air is collected from the sides. To facilitate maintenance, the racks hang from the ceiling. To lower power consumption, the container has been designed very carefully to ensure that there are no spots without air circulation. This allows the cold air to be kept at 81°F instead of the conventional 71°F. In less carefully designed data centers, the cold air has to be cooled intensely so that even spots without air circulation do not get too hot to harm servers. But this safety comes at a very high operational cost for cooling. By eliminating dead zones and by locating servers in cooler climates, such as Oregon, many companies are even trying to lower costs by eliminating air conditioning altogether.

While Google has attracted a lot of attention for its data center innovations, it is by no means the only company at the forefront of these innovations. In 2011, Facebook announced power efficiencies that greatly surpassed those reported by Google.[25]

Cold Scandinavian locations are hot for data centers

Data centers accounted for about 1.5% of worldwide electricity use in 2010. Scandinavian locations can have mean annual temperatures of 36°F, with less than an hour each year when the temperature exceeds 86°F. They also have stable politics, network connectivity, and very reliable hydroelectric power with no power outage in 45 years. This has made them very attractive destinations for large data centers serving European and Russian users.[26]

The tech firms also are investing billions in captive network capacity to carry traffic from their data centers to population centers.[27]

Cash-mobs, a social media business innovation

This book has focused on technical innovation and Internet firms have been at the center of attention in this chapter. However, the technical services enabled by these innovations have, in turn, enabled other inspiring innovations, one of them being cash-mobs. Cash-mobs are self-organized events where people identify a local business and use social media to encourage their friends to gather at the business at a specific time with the intent of spending money at the business.

The first cash-mob is credited to Chris Smith, in Buffalo, NY, who used Twitter and Facebook to gather more than 100 people to purchase wine at City Wine Merchant on August 5, 2011, about triple the average daily number of buyers at the business.[28]

---

25 https://code.facebook.com/posts/272417392924843/open-sourcing-pue-wue-dashboards/ (accessed June 29, 2014).

26 Sven Grundberg and Niclas Rolander, "For data center, Google goes for the cold," *Wall Street Journal*, Nov. 12, 2011.

27 Drew Fitzgerald and Spencer Ante, "Tech firms push to control web's pipes," *Wall Street Journal*, Dec. 16, 2013.

28 Emily Maltby, "Cash mobs help ignite buy-local effort," *Wall Street Journal*, Dec. 27, 2011.

Cloud Computing

Based on the efficiencies of WSCs, the all-inclusive computing cost for a WSC has been estimated at $0.11/server/hour. This is substantially lower than the fully amortized computing costs for most enterprises. For this reason, large IT vendors have entered into the space of offering computing as a service, which is conventionally called cloud computing. NIST (publication 800–145) defines it as follows: *Cloud computing is a model*[29] *for enabling ubiquitous, convenient, on-demand network access to a shared pool of configurable computing resources (e.g. networks, servers, storage, applications, and services) that can be rapidly provisioned and released with minimal management effort or service provider interaction.* In simple terms, cloud computing is the use of computing resources on rent.

Characteristics

According to NIST, cloud computing has five essential characteristics:

1. On-demand self-service: The consumer[30] should be able to provision infrastructure, such as CPUs and storage, without requiring human interaction with the service provider.
2. Broad network access: Computing capabilities are accessed over the network using standard, non-proprietary mechanisms over a medium of the customer's choice (e.g. mobile phones, tablets, laptops, and workstations).
3. Resource pooling: Customers share a common infrastructure from the provider using a multi-tenant model. Customers generally have no knowledge about the exact location of the provided resources but may be able to specify preferences such as country or state.
4. Rapid elasticity: Required computing capacity can be rapidly grown or shrunk as much as necessary to stay commensurate with demand.
5. Measured service: Computing usage can be transparently metered at a level of granularity appropriate to the service (e.g. storage capacity, processing speed, bandwidth usage, and user account counts).

Access

In terms of access, NIST has identified four possibilities:

1. Private cloud: The cloud infrastructure is provisioned for exclusive use by a single organization comprising multiple internal consumers (e.g. business units). It may be owned, managed, and operated by the organization's IT group, or by an external provider, and it may or may not be located off premises. Many companies are beginning to use private clouds for shared file storage. USF's IT group has created a private

29 There is a humorous quote relevant to models, hence it is reproduced here: "The mathematician tends to idealize any situation with which he is confronted. His gases are 'ideal,' his conductors are 'perfect,' his surfaces 'smooth.' He calls this 'getting down to essentials.' The engineer is likely to dub it 'ignoring the facts.'" Thornton C. Fry, "Industrial Mathematics," *Amer. Math. Monthly*, 48 (1941): 138.

30 The consumer here refers to the consumer of the service, whether an individual or an organization.

cloud to host most scientific and desktop applications, eliminating the need to maintain student computer labs.

2. Community cloud: The cloud infrastructure is provisioned for exclusive use by a specific community of consumers from organizations that have shared concerns (e.g. mission, security requirements, policy, and compliance considerations).
3. Public cloud: The cloud infrastructure is provisioned for open use by the general public. It may be owned, managed, and operated by a business, academic, or government organization, or some combination of them. It exists on the premises of the cloud provider. Examples of public clouds include Amazon AWS, Microsoft's Azure, and Rackspace's Open Cloud.
4. Hybrid cloud: The cloud infrastructure is a composition of two or more distinct cloud infrastructures (private, community, or public) that remain unique entities but are bound together by standardized or proprietary technology that enables data and application portability. Example applications include using a public cloud to augment the private cloud's capacity during periods of high activity.

Services

In terms of the services offered, NIST has classified cloud services into three categories:

1. Software as a Service (SaaS): Consumers get access to ready-to-use applications running on a cloud infrastructure. Consumers have minimal control over the application configuration beyond user-specific settings. Gmail, Salesforce, WebEx are examples of SaaS.
2. Platform as a Service (PaaS): Consumers get the ability to deploy their own application onto the cloud infrastructure. The applications need to be created using programming languages and tools supported by the provider. The best-known PaaS is Microsoft's Azure cloud, which can be used to host applications developed using Microsoft's development environment. To promote Azure usage, Microsoft has progressively simplified deployment to the Azure cloud directly from within Visual Studio.
3. Infrastructure as a Service (IaaS): The consumer is allowed to specify core computing resources such as processor, memory, storage, and network bandwidth. Consumers can deploy any software of their choice, including operating systems. The consumer has complete control over the components he deploys, including the operating system, storage, and any deployed applications. Amazon's AWS and Rackspace's Open Cloud are the best known examples of IaaS.

Security Concerns on the Cloud

Cloud providers generally take great care to secure their services because that is essential to the survival of their business. However, consumers of cloud services need to be aware that IaaS and PaaS cloud providers generally make information security the responsibility of their customers. This is justified by the idea that customers have tremendous latitude in configuring how they use cloud services, and therefore consumers should be held responsible for the security issues they create. Since there is no client-side data in SaaS, security is typically the responsibility of the cloud provider in SaaS. Customers also need to

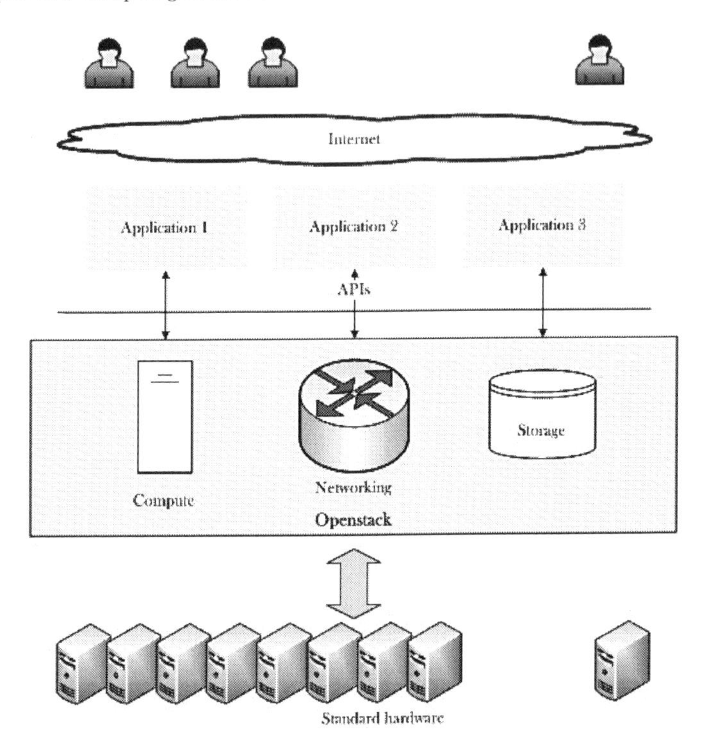

Figure 12.14: Cloud architecture with OpenStack

be aware that when using cloud services, they are generally placing their data in the custody of a third party, bringing with it all the trust issues associated with the arrangement.

Users can take various measures to mitigate risks. PaaS providers expose functionality to users and developers through APIs. When using PaaS, developers should try to use as many APIs provided by the cloud provider as possible. Providers will generally be in a better position to implement the correct cryptographic and other protocols in their APIs than individual developers. Developers should also try to integrate as many security services between their own organization and the PaaS provider as possible. This includes identification and authentication frameworks so that there are as few identification and authentication systems in use as possible. When using IaaS, organizations should minimize the number of users with administrative access to the cloud.

Implementation

Technologies to build and deliver cloud services are in a rapid state of flux and they are perhaps the area of greatest infrastructure technology development today. This section may be obsolete by the time you read it. However, at a high level, the goal of these cloud technology developments is to hide the underlying hardware and operating system details from applications. Cloud users deploy their applications to the cloud and the cloud software identifies available processors and other hardware resources to run these applications. Application developers are unaware of the identities of the processors working for them.

In some sense, this goal of cloud computing is the opposite of the goals of virtualization. Whereas virtualization makes one computer appear as multiple computers, cloud technologies make multiple computers appear as one. OpenStack is a popular framework for cloud deployments. Figure 12.14 shows the OpenStack model for cloud deployments.

Acknowledgment

This chapter is heavily influenced by two books:

1. Abraham Silberschatz et al., *Operating Systems Concepts*, 8th edition (Wiley, 2008).
2. John A. Hennessy and David Patterson, *Computer Architecture, A Quantitative Approach* (Morgan Kaufmann, 2011), especially chapter 6. The interested reader is also encouraged to look at the article by Barroso and Holzle on Warehouse Scale Computers.[31]

Summary

This chapter introduced the elements of the infrastructures that make up the computational part of modern IT infrastructures. We started with the basic design of all modern computers, tracing their architecture to the original von Neumann architecture. We then looked at operating systems, the critical software component of this architecture that orchestrates the functioning of all activities on a computer. Activities on a computer are organized as processes, which are moved to and from the processor by the operating system as needed.

Modern computer applications frequently have processing needs that are beyond the capabilities of an individual computer. Various architectures have been proposed to address this problem, and in this chapter we looked at two of these architectures: virtual machines and warehouse scale computers. Virtual machines are used to partition large computers into multiple smaller computers, and warehouse scale computers aggregate regular desktop-style computers into large processors. There is now an emerging industry to provide computing capability as a standardized service, much like electricity and the Internet. We looked at the basic models of this emerging cloud computing industry.

In each section, we also looked at the important security concerns that arise in contemporary use. In the next chapter, we will see how this infrastructure is used to provide services to end users.

About the Colophon

We have come a long way. In March 1975, Paul Allen visited Ed Roberts of MITS Computer in Albuquerque, New Mexico, to demonstrate their version of BASIC on the Altair 8800 computer that MITS had begun to sell. The entire BASIC program was punched on paper tape. Neither Paul Allen nor Ed Roberts had any hope that the computer or the BASIC

31 http://www.morganclaypool.com/doi/pdf/10.2200/s00193ed1v01y200905cac006 (accessed June 29, 2014).

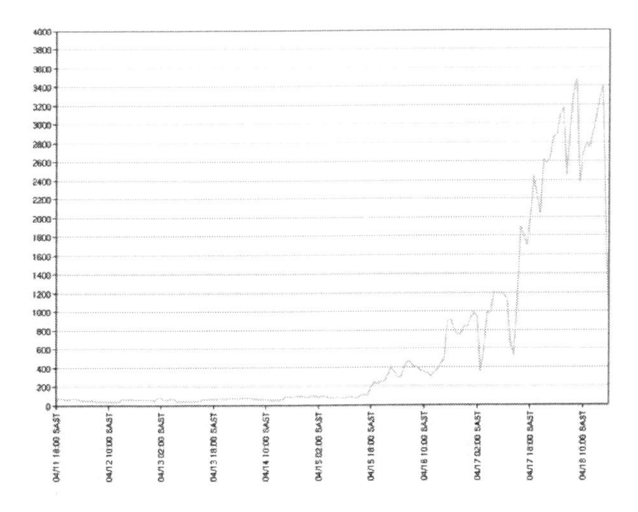

Figure 12.15: Cloud computing usage at Animoto

program would actually work. So, after the 15 minutes it took to read the BASIC interpreter, when the computer responded with the command MEMORY SIZE?, Allen and Roberts were amazed. For both, it was great enough that the combination of BASIC interpreter and Altair 8800 computer actually typed something on the screen.

Ed Roberts (1941–2010) is widely considered the father of the personal computer. Microsoft was founded in Albuquerque, NM, on April 4, 1975, and was located there until 1979, when it moved to Bellevue, Washington.

EXAMPLE CASE—40 servers to 5,000 in 3 days

One well-known case study highlighted by Amazon about the advantages of cloud computing relates to a start-up company, Animoto. Figure 12.15 shows how their IT needs evolved over a four-day period.

References

1. Case studies: https://aws.amazon.com/solutions/case-studies/ (accessed Apr. 9, 2016).
2. Animoto: https://aws.amazon.com/solutions/case-studies/animoto/ (accessed Apr. 9, 2016).

REVIEW QUESTIONS

1. What was the von Neumann report? What is its significance to the modern IT industry?
2. What are the five significant parts of a computer? Briefly describe each part and its role in computing.

3. In what ways does the structure of a modern computer create challenges for students trying to enter the professional workforce?
4. What is the role of an operating system in modern computers?
5. What is a *system call*?
6. Why are some tasks reserved for the operating system?
7. What is the *dual mode of operation* of computers? Why is it useful?
8. What is *multitasking*? Why is it useful for operating systems? Why is multitasking increasingly being considered harmful for people to attempt in their personal lives?[32]
9. What is a *computer process*? What are the contents of a process?
10. What is a *process state*? What are the different states in which a process can be in a computer?
11. What are *threads*? How are they different from processes?
12. How are threads useful on the server end? How are they useful at the client end?
13. What is *memory*? Why is it necessary on a computer?
14. What is *virtual memory*? Why is it useful? Why is it useful to minimize the use of virtual memory?
15. What is the *file system*? What services does it provide?
16. What are the important fundamental security design principles used to protect operating systems from harm?
17. What is the sequence of operations involved when a computer is powered on?
18. What is *big data*? How does it affect organizational IT infrastructures?
19. What is *virtualization*? Why is it useful? What are the two types of virtualization?
20. What are the important security concerns with virtualization?
21. What is a *warehouse scale computer*? How is a WSC different from high-performance computers (HPCs)?
22. What are the important requirements of WSCs?
23. What are some important economic advantages obtained from WSCs?
24. What is *cloud computing*? Why is it useful? How is cloud computing related to WSCs?
25. What are the important characteristics of cloud computing?

EXAMPLE CASE QUESTIONS

1. What service does Animoto offer?
2. What was the IT challenge that Animoto faced, as introduced in the case?
3. How did cloud computing help Animoto address the challenge?
4. In the absence of cloud computing, what would have been the outcome for Animoto?

HANDS-ON EXERCISE—perfmon

This chapter lends itself to numerous hands-on activities. The one chosen here gives you visibility into some of the new concepts introduced in this chapter that you may not have

32 To answer the last part of the question, you may have to turn to the Internet for research on the subject. One relevant article is by Christopher Mims, "Say no to the distraction-industrial complex," *Wall Street Journal*, June 30, 2014.

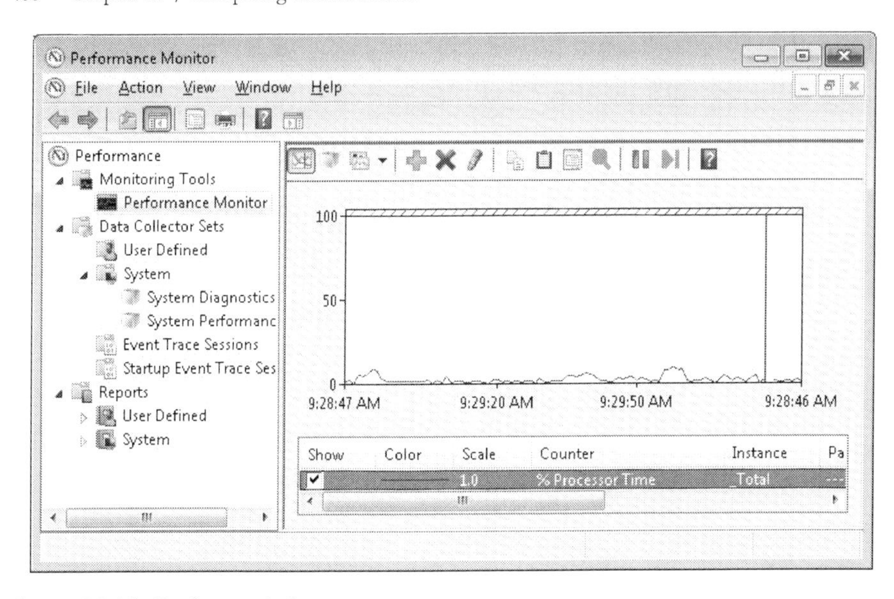

Figure 12.16: Perfmon window

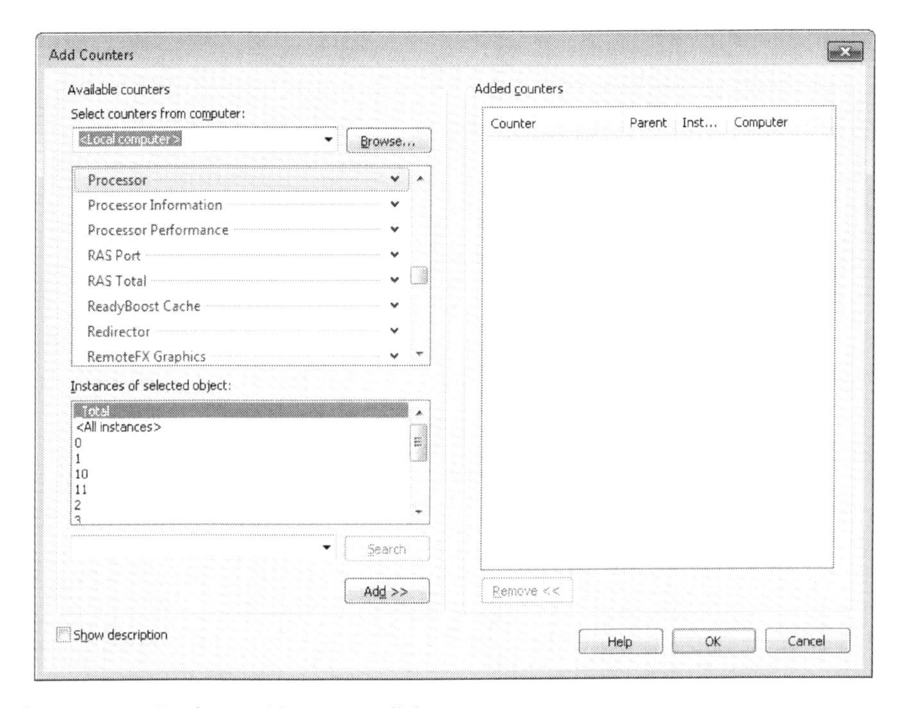

Figure 12.17: Perfmon add counters dialog

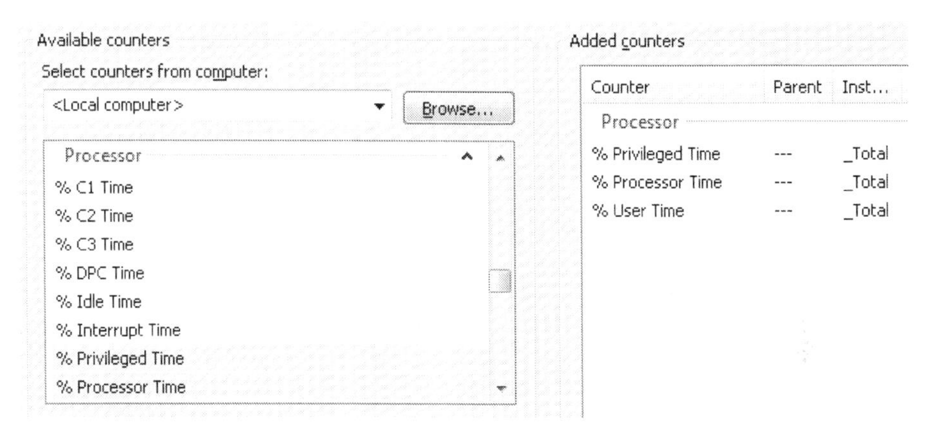

Figure 12.18: Perfmon counters added

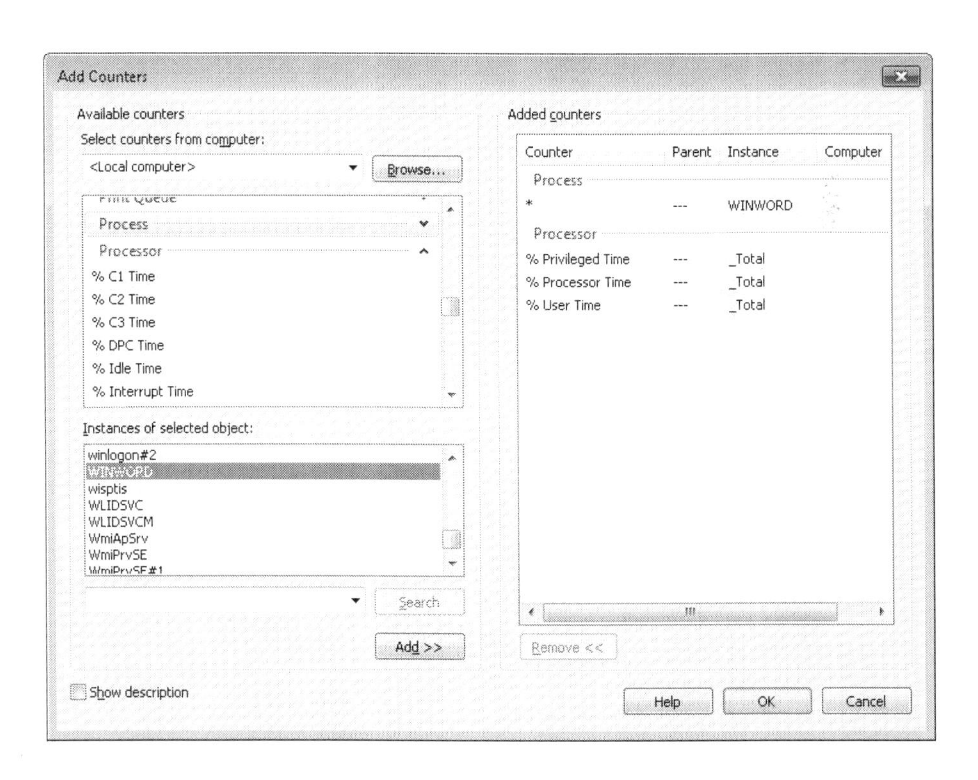

Figure 12.19: Perfmon counters for the exercise

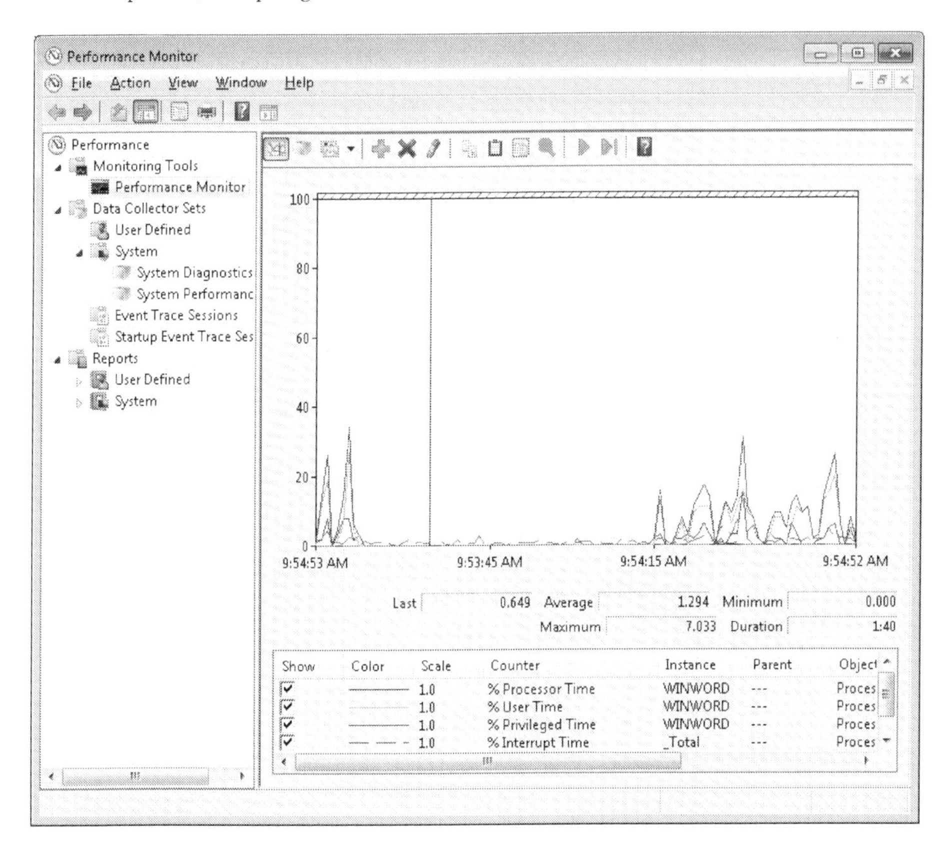

Figure 12.20: Perfmon counters activity capture

encountered before, and may help you diagnose issues with applications you develop. We will use the perfmon utility that comes with Windows to look at how the processor responds to common operations invoked on the computer. You will monitor some basic CPU statistics while performing some common operations.

To start perfmon, go to Start → Run and type `perfmon`.[33] The perfmon window is shown in Figure 12.16. To collect statistics of interest, you need to add some counters and filter for processes of interest. To do so, click on the Add button (+) to bring up the "Add counters" dialog shown in Figure 12.17. You may notice that of the numerous available filters sorted alphabetically, Windows brings you to the middle of the list where you have one of the two most useful filters for this exercise: processor. Expand the processor node to see the available counters and add the three counters shown in Figure 12.18.

After adding the counters, scroll up in the available counters to select the process item to show the available processes to sample. In the "instances of selected object" list, select

33 http://blogs.msdn.com/b/securitytools/archive/2009/11/04/how-to-use-perfmon-in-windows-7.aspx (accessed July 1, 2014).

a process of interest and add it to the selection. For this exercise, I have selected the "WINWORD" process for MS Word. The result is shown in Figure 12.19. Select OK.

Now perform some operations on the selected software. For example, I am writing this sentence, saving the file, cutting an image, saving the file, pasting the image back, and saving the file again. The result is the capture shown in Figure 12.20. Please perform a few common operations and see for yourself how the processor responds to these user commands.

HANDS-ON ACTIVITY QUESTIONS

1. What is the *perfmon* utility? What is it typically used for?
2. Show a screenshot of the counters for your case.
3. Interpret the perfmon results. What activities generate the most privileged time? User time? Interrupt time?

CRITICAL-THINKING EXERCISE—Nano Scale Computers

Implicitly, this chapter has assumed that development in computers is towards bigger and faster computers. But development can also occur in the other direction towards smaller and more power-efficient computers. Smart watches are a manifestation of this development. Toolkits like Arduino are another. Imagine you have access to the world's best miniature computer design and fabrication facility.

1. What application(s) do you think would be very profitable for you to create?
2. How would you price your invention(s) in Question 1?

IT INFRASTRUCTURE DESIGN EXERCISE—Using the Cloud

The CIO of TrendyWidgets has decided to use AWS for some experimental projects. Update your infrastructure diagram to indicate a connection to the AWS cloud.

Hint: you will access AWS through the Internet, not through your WAN.

CHAPTER 13

Services Delivery

People sacrifice the present for the future. But life is available only in the present.

Thich Nhat Hanh

Introduction

The computing infrastructure introduced in the previous chapter, together with the communications infrastructure discussed in prior chapters, is used to deliver technology services to end users. While the previous chapters have focused on the technology design choices to maximize efficiency in such service delivery, managerial design choices are equally necessary to maximize efficiency. In recent years, organizations have increasingly focused on these types of choices. This chapter introduces the key managerial design choices used in industry for the efficient delivery of technology services to end users. At the end of this chapter you should know:

- the key managerial design choices used in delivering technology services to end users
- the popular design choices for maintaining high availability of services
- the essential practices used to maintain service delivery during extreme events

IT Services Management

IT services management (ITSM) is the set of activities performed by an organization to deliver IT services to end users. ITSM aims to focus the IT organization on using technology to meet the organization's business needs. ITSM is a relatively new domain of research and practice and reflects managerial design choices in services delivery.

Careful consideration and adaptation of ITSM recommendations can be very useful. Most students reading this text will have no control over their technology design choices. For example, students of this text are very unlikely to develop a new network layer protocol. However, most students will have substantial control over their managerial design choices, on budgeting, training, and capacity planning, for example. Poor choices can cause significant delivery inefficiencies in spite of deploying the best technology infrastructure.

Therefore, while knowledge of the underlying technology design principles is important, knowledge of ITSM recommendations can help students actually influence the design of technology solutions.

ITSM Frameworks

A number of frameworks have been developed over the years for technology services delivery, each reflecting the perspectives of the leading sponsors of the framework. These frameworks include COBIT, ISO 20000, and ITIL. COBIT (Control Objectives for IT) is sponsored by ISACA (Information Systems Audit and Control Association) and, reflecting the perspective of auditors, has a strong focus on compliance and governance of the IT function. ISO 20000 has been developed by the International Standards Organization (the same organization that developed the OSI model) and has a strong focus on defining the requirements that an IT service provider should meet. ITIL (IT infrastructure library) was initially developed by the UK government as best practice recommendations so that the various departments of the government could manage their IT functions consistently and efficiently. ITIL therefore has a focus on providing detailed guidance on implementing processes for services delivery. ITIL has become increasingly popular as a framework for discussions between business and IT, and the recommendations introduced in this chapter are inspired by ITIL.

IT Services Delivery Background

Historically, business users, solution developers, and the IT operations team worked independently. Business users demanded applications, solution developers developed these applications to the best of their ability, and the operations team did what it could to keep the applications running without interruption over the life of the application. Application developers were often unaware of the capabilities and limitations of the IT infrastructure maintained by the operations team until the application was ready to be deployed. This was acceptable until around 2000 when computer applications were largely internal, and most applications had very few users.

However, beginning in around 2000, applications went online, and popular applications began to attract hundreds of thousands of users each day. Most application developers, who generally tested their applications on their local machines, had no idea of concepts such as threading, memory management, and security that were needed for their applications to serve this many users. The old model became indefensible in this new internet-driven environment. It became necessary for the systems development function to be aware of the operational prerequisites and to embed these prerequisites in every new systems build.

The ITSM recommendations aim to make the IT organization strictly focus on client needs, using well-designed IT processes. As a result, the IT organization concentrates on the services required by the customer, rather than focusing on the technology fashions of the day. In the next section, we look at some key managerial design recommendations for efficient service delivery. These have been drawn from ITILv2 and focus on the steady state operation of services, which (in the opinion of the authors) is conceptually simpler for fresh graduates to understand. ITILv3, introduced in 2007, replaced this model with a life cycle model, focused on ongoing improvement of services. If your work involves services delivery, you will learn about these issues in great detail at work. This chapter is intended to introduce you to services delivery and its concerns.

Bill Gates's trustworthy computing memo and Microsoft's code moratorium

On January 18, 2002, Microsoft's then chairman, Bill Gates, wrote a famous memo titled "Trustworthy Computing."[1] In the memo, he indicated the need for Microsoft's products to focus on security to the same extent that they historically focused on adding new features. Later, in February 2002, Microsoft put a moratorium on new code development in Windows in order to train its developers on security.

The overall cost of this effort exceeded $100 million. It is considered an important landmark in the evolution of technology development to integrate operational details at the time of technology development.

Service Delivery Disciplines

The five areas of managerial discipline identified by ITILv2 are listed below and briefly described later in this section. The general idea is that an entity at the service provider (called the service desk) enforces these disciplines.

- service level management
- capacity management
- continuity management
- availability management
- IT financial management

Service Level Management

Service level management refers to the maintenance of a catalog of all services offered, together with binding agreements with both the provider and the customer for performance. Effective service level management reflects technological feasibility and client budgets and can minimize conflicts. The agreements with providers and customers are called service level agreements (SLAs).

SLAs are output-based contracts between customers and providers that define the service the customer can expect to receive in return for the fees paid. A typical SLA includes definitions for the service under consideration, the reliability and other performance metrics the user can expect, the procedures for reporting problems, the responsiveness a client can expect when reporting a problem, the consequences for not meeting service levels, and any escape clauses that

1 https://www.microsoft.com/mscorp/execmail/2002/07-18twc.mspx (accessed Feb. 2016).

may invalidate the SLA. Typical escape clauses include extreme events such as floods, earthquakes, terrorist events, etc.[2]

Developing SLAs can lead to other benefits beyond reduced conflict between client and vendor. During the course of developing the SLA, both client and vendor can better understand each other's priorities, especially any priorities that are unique to the other party. Also, clients often better understand the importance of quantitatively measuring all aspects of their services in order to be able to report problems.

Capacity Management

Capacity management is the managerial discipline for maintaining IT infrastructure at the right size to meet current and anticipated business needs in a cost-effective manner. Capacity management requires rigorous performance measurements of current services to detect bottlenecks and unused resources. These measurements can help reduce the capacity-planning uncertainties related to anticipated growth.

Global Crossing

One of the best-known cases of capacity management gone bad in any industry relates to the telecom firm Global Crossing. Between 1997 and 2002, the firm incurred more than $12 billion in debt to build out a global submarine optical fiber network in anticipation of growth in data and voice traffic resulting from the growth of the Internet. When the demand did not materialize, the firm had to declare bankruptcy.[3] Thousands of technology executives were affected.

There was likely an element of greed in the rushed buildout, but unwise emotions often accompany poor capacity planning.

Continuity Management

Continuity management is the discipline of planning to ensure that IT Services can recover and continue should a serious unexpected incident occur. Continuity management includes proactive measures to reduce the likelihood of unexpected incidents in addition to reactive measures to be taken when the incidents occur. As unexpected incidents such as terrorist attacks and financial setbacks become more frequent, most large organizations require continuity management from IT providers before doing business with them.[4]

An entire section of this chapter is devoted to key elements of continuity management—business continuity and disaster recovery. At a high level, however, continuity management involves prioritizing the business processes to be recovered if an incident occurs, minimizing

---

2 Sample SLAs can be found on the Internet. An example is http://www.itdonut.co.uk/it/it-support/it-support-contracts/sample-service-level-agreement (accessed Mar. 2016).

3 For a story on the heady start of the company, please see http://www.forbes.com/forbes/1999/0419/6308242a.html (accessed Feb. 2016).

4 FEMA has a website with templates for many disciplines in services delivery: https://www.fema.gov/planning-templates (accessed Mar. 2016).

the risks facing high-priority services, and developing options for recovery under different circumstances. Once developed, these plans are tested and revised periodically.

Availability Management

Availability management is the discipline of ensuring that resources such as IT infrastructure and personnel are appropriate for meeting the service-level agreements in place. On a day-to-day basis, end users are generally most concerned about the availability of IT resources, therefore weaknesses in availability management can be extremely stressful to both the client and the service provider. This puts a premium on developing maintainable software and components so problems can be remedied quickly and designing resilience into the technology so that common usage scenarios do not cause the system to fail.

IT Financial Management

IT financial management is the discipline of accurate accounting of IT services and using this information to deliver IT services in the most cost-effective manner possible. IT financial management improves the viability of the infrastructure supporting service delivery by establishing sound client prices, and it also helps in identifying areas where productive investments are possible. Financial management is usually not very interesting to IT staff, and so it is, unfortunately, neglected, which often results in technology under-investment.

The financial management exercise can help organizations identify costs that may have escaped notice, such as building costs, depreciation costs, external service-provider costs, and software-licensing costs.

These disciplines are evolving with technology and end-user expectations, as well as with the industry's experiences with deploying these disciplines. It is anticipated that with the right managerial discipline, the IT organization can efficiently deliver services to agreed standards. The standards themselves would be developed in dialog with customers. Changes in end-user needs would be met with minimal disruptions to ongoing services and would leverage existing infrastructures. Suppliers would be kept in the loop as changes are planned so that suppliers could build and deliver the necessary components on time, with minimal disruptions to their own production schedules.

The next two sections in this chapter take a deeper look at two components of service delivery in more detail—high availability and business continuity. These components have been chosen for their salience to clients on a day-to-day basis.

High-availability Concepts

Today's businesses demand 24/7 availability of information in order to serve customers, make smart decisions, push innovation, take advantage of opportunities, and stay one step ahead of the competition. *High availability is the ability of a system to remain operational for a duration significantly higher than normal.* High availability is an important goal of service delivery, and in this section we look at some of the important ways in which high availability is achieved.

Characteristics of High Availability

IT systems that support business must be online and accessible through both planned and unplanned events. However, organizations need to achieve high availability within budgetary constraints, carefully determining the right level of availability for each part of the IT infrastructure, and striking a balance between costs and the risks associated with non-availability of information. Ongoing availability management aims to optimize this balance.

High availability (HA) is assessed from the perspective of an application's end user. End users are disappointed when needed data is unavailable, and they can easily switch their loyalty to a competitor's brand when system availability becomes unacceptable, for whatever reason. Availability failures due to higher-than-expected usage create the same adverse outcomes as failures due to outages in critical components in the solution.

High-availability system designs typically share four characteristics: reliability, recoverability, error detection, and continuous operations. Reliable hardware and software components facilitate the design of HA solutions. Software components include the database, web servers, and applications. Recoverability involves the design choices that facilitate recovering from a failure if one occurs. This involves anticipating important failures and planning to recover from those failures in the time that meets business requirements. For example, if a critical table is accidentally deleted from the database, how would you recover it? Error detection refers to quick awareness of a problem so recovery procedures can be adopted. Continuous operations refer to keeping all maintenance operations invisible to the end user.

High-availability Requirements

Since IT systems run throughout the year, even small failures of availability add to downtime significantly, as shown in Table 13.1.

Table 13.1: Downtime as a function of availability

Availability percentage	Approximate downtime per year
95%	18 days
99%	4 days
99.9%	9 hours
99.99%	1 hour
99.999%	5 minutes

While 99% performance would be adequate in most contexts, it leads to four days of downtime in a year, an unacceptable level of downtime for any business. However, implementing high-availability solutions is expensive. High-availability implementations involve more fault-tolerant (i.e. expensive) and redundant systems for business technology components and also require greater ongoing expenses in IT staff, processes, and services to reduce downtime. The transition to high availability can involve tasks such as:

- retiring legacy systems
- investing in more sophisticated and robust systems and facilities
- redesigning the overall IT architecture to adapt to a high-availability model
- redesigning business processes to support high availability
- hiring and training personnel

Therefore, businesses start by identifying technology components that require high availability. An analysis of the business requirements for high availability and an understanding of the accompanying costs help businesses achieve availability within budgetary constraints.

Framework for Determining High-availability Requirements

Two considerations are commonly used to identify the business and technology components that most benefit from high availability: business impact analysis and cost of downtime.

Business impact analysis A rigorous business impact analysis identifies the critical business processes within an organization, calculates the quantifiable loss risk for unplanned and planned IT outages affecting each of these business processes, and outlines the less tangible impacts of these outages. It takes into consideration essential business functions, people and system resources, government regulations, and internal and external business dependencies. This analysis is done using objective and subjective data gathered from interviews with knowledgeable and experienced personnel, reviewing business practice history, financial reports, IT systems logs, and so on.

The business impact analysis categorizes business processes based on the severity of the impact of IT-related outages. For example, at a semiconductor manufacturer, with chip design centers located worldwide, an internal corporate system providing access to human resources, business expenses, and internal procurement is not likely to be considered as mission-critical as the customer-facing website. Any downtime of the customer-facing website is likely to severely affect the ability of the organization to get customer orders, which in turn can have material financial impact on the company. At a consulting organization, on the other hand, the internal HR system is mission-critical since a problem with the HR system at the organization will prevent the organization from assigning and tracking its consultants working on different projects. The customer-facing website may be less critical in the short run. This leads us to the next element in the high-availability requirements framework—cost of downtime.

Cost of downtime A well-implemented business impact analysis provides insights into the costs that result from unplanned and planned downtimes of the IT systems supporting the various business processes. Understanding this cost is essential because this has a direct influence on the high-availability technology chosen to minimize the downtime risk.

Various reports have been published documenting the costs of downtime across industry verticals. These costs range from millions of dollars per hour for brokerage operations and credit card sales, to tens of thousands of dollars per hour for package shipping services.

While these numbers are staggering, the reasons are quite obvious. The Internet has brought millions of customers directly to the businesses' electronic storefronts.

Critical and interdependent business issues such as customer relationships, competitive advantages, legal obligations, industry reputation, and shareholder confidence are even more critical now because of their increased vulnerability to business and technology disruptions. The higher the downtime costs arising from a business process, the greater the justification for implementing high-availability technology solutions to support the business process.

Upon completion of the business impact analysis, businesses can define service-level agreements (SLAs) in terms of high availability for critical aspects of their business. Commonly, business processes are categorized into several HA tiers:

- Tier 1 business processes have maximum business impact. They have the most stringent HA requirements, and the systems supporting them need to be available on a continuous basis. For a business with a high-volume e-commerce presence, this may be the web-based customer interaction system.
- Tier 2 business processes can have slightly relaxed HA requirements. The second tier of an e-commerce business may be their supply chain/merchandising systems. For example, these systems do not need to maintain 99.999% availability. Thus, the HA systems and technologies chosen to support tier 1 and tier 2 are likely to be different.
- Tier 3 business processes may be related to internal development and quality assurance processes. Systems supporting these processes need not have the rigorous HA requirements of the other tiers.

High-availability Architectures

Contemporary high-availability architectures can be categorized into local high availability solutions and disaster-recovery solutions. Local high-availability solutions provide high availability at a single location, typically a single data center. Local high-availability solutions can protect against threats such as process, node, and media failures, as well as human errors. Disaster recovery solutions are usually geographically distributed and provide high-availability during disasters such as floods, hurricanes, or regional network outages. Disaster recovery solutions can protect against local disasters that affect an entire data center.

A number of technologies and best practices are used to achieve high availability. The most common mechanism, though, is redundancy, whereby redundant systems and components are used to process user requests. *When all necessary resources are organized in the manner most suitable to deliver user services, this is called a system instance.* High-availability solutions are built by organizing system instances appropriately. Active system instances are instances involved with handling user requests. Passive system instances are fully configured instances, but which are not currently handling user requests. Local high-availability solutions are categorized into active-active solutions and active-passive solutions by their level of redundancy (Figure 13.1):

- Active-active solutions deploy two or more active system instances and can be used to improve scalability as well as provide high availability. In active-active deployments, all instances handle requests concurrently.

- Active-passive solutions deploy an active instance that handles requests and a passive instance that is on standby. In addition, a heartbeat mechanism is set up between these two instances. The heartbeat mechanisms are vendor-specific operating system technologies to automatically monitor and failover between cluster nodes, so that when the active instance fails, an agent shuts down the active instance completely, brings up the passive instance, and application services can successfully resume processing. As a result, the active-passive roles are now switched. The same procedure can be done manually for planned or unplanned downtime. Active-passive solutions are also called cold failover clusters.

Achieving High Availability

High availability is commonly achieved from the hardware. Additionally, operating systems and applications can help with achieving high availability.

The most basic high-availability strategy is to ensure that the hardware is as robust as possible, minimizing failures in the first place. Application servers also usually have many built-in high-availability features. For example, servers can provide the means to replicate data between multiple instances of applications to maintain service and data availability. Enterprise server operating systems have features to provide failover clustering, which keeps both applications and operating systems highly available.

Achieving High Availability at the Hardware Level

High availability generally begins with the hardware. An effective hardware strategy can significantly improve system availability. Hardware strategies can range from simply adopting commonsense practices with inexpensive hardware to using expensive fault-tolerant equipment.

High Availability at the Hardware Level using Robust Hardware

Unplanned downtime can be reduced significantly with high-quality, reliable components that are less likely to fail in the first place. Redundant components can then be added to

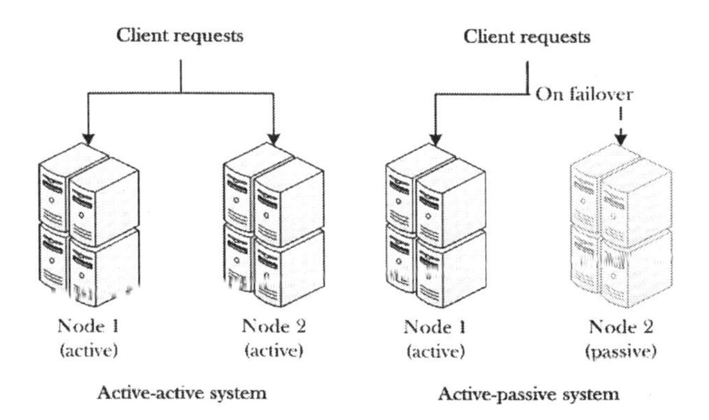

Figure 13.1: Comparing active-active and active-passive high-availability solutions

take over in case of a hardware failure. An effective hardware strategy involves standardized components, accessible spares, and careful maintenance of the environment.

Fault-tolerant servers Fault-tolerant servers have high or complete redundancy across all hardware components. This includes the power supplies, fans, hard disk, memory, and CPUs. When a component such as a power supply fails, secondary components continue to seamlessly serve user workloads. As such, fault-tolerant systems "operate through" a component failure without loss of data or application state.

Server equipment varies in its level of fault-tolerance. Most high-end servers employ at least some redundant components, especially to eliminate common points of failure, but they will still fail when a component that is not redundant, such as a microprocessor or memory controller, fails. True fault-tolerant servers use complete redundancy across all system components, ensuring that no single point of failure can compromise system availability. Some fault-tolerant server designs extend this level of redundancy across data-center boundaries by letting the server's redundant subsystems be installed in separate, yet connected, locations. Support for fault tolerance in Enterprise Servers is handled completely at the operating system kernel and hardware abstraction layer—a method that makes it transparent to applications.

Dynamic hardware partitioning We have seen earlier (in the discussion of virtual machines) that contemporary server hardware and software can generally be configured into one or more isolated hardware partitions where each hardware partition runs its own instance of the operating system and associated applications. Hardware resources on any partition are isolated from the other partitions on the same server. These partitions can even be changed dynamically. *Dynamic hardware partitioning is the capability of servers that allows system administrators to change the allocation of hardware partition units to each partition while the servers are still running.*

Dynamic hardware partitioning further enhances server availability and fault tolerance. On servers that can be dynamically partitioned, administrators can hot replace or hot add additional processors and memory to partitions without restarting the operating system or applications running on the hardware partition as needed.[5] This significantly increases the reliability, availability, and serviceability of servers. For example, memory chips that show signs of failing can be replaced, or spare processors can be added to partitions as demand increases.

High Availability at the Hardware Level Using Unreliable Commodity PC Architecture

High reliability also can be achieved by creating a reliable computing infrastructure from clusters of unreliable commodity PCs. This architecture was originally promulgated by popular technology leaders such as Google by replicating services across many different low-cost PC machines (instead of high-cost fault-tolerant servers) and automatically detecting and handling failures.

5 In electronics, the term *hot* refers to systems that are currently powered or active. By contrast, *cold* systems are not currently powered or active.

In this architecture, the cost advantages of using inexpensive, Intel PC-based clusters over high-end multiprocessor servers can be quite substantial, at least for applications that can be executed in parallel. A common feature of these applications is that they use unstructured data and primarily perform read operations on the database. The architecture of these designs that leverage commodity PCs follows a few key design principles, including the following.[6]

Software reliability Instead of fault-tolerant hardware features such as redundant power supplies and high-quality components, the software used in this architecture detects failures and directs requests appropriately. The architecture leverages the fact that most commodity PC hardware can now be assumed to be quite reliable. You may observe parallels between this assumption and the assumption made when using cyclic redundancy check (CRC)—that most modern networks can be assumed to be quite reliable so it is acceptable for the underlying network to discard defective packets.

Replication When commodity PCs are used to serve large workloads, a large number of machines are inherently necessary to process the user requests. Therefore, replication is already built into the system design from the ground up. When the software can detect and respond to failures, high availability is achieved at almost no additional cost.

Price-focused design is preferable to performance-focused design Hardware that provides the best computing performance per unit price is preferable to hardware that provides the best absolute performance. Generally, this allows more computational resources to be directed at user tasks for a given budget.

Achieving High Availability at the Application/Middleware Level

While the high-availability hardware designs prevent catastrophes that can take longer times (days or weeks) to recover from, high-availability designs at the application and database level prevent data loss in running applications. High-availability design at the application and middleware level includes paying attention to network disconnections between the application server and end users, application server failure, and database server failure.

Application Server High-availability Architecture

Redundancy continues to be the basic premise of high-availability application server architectures. Redundancy is commonly achieved by using clusters consisting of redundant application server nodes with failover policies like "1 of N," whereby the application server runs on 1 out of N available server machines; or "static," whereby the application server runs on a dedicated server machine. *A cluster is a collection of servers (called nodes), any of which can run the workloads common to the cluster.*

Within the cluster, application servers provide continuous replication, which provides data availability for the system. This is done by automatically backing up transaction log files

6 L.A. Barroso, J. Dean, and U. Holzle, "Web search for a planet: The Google cluster architecture," *IEEE Micro*, 23(2) (Mar.-Apr. 2003): 22–28.

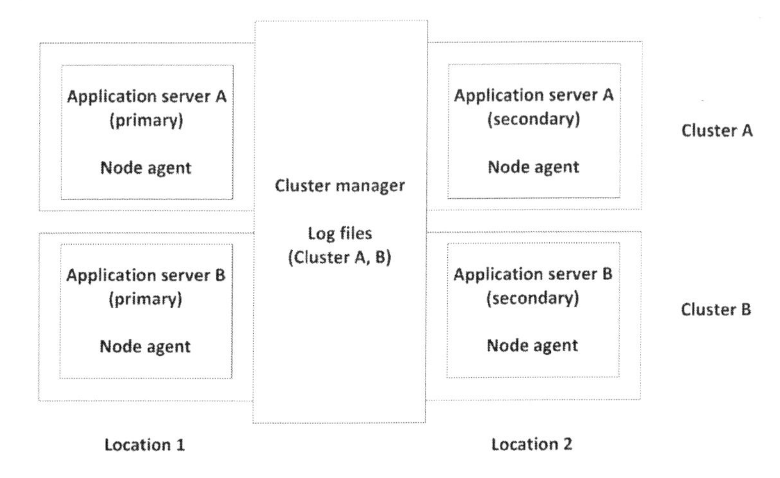

Figure 13.2: High-availability application server configuration example

to one or more copies located on a second set of disks, preferably on another server that is preferably in another location. The copies of the database transaction log files are used to replay the completed transactions on a copy of the database, thereby keeping the databases consistent without having to be synchronous with each other.

When a server in the cluster experiences a hardware or software failure, the management software in the cluster detects it and starts this service on another node in the cluster. Applications running on clusters need to be cluster-aware and capable of responding to these signals from the management software. They need to be capable of taking over immediately without any need for server restarts and replaying transactions, other than those that were "in-flight" at the point in time the primary server failed.

A representative architecture for achieving high-application server availability is illustrated in Figure 13.2.[7] Five nodes are organized in two clusters and one server running both the cluster manager and logs. The nodes on the left are the primary application servers while the nodes on the right serve as backups. The shared server is mounted to all nodes so that the transaction log files are accessible from all nodes, with appropriate permissions. Provided that both nodes of a cluster do not fail simultaneously, and that each node is independently capable of serving user needs for the applications served by the cluster, the architecture will improve availability significantly.

Database Server High-availability Features

A typical high-availability configuration for a web application is shown in Figure 13.3. We see that a failure in the database can independently hurt availability even if the application servers are available. Therefore, enterprise database servers have a comprehensive set of HA capabilities. High-availability features in the database typically include data protection,

---

7 Udo Pletat, "High availability in a J2EE enterprise application environment," http://ceur-ws.org/Vol-141/paper13.pdf (accessed Jan. 2016).

disaster recovery, real-time data integration and replication, and support for multiple hardware and operating-system platforms.

Three mechanisms commonly support high-availability database operations: database mirroring, failover clustering, and replication.

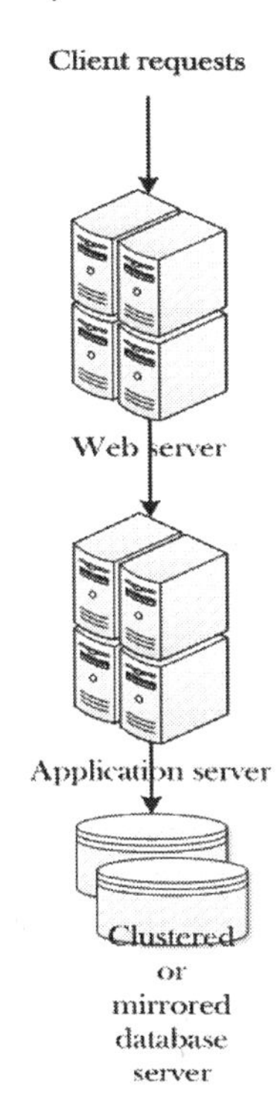

Figure 13.3: Typical web application architecture for high availability

Database mirroring *Database mirroring is a software solution for providing almost instantaneous failover with no loss of committed data.* It is now a standard feature of most enterprise database server software to provide high availability. Database mirroring can be used to maintain a single standby database (mirror database) for a corresponding production database (principal database). The mirror database is often also used for various reporting needs without disturbing the customer-facing principal database. Mirror databases are also for snapshots, to obtain read-only access to data as it existed at the time when the snapshot was created.

In industrial use, database mirroring is run in either synchronous operation in high-safety mode, or asynchronous operation in high-performance mode. In high-performance mode, the transactions commit without waiting for the mirror server to write the log to disk, which maximizes performance. In high-safety mode, a transaction is completed only when it is committed on both partners, but this usually reduces the transaction speeds of the database.

Failover clustering *Failover clustering is an architecture that provides redundancy through a configuration in which other servers essentially act as clones of the main production system.* A failover cluster comprises one or more servers (nodes) with a set of shared cluster disks. The shared disk array is configured to allow all nodes access to the disk resources, but with only one node actively processing data. When a server node fails, the failover cluster automatically moves control of the shared resources to a working node. This configuration allows seamless failover capabilities in the event of a CPU, memory, or other hardware failure that does not affect storage.

Replication *Database replication allows two or more database servers to stay "in sync" so that the secondary servers can answer queries and potentially actually change data.* If data on the secondary servers is changed, it is merged during synchronization.

Replication can be used to allow "slices" of a database to be replicated between several sites. The "slice" can be a set of database objects (i.e. tables) or even parts of a table, such as only certain specific rows (horizontal slicing) or only certain columns. While replication is primarily a technology for making data available at off-sites and to consolidate data to central sites, it also can be used for high availability or for disaster recoverability.

High Availability at the Operating System Level

Failover clusters, a technology we saw in the context of databases, also can be provided by the operating system for high availability by reducing single points of failure. Failover clusters are often used for key databases, file sharing on a network, business applications, and customer services such as e-commerce websites. There are two basic types of clustering technologies at the operating system (OS) level: network load balancing clusters and failover clusters. The choice of which technology to use generally depends upon the service being delivered by the cluster. For the purpose of availability, services may be characterized as stateless or stateful. *Stateless transactions are self-contained transactions, requiring no awareness by the server of the prior history of transactions. Stateful workloads require awareness of the history of the transaction to complete successfully.* Since stateful services require an awareness of the history of a transaction, they are more challenging for availability.

Network Load Balancing for Stateless Workloads

Network load balancing is an effective, scalable means for achieving high availability for stateless server workloads. Client requests handled before a given client request have no impact on that current transaction. The standard example of a stateless transaction is a web request. The http protocol has been designed such that each request for a web page is self-contained, and when responding to the request, the web server gathers all necessary information to present the page to the client. Upon delivering the response, the server discards all these resources collected to respond to the request. When the user clicks on a link on the page to continue the transaction, the click contains all necessary information for the web server to respond, and it is treated as a new stand-alone request from the client. In this situation, since each request supplies all the information needed by the server to fulfill the request, any given request can be processed by any instance of a server with access to the same underlying data.

Network load-balanced clusters leverage this flexibility to automatically distribute incoming requests among available computers. If a server in a NLB cluster fails unexpectedly, only the active connections to the failed server are lost. However, more commonly, NLB greatly simplifies bringing hosts down intentionally for planned maintenance. Upon completion of maintenance, servers can be added back to the cluster to resume sharing workloads.

Failover Clusters for Stateful Workloads

The prototypical example of a stateful workload is a database transaction. When responding to a database request, the server could potentially alter the data. However, since reading and writing information from and to storage is slow, a request involving a series of transactions is first completed in memory and the final results are recorded back in storage. If there are problems during the sequence, all transactions can be reverted. Thus, previous client requests in a sequence can influence subsequent transactions if all transactions in the sequence are to be consistent with each other.

Operating systems can facilitate failover clusters. If the primary node in a clustered application fails or that node is taken offline for maintenance, the cluster manager will start the clustered application on a backup cluster node. The operating system can now immediately redirect requests for resources to the backup cluster node to minimize the impact of the failure.

Multi-site Clusters

While failover clusters improve availability, servers in close proximity to each other are vulnerable to natural disasters, power failures, and wide-area network (WAN) outages (which can themselves be caused by construction accidents). In these situations, all cluster nodes can be disabled at once. Multi-site clusters can mitigate this risk. Cluster nodes can be connected through LAN or WAN links, providing high availability and disaster recovery. If any given node in a multi-site cluster fails, subsequent requests are directed to a node at one of the available sites.

Business Continuity and Disaster Recovery

"Snow Blizzard Shutting Down NYC"; "Earthquake in California Damaging Buildings"; "Super Storm Sandy Wiping out the New Jersey Boardwalk." These headlines are all too common these days, and the media narrative makes it appear that the storms are getting larger, more frequent, and more destructive. How does this affect you as an IT professional?

As an IT professional, one of your primary responsibilities is IT service delivery, as we discussed earlier in this chapter. IT is in every corner of just about every organization today. In small businesses, IT may be as simple as a router provided by the ISP, some servers for data storage, and a handful of desktops or laptops and printers. In larger organizations such as your university, IT can include hundreds of applications running on hundreds of servers across multiple load-balanced locations.

Regardless of how simple or complex your IT environment is, you need to plan for business disruptions. Without such a plan, events such as a local power outage or a tornado, hurricane, or earthquake can keep you disconnected from the market for extended periods, while your clients shift to your better-prepared competitors, hurting your business. Unfortunately, the data suggests that companies may be unprepared for disasters. For example, almost 75% of respondents in a recent survey had no backup plan in case their phone lines went down.[8] There are many reasons for this, including a lack of awareness, time, resources, or sense of urgency. Hopefully by the end of this chapter, you will understand the importance of being better prepared.

Business continuity planning (BCP) and disaster recovery are processes used by IT organizations to prepare for emergencies. *Business continuity planning is the methodology used to create and validate a plan for maintaining continuous business operations before, during, and after disruptive events.* The subset of BCP that is used to restore the affected services is called disaster recovery. *Disaster recovery is the set of procedures used to restore technology services that were disrupted during an extreme event.*

BCP is an important component in evaluating competing technology choices in many high-value industries. For example, large financial institutions, utility companies, healthcare

8 http://www.information-age.com/it-management/risk-and-compliance/123458335/two-thirds-companies-dont-have-telecoms-back-plan (accessed Feb. 2016).

organizations, credit card processing companies, mainstream media companies, and high-volume online retailers may decide that they cannot tolerate even a few minutes of downtime under any circumstances. Such downtime could cause large losses for businesses, or could put lives at stake at the hospital. These operational requirements will therefore justify the costs of fully redundant systems as part of BCP.

Disaster recovery usually involves several discreet steps in the planning stages, though those steps blur quickly during implementation because the situation during a crisis is almost never exactly as planned. Disaster recovery involves stopping the effects of the disaster as quickly as possible and addressing the immediate aftermath. This might include shutting down systems that have been breached, evaluating which systems are impacted by a flood or earthquake, and determining the best way to proceed.

Figure 13.4 shows the timeline for business continuity and disaster recovery activities around a disaster event. Business continuity activities are ongoing throughout the event. During active disaster recovery, business continuity activities include determining where to set up temporary systems, how to procure replacement systems or parts, and how to set up security in a new location. Finally, at the end of the event, once normal operations are resumed, the lessons learned are used to revise the continuity and recovery plans for more efficient response during the next event.

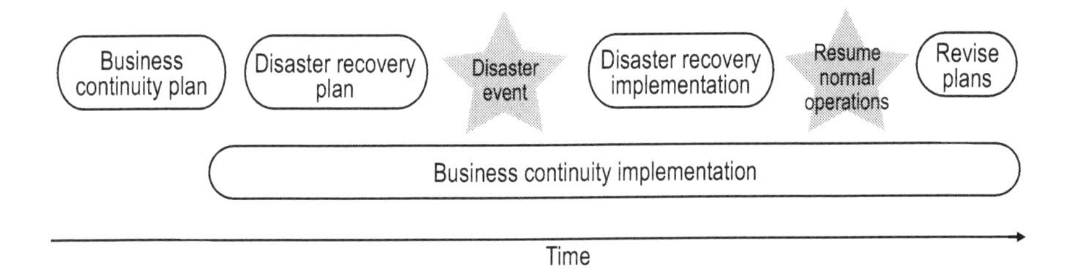

Figure 13.4: Business continuity and disaster-recovery cycle

Planning for Business Continuity and Disaster Recovery

The role of IT professionals is unique in BC/DR. On the one hand they are not responsible for the company's comprehensive BC/DR planning, but on the other hand technology is so integral to most corporate operations that continued IT operations is one of the core concerns of BC/DR. As a result, a holistic view of the organization allows IT to determine the most appropriate role of the IT group within the organization. Many elements of a BC/DR plan extend beyond the walls of the IT department. For example, corporate communications will keep stakeholders informed, the power company can try to restore power to critical buildings, and so on. The BC/DR project team therefore needs expertise in several areas. Some typical areas are shown in Table 13.2 below.

Table 13.2: Subject matter expertise needed for BC/DR planning

IT	Facilities	Specialty
Data centers	Office spaces	Off-site data storage
IT infrastructure	Production facilities	Critical data/records
End-user IT (laptops)	Manufacturing facilities	Critical equipment
Voice and data communications	Inventory storage areas	

The BC/DR plans are usually saved as documents and include the anticipated scenarios, contact information for the key personnel involved, detailed recovery procedures for the identified scenarios, and target recovery times.

Understanding BC/DR Requirements

When you are ready to develop a BC/DR plan, you can invoke ideas from your systems analysis and design class. An effective plan starts with understanding of the client's requirements. The BC/DR requirements rely strongly on the organization's specific location(s), industry, and operations-specific details to shortlist the types of disasters and events most relevant to the plan. As the plan is being developed, information that is likely to be helpful includes the following:

- Which system functions are critical?
- How much downtime is tolerable?
- How much data loss is tolerable in the event of a server or facility failure (if logs that would enable point-in-time recovery are lost)?
- Is the client willing to pay for an off-site recovery site that's always available?
- Could the client accept only critical functions being available in the case of a serious disaster? Which ones and for how long?

You may already know many of these answers from organizational documentation, but often, the BC/DR planning exercise is the first time the organization develops clarity on these concerns. An open dialog with the client can be helpful in developing this information.

BC/DR Plan Maintenance

Once the high-level BC/DR plan is developed, the disaster recovery components can be developed, which include detailed information that will likely be needed during a disaster (e.g. backup and recovery scripts, parameter files, etc.). The full package (backup/recovery plan plus recovery components) should be jointly reviewed by business stakeholders and IT to ensure that the BC/DR package is consistent and coherent. The package can then be used when required for a real emergency.

The BC/DR package should be tested at initial implementation time and regularly thereafter for as many scenarios as possible. The tests often require cooperation from other teams, so they are complex exercises that cannot be done frivolously.

Summary

In this chapter, we looked at the core concepts underlying industry-scale delivery of IT services. Some popular frameworks were introduced and the underlying ideas behind a popular service management framework—ITIL—were discussed. We identified high availability as a key concern for businesses and looked at how high availability is commonly achieved at the hardware, middleware, and operating system levels.

Finally, we introduced the idea of planning for disasters during peacetime. Business continuity planning is an ongoing activity that aims to help the organization continue to serve customers when unanticipated problems happen.

About the Colophon

The quote is attributed to a widely respected philosopher and teacher. It captures the idea so important to IT service providers that they may have no tomorrow if performance today is short of flawless. On a personal level, the philosopher probably asks people to pause and reflect. But for the IT professional, this quote is a call to alertness.

EXAMPLE CASE—Chaos Monkey at Netflix

It is now common knowledge that many of the most interesting consumer services that have emerged in recent years have been developed by technology companies that have brought together some of the smartest minds on the planet to build products and services used by tens of millions of users around the world. Google, Facebook, and Amazon are examples of these organizations.

Less well known is the fact that many of these companies also maintain interesting blogs that document the engineering that goes behind making these services highly available. These blogs are very detailed and students reading this book will find them very interesting. Since they are generally so well written, we will, for the most part, simply refer students to these blogs and limit this discussion to providing context where appropriate.

Chaos Monkey sounds like an odd name for a technology designed to improve availability. But this is the name given by Netflix to technology it has developed to deliberately fail components in live customer-facing systems and observe the behavior of their infrastructure in response. Chaos Monkey helps Netflix discover problems in its infrastructure and anticipate availability problems before they occur. The team summarizes the technology with the line, "Do you think your applications can handle a troop of mischievous monkeys loose in your infrastructure? Now you can find out." You can read more about Chaos Monkey on the Netflix tech blog.[9]

REVIEW QUESTIONS

1. What is *IT services management*? What is its objective?

9 http://techblog.netflix.com/2012/07/chaos-monkey-released-into-wild.html (accessed Jan. 2016).

2. What are some common frameworks that can be used for ITSM? What are their principal advantages and disadvantages?
3. What are the five service delivery disciplines as defined by ITILv2? In your opinion, which of these disciplines is most important for efficient service delivery? Why?
4. What is *service level management*? Why is it important?
5. What is a *service level agreement*? Why is it useful?
6. What are some of the important components of a SLA? In your opinion, which of these components is most important for effective relations between provider and client? Why?
7. What is *capacity management*? Why is it important?
8. What is *continuity management*? Why is it important?
9. What is *availability management*? Why is it important?
10. What is IT financial management? Why is it important?
11. What is *high availability*? Why is it important?
12. What are the four characteristics of high-availability system designs?
13. What are some of the important costs associated with transitioning to high-availability systems?
14. What is *business impact analysis*? How does it affect high availability?
15. What is cost of downtime? How does it affect high availability?
16. How are local high-availability solutions different from disaster-recovery solutions? How are they similar?
17. What are *active-active systems*? When would you use an active-active high-availability system over an active-passive high-availability system?
18. What are *fault-tolerant servers*?
19. What is an *application server cluster*? How does it improve availability?
20. What are the common ways of achieving high-availability database operations?
21. What is *business continuity planning*?
22. What is *disaster recovery*? How is it related to business continuity?
23. What is the distinction between stateful and stateless workloads? How do they affect high-availability design?
24. What are *multi-site clusters*? How do they improve availability?
25. What is the natural disaster most likely to occur in your area? If you were preparing a business continuity plan for your organization, what are some of the most useful items of information you can find?

HANDS-ON EXERCISE—Device Uptime

In this simple exercise, we will use command line utilities to obtain the uptime for common operating systems. In Windows, use the command `net stats srv` and on the Mac, use the command `uptime`

1. What is the uptime on your machine?

CRITICAL-THINKING EXERCISE—Personal High Availability

Are you personally prepared for high availability? Let us complete this exercise with the goal of helping you achieve your most important current priority. For most readers of this chapter, this priority is likely to be to find a job that can lead to a productive career. Use the following questions as a framework to minimize hurdles in achieving this goal.

1. What is your most important priority right now?
2. What are the top three hurdles you can foresee in achieving this priority?
3. What would be the most effective and cost-effective ways by which you can minimize these hurdles?

IT INFRASTRUCTURE DESIGN EXERCISE—Including Active Replication[10]

Initially, TrendyWidgets assumed that AWS guaranteed high availability. But after its CIO read about AWS's NoSQL outage,[11] it decided to maintain a two-region active-active replication of its data. Answer the following questions and update your infrastructure diagram as suggested.

1. What are the locations of some AWS regions?
2. Say TrendyWidgets decides to locate its servers in the US-East and US-West locations. Update your infrastructure diagram to indicate these locations. For simplicity, show a direct connection from TrendyWidgets to the US-East location and use any informal representation (say dotted lines) to indicate the internal AWS network.

EXAMPLE CASE EXERCISE

1. What is *Chaos Monkey*? What are some of its features?
2. How can a technology such as Chaos Monkey help improve service delivery?
3. Look at the tech blog at one of the leading Internet technology companies that maintains such a blog. Briefly summarize the most recent article on the blog.

---

10 A popular resource for information on data centers, an essential part of modern IT infrastructures, is the data center knowledge blog, http://www.datacenterknowledge.com (accessed Apr. 9, 2016).

11 http://www.techrepublic.com/article/aws-outage-how-netflix-weathered-the-storm-by-preparing-for-the-worst/ (accessed Feb. 2016).

CHAPTER 14

Managerial Issues

Let us raise a standard to which the wise and the honest can repair.

George Washington

Overview

Most networks you will encounter in the professional world extensively use the technologies such as Ethernet, IP, and WANs covered in this book. In addition to the technical issues covered in earlier chapters, there are nontechnical issues in using these technologies that are important to senior managers you will work with. Managers need to design and implement networks in ways that meet user needs and budgets. Networks need to be maintained to minimize downtime. Investment decisions for networking technologies will depend upon the maturity of technology standards to improve the interoperability of selected technologies with other technologies. Finally, many organizations, particularly the telecom carriers, are deeply affected by government policy and legal rulings. As a networking professional, you will be affected by all of these issues. At the end of this chapter you should know about:

- a high-level overview of designing computer networks
- the important concerns in network management
- the process by which computer networking standards are developed
- the role of government and legal process in computer networks

Introduction

Almost every student reading this book is likely to be the person responsible for maintaining their home computer network. These networks are small, built around a simple and inexpensive wireless router that also acts as a DNS and DHCP server for the internal home network as we saw in Chapter 7. The typical home network may have 1–10 devices including desktops, laptops, printers, etc.

Such a network is simple to maintain and generally requires no special knowledge on the part of home users to operate. As technology or user needs evolve at home, new components can be bought at big-box retail stores and added to the network in an ad hoc manner.

Since these components are generally quite inexpensive, these expansions are usually done without serious consideration of costs, design, manageability, or future network expansion.

If the above is the extent of a person's involvement with computer networks, the information in the earlier chapters is enough to help them understand and operate such networks. However, large networks cannot be operated in this manner. Ad hoc implementation and expansion of organizational networks will lead to very expensive and unreliable networks. Without good IT infrastructure design and implementation, if a user in a large network, such as a typical college or university, experienced an outage, it would become very difficult to locate the fault and fix the problem. Therefore, many managerial practices have evolved that help in cost-effective creation and maintenance of computer networks.

In earlier chapters, we placed great emphasis on highlighting the efforts made to efficiently transfer data. Packetizing, layering, broadcasting, multiplexing, multi-part addressing, port-addressing, DNS, DHCP, and NAT are all examples of technological innovations that help in the efficient transfer of data over computer networks. It is only appropriate that we now also look at the nontechnical developments that help cost-efficient network operations. This chapter will focus on four important nontechnical innovations in computer networks that help computer networks meet user needs while keeping costs to the minimum: network design, network management, network technology standards, and legal issues.

An area that is likely to gain increasing importance, but that is not covered here, is techno-stress. Ubiquitous technology has changed how we work and relate to friends and family. In recent research, work overload and role ambiguity were identified as important stressors, whereas intrusive technology characteristics, in turn, predicted the existence of stressors.[1] As managers, you are likely to see more discussion of these issues in your career.

Network Design[2]

Thus far in this book, we have always assumed that a computer network is in operation and the focus is primarily on understanding how the network works. An example of such a network would be the computer network at a typical small business. If you work in IT, or more specifically, in the networking group of a sizeable organization, eventually you will be involved in designing and installing a new network, or adding to an existing network. Typically, this would happen if the organization added a building or department, or merged with another organization. A formal design process improves the likelihood that the new network meets current and future user needs, is the most cost-effective solution available, uses the most appropriate technology, and addresses any organizational constraints. This section introduces the basic concepts of network design and implementation that you may find useful in such situations.

---

1 Ramakrishna Ayyagari, Varun Grover, and Russell Purvis, "Technostress: Technological antecedents and implications," *MIS Quarterly*, 35(4) (Dec. 2011): 831–858.

2 For an article that describes how this was done at USF for the deployment of Gigabit Ethernet, see Joe Rogers, "Using Fast/Gigabit Ethernet to Satisfy Expanding Bandwidth Needs," EDUCAUSE '98 (Dec. 10, 1998), https://net.educause.edu/ir/library/html/cnc9805/cnc9805.html. Though some years old, the process description is very detailed and perennially relevant.

A design is a detailed description of a product or service. Like any design project, the goal of network design is to meet current and foreseeable end-user needs, in a manner that minimizes costs over a specified time period. Costs include the costs of setting up the network initially and maintaining the network over the specified time period. If the network is too slow or unreliable, there are also costs associated with user downtime, when users are unable to perform their jobs because they cannot access the network.

Rather than develop new design principles for networks, the basic principles of systems analysis and design have been adapted for network design. These principles are taught in the systems analysis and design class you may have taken. A typical network design project will go through two stages: requirements analysis and physical design.[3]

Requirements Analysis

Like any design and development project, the creation of a network begins with identifying the needs of end users. Most network users in professional offices and university campuses have typical network needs—web browsing, e-mail, and access to some enterprise applications. However, there usually are also pockets of users with intensive network needs—for example, graphic designers, who may need to exchange large media files or medical imaging applications that create and read very large high-resolution image files. Enterprise applications such as web servers also need high-data-rate connections to the network because they are accessed by a number of users worldwide. At many universities, a good example of such an enterprise application that needs high-speed connectivity is the course portal where students access course materials, grades, etc.

During the requirements-analysis phase, the goal of network designers is to identify the different categories of users and applications. It is also necessary to determine their office locations within the organization so that the network can support high bandwidth at the required locations.

Fortunately, since most organizations already have computer networks in some form, data-traffic needs are usually very well understood. In addition to determining current needs, during the requirements-analysis phase, network designers also need to predict future network needs. Prior experience can provide baseline information on the growth rate of traffic. In addition, traffic growth comes from applications that are likely to be deployed in the future. For example, Voice-over IP and desktop video conferencing are not very common right now, but they are expected to soon become very popular and to generate significant volumes of network traffic. The requirements-analysis phase is a good time to anticipate new networking applications that are likely to be deployed in the organization and plan for the network traffic that is likely to be generated by these new applications.[4]

At the end of the requirements-analysis phase, it is a good idea to draw the logical network design. *The logical network design indicates the layout of the network, the subnets in the*

---

3 A classic article on system design, one that encourages an end-to-end look at system design, is J.H. Saltzer, D.P. Reed, and D.D. Clark., "End-to-end arguments in system design," *ACM Trans. Comput. Syst.* 2(4) (Nov. 1984): 277–288.

4 A very good resource is the work on the early design of the Internet. The 35th-anniversary celebration at UCLA has great presentations, including Kleinrock's introduction at http://internetanniversary.cs.ucla.edu/Technical_Sessions.html (accessed Dec. 2015).

network, IP addressing and naming schemes used in the network, and management strategies used in the network. Drawings of the logical network design provide a visual guide and facilitate group discussion. Drawings are a very effective tool for facilitating communication between end users and technical analysts. If the network does not adequately meet end-user needs, fixing the deficiencies later can be very expensive. It is therefore useful to carefully discuss and clarify current and future network needs with end users. If you don't, the end users will have to live with a suboptimal network for a long time.

To ensure that the logical design captures end-user needs, the logical diagram should show the total number of general-purpose desktops, servers, shared devices such as printers, and any devices with special network needs for every area of the network. Subnets should be indicated for each subnet, to help analysts determine the number of IP addresses available in each subnet for future expansion. The logical design also identifies network services such as DNS, DHCP, and network security and management utilities such as firewalls.

An example of a logical network design diagram for a small business specializing in video production is shown in Figure 14.1. In the example, the editing group needs high-bandwidth connectivity (1 Gbps) to its storage network, but the other groups only need plain desktop connectivity (100 Mbps). There are some extra network ports provisioned in every subnet for future expansion. The figure shows networked devices such as printers and portal servers. The connection speed to the external network is also shown. With a figure like this, all stakeholders can easily see the current and future capabilities of the proposed network. This facilitates identification of any missed requirements.

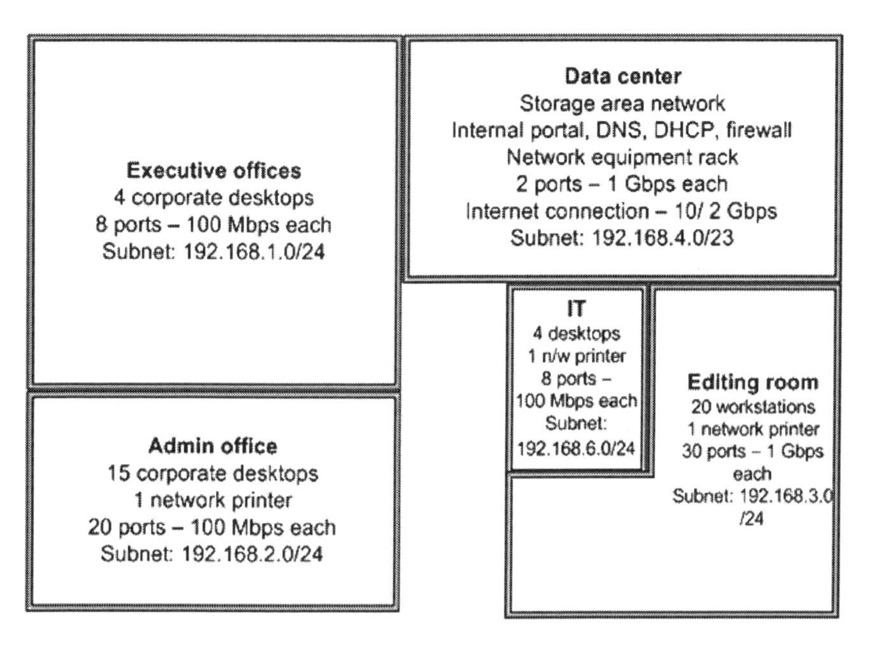

Figure 14.1: Logical network design example

Drawings serve another important function. They are repositories of organizational memory. It is said, "A picture speaks a thousand words." Long after the network has been implemented and the analysts and users who participated in the network design have left the organization, the drawings left behind are one of the most useful pieces of information to help future analysts and users in updating the network.

Physical Design

Once the logical design has been finalized, a physical network design is developed to deliver the functionality specified in the logical network design. *The physical network design indicates the technologies (for example, copper/fiber media for cabling; switches; routers; and data-link layer technologies such as Ethernet) that will be used to implement the logical design.*

Since the rooms in the example are adjacent to each other, all the links are likely to be shorter than 100 meters long. Therefore, we can use LAN technologies to create the network. LANs were covered in Chapter 3. From that chapter, we know that the most popular LAN technology is Ethernet. Therefore, we could use an Ethernet in each subnet. There would be suitably placed Ethernet outlets in each room, one for each desktop or workstation. The cable from each such outlet would terminate in a port on the Ethernet switch. The storage area network and video workstations could be located in one subnet served by a high-speed switch to isolate the high-speed traffic within the subnet. To create the subnets, we could use a router in the internal network. A firewall would provide the necessary protection for the internal network. With these choices, the physical network design for the example will look as in Figure 14.2. Standard icons are used to denote switches, routers, firewalls, and the Internet. The legend indicates the symbols used to show the different LAN and WAN links.[5]

Implementation

It is commonly seen that when organizations implement computer networks, they customize the network design in each part of the network. This is driven by user needs as well as the preferences of network administrators. It is also common for networks to be built using equipment from a number of vendors. However, over time it becomes cumbersome and expensive to maintain expertise in using technologies from a number of vendors. Therefore, organizations frequently standardize, using one vendor for network hardware. Organizations also discover some common patterns and data-link layer technologies that are used to implement the networks. Rather than try to create unique network designs for each building or department, they find it very convenient to implement networks using these standard patterns and technologies. Once a design is standardized, the network administrators can focus on developing deep expertise in the chosen technologies. This expertise helps them manage large networks with thousands, or even hundreds of thousands, of routers, switches, and network devices.

---

5 One of the most influential early papers on the Internet built on this scenario to highlight the role computers could play in communication. Please see J.C.R. Licklider and Robert W. Taylor, "The computer as a communication device," 1968. http://memex.org/licklider.pdf (accessed Apr. 9, 2016).

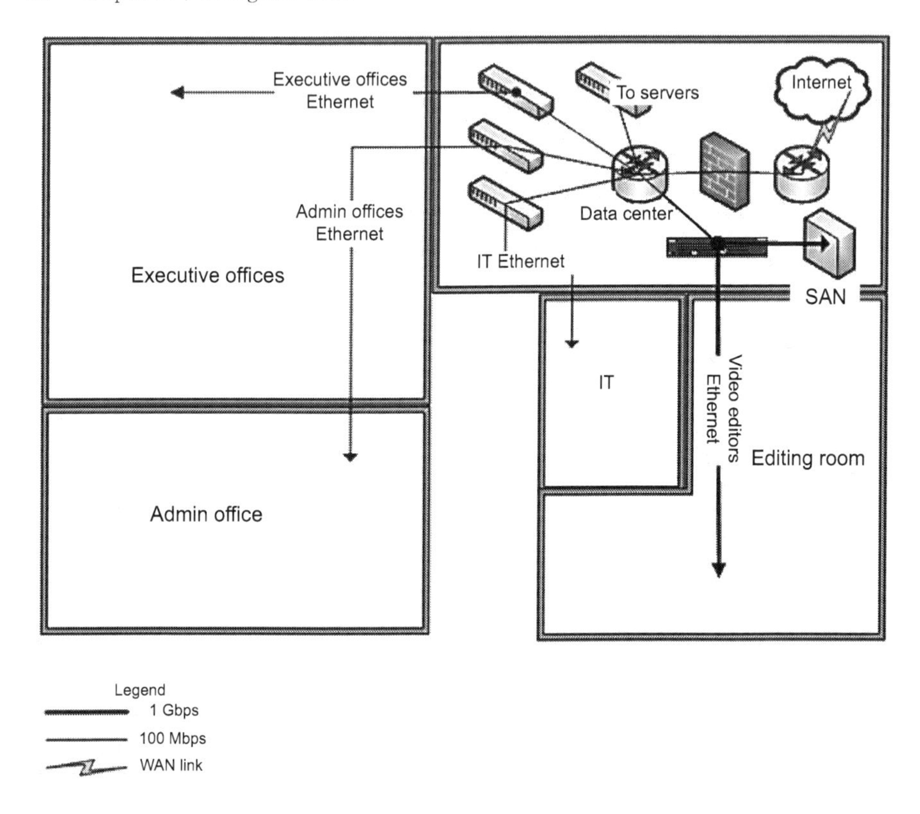

Figure 14.2: Physical network design example

The data communications timeline in Chapter 1 shows a diagram of the first Internet. Is that a detailed diagram, or a logical diagram?

For example, our network design example in Figure 14.2 uses two different Ethernet technologies to implement the network: 100 Mbps and 1 Gbps. Since the costs of Ethernet technologies have come down fairly significantly, it would also be possible to implement the entire network using 1-Gbps Ethernet. If we made this choice, the entire network would only use one switch model instead of two switch models. In fact, USF has indeed standardized on 1-Gbps outlets on all new and upgraded networks for this reason.

Our network design as shown is fairly simple. These simple examples have been used to introduce the components of logical and physical network designs. In general, to deal with the requirements of organizations of different sizes, network designs fall under three categories: building design, campus design, and enterprise design.

The simplest network consists of a few networked devices, all located within a single building. *A building network is a network that connects devices located within a single building.*

Our example was a building network. If the organization is successful, it grows and expands to multiple buildings. When this happens, the data network needs to span multiple buildings. *A network that spans multiple buildings is called a campus network.* The network of a typical university is an example of a campus network. Since this market is large, the most recent updates to Ethernet have defined versions of Ethernet that can span up to 30 miles. Therefore, building and campus networks can now typically be built using LAN technologies such as Ethernet.

Most networks begin their lives as building networks. It is not unusual for an organization's earliest network to be patched together using off-the-shelf components bought from a computer retailer. However, as the network grows, it becomes important for the network to be designed to keep costs low (efficiency) and to allow future growth. A common method to accomplish this is to perfect the design for a building network and repeat the design in every building in order to create the campus network in a modular fashion. Such an approach is called a modular or building-block approach to network design.

The campus network is typically organized into three layers: the core layer, the distribution layer, and the access layer. An example of a layered campus network is shown in Figure 14.3.

In a campus network, *the core layer is the layer that is connected to all parts of the campus network and is responsible for fast and reliable transportation of data across the different parts of the network.* The core layer is also called the network backbone. All traffic between different parts of the network passes through the core. The network core is optimized for fast and efficient packet handling. *The distribution layer is the administrative layer of the network.* It organizes the network into subnets to minimize traffic that needs to be handled by the core. The distribution layer also uses policies to filter traffic, for example by using firewalls and authentication to deny network access to unauthorized users. WAN access is also usually provided by the

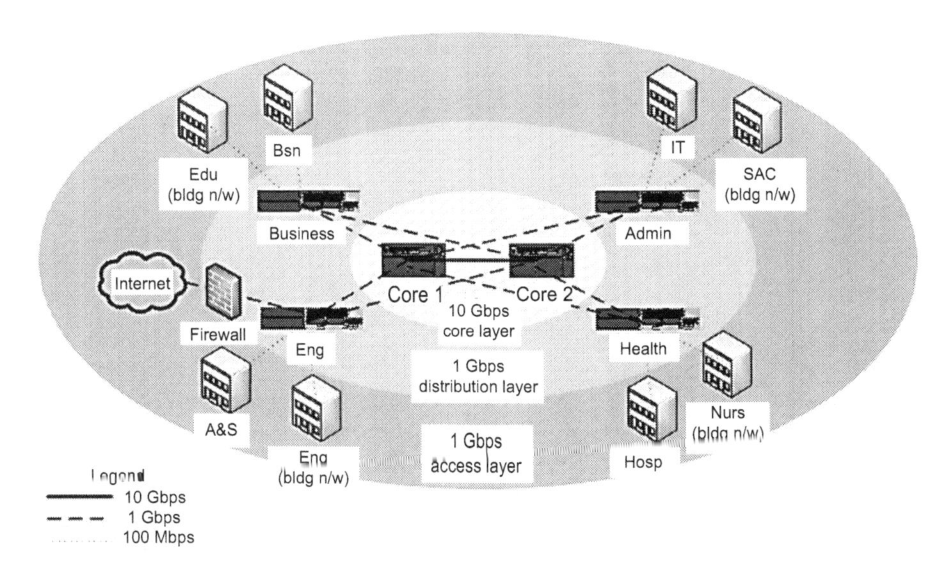

Figure 14.3: Layered campus network example

distribution layer. End users are typically not connected to the distribution layer, but most of the support services (e.g. DHCP, DNS, Time) are connected to this layer. Typically, once the distribution layer determines what service is desired by the end user and determines that the end user is authorized to access the service, it passes the packet to the core layer. The core layer then forwards the packet to the location in the distribution layer that provides that service.

The access layer is responsible for providing network ports to end users. It typically comprises hubs and switches that organize the devices in a building into Ethernets. The aggregated traffic from these Ethernets is passed to the distribution layer for processing. The typical building network is at the access layer.

In the campus network example of Figure 14.3, the network has two high-speed routers at the core. Both routers provide identical services and are configured to back each other up in case one of the routers fails. Each device in the distribution layer is connected to both routers so that the network is not affected by a failure in one of the core routers. The Internet connection is at one of the nodes at the distribution layer where a firewall can block unwanted connections. Buildings are also connected to the distribution layer.

Typically, there is redundancy in the core to ensure availability. But it is very expensive to create redundancy at the distribution and access layers because of the large number of devices at these layers. Therefore, lack of redundancy at the distribution and access layers is an acceptable trade-off because a failure in one of the devices at these layers only affects network connectivity within an isolated part of the campus network.

If the organization grows even larger and establishes offices in multiple locations around a state, country, or even the world, its network needs to expand beyond a campus network. The organization now needs WAN links to create a network that connects all the different offices. The network is now called an enterprise network. *An enterprise network is one organization's geographically scattered network.* The enterprise network connects campus networks at the organization's various locations. An ISP provides WAN connectivity between campuses. It is quite common for organizations to have offices in locations outside the service area of their ISP. In such cases, the ISP typically subcontracts with other ISPs to provide connectivity to all the campuses.

Figure 14.4 shows a typical enterprise network. As seen in the figure, most large organizations have a few large campuses with multi-tiered campus networks and a large number of smaller branches with smaller networks. All these sites are connected by a WAN operated by one or more telecom carriers (ISPs).

Maintenance

After data networks are designed and implemented, they need ongoing maintenance. Like all machines, network equipment can fail, or degrade in performance. For example, hard drives crack after spinning at high speeds for a long time. Power supplies get overheated from power surges. Components also fail randomly. Other maintenance issues arise from the need to update software. When security vulnerabilities are identified in software, network administrators need to ensure that the updates supplied by vendors are tested and deployed to all network devices at the earliest opportunity. Finally, users often install software on

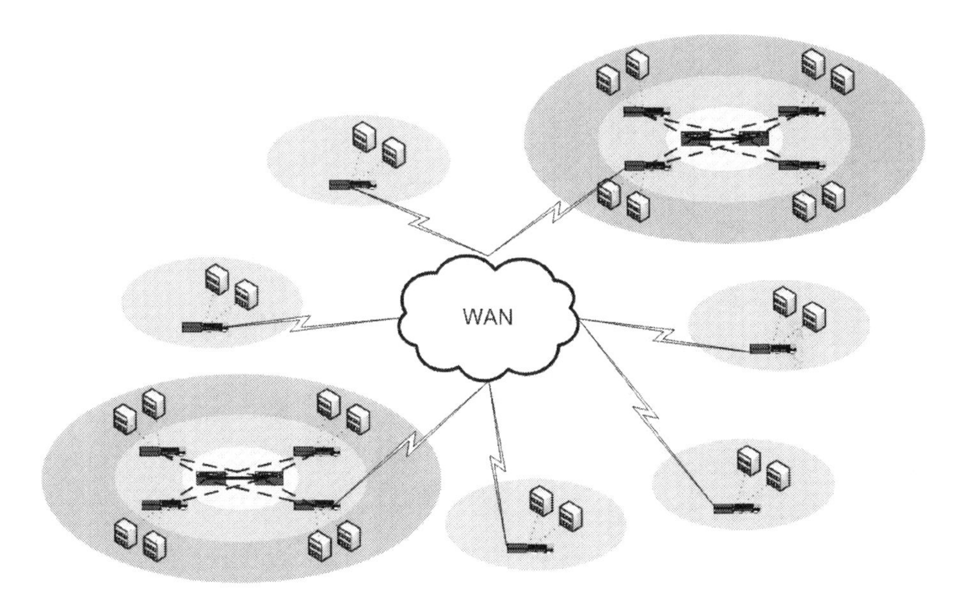

Figure 14.4: Enterprise network example

their local computers and ignore applying updates to such locally installed software. If this software has security vulnerabilities, remote attackers can exploit the vulnerability and put the organization at risk. Network maintenance attends to all these ongoing issues.

Network maintenance is the set of activities performed to keep networks in a serviceable condition or to restore them to serviceability. It includes activities such as inspection, testing, and servicing. Broadly speaking, therefore, network maintenance requires regular monitoring and updating of both hardware and software on all the computer equipment in an organization. As organizations get large, it becomes prohibitively expensive to perform such monitoring and updating manually. For example, the network of a typical state university is likely to have more than 1,000 network devices such as switches and routers and more than 10,000 desktops and laptops. If all these devices were to be scanned periodically by a human administrator for maintenance, most universities would have to hire too many IT administrators, seriously hampering the universities' ability to invest in other innovative applications. Therefore, tools have been developed to perform most of these activities automatically. Broadly speaking, these tools fall into two categories: tools to identify problems in network hardware and tools to identify software vulnerabilities.

Maintaining Network Hardware

Remote monitoring is the standard procedure used to identify problems in network hardware. The technology industry has long been aware of the need to facilitate remote monitoring of computer hardware. An important development in this area was the development of network management protocols, specifically the simple network management protocol (SNMP). *SNMP is a protocol used to manage and control IP (Internet protocol) devices.* SNMP was

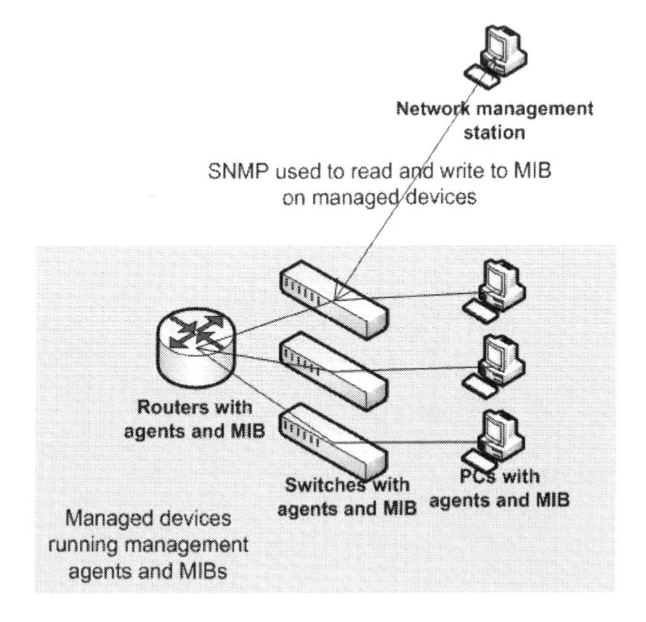

Figure 14.5: SNMP architecture

introduced in RFC 1157 in 1990 based on the idea that "management information for a network element may be inspected or altered by remote users." Devices that can be managed store configuration information in a database called the management information base (MIB). *A management information base (MIB) is a store of data regarding the entities in a communications network.* Example information found in the MIB includes network addresses, startup time, packet count, and power consumption. Only information that is essential for fault management or configuration management is included in the device's MIB.

The SNMP architecture is shown in Figure 14.5. A managed network has one or more network management stations and a large number of managed network elements such as hosts, switches, and routers. Managed network elements include management agents. Management agents are the software responsible for performing the network management functions requested by the network management stations. SNMP communicates management information from the MIB between the network management stations and the agents in the network elements. Management stations run management applications such as Spiceworks to monitor and control the managed network elements.

A network with about 1,000 managed elements (like a typical college), with about 100 information items per element, has about 100,000 items of information that need to be monitored by the network administrator. With SNMP, the management application on the network management station can poll all the managed elements, say, once per hour, and observe the status of all the information items. If any discrepancy is observed, the management application can alert the administrator by e-mail, SMS, phone, page, or another mechanism. Thus, by using SNMP, network administrators can focus on their work and attend only to configuration problems and failures when they arise. The user interface of the management application on the network management station can help the administrator quickly identify issues that need attention. These identified issues can be dealt with manually.

Maintaining Software

The primary concern with maintaining enterprise software is to ensure information security in the organization. Therefore, network administrators are interested to know the vulnerability status of all operating system and application software installed on all the devices in the network. Vulnerabilities that are thus located need to be fixed to prevent attacks on the network.

A common approach to identifying vulnerabilities on the network is to use network vulnerability scanner software to scan all devices on the network. Such scanners can scan all specified devices on a network for vulnerabilities. Like virus scanners, network vulnerability scanners use rules to identify vulnerabilities on the scanned devices. A popular vulnerability scanner is Nessus. A sample report from a Nessus scan on a host is shown in Figure 14.6. The report shows vulnerabilities in applications listening on ports 898 and 161. It also indicates open ports (22) and warnings on certain applications (ports 1241 and 80).

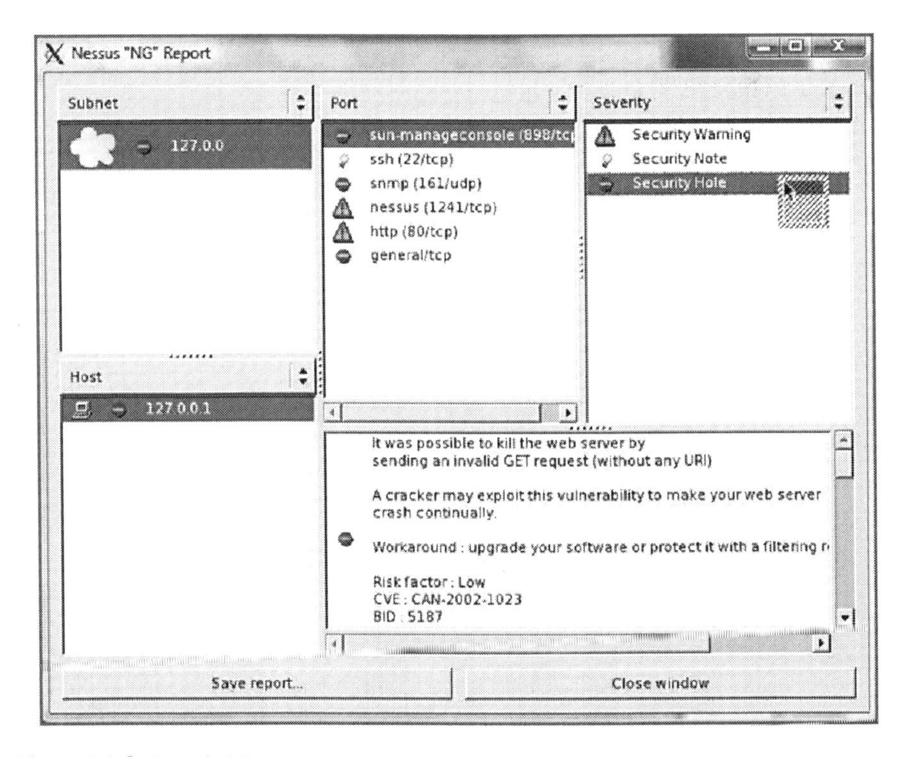

Figure 14.6: Sample Nessus report

As seen in the figure, to help administrators, where possible, vulnerability scanners point to the common vulnerabilities and exposures (CVE) database.[6] The CVE database is a database maintained by the IT industry to centralize information on all identified software vulnerabilities. The CVE database helps users and vendors gather information about software vulnerabilities and disseminate information to fix these vulnerabilities. Here users can gather more information about the vulnerability.

Another important concern is license management, to ensure that all software being used in the organization is duly licensed. License manager software utilities operate similarly, scanning all the machines in the organization to identify any software that offends the organization's licensing policies.

Standards

An important administrative development on the Internet is the use of technology standards. In all your experience on the Internet, most likely you have never had problems using any Internet service from anywhere in the world on any computer running any operating system. Because of the use of standardized technologies on the Internet, when you send e-mail, you are free to use any e-mail client of your choice, without concern for the operating system or brand of the e-mail server software. Similarly, when you visit websites, you are able to visit websites in any country, without concern for the operating system or brand of web server running the website. If you use wireless networks for connectivity on your laptop, you are able to move seamlessly from your home wireless network to the wireless network at school without any changes to your network hardware.

Compare this easy connectivity on the Internet to some other comparable situations you encounter. You probably have a different charger for each powered electronics device you use. Most likely, each such device also uses a different battery. In many cases, each device also uses a different cable.

The lack of uniformity is pervasive. If you drive internationally, you will find that people in many countries drive on the left side of the street. Power outlets look different in each country and even supply power at different voltages, 220 volts being quite popular. If you travel abroad, you will need to carry special adapters to power electronic equipment purchased in America. In video, some countries use the PAL TV standard for conventional TV broadcast, whereas the United States uses NTSC. Therefore, you cannot play DVDs purchased in many countries abroad on DVD players in the United States.

By comparison, Internet applications are remarkably well behaved. You can use a PC, a Mac, a Linux laptop, or even a smart phone and connect to almost any networked application hosted anywhere in the world. All this is possible because of the existence of technology standards on the Internet and the adherence of software and hardware vendors to these standards. In fact, the Internet may be the most standards-compliant global system ever created.[7]

6 http://nvd.nist.gov/.

7 The phone system is probably the only other system that is comparably standardized across the globe.

Standards are documents, established by consensus and approved by a recognized body, that provide rules, guidelines, or characteristics aimed at achieving order in a given context.[8] More simply, standards are rules that ensure interoperability. Standards help developers, working independently of each other, to create products that are guaranteed to interoperate with other standards-based products created by other developers.

Standards battles

Most communication standards have emerged through the standards development process. Therefore there is a high degree of interoperability in communication technologies. However, it is important to remember that many important computing standards have developed due to the efforts of individual companies. For example, Windows, Java, and Office are important computing standards whose development has been spearheaded by individual companies or a consortium of companies.[9]

If a company can create a non-standardized technology that becomes extremely popular, it can be highly profitable because customers interested in using the technology have no choice but to use the products supplied by the company. This gives the company monopoly pricing power. Customers of the technology can also get locked-in with the company's products, requiring them to buy upgrades and other connected products from the same company. Therefore, standardization often creates battles between competing firms or groups of firms, with dominant firms trying to resist the standardization process and other firms promoting standardization.

Standardization on the Internet has benefitted both customers and the industry. Standardization allows multiple vendors to enter the market, creating devices that are guaranteed to be compatible with devices sold by other vendors. The existence of multiple vendors creates competition in the industry, which generally lowers prices. Lower prices make technology affordable to a larger group of customers, creating an even larger market for vendors.

Many organizations have been involved in developing standards used on the Internet. Two of these are extremely important to most end users of the Internet: IEEE and IETF. The Institute for Electrical and Electronics Engineers (IEEE) develops standards used at the data-link layer in Ethernet and wireless LANs. Examples of such standards include 802.3 (Ethernet) and 802.11 (wireless). The Internet Engineering Task Force (IETF) develops standards used at the network, transport, and application layers. Examples of these standards include IP, TCP, UDP, HTTP, SMTP, POP, IMAP, and SNMP. The International Telecommunication Union (ITU) takes the lead in developing WAN standards.

---

8 This is an adaptation from the more complete definition at the ATIS glossary, "Standards are documents, established by consensus and approved by a recognized body, that provide, for common and repeated use, rules, guidelines, or characteristics for activities or their results, aimed at the achievement of the optimum degree of order in a given context."

9 Standards or conventions may also emerge simply because they catch on. For example, the @ symbol on Twitter was first used by Apple designer Robert Andersen on Nov. 2, 2006, but it quickly caught on. Source: Nick Bilton, *Hatching Twitter* (Portfolio, 2014).

President Lincoln and railroad standards[10]

President Lincoln is a revered president of the United States. However, prior to his political prominence, he was a prominent railroad attorney and was familiar with the importance of nationwide railroad standards to the growth of the industry.

On January 21, 1863, President Lincoln signed an order declaring that all railroads built pursuant to the Pacific Railroad Act of 1862 would be 5 feet in width. However, two months later, on March 3, 1863, Congress superseded the order and specified 4′ 8″ as the gauge for the railroad.

The development of standards usually follows a well-established procedure. Standards development begins with the identification of a user need that is not met by current technologies. There is general agreement in the IT industry about the organization that is in the best position to lead the development of a standard for the identified need. For example, the IEEE takes the lead in developing data-link layer standards for LANs. The IETF takes the lead in developing standards for protocols at the network, transport, and application layers. The lead organization for the standard establishes a working group to specify requirements for the proposed standard. These requirements are such that they can be implemented with current technologies at reasonable costs. Companies and expert groups next propose technical solutions to implement the proposed requirements. Members of the lead organization then vote on the various proposed solutions to identify a solution that wins an overwhelming majority of the vote. If no solution obtains an overwhelming majority, voting is conducted in multiple rounds, candidates with very few votes are eliminated, and the process continues until one solution emerges as a clear winner. This solution is published as the standard for the technology. Vendors then bring products based on the standard to the market.

A very good example of the standards development process is the development of the IEEE 802.11n wireless standard.[11, 12] After the IEEE approved the development of the standard, proposals were invited to implement the standard. In the first round, four complete and 28 incomplete proposals were received. At the working group's meeting in November 2004, all four complete proposals obtained the required number of votes for further consideration, and the 28 incomplete proposals were eliminated from further consideration. At the group's next meeting, only three of the four proposals were presented, and one of the three proposals was voted down, leaving two candidate proposals. In March and May 2005, one of the two proposals secured a majority, but not the overwhelming majority that is required by the IEEE for the acceptance of a proposal. As per procedure, this reinstated all three of the previously rejected proposals into the voting process. Eventually, the advocates of three proposals decided to merge their efforts. The resulting proposal secured the desired majority and work began on refining the initial draft by incorporating the changes suggested by the working group. The 802.11n standard was finalized in September 2009. Standards development takes time.

---

10 http://quod.lib.umich.edu/l/lincoln/lincoln6/1:125?rgn=div1;view=fulltext (accessed Dec. 2015).

11 http://standards.ieee.org/board/nes/projects/802-11n.pdf.

12 http://grouper.ieee.org/groups/802/11/Reports/tgn_update.htm.

RAND licensing for SEP[13]

Most technical standards use a mix of publicly available technologies and patented technologies. *Standards essential patents (SEP) are patents that are required for implementing a technology standard.* Technologists working on a technology are usually unaware of the patent implications of their work, and courts have held that SEP should be made available by patent holders at reasonable and non-discriminatory rates (RAND) to anyone implementing the standard.

In February 2014, Cisco, the maker of infrastructure gear, settled a lawsuit against Innovatio over RAND. Innovatio had bought old patents from Broadcom and sent letters to 13,000 individual hotels and coffee shops asking for $2,300–$5,000 in license fees for using off-the-shelf Wi-Fi routers that had incorporated technologies covered by Broadcom's patents. Cisco defended its buyers in court, spending $13 million in legal expenses, and Innovatio settled for 3.2 cents per device, for which Cisco will pay Innovatio $2.7 million.[14]

Government Involvement, Legal Issues

Throughout the book, we have focused primarily on the role of technical institutions in the development of the Internet and related technologies. We have paid very little attention to the role of government and the judicial system in influencing the development and adoption of communication technologies.

In general, developments on the Internet have been led by commercial interests. The government has not played a significant role in the day-to-day evolution of Internet technologies. However, the government has played a very constructive role at critical points in the development of the technologies. This section highlights some of these developments.

Eight years in the making—digital projection standards

How long can standards development take, even if the payoffs to such standards development are very significant, and all parties around the table are seasoned professionals? In the movie industry, standards development and other related developments to move from film distribution to digital distribution of movies took almost eight years.[15] The team that worked on the project called it the Kasima project—Kicking and screaming into the modern age.

13 A good read is the verdict in the dispute between Microsoft and Motorola. The annotated version is an easier read. http://essentialpatentblog.wp.lexblogs.com/wp-content/uploads/sites/234/2013/05/2013.04.25-Microsoft-Motorola-RAND-Decision-Annotated.pdf (accessed Feb. 2016).

14 Joe Mullin, "Wi-fi patent troll will only get 3.2 cents per router from Cisco," *arsTechnica*, Feb. 6, 2014.

15 https://en.wikipedia.org/wiki/Digital_cinema (accessed Feb. 2016).

The Development of Packetization and TCP/IP led by the Department of Defense

It is quite possible that the Internet as we know it today would not have existed without the efforts of the US Department of Defense (DoD) to develop a robust communication system that would continue to function in the event of a large-scale war. This effort led to funding for the development of packetization and related packet-transfer technologies—TCP and IP. Similar efforts also put in place other building blocks of the modern Internet, such as DNS.

Being a government body, and not treating TCP/IP as a classified technology, the US Department of Defense made these taxpayer-funded technologies freely available for general use, without a profit motive. This allowed early UNIX implementations to use TCP and IP for networking, bringing the technologies within reach of a wider audience. The DoD also insisted that TCP/IP should be used for network functionality in all computer software sold to the DoD. To improve TCP/IP technology, the DoD also funded the development of a communication network, the ARPANET, which functioned from 1969 to 1990.

Eventually, all computer manufacturers replaced their proprietary network layer and transport layer technologies with TCP and IP, enabling TCP/IP to provide a standard platform for computer communication across the globe.

The Early Internet, Funded by the National Science Foundation

The National Science Foundation (NSF) is the primary agency of the US government that funds research in nonmedical disciplines. NSF initiated a program in 1985 to fund the development of a nationwide network, NSFNET, to serve the entire academic community in the United States. NSF required US universities requesting NSF funding for an Internet connection to make the connection available to all departments on campus. This initiative liberated the Internet from the confines of Computer Science departments and enabled nontechnical departments to get access to the Internet. NSF chose to use TCP/IP for the NSFNET program, thereby introducing the Department of Defense's TCP/IP to the nontechnical community. In some sense, NSFNET may be considered the earliest version of the Internet.

NSFNET functioned until 1995, by which time the Internet had become wildly popular and commercial service providers had emerged to offer the functionality provided by NSFNET, even to home users. In April 1995, funding for NSFNET was stopped, and universities began to use commercial ISPs for Internet connectivity.[16]

The Development of the Web Browser at the National Center for Supercomputing Applications

TCP and IP were very useful for computer communication. However, early applications that used TCP and IP were not very user friendly. The primary mechanism for interacting with these applications was through the command line interface (CLI). The DOS prompt that

---

16 B.M. Leiner, V.G. Cerf, D.D. Clark, R.E. Kahn, L. Kleinrock, D.C. Lynch, J. Postel, L.G. Roberts, and S. Wolff, "A Brief History of the Internet," http://research.microsoft.com/en-us/um/people/padmanab/cse561/papers/internet-history.htm.

you have used in the hands-on exercises of earlier chapters is an example of the command line interface. Most users are very uncomfortable with the command line. Therefore, the Internet did not become useful for the masses until the graphical web browser was developed. This happened at the National Center for Supercomputing Applications (NCSA) at the University of Illinois. Two students at the center, Marc Andreessen and Eric Bina, developed a graphical web browser for UNIX computers. The first version of the browser was released on January 23, 1993, as an application that ran on X-Windows, the UNIX GUI. Soon thereafter, the browser led to the boom in the PC industry, e-commerce, the dot-com era, and related developments. The dot-com boom ended badly for many investors, but the user-friendly web has changed our lives forever.

NCSA and the students who developed the first browser were largely funded by the US National Science Foundation.

Wireless Spectrum

The government, in its coordinating role, has identified some wireless bandwidth for free use by wireless LAN and other technologies (more details are in the supplement). This allocation of bandwidth has enabled the development of many generations of 802.11 WLAN technologies. This, in turn, has enabled free wireless Internet access at many offices and commercial establishments, allowing professionals to become more mobile using laptops. Without the availability of this free ISM bandwidth, cell phones would probably have been used for wireless LANs. This would have raised the costs of wireless Internet access because cell phone service is not free. It would not have been surprising if the costs of wireless Internet access would have been comparable to the costs of using a cell phone.

AT&T Split

The government took an active role in breaking the monopoly of AT&T in the phone business in the United States (more details are in the supplement). In 1982, the US government and AT&T settled on a judicial agreement that generated competition in long-distance phone calls. The government initiates such antitrust proceedings when it believes that a company is using its monopoly position in an industry in an unfair manner. Antitrust proceedings are contentious and often controversial. However, the authority of the government to intervene and prevent the misuse of monopoly power is well recognized. The AT&T divestitures are a good example of the government's role in preventing abuse of monopoly power. Today, two decades after that judgment, competition and technology development has made long-distance phone service almost free, even to many international destinations.

We also saw how the courts and the legislature were involved in the process of enabling competition in the phone industry, ultimately changing the law that governed how telecommunication companies operated in the United States. In 1996, the Telecommunications Act was passed to open all forms of communication to competition. As a result, today we are beginning to see the web emerge as a viable competitor to cable TV service.

Balancing regulation and innovation

As customers disconnect their landlines and embrace cell phones and telecom carriers respond to these changes, some benefits of regulation have come to the fore. As carriers experiment with all-IP cellular-based phone networks in small cities as a precursor to national roll-outs, disagreements in how to handle traffic from competing carriers are emerging. In Illinois in 2014, for example, Sprint was converting its IP traffic to switched traffic so that AT&T would accept it under the terms of a regulatory agreement. Though AT&T subsequently converted the traffic back to IP traffic for distribution over its own networks to its customers, it saw no obligation to accept the same traffic directly as unregulated IP traffic.[17]

Patents

Technology development requires large investments and the government encourages such investments through patent protection. Patents grant property rights to the inventor, and are issued by the Patent and Trademark Offices in the country where the protection is sought. In the US, patents grant the right to the inventor to exclude others from "making, using, offering for sale, or selling" the invention in the United States or "importing" the invention into the United States. The patent office does not enforce the patent, which is the responsibility of the patentee. In the US, a new patent generally protects the invention for 20 years.

The patent law specifies the requirements for obtaining a patent. According to the law, any person who "invents or discovers any new and useful process, machine, manufacture, or composition of matter, or any new and useful improvement thereof, may obtain a patent," subject to the conditions and requirements of the law. The word "process" is defined by law as a process, act, or method, and it primarily includes industrial or technical processes. The term "manufacture" refers to articles that are made and includes all manufactured articles. The term "composition of matter" relates to chemical compositions and may include mixtures of ingredients as well as new chemical compounds. These classes of subject matter taken together include practically everything that is made by man and the processes for making the products. The patent law also specifies that the subject matter must be "useful." Issued patents can be searched at the patent office.[18]

Interpretations of the statute by the courts have defined the limits of the field of subject matter that can be patented, and the laws of nature, physical phenomena, and abstract ideas are considered not patentable. In a famous case from 1853, O'Reilly vs. Morse, Samuel Morse was granted the patent for the telegraph machine that delivered the famous message "What Hath God Wrought?" However, Morse was trying to obtain a patent for the idea of sending messages electronically across great distances. The Supreme Court did not allow that on the grounds that ideas alone could not be patented.[19]

---

17 Ryan Knuttson, "AT&T's plan for the future: No landlines, less regulation," *Wall Street Journal*, Apr. 7, 2014.

18 In the US, this site is http://www.uspto.gov/patents-application-process/search-patents.

19 L. Gordon Crovitz, "Could Morse have Patented the Web?" *Wall Street Journal*, Mar. 25, 2012.

The perils of rushing to regulate a new technology[20]

It's important to consider the potential ill effects of a new technology, say, for instance, Google Glass. However, we shouldn't be so quick to regulate it. Unfortunately, history shows that society tends to ignore that advice. When automobiles first arrived on the roads, a safety law required cars to be preceded by a person carrying a red flag. Such stories are useful to remember when we argue about how new technologies could possibly hurt us. Irrational fears are not very helpful.

EXAMPLE CASE—Telework, Telemedicine

The cases in the previous chapters have looked at how computer networks have affected various industries. As we come to the end of the book, it is time to see how computer networks are helping the people who work in these industries. Computer networks are enabling new work practices, changing old practices, and lowering business costs across industries. With the help of computer networks, many professionals are able to work from home and central offices, saving commute time and bringing world-class services to the remotest corners of the world.

Consultant Jack Niles is credited with coining the terms "teleworking" and "telecommuting" in 1973. *Teleworking is any substitution of information technologies for work-related travel.* Teleworking moves work to the workers instead of moving the workers to work. Telecommuting is a form of teleworking where workers work away from their primary office space, at least one day per week. Telecommuters may work at home, client sites, or at a telework center. Telecommuting substitutes information technologies for the commute to work. The preferred terms now are distributed work, mobile work, or remote work.

Telecommuting is gaining popularity in the United States and around the world. In 1997, approximately 3.6 million people were telecommuting, and in 2015, approximately 25% of the workforce telecommutes.[21]

How far we have come. Just five generations ago, the average American worked 60 hours a week, took no vacations, and earned less than the modern-day equivalent of $6,000 a year. He or she rarely traveled more than a few miles from home, had no central heat or running water, and died at age 50.[22]

Telecommuting offers many advantages and some risks. Real estate cost savings can be considerable. Cisco estimates annual savings of $277 million (compared to annual profits of $6 billion) from telework and telecommuting. During the workday, employees generally find fewer distractions at home than at work and can therefore be very productive. Employees report significant increase in productivity and overall satisfaction as a result of their ability to telework. There are also other social benefits that are not captured by the organization, such as reduced traffic congestion and lower fuel consumption.

20 Based on HBR Daily idea, June 26, 2013, adapted from Larry Downes, "What Google Glass Reveals About Privacy Fears," *Harvard Business Review*, May 23, 2013.
21 http://globalworkplaceanalytics.com/telecommuting-statistics.
22 Steven E. Landsburg, "How the Death Tax Hurts the Poor," *Wall Street Journal*, Oct. 29, 2011, A15.

Telecommuting has risks. Many workers are not able to separate their work and personal lives effectively. Telecommuters can feel left out of happenings in the workplace. The productivity of workers involved in activities that need face-to-face interaction can be dramatically reduced with telecommuting. While technology can facilitate telecommuting, it cannot address these risks. The risks have to be managed through well-developed managerial practices.

The field of medicine has successfully developed variations of telework to reduce medical costs and to bring medical expertise to remote regions of the world. This is called telemedicine. *Telemedicine is the application of computer networks to transfer medical information through the phone or Internet for medical consultation, and even to remotely provide medical examinations and procedures.*

Telemedicine traces its origins to NASA's space program, which developed procedures in the 1960s to serve astronauts in outer space. In the best-known examples of telemedicine, medical data like medical images and bio-signals (such as ECGs and EEGs) are transmitted to doctors or medical specialists for assessment. Even in traditional medicine, such assessment is not done face-to-face and the information required for such assessment can be completely digitized. Therefore, such assessments are excellent candidates for telemedicine.

In another category of telemedicine, medical professionals use networking technologies and applications to remotely monitor patients with chronic diseases or specific conditions, such as heart disease, diabetes, or asthma. Not only does remote monitoring reduce the need for patients to commute to hospitals and clinics, it can be cost-effective and provide health outcomes comparable to traditional medical visits.

In the most advanced category of telemedicine, it can provide live interactions between patients and providers. Psychiatric evaluations and ophthalmology assessments are considered prime candidates for interactive telemedicine.

Advantages of Telemedicine

Telemedicine offers many advantages. Rural populations that typically do not have the resources to support specialty-care clinics can be served by telemedicine using video consultations. Such technical solutions can significantly reduce costs for patients. Additionally, healthcare workers in isolated or low-income areas can undergo regular continuing education programs using simple video technologies.

Barriers to Telemedicine

As with most technology-driven advancements, barriers continue to exist that prevent rapid adoption of telemedicine. Many states require a license in order to provide medical consultations to patients in the state. Physicians have to consider the possibility of malpractice suits if a patient is dissatisfied with the treatment. Many physicians are not comfortable with information technology and have strong preferences for personal interactions with patients. Reimbursement practices of many private insurers and Medicare may discourage telemedicine. Many target sites lack appropriate telecommunications infrastructure because rural areas still generally lack access to high-bandwidth data networks to transmit video and medical images. Regular telephone lines are available everywhere but they provide insufficient bandwidth for telemedicine applications. Unfortunately, therefore, telemedicine may be least available to those who could benefit most from it.

Telemedicine Success

To end the case, chapter, and text on a positive note, let us look at a recent success story in telemedicine. After the earthquake in Haiti on January 12, 2010, the University of Miami established a 240-bed tent hospital near the airport in Port-au-Prince, Haiti. Volunteer surgeons from the United States have performed more than 1,000 on-site surgical procedures at the site. For ongoing operations, experts who helped develop telemedicine applications for space shuttles have installed systems to enable on-site doctors to consult with specialists in Miami and other medical centers via satellite. The satellite connection will provide sufficient bandwidth for telemedicine consultations and is unlikely to be affected by terrestrial phenomena. While it is difficult for busy physicians to stay for long at disaster sites, after a visit they are usually available for telemedicine consultations. If video sessions are necessary, e-mail and secure Internet messaging is used to exchange pictures, x-rays, and pathology reports beforehand to optimally use physician time during the video session.

References

1. Cisco. "Cisco Study Finds Telecommuting Significantly Increases Employee Productivity, Work-Life Flexibility and Job Satisfaction," https://newsroom.cisco.com/press-release-content?articleId=5000107 (accessed Apr. 9, 2016).
2. Freudenheim, M., "In Haiti, Practicing Medicine from Afar," *New York Times*, Feb. 9, 2010, D5.
3. Healy, M. "Wounded Soldier's Shattered Pancreas Gets Replaced in a Whole New Way." *Los Angeles Times*, Dec. 15, 2009.
4. Jala International, www.jala.com.
5. Jennings, J. *Less Is More: How Great Companies Use Productivity*. Portfolio, Penguin Putnam, 2002.
6. Mariani, M. "Telecommuters." *Occupational Outlook Quarterly*, Fall 2000.
7. Nilles, J.M. "Telecommunications and Organizational Decentralization." *IEEE Transactions on Communications*, 23 (10) (Oct. 1975).
8. Schadler, T. and M. Brown. "US Telecommuting Forecast, 2009 to 2016." www.forrester.com.
9. US Bureau of Labor Statistics, Table A-1. http://www.bls.gov/news.release/empsit.t01.htm (accessed Apr. 9, 2016).
10. Wikipedia on telemedicine.

Summary

This chapter introduced the managerial and procedural issues in computer networking. Computer networks are designed following standard systems analysis and design procedures. Some standard designs have evolved to build networks. These standard designs are used in a modular manner to create larger networks. Once deployed, hardware and software in networks need to be monitored and maintained. Protocols such as SNMP have been developed to automate many of these maintenance functions.

Computer networks have benefitted immensely from the development of standards for communication. As a result, users can use any network application running on any operating system and hardware to communicate with applications running on any other operating system and hardware. Many organizations, including the IEEE, ITU, and IETF, are responsible for the successful development of these standards.

Over the years, the government has played a very significant role in funding the development of many critical components of the Internet. These include the TCP and IP protocols, the early Internet, and the early Internet browser. The regulatory role of the government has also become increasingly important in facilitating the orderly development of business data communication technologies.

About the Colophon

Standards are extremely important elements in ensuring the smooth operation of a community. This book has focused on technical standards that are important for data communications. Standards exist in many other technical domains as well, such as the voltage levels on a wire.

But standards can have even wider impacts. As the colophon to the chapter indicates, the Constitutional Convention started with the goal of creating a standard documenting the highest ideals for a society. The resulting standard is the Constitution of the United States. For years, this standard has served as a guiding light for lawmakers in creating laws, rules, and conventions that govern life in the United States. Standards can help guide activities whenever coordination among multiple entities is required to accomplish a goal.

REVIEW QUESTIONS

1. What are the goals of network design and implementation?
2. What are some of the important pieces of information you should gather during the requirements-analysis phase of network design?
3. What is a *logical network design*?
4. Why is it useful to represent the logical network design as a drawing?
5. What are some important pieces of information you should show in a drawing of the logical network design?
6. What is a *physical network design*?
7. What is a *building network*? What data-link layer technology are you most likely to see in a building network?

8. Why is it useful to develop a standardized design for the building network and to use it as a building block to network every building, even buildings that have much lower network demands?
9. What is a *campus network*?
10. What is the *core layer* in a campus network? What is the role of the core layer?
11. What is the *distribution layer* in a campus network? What are the roles of the distribution layer?
12. What is the *access layer* in a campus network? What service is offered by the access layer?
13. Which layer in the campus network is most suitable to providing Internet connectivity? Why?
14. What is an *enterprise network*?
15. What is *network maintenance*? What are the important activities in network maintenance?
16. What is *SNMP*? Briefly describe how it is used to maintain network hardware.
17. What is the *management information base* (MIB)? What are some pieces of information you are likely to find in an MIB?
18. What are the typical concerns in software maintenance on the network?
19. What are *standards*? How are they useful?
20. What is the typical procedure by which a standard is developed?
21. How did the government facilitate the development of the technologies (TCP and IP) on which the Internet is based?
22. What are *standards essential patents*? What are some regulations on their use?
23. How did the government facilitate the creation of the Internet?
24. What has been the role of the government in the evolution of the phone industry?
25. What are the general ways in which the government influences the development of the data communications industry?

EXAMPLE CASE QUESTIONS

1. What is your desired career path? Look at the characteristics of tasks for which you could telecommute, found on pages 15–16 of the article by Matthew Mariani. (Please see references above; the article is available online.) Which of these characteristics does your dream job possess? Based on these characteristics, how suitable is your career path for telecommuting?
2. Use the "Telecommuter self-assessment" in the same article (pages 16–17) to assess your ability to succeed at telecommuting. What is your score on the self-assessment? In what areas do you need to improve to be able to telecommute successfully?
3. What practices would you recommend to a manager who is supervising teleworkers, to optimize their professional performance?
4. What is the medical research center that is closest to you? Look up its website to see what telemedicine services the center provides, or plans to provide.
5. Recall a medical procedure that was performed on someone you know. If telemedicine was used during any stage of the disease, briefly describe how telemedicine was used in the case. If not, briefly describe how telemedicine may have been used in the case. (Assign anonymous names, such as A, B, etc., to prevent revealing personal information.)

HANDS-ON EXERCISE—Standards Development Review

In this exercise you will do a quick review of the standards development process, by following the timeline for the development of a contemporary technology standard. You may pick a standard of your choice; the instructions below provide links to the 802.11ac Wi-Fi standard.

1. What are the start and end dates for the development of the standard? For 802.11ac, please use the link http://grouper.ieee.org/groups/802/11/Reports/tgac_update.htm (accessed Feb. 2016).
2. What are some major milestones in the timeline for the development of the standard?
3. Provide evidence of the attention to detail in the development of the standard. For the 802.11ac standard, download the comments spreadsheet at the timeline and briefly describe one comment and how it was resolved.

CRITICAL-THINKING EXERCISE—Patents

Critics have said that Samuel Morse's patent for the idea of sending messages electronically across great distances could have been a patent on the web. Do you agree? Why or why not?

CRITICAL-THINKING REFERENCES

1. Crovitz, L. Gordon. "Could Morse have patented the web?" *Wall Street Journal*, Mar. 25, 2012.
2. Mossoff, Adam. O'Reilly v. Morse. George Mason University School of Law, https://en.wikipedia.org/wiki/O%27Reilly_v._Morse.

IT INFRASTRUCTURE DESIGN EXERCISE—Hire Yourself

Your network is ready to use in most respects. To maintain the network, TrendyWidgets has decided to hire a network administrator. Answer the following:

1. Create a job ad that TrendyWidgets can use to attract the right individual to the job. In the ad, include all the relevant information to maximize the likelihood that the most suitable individuals will be attracted to the opening. A good place to start is to look at current job openings for people with the CCIE certification (this is the premier certification offered by Cisco, one of the leading vendors of networking equipment). Select one such job advertisement and adapt it to TrendyWidgets's context, trying to remain as vendor-neutral as possible.
2. Recommend three media outlets to advertise the opening.

Appendix: Networking Careers

I don't think of my life as a career. I do stuff. I respond to stuff. That's not a career—it's a life.

Steve Jobs, *Time*, April 12, 2010

Hopefully this book has gotten you interested in exploring opportunities in the IT sector that relate to networking. To help in your job search in this sector, a short summary of the jobs performed by networking professionals is provided here. There are many ways of classifying networking jobs. The approach used here parallels the chapter structure of the book. In general, each layer creates professional opportunities.

Many entrepreneurs get started with cabling. Schools and small and large businesses are constantly investing in installing and upgrading networks, which usually requires running Cat5 or Cat6 cabling through the structures. This job mostly involves using elbow grease and may be considered the most easily attainable job opportunity in IT.

Moving up to the data-link layer, small businesses start with buying a switch to set up an Ethernet LAN to share an Internet connection, and hardware such as printers, among the computers in the organization. Most such organizations need someone with at least rudimentary expertise to set up equipment, troubleshoot problems with network connectivity, isolate faults, and generally ensure that the network is operating without problems. A popular certification for professionals who attend to these networks is offered by Cisco, and is called CCNA (Cisco Certified Network Associate).

When organizations are successful, they grow larger and establish presence in multiple locations. Each of these locations typically has a LAN. Routers are used to exchange traffic between these LANs. Layer-3 (network layer) technologies form the core of the network in these organizations. The organization now interfaces with telecom carriers and needs to select suitable network technologies that will provide the most cost-effective solution for its data transport needs. The organization now needs access to expertise in operating and maintaining routers and designing and managing the network. One of the certifications that professionals seek to obtain to demonstrate expertise in this area is the CCIE (Cisco Certified Internet Expert).

Transport layer details are generally hidden from end users, but two categories of professionals are interested in transport layer details: network-security administrators and application developers. Network-security professionals are very interested in ensuring that there are no unnecessary ports open on computers on the network. Unnecessarily open ports represent applications that may expose software vulnerabilities. Security professionals routinely scan computers on the organization's network to ensure that all non-essential computer ports are closed. Web-application developers are interested in developing session-aware applications to improve the customer experience. Most high-level applications-development frameworks such as .NET and JSF support the development of session-aware applications.

The application layer has revolutionized business and has created professional opportunities even in non-technical functional areas of the business. As an example, search-engine optimizers help marketing and are widely quoted as an example of a completely new category of professionals that did not exist a decade ago, and would not have existed were it not for the Internet.

A lot of the day-to-day network-operations work involves the support functions. Network professionals who maintain the WAN typically also maintain network services such as DHCP and DNS. These professionals also periodically are involved with carving out new subnets to accommodate organizational growth.

More and more organizations are using wireless LANs instead of wired Ethernet LANs. Network technicians who maintain LANs also typically maintain the wireless LANs. In larger organizations, the creation of a wireless network is typically accompanied by the creation of new subnets and NAT domains to accommodate wireless users.

WANs are almost exclusively operated by the telecom carriers. These organizations (e.g. AT&T, Verizon, and Comcast) specialize in long-distance data transport and employ large cadres of network experts to manage all parts of their networks. These experts are responsible for routing and optimizing the paths taken by packets through their networks. These experts also design networks to accommodate traffic growth created by video and cell phones (recall the statistics from Chapter 1).

Landline phones are now a declining business. But cell phone networks are growing rapidly. Operators of these networks need experts to keep the networks running even as traffic volumes grow rapidly. Also, network security is experiencing rapid growth. These professionals ensure that the organization complies with information-security requirements specified by law. They also deploy solutions such as single sign-on that simplify network security.

Finally, all keen IT professionals have the opportunity to provide their input on network standards through voting memberships in organizations such as IEEE.

About the Colophon

Here, Steve Jobs reminds us that, ultimately, a career is only one component of a larger expedition—life.

Glossary

Proper words in proper places, make the true definition of a style.

Jonathan Swift

Chapter 1: Introduction

- **Business data communications:** The movement of information from one computer application on one computer to another application on another computer by means of electrical or optical transmission systems.
- **Data network:** A transmission system that enables computer networking.
- **Multiplexing:** The ability to combine multiple channels of information on a common transmission medium.
- **Switching:** Transmitting data between selected points in a circuit.
- **Packet switching:** The process of routing data using addressed packets so that a channel is occupied only during the transmission of the packet.
- **Circuit:** An electronic closed-loop path among two or more points for signal transfer.
- **Routers:** Devices used to interconnect two or more networks.
- **Packetization:** The process of breaking down user data into small segments and packaging these segments appropriately so that they can be delivered and reassembled across the network.
- **Burstiness:** Short periods in which large volumes of data are uploaded or downloaded by the user followed by long periods of minimal activity.
- **Layering:** The practice of arranging functionality of components in a system in a hierarchical manner such that lower layers provide functions and services that support the functions and services of higher layers.
- **Hop:** Each link from one router to the next router.
- **OSI model:** A logical structure for communications networks standardized by the International Organization for Standardization (ISO).
- **Protocols:** A set of rules that permit information systems (ISs) to exchange information with each other.

Chapter 2: Physical Layer

- **Signaling:** Providing transparent transmission of a bit stream over a circuit built from some physical communications medium.
- **Signals:** Detectable transmitted energy that can be used to carry information.
- **Physical medium:** The transmission path over which a signal propagates.
- **Optical fiber:** A thin strand of glass that guides light along its length.
- **Data:** Numbers, letters, or other representation of information that can be processed by people or machines.

- **Digital signals:** Signals in which discrete steps are used to represent information.
- **Bit period:** The amount of time required to transmit one bit of data.
- **Analog signals:** Signals that have a continuous nature rather than a pulsed or discrete nature.
- **Noise:** Any disturbance that interferes with the normal operation of a device.
- **Bit:** Unit of information that designates one of two possible states of anything that conveys information.
- **Coding:** The transformation of elements of one set to elements of another set.
- **Smart grid:** The system that delivers electricity from suppliers to consumers using digital technology to save energy, reduce cost, and increase reliability and transparency.

Chapter 3: Data-link Layer

- **Broadcasting:** The transmission of signals that may be simultaneously received by stations that usually make no acknowledgement.
- **Collision:** The situation that occurs when two or more demands are made simultaneously on a system that can only handle one demand at any given instant.
- **Medium access control:** The method used to determine who gets to send data over a shared medium.
- **Multiple access:** A scheme that gives more than one computer access to the network for the purpose of transmitting information.
- **Carrier sensing:** An ongoing activity of a data station in a multiple access network to detect whether another station is transmitting.
- **Collision detection:** The requirement that a transmitting computer that detects another signal while transmitting data, stops transmitting that data.
- **Cyclic redundancy check (CRC):** An error-checking algorithm that checks data integrity by computing a polynomial algorithm-based checksum.

Chapter 4: Network Layer

- **Routing:** The process of selecting a path on the Internet that can be used to deliver data to a destination.
- **Best-effort delivery:** A network service in which the network does not provide any guarantee that data will be delivered.

Chapter 5: Transport Layer

- **Transmission control protocol (TCP):** A highly reliable host-to-host transport-layer protocol over packet-switched networks.
- **Flow control:** Slowing the transmission rate by a sender terminal so that the data can be received conveniently by the receiver.
- **Window size:** The amount of data the receiver is capable of processing.
- **Sliding-window flow control:** A flow-control mechanism that uses a variable-length window to specify the number of data units that can be transmitted before an acknowledgement is received.
- **Process:** A program and associated data ready to be executed by the computer.

Chapter 6: Application Layer

- **HTTP:** The protocol that facilitates the transfer of files between local and remote systems on the World Wide Web.
- **An internet:** Any interconnection among or between computer networks.
- **The Internet:** A worldwide interconnection of individual computer networks that provide access to all other users on the Internet.
- **World Wide Web (web):** The information system that displays pages containing hypertext, graphics, and audio and video content from computers located anywhere around the globe. The web is the part of the Internet where information is accessed using HTTP.
- **Search advertising:** The placement of advertisements on web search results.
- **Inlinks:** Links on other web pages that point to a page.
- **Hypertext:** Text that includes navigable links to other hypertext.
- **Uniform resource locator (URL):** A character string describing the location and access method of a resource on the Internet.
- **E-mail:** An electronic means for communication in which information—including text, graphics, and sound—is sent, stored, processed, and received. Messages are held in storage until called for by the addressee.
- **Simple mail transfer protocol (SMTP):** The protocol used to transfer e-mail between mail servers.
- **Post office protocol (POP):** The protocol that allows a user to access a mailbox on an e-mail server and perform useful actions on the contents of the mailbox.
- **Internet message access protocol (IMAP):** The protocol that allows a client to access a mailbox on an e-mail server and manipulate messages located on the server as conveniently as they could be manipulated locally.
- **File transfer protocol (FTP):** The Internet protocol for transferring files from one computer to another, regardless of the hardware and software configurations of the two computers.
- **Instant messaging (IM):** An application-layer protocol that allows users to send short, quick messages to each other.
- **Presence:** The ability of users to subscribe to each other and be notified of changes in state such as being online or busy or away.

Chapter 7: Support Services

- **Dynamic host configuration protocol (DHCP):** A technology that enables automatic assignment and collection of IP addresses and other network configuration parameters of networked computers.
- **Lease time:** The duration for which an IP address is provided.
- **Network and port address translation (NAPT):** The method by which IP addresses are mapped from one address block to another, providing transparent routing to end hosts.
- **Address resolution protocol (ARP):** A protocol that dynamically determines the network-layer IP address associated with a data-link-layer physical hardware address.
- **Domain name system (DNS):** The set of databases that performs the correspondence between the domain name and its IP address.

Chapter 8: Routing

- **Subnetting:** A way of breaking down large blocks of IP addresses into smaller address blocks.
- **Subnets:** Logically visible subsections of a single Internet network.
- **Subnet mask:** A number that tells the host what bits in an IP address constitute the network ID and subnet ID of the network.

Chapter 9: Subnetting

- **Routing:** The process of selecting paths to move information across networks from the source network to the destination network.
- **Routers:** Devices used to interconnect two or more networks.
- **Autonomous system:** A collection of routers that fall under one administrative entity.
- **Routing table:** A collection of paths that can be reached from the router along with information about the path.
- **Routing protocols:** The mechanisms used by routers on the Internet to maintain routing tables.
- **Route aggregation:** The combination of two or more IP address blocks to one larger address block.
- **Multi-protocol label switching (MPLS):** A packet-forwarding mechanism that uses predefined labels to determine how to deliver packets.
- **Forwarding-equivalence class:** A group of IP packets that are forwarded in the same manner, for example, over the same path.

Chapter 10: Wide-area Networks

- **Wide-area networks (WANs):** Networks that provide data communications to a large number of independent users. These users are usually spread over a larger geographic area than a LAN.
- **Statistical multiplexing:** Allocating network resources according to need (the name arises because the need may typically be modeled using statistical distributions).
- **Virtual circuit:** A communications arrangement in which data from a source user may be passed to a destination user over various real circuits.

Chapter 11: Network Security

- **Information security**: Information security is the provision of confidentiality, integrity, and availability to information.
- **Confidentiality**: Confidentiality is to preserve authorized restrictions on information to protect personal privacy and proprietary information.
- **Integrity**: Integrity is to guard against improper modification or destruction of information and ensure authenticity of information.
- **Availability**: Availability is to ensure timely and reliable use of information.
- **Vulnerabilities**: Vulnerabilities are weaknesses in an information system that could be exploited by interested hackers to compromise the security of the information held in the information system.

- **Threats**: Threats are capabilities, intentions, and attack methods of adversaries to cause harm to information.
- **Controls**: Controls are measures taken to mitigate the dangers arising from information security threats.
- **Network security**: Network security is the provision of information security in the presence of dangers created by computer networks.
- **Patches**: When software weaknesses become known, developers quickly issue updates to fix problems. These updates are called patches, and the process of applying updates is called patching.
- **Authentication**: Authentication is the verification of a claimed identity.
- **Authorization**: Authorization is the granting of rights to a user to access, read, modify, insert, or delete certain data, or to execute certain programs.
- **Firewall**: A firewall is a computer that lies between two networks and regulates traffic between the networks in order to protect the internal network from electronic attacks originating from the external network.
- **Access list**: An access list is a list of permissions associated with specified objects.
- **Demilitarized zone**: The DMZ is a network that contains the organization's external services and connects them to the Internet.
- **Denial-of-service attack**: A denial-of-service (DOS) attack is when an attacker consumes the resources on a computer or network for things it was not intended to be doing, thus preventing normal use of the resources for legitimate purposes.
- **Encryption**: Encryption is the process of rendering plain information unintelligible in such a manner that it may later be restored to intelligible form. The output from encryption is called ciphertext. Decryption is the process of converting ciphertext to plaintext.
- **Encryption algorithm**: An encryption algorithm is a mathematically expressed process to create ciphertext.
- **Encryption key**: An encryption key is a sequence of symbols that controls the operation of the encryption algorithm.

Chapter 12: Computing Infrastructures

- **System call**: A system call is an invocation by a user program of a function provided by the operating system.
- **Process**: A process is a computer program that has been loaded into memory and is executing.
- **Threads**: Threads are the smallest sequence of instructions that an operating system can manage independently.
- **Memory**: Memory is the storehouse of data on the computer that can be quickly and directly accessed by both the CPU and the I/O (input/output) devices on a computer
- **Virtual memory**: Virtual memory is a technique to take up space on the hard drive and use it as an extension of the main memory.
- **File system**: The file system is the component of the operating system that provides efficient and convenient access to storage devices by allowing data to be stored, located, and retrieved easily.

- **API**: An application programming interface (API) is a formalized set of software function calls that can be referenced by user programs in order to access supporting services.
- **Big data**: Big data is the general term used to describe data that is difficult to process using traditional database management systems.
- **Warehouse scale computer**: A WSC is a set of many (tens of thousands) hardware and software resources working in concert to appear as one large computer that efficiently delivers desired levels of Internet service performance.
- **Cloud computing**: Cloud computing is a model for enabling ubiquitous, convenient, on-demand network access to a shared pool of configurable computing resources (e.g. networks, servers, storage, applications, and services) that can be rapidly provisioned and released with minimal management effort or service provider interaction.

Chapter 13: Services Delivery

- **IT services management**: IT services management (ITSM) is the set of activities performed by an organization to deliver IT services to end users.
- **Service level management**: Service level management refers to the maintenance of a catalog of all services offered for performance, together with binding agreements with both the provider and the customer.
- **SLA**: SLAs are output-based contracts between customers and providers that describe the service the customer can expect to receive in return for the fees paid.
- **Capacity management**: Capacity management is the managerial discipline for maintaining IT infrastructure at the right size to meet current and anticipated business needs in a cost-effective manner.
- **Continuity management**: Continuity management is the discipline of planning to ensure that IT services can recover and continue should a serious unexpected incident occur.
- **Availability management**: Availability management is the discipline of ensuring that resources such as IT infrastructure and personnel are appropriate for meeting the service-level agreements in place.
- **IT financial management**: IT financial management is the discipline of maintaining accurate accounting of IT services and using this information to deliver IT services in the most cost-effective manner possible.
- **High availability**: High availability is the ability of a system to remain operational for a duration significantly higher than normal.
- **Cluster**: A cluster is a collection of servers (called nodes), any of which can run the workloads common to the cluster.
- **Database mirroring**: Database mirroring is a software solution for providing almost instantaneous failover with no loss of committed data.
- **Database replication**: Database replication allows two or more database servers to stay "in sync" so that the secondary servers can answer queries and potentially actually change data.
- **Stateless transactions**: Stateless transactions are self-contained transactions, requiring no awareness by the server of the prior history of transactions.
- **Stateful workloads**: Stateful workloads require awareness of the history of the transaction to complete successfully.

- **Business continuity planning**: Business continuity planning is the methodology used to create and validate a plan for maintaining continuous business operations before, during, and after disruptive events.
- **Disaster recovery**: Disaster recovery is the set of procedures used to restore technology services that were disrupted during an extreme event.

Chapter 14: Managerial Issues

- **Design:** Detailed description of a product or service.
- **Logical network design:** A representation of the layout of the network, the subnets in the network, IP addressing and naming schemes used in the network, and management strategies used in the network.
- **Physical network design:** A representation that indicates the technologies (for example, copper/fiber media for cabling; switches; routers; and data-link-layer technologies such as Ethernet) that will be used to implement the logical design.
- **Building network:** A network that connects devices located within a single building.
- **Campus network:** A network that spans multiple buildings.
- **Core layer:** Network components that are connected to all parts of the campus network and are responsible for fast and reliable transportation of data across the different parts of the network.
- **Distribution layer:** The administrative layer of the network.
- **Access layer:** Network components responsible for providing network ports to end users.
- **Enterprise network:** One organization's geographically scattered network.
- **Network maintenance:** The set of activities performed to keep networks in a serviceable condition or to restore them to serviceability. It includes activities such as inspection, testing, and servicing.
- **Simple network management protocol (SNMP):** A protocol used to manage and control IP (Internet protocol) devices.
- **Standards:** Documents, established by consensus and approved by a recognized body, that provide rules, guidelines, or characteristics aimed at achieving order in a given context.
- **Teleworking:** Any substitution of information technologies for work-related travel.
- **Telemedicine:** The application of computer networks to transfer medical information through the phone or Internet for medical consultation, and even to remotely provide medical examinations and procedures.

Supplementary Chapter 15: Wireless (eTextbook only)

- **Wireless networks**: Wireless networks are computer networks that use the ISM wireless frequency bands for signal transmission.
- **ISM frequencies**: ISM frequencies or ISM bands are radio frequencies available internationally for free use for industrial, scientific, and medical applications.
- **Wireless access point**: A wireless access point is a device that allows wireless hosts to connect to a wired network using wireless LAN technologies such as Wi-Fi.

- **Basic service area**: The area covered by an access point is called a basic service area (BSA).
- **Basic service set**: The basic service area and the access point covering that area together are called a basic service set (BSS).
- **Personal area networks**: Personal-area networks are computer networks designed for data transmission among devices in close proximity, typically owned by the same individual.
- **Piconet**: A piconet is a collection of devices connected to each other using Bluetooth.
- **Wireless metropolitan-area network**: A wireless metropolitan-area network (MAN) is a moderately high-speed computer network that usually spans a city or large enterprise campus.

Supplementary Chapter 16: Phone Networks (eTextbook only)

- **Circuit switching**: A process that, on demand, connects two or more communicating devices and permits the exclusive use of a data circuit between them until the connection is released.
- **Local loop**: A circuit from the customer premises to the last switch of the phone company's network.
- **End office (central office)**: The location where the phone company operates equipment that is responsible for providing the customer's dial tone.
- **Inter-exchange carriers (IXCs)**: Networks that carry traffic between end offices.
- **Hertz**: A unit of frequency that is equivalent to one cycle per second.
- **Digital subscriber line (DSL)**: Technology that provides full-duplex service on a single, twisted, metallic pair of phone wires at a rate sufficient to support basic high-speed data service.
- **Cellular telephony**: A mobile communications system that uses a combination of radio transmission and conventional telephone switching to permit mobile users within a specified area to access full-duplex telephone service.
- **Mobile telephony switching office (MTSO)**: The switching office that connects all the individual cell towers to the Central Office (CO) and switches conversations to the cell towers with the best possible reception.
- **Handoff**: The process of transferring a phone call in progress from one cell transmitter and frequency pair to another cell transmitter and receiver, using a different frequency pair without interruption of the call.
- **Code division multiple access (CDMA)**: A coding scheme in which multiple channels are independently coded for transmission over a single wideband channel.

Index

Note: Page numbers higher than 456 refer to the two supplementary chapters available only in the eTextbook.

Access layer, 440
Access list, 355
ACID (atomicity, consistency, isolation, and durability), 398
ACK field, 179
Acknowledgment number, 181
ACM (Association for Computing Machinery), 290
Active replication, 432
Address resolution protocol (ARP), 247–249, 517
Addresses
- access-point, 467
- Ethernet, 102–106
- leasing, 238
- multi-part, 135–139
- vs. names, 90, 126
- non-routable, 242–243
- *See also* IP addresses; MAC addresses

Advanced Encryption Standard (AES), 362–363
Advertising, online, 223–224
AdWords, 200
Aerotropolis, 289–290
AES (Advanced Encryption Standard), 362–363
Africa
- mobile telephony in, 506
- Wi-Fi in, 505

Aggregation, 17
Agriculture, use of computer networks in, 109–110
Akamai content-delivery network, 293
Algorithms, 98, 184
- encryption, 361–363
- error-detecting, 98, 369
- Luhn's 136
- MD5, 370
- Nagle, 184
- SHA, 370
- spanning tree, 107–108
- for subnet masks, 317

Alibaba, 506
ALL CAPS, 225
Allen, Paul, 405
Alliance for Telecommunications Industry Solutions (ATIS), 41–42
Alphabet symbols, 43
Altair 8800 computer, 405–406
AM (amplitude modulation), 61–62
Amazon, 205, 257
Amazon AWS, 403
American National Standard, 41
American Standard Code for Information Interchange. *See* ASCII coding
Amplitude, 60–61
Amplitude modulation (AM), 61–62
Amplitude Shift Keying (ASK), 63, 80–82
Analog data, 73
Analog signals, 59, 60–62
Andreessen, Marc, 449
Anti-trust lawsuits, 492
Antivirus software, 359
AOL instant messaging, 222. *See also* Instant messaging
Apache Web Server, 263
API (application programming interface), 392
Application layer, 193–194
- diagram, 195

Application programming interface (API), 392
Application server high-availability architecture, 423–424
ARP (address resolution protocol), 247–249, 517
- cache, 248
- packets exchanged, 248
- sequence of operations, 247

ARPANET 448
AS (autonomous systems), 273–275
ASCII coding, 8–9, 68–70, 364
ASK (Amplitude Shift Keying), 63, 80–82
Association for Computing Machinery (ACM), 290
Asymmetric key encryption, 363–365, 369
Asynchronous transfer mode (ATM), 336–337
AT&T, 491, 502, 506
- company split, 449–550

ATIS (Alliance for Telecommunications Industry Solutions), 41–42
ATM (Asynchronous Transfer Mode), 336–337

Authentication, 353–355
Authorization, 354–355
Automatic allocation, 240
Autonomous systems (AS), 273–275
Availability, 347, 348, 352, 358–360, 371–372
Availability management, 417
Azure 403

Backlinks, 200
Backus, John, 380
Bandwidth
data rates and, 66
efficient utilization of, 57–58
Barger, Jorn, 199
Barksdale, Jim, 185
BASIC, 405–406
Basic service area (BSA), 464
Basic service set (BSS), 464
Batteries, 58
BC/DR. *See* Business continuity and disaster recovery
BCP (business continuity planning), 427–428
Bell, Alexander Graham, 491, 492, 507–508
Berners-Lee, Tim, 198
Best-effort delivery, 121–122
Bezos, Jeff, 205
BGP (Border Gateway Protocol), 280–281
BGPlay, 274, 286, 296–297
Bhushan, Abhay, 220
Big data, 394–395
Bin Laden, Osama, 44
Bina, Eric, 449
Binary digits (bits) 65–66, 66n17, 128
Binary numbers, 64, 127–129
conversion to and from decimal 129–131
Binary representations, 69
Binary signals, 63–66
BIND, 258–259
Bit periods, 60
Bit Torrent, 219
Bits (binary digits), 65–66, 66n17, 128
Blogs, 199
Bluetooth, 180, 462, 471–472, 475
categories of, 476
Bluetooth architecture, 472–473
Bluetooth cards, 113–114
Bluetooth device discovery, 474–475
Bluetooth frame structure, 473–474
Boggs, David, 111
Boot-up, 393–394
Border Gateway Protocol (BGP), 280–281
Broadband signals, 494
Broadcast mechanism, vs. broadcast address, 92
Broadcasting, 89–90, 270
BSA (basic service area), 464
BSD UNIX, 120–121
BSS (basic service set), 464
Building network, 438–439
Burstiness, 16, 332–333
Business continuity, 427–429
Business continuity and disaster recovery (BC/DR) planning, 428–429
plan maintenance, 429
requirements for, 429
Business continuity planning (BCP), 427–428
Business data communications, 1. *See also* computer networking
Business impact analysis, 419
Bytes, 66

Cable TV, 494. *See also* Television
Cables
Cat3, 47–48, 495
Cat5, 47–48, 51, 67, 513
Cat5e, 48–50, 51
Cat6, 49, 513
Cat7, 49
connectors for, 49
optical fiber, 45–46, 51–55
RJ45 jacks, 49–50
shielded, 47
straight-through vs. crossover, 51
submarine, 53
terminating, 48
UTP (unshielded twisted pair), 47–49
Campus network, 438
Campus-wide wireless LAN, 465
Capacity management, 416
Careers, in computer networking, 5–6, 513–514
Carrefour, 149
Carrier frequencies, 72
Carrier pigeons, 329
Carrier sense and sensing, 92, 516
Carrier-sense multiple access with collision detection (CSMA/CD), 85, 91–95, 463
Carriers, value of, 82
Cash-mobs, 401
CCMP (Counter-mode with cipher-block chaining message authentication code protocol), 367
CDMA (code division multiple access), 499, 504–505, 507, 510–511
Cell phone networks, 506–507, 514. *See also* Telephone networks
Cell phone technology
second-generation (2G), 497, 499
third-generation (3G), 497–499
fourth-generation (4G; 4G LTE), 499
evolution of, 497–499

system architecture, 499–500
Cell phone towers, 58, 500, 501, 502–3
accidents, 501
locations, 500, 502–3
Cell phones, 58, 496.
expenses, 3–4
frequency reuse, 500–502
and global development, 505–506
See also Telephones
Cellular telephony, 496. *See also* Cell phone technology; Cell phones
Celtel, 505
Censorship, 281, 297
Central Index Key (CIK), 221
Central office (CO), 489, 503
Chaos Monkey, 430
Chat, 208
Checksum
IP header, 124, 145
TCP, 167, 181–182
UDP, 183
See also Cyclic Redundancy Check (CRC)
Chevron, 479–481
Chilean mining incident, 46
China, mobile telephony in, 506
China Telecom, 280
CIDR (classless inter-domain routing), 141–143, 152
CIK (Central index key), 221
CIOs, 6
Ciphertext, 361
Circuit switching, 13–14, 488
vs. packet switching, 16–18
Circuits, 334
virtual, 333–336, 339, 342
Clarke, Arthur C., 31
Classless inter-domain routing (CIDR), 141–143, 152
Cloud computing, 401–405
access, 402–403
characteristics, 402
implementation, 404–405
security concerns, 403–404
services, 403
usage at Animoto, 406
Clusters, 423
failover, 426–427
multi-site, 427
CNN, strategies post-9/11, 293
CO (Central office), 489, 503
COBIT (Control Objectives for IT), 414
Code division multiple access (CDMA), 499, 504–505, 507, 510–511
Code Red virus, 359
Code talkers, 361
Coding, 8, 68–69
Collisions, 92–93
avoidance of, 463
detection of, 92–93
Command line interface (CLI), 448
Common vulnerabilities and exposures (CVE) database, 444
Communication
global, 52
as killer app, 207, 208
Communication sockets, 168. *See also* Sockets
Community cloud, 403. *See also* Cloud computing
Competitive local exchange carriers (CLECs), 494
Compounding, 132
Computer networking, 1–2
dial-up, 328–329
related careers, 5–6, 513–514
utility of, 3–5, 30
wireless, 457–568
See also Business data communications
Computer networks, 29–30
Computer Science and Telecommunication Board (CSTB), 290
Computer software. *See* Software
Computers
desktop, 2
nano scale, 411
warehouse scale (WSCs), 394, 397–401, 405
Computing infrastructures, 379
cloud computing, 401–405
operating system, 381–394
origins, 379–381
VMs (virtual machines), 394, 395–397
WSCs (warehouse scale computers), 394, 397–401
Confidentiality, 347, 348, 352–355, 360–369
Confidentiality, integrity, and availability (CIA), 348, 352
Connection establishment, 177–180
ConstantContact, 371
Continuity management, 416–417
Control bits, 181
Control objectives for IT (COBIT), 414
Controls, 348, 350–351
authentication and authorization, 353–354
for availability, 358–360, 371–372
for confidentiality, 352–355, 360–369
denial-of-service (DOS) attacks, 360
firewalls, 355–358
for incoming information, 352–360
for integrity, 355–358, 369–370
for outgoing information, 360–372
patching, 353
physical, 350–351
procedural, 350–351

technical, 350–351
viruses and worms, 358–359
Copper, and the S&P 500, 45
Copper wire, 43–51, 76
compared to optical fiber, 55
Core layer, 439
Counter-mode with Cipher-block chaining message authentication code protocol (CCMP), 367
Craig's List, 223
CRC (Cyclic redundancy check), 95–101
Credit card numbers, 136
Critical-thinking exercises
Broadcast and search, 114–115
Flow control of distractions, 190
Genetic code, 159
Identifying threats, 377
Internet censorship, 297
Nano scale computers, 411
Nissan Computer Corp., 267
Other Three Billionaires, 512
Patents, 456
Personal high availability, 432
Power of universal exchange formats, 37
Professional WAN, 345
Protesting SOPA, 233–234
Subnet design, 323
Value of carriers, 82
Wi-Fi, 485
Crocker, Steve, 121
Cross-docking, 150
CSMA/CD (carrier-sense multiple access with collision detection), 85, 91–95, 463
advantages and disadvantages of, 94–95
CSTB (Computer Science and Telecommunication Board), 290
Curie, Marie, 460
Cybersecurity 4, 372–374
Cyclic redundancy check (CRC), 95–99
example, 99–101

Data
analog, 73
binary representation of, 69
coding, 68–70
decoding, 70
defined, 56
digital, 73
estimating requirements for, 190–191
meta-, 96–97, 96n10
vs. signals, 56
transmission and reception of using signals, 67–70, 73–75
Data centers, in Scandinavia, 401
Data communications timeline, 12
Data Encryption Standard (DES), 362
Data network technologies, 2
Data networks 1
Data offset, 181
Data packets, 14
compared to letters, 14–15
names for, 164
Data rates
and bandwidth, 66
for 802.16, 477
Data storage, 372, 391
Data transfer
bidirectional 178
via Ethernet 87–89
three-way handshake, 179
Data transmission using signals
coding data, 68–69
decoding data 70
demodulating signal, 70, 74–75
modulating carrier, 69–70, 73
Database mirroring, 425
Database server high-availability features, 424–425
Datagrams, 164, 166–167
Data-link layer, 110–111, 125
CSMA/CD, 91–95
cyclic redundancy check (CRC), 98–101
error detection and correction, 95–97
Ethernet, 85–91
Ethernet addresses, 103–105
Ethernet frame structure, 101–103
Ethernet LANs, 108
functions of, 83–85
overview, 83
spanning tree protocol, 107–108
switched Ethernet, 106–107
DC power, 51
Decimal numbers, 129
conversion to and from binary, 129–131
Demilitarized zone, 356, 357
Demodulating signal, 70, 74–75
Denial-of-service (DOS) attacks, 360
Denning, Peter J., 392
Dense wavelength division multiplexing (DWDM), 325, 328, 338
Department of Defense (DoD), 448
Depository Trust and Clearing Corporation (DTCC), 184
Design, 435
Device discovery, 474–475
Device uptime, 431
DHCP (dynamic host configuration protocol), 235–238
allocation schemes, 240–241
basic operation, 238–240

configuration, 242
and non-routable addresses, 242–243
operation timeline, 239
sample server-configuration file, 241
DHCP lease-time, 238
Dial-up networking, 328–329
Differential quadrature-phase shift keying (DQPSK), 78
Digital data, 73
Digital living network alliance (DLNA), 470
Digital projection standards, 447
Digital signals, 59–60
impact of noise on, 64
Digital subscriber line (DSL), 491, 494–495, 506
Disaster recovery, 427–429
Disaster response, 290
9/11, 290–292
Hurricane Katrina, 290
Internet traffic patterns on 9/11, 292–293
Distance-vector protocols, 281
Distribution layer, 439
DMZ (demilitarized zone), 356–357
DNS (domain name service/system), 249–250
and ipconfig, 257
lookup, 253–254
and the phone system, 256
recursive query resolution, 254
and router configuration error, 258
sample BIND DNS server configuration file, 259
structure, 251
tracing a query, 256
typical query, 255
and virtual hosts, 261–264
DNS poisoning, 251
DoD (Department of Defense), 448
Domain name hierarchy, 252
Domain name service. *See* DNS
Domain names, caching of, 259
Domain separation, 392
Domains, 250
open, 252
top-level, 251
See also DNS (domain name service/system)
Domino's Pizza, 31–33
Domke, Scott, 290
DOS (denial-of-service) attacks, 360
Dotted decimal notation, 133
Downtime, 418–419
cost of, 419–420
Doyle, Patrick, 33
DQPSK (Differential quadrature-phase shift keying), 78
Driverless cars, 114
DSL (digital subscriber line), 491, 494–495, 506
DS-signals, 329–330
DTCC (Depository Trust and Clearing Corporation), 184
Dual code operation, 383–384
Duration/ID, 466
DWDM (dense wavelength division multiplexing), 325, 328, 338
Dynamic allocation, 240–241
Dynamic hardware partitioning, 422
Dynamic host configuration protocol. *See* DHCP

Eckert, J. Presper, 379
Economic development, and the telecommunications infrastructure, 4
EDGE (enhanced data rates for global evolution), 499
Edison, Thomas, 6, 9
Edwards, Douglas, 200
Electric power infrastructure, 77
Electronic dead drops, 218
Electronic Numerical Integrator and Calculator (ENIAC), 379, 381
E-mail, 2, 207–208
as communication medium, 208–210
outages at USF, 218
"push," 208
suspicious, 359
system architecture, 211
E-mail etiquette, 225–226
E-mail protocols, 211–212
IMAP (Internet message access protocol), 212, 216–218
POP (post office protocol), 212, 214–215
SMTP, 211–214
Encryption, 361, 376–377
asymmetric key, 363–365
example, 378
in practice, 365–366
SSH (secure shell), 366–369
symmetric key, 362–363
in wireless networks, 367
Encryption algorithms, 362
Encryption keys, 362
End office (central office), 489
Enhanced data rates for Global Evolution (EDGE), 499
ENIAC (Electronic Numerical Integrator and Calculator) machine, 379, 381
Enterprise network, 440
Ericsson, 472
Error control 20
Error correction, 84, 100
example, 97

Error detection, 84, 95–97
  cycle redundancy check (CRC), 95–99
  in LANs, 464
Ethernet, 8, 24, 27, 51, 85–91
  addresses, 102–106
  advantages and disadvantages of using CSMA/CD in, 94–95
  broadcast in, 89–91
  connections, 84–85
  and CSMA/CD, 91–94
  development of, 85–87, 111
  early diagram, 86
  error detection in, 95–101
  frame structure, 101–103
  hub-based, 87
  in LANs, 108
  naming, 107
  networks using, 85–87
  operation of, 87–89
  packet receipt in, 91
  as part of larger network, 109–110
  spanning tree protocol, 107–108
  switched, 106–107
  transmitter-receiver, 87
  vs. wireless LANs, 462–463
Ethernet ports, 84
Extended Carrier Route Walking Sequence Saturation (ECRWSS), 90–91
Extended service set, 464–465

Failover, 298
Failover clusters and clustering, 425–427
Fault-tolerant servers, 422, 431
Favicons, 262, 266
FCC (Federal Communications Commission), 459, 493
FCC spectrum auctions, 59
FCS (frame-check sequence) field, 101
FDM (frequency-division multiplexing), 71, 504–505
FDM (frequency-division multiplexing) WANs, 337–338
Federal Communications Commission (FCC), 459, 493
Feinler, Elizabeth, 251
Feynman, Richard, 460
File system, 391
File Transfer Protocol (FTP), 193, 195, 219–221
Filezilla, 219
Financial industry, 183–184
  latency in, 184–185l
Firefox, 198
Firewalls, 240, 355–358
Flags, 216
Flaming, 226
Flat-pack methodology, 16
Flickr, 220
Flow control, 174–176
  sliding-window, 177–178
  stop-and-wait, 175–176
  stop-and-wait with ISN, 178
  window size, 176–177
Ford, Henry, 321
Fortran, 380
Forwarding-evidence class, 287–288
Frame control, 466
Frame Relay, 336
Frame-check sequence (FCS) field, 101
Franklin, Rosalind, 460
Frequency, 60–61
Frequency allocation chart, 72
Frequency hopping, 473
Frequency reuse, 500–502, 507
Frequency-division multiplexing (FDM), 71, 504–505
Frequency-division multiplexing (FDM) WANs, 337–338
FTP (file transfer protocol), 193, 195, 219–221
  financial sector example, 221
  vs. HTTP, 220
  and SMB, 221

Gates, Bill, 86n3, 415
General information risk management model, 349
General Packet Radio Service (GPRS), 499
Glass, David, 150
Global Crossing, 416
Global development, cell phones and, 505–506
Gonzalez, Albert 351, 373
Google, 200, 223, 289, 371, 399–401
Google Ads, 222–224
Google Glass, 451
Government involvement
  in the AT&T split, 449
  development of packetization, 448
  development of TCP/IP, 448
  development of the web browser, 448–449
  early internet, 448
  patents, 450
  in the wireless spectrum, 449
GPRS (general packet radio service), 499
Graphical web browser, 449
The Grid, 77
Groupon, 209
Guzman, Joaquin “El Chapo,” 44

Hackers, 351, 352–353, 355, 358, 458
Handoff, 504
Hands-on exercises
  AirPCap Wireshark Captures, 483–485

Amplitude Shift Keying (ASK), 80–82
bgplay, 296–2987
CDMA, 510–511
Device Uptime 431
https, 376–377
ipconfig and ping 154–158
netstat, 188–190
nslookup, 266–267
OUI Lookup 112–114
perfmon, 407–411
standards development, 456
Subnet Mask, 322–323
Traceroute, 35–37
Web Page Debugging, 344
Wireshark, 227–233
Hardware maintenance, 441–443
Header information, 14, 21, 22
Headley, David, 218
Heartland Payment Systems, 351, 352, 353, 373
Hertz (Hz), 67, 490
Hexadecimal notation, 105
Hibernian Express submarine cable, 185
High availability
database server features, 424–425
personal, 432
High-availability architectures, 420–421
at the application/middleware level, 423–425
at the hardware level, 421–423
at the operating systems level, 426–427
web application, 425
High-availability concepts, 417–418
High-performance-computing (HPC) systems, 397–398
Home networking, 259–261
Home PC IP configuration, 260
Home router security settings, 368
Homegrid/ITU G.hn, 78
Homeplug, 78
Hops, 21
HTML (Hypertext Markup Language), 198, 200–201
HTML 5, 201
HTTP (hypertext transfer protocol), 193–205
vs FTP, 220
GET command, 203
POST command, 203
HTTPS, 376–377
Hungarian notation, 86n3
Hurricane Katrina, 290
Hybrid cloud, 403
Hypertext, 201
Hypertext Markup Language (HTML), 198, 200–201
Hypertext Transfer Protocol. *See* HTTP
IaaS (Information as a Service), 403
IANA (Internet Assigned Numbers Authority), 143, 171, 251, 252
IBM, 382
Ibrahim, Mo, 505
ICMP (Internet Control Message Protocol), 156
IDEA (International Data Encryption Algorithm), 362
IEEE (Institute for Electrical and Electronics Engineers), 445–446
IEEE standards, 78, 462, 464, 476–477
IETF (Internet Engineering Task Force), 445–446
IIS (Internet Information Server), 199
IM (instant messaging), 193, 221–222, 292
IMAP (Internet message access protocol), 212, 216–218
Incumbents local exchange carriers (ILECs), 494
Information, 56
Information-exchange industry, 4
Information redundancy, and multitasking, 95
Information security, 347–348. *See also* Network security
Infrastructure as a Service (IaaS), 403
Initial sequence number (ISN), 178–180
Inlinks, 200
Innovation, balancing with regulation, 450
Instant messaging (IM), 193, 221–222, 292
Institute for Electrical and Electronics Engineers (IEEE), 78, 445–446, 462, 464, 476–477
Integrated services digital network (ISDN), 331
Integrity, 347–348, 352, 355–358, 369–370
Intelsat, 31
Interexchange carriers (IXCs), 489–490
International Data Encryption Algorithm (IDEA), 362
International Organization for Standardization (IOS), 25
International Standards Organization (ISO), 24–25
International Telecommunication Union (ITU), 445, 497
Internet, 30–31
early development, 448
estimated size of, 198
and Hurricane Katrina, 290
vs. internet, 196
and 9/11, 290–293
and the print media, 222–224
Internet Assigned Numbers Authority (IANA), 143, 171, 251, 252
Internet censorship, 281, 297
Internet Control Message Protocol (ICMP), 156
Internet Engineering Task Force (IETF), 445–446
Internet Explorer, 198
Internet Information Server (IIS), 199

Internet message access protocol (IMAP), 212, 216–218
Internet of Things, 289
Internet Protocol (IP), 120–122. *See also* IP addresses; IP configuration; IP header
Internet protocols, principles of, 27–29
Internet service, 493
Internet Service Providers (ISPs), 29, 197, 487
Internet traffic, 4
Investigative journalism, 222
IOS (International Organization for Standardization), 26
IP (Internet Protocol), 120–122
IP addresses, 118–120, 126–127, 154
  classes of, 139–141
  and DHCP, 235–236
  dotted decimal notation for, 133
  of home network, 260–261
  vs. MAC addresses, 235
  non-routable, 242–244
  obtaining, 143–144
  special, 134–135
  three-part, 303–305
  structure of, 134–135
  and subnetting, 299–302
IP configuration, home PC, 260
IP header, 122
  flags, 124
  fragment offset, 124
  header checksum, 124
  header length, 123
  identification, 123
  options, 125
  padding, 125
  protocol, 124
  source and destination addresses, 125
  time to live, 124
  total length, 123
  type of service, 123
  version field, 122
IP traffic, 4–6
Ipconfig, 114, 154–155, 257, 260
IPSec, 366
IPv6, 120, 144–146, 237
  addressing example, 147–148
  auto-configuration, 242
  header fields, 146
  subnetting in, 318
I-root server, 258
ISM frequency bands, 459–462
ISO (International Standards Organization), 24–25
ISO 20000, 414
ISPs (Internet Service Providers), 29, 197
IT financial management, 417
IT Infrastructure Design Exercises. *See* TrendyWidgets
IT infrastructure library (ITIL), 414
IT infrastructures, 1
IT services management (ITSM), 413–415
  delivery background, 414–415
  frameworks, 414
ITIL (IT infrastructure library), 414
ITSM. *See* IT services management
ITU (International Telecommunications Union), 445, 497
IXCs (interexchange carriers), 489–490

Jacks, RJ45, 49–50
JavaScript, 344

Kamprad, Ingvar, 16
Kao, Charles Kuen, 54
Karmazin, Mel, 223
Kernel mode, 383
Kindleberger, Charles, 207
Kmart, 149, 150–151
Kresge, Sebastian S., 149

Landline phones, 499, 514
LANs (local area networks), 85, 513–514
  compared to road network, 326–328
  device discovery in, 474
  home LAN with wireless router, 260
  vs. WANs, 325–326
Latency, 184–185
Layering, 18–19, 30
  in organizations, 19
  and packetization, 20–21
  in software, 18–20
  summary of tasks, 25–26
Learning to read, 43
Lease-time, 238
Least privilege, 392–393
Length, 183
Licensed frequencies, 478
LIDAR (light radar), 114
Lincoln, Abraham, 446
LinkedIn network, 345
Links, 200
Link-state protocols, 281
Local area networks. *See* LANs
Logical network design, 435–437
Loops
  local, 489
  in networks, 108
Luhn's algorithm, 136
Lundgren, Gillis, 16

MAC (medium-access control), 92, 463
MAC addresses, 90, 107, 112, 126, 154
compared to IP addresses, 235
virtualization of, 104
MAC layer frame format, 466–467
Mail exchange (MX), 258
Mail-transfer agent (MTA), 211
Mail-user agent (MUA), 211
Maintenance, 50, 440–441
of network hardware, 441–443
of software, 443–444
Management information base (MIB), 442
Managerial issues, 433–434
government involvement and legal issues, 447–451
maintenance, 440–444
network design, 434–440
standards, 444–447
See also Supply-chain management
Manual allocation, 240
Marconi, Guglielmo, 459
Mauchly, John, 380
Medium access, 92
Medium access control (MAC), 92, 463. *See also* MAC addresses
Memory, 390–391
Merger and acquisitions (M&A), 502
Merholz, Peter, 199
Mesh network, 481
Message IDs, 216
Meta-data, 96–97, 96n10
Metcalfe, Bob, 85–86, 86n3, 111
Metro zones, 477
Metropolitan-area networks, 462
MH 370 disaster, 475
MIB (management information base), 442
Microsoft, 403, 406
Microwave ovens, 461
Mobile-telephone switching office (MTSO), 499, 503–504
Modulating carrier, 69–70, 73
Monaghan, James, 31
Monaghan, Thomas, 31
Monopolies, 491
Moore's law, 18
Morse, Samuel, 6, 450
Morse code, 6–9, 7n9, 8nn12–13
vs. text messaging, 9
Mozilla, 198
MPLS (multiprotocol label switching), 285–288
MSI Cellular Investments (Celtel), 505
MTA (mail-transfer agent), 211
MTC Kuwait, 505
MUA (mail-user agent), 211
Multi-home logistics, 298
Multi-part addresses, 135–139
Multiple access, 92, 93
Multiplexing, 9–11, 57, 70–75, 168–174
AM example, 73–75
frequency-division (FDM), 71, 504
with LANs, 463
in TCP, 168–174
and TCP ports, 167
time-division (TDM), 71, 504
in wide-area networks (WANs), 331–338
Multiprotocol label switching (MPLS), 285–288
Multi-site clusters, 427
Multitasking, and information redundancy, 95
MX (mail exchange), 258
MySQL, 207

Nagle algorithm, 184
Names, vs. addresses, 90
Nano scale computers, 411
Nanpa (North American Numbering Plan Administration), 241
NAPT (network address port translation), 244–247
and IP address assignment, 244
NAT (network address translation), 244, 245. *See also* NAPT
National Center for Supercomputing Applications (NCSA), 448–449
National Institute for Standards and Technology (NIST), 78, 362
National Science Foundation (NSF), 448–449
Near-field communication (NFC), 475
Netcraft, 199
Netflix, 430
Netscape Navigator, 198
Netstat, 172, 188–190
security use of, 172–174
Network address port translation (NAPT), 244–247
Network address translation (NAT), 244, 245. *See also* NAPT
Network design, 434
implementation, 437–440
maintenance, 440–441
physical design, 437, 438
requirements analysis, 435–437
Network File System (NFS), 221
Network interface card (NIC), 104
Network interfaces, 113
Network layer, 117
classless inter-domain routing (CIDR), 141–143
functions of, 117–120
IP addresses, 126–141
IP headers, 122–125
IPv6, 144–148

obtaining IP addresses, 143–144
overview of the Internet protocol (IP), 120–122
Network layer tasks, 21
Network load balancing, 426
Network maintenance. *See* Maintenance
Network redundancy, 371, 372, 423
Network security, 374
and cloud computing, 403–404
cybersecurity 4, 372–374
general information risk management model, 349
on home routers, 368
importance of, 351–352
introduction, 347–350
at Microsoft, 415
and the operating system, 391–392
overview, 347
procedural, physical and technical controls, 350–351
threats and vulnerabilities, 4, 349
and virtual machines, 397
wireless, 367
See also Controls
Network traffic, aggregating from copper to fiber, 52
Networking
political impacts of, 119
See also Computer networking; Wireless networking
Networks
big data applications 109–110
corporate-sponsored, 23
dedicated, 28
loops in, 108
topologies, 106
See also Cell phone networks; Telephone networks; Television networks; Wireless networks
Newspaper circulation, 222–224
NFC (near-field communication), 475
NFS (network file system), 221
NIC (network interface card), 104
Niles, Jack, 451
Nimda virus, 291
9/11 terrorist attacks, 290–293
Nissan Computer Corporation, 267
Nissan Motors Corporation, 267
NIST (National Institute for Standards and Technology), 78, 362
Nodes, 423
Noise, 57, 63–66
North American Numbering Plan Administration (Nanpa), 241
NSF (National Science Foundation), 448–449
NSFNET, 448
Nslookup, 255, 266–267
Number systems, 132
binary, 64, 127–131
decimal, 129–131

O3B, 505, 512
Oil industry, 479–481
Open Cloud, 403
Open domains, 252
Open shortest path first (OSPF) protocol, 280, 281
Open Standards Interconnect. *See* OSI model
OpenFlow, 288–289
OpenStack, 405
Operating system
components, 384–385
installation and startup, 393–394
origins, 381–383
process state, 385–390
security protections, 391–393
system calls, 383–384
Optical fiber, 45–46, 51–55, 76
compared to copper wire, 55
construction of, 54–55
how it works, 52–53
multimode, 54
single-mode, 54
types of, 54
Options, 181
Orthogonal frequency-division multiplexing (OFDM), 78
OSI (Open Standards Interconnect) model, 24–27
compared to TCP/IP, 26–27
current configuration, 26
mnemonic, 25
network layer names and tasks, 25
OSPF (open shortest path first), 280, 281
OUI lookup, 112–114
Outlook, 211

PaaS (Platform as a Service), 403, 404
Packet delivery, 101
postal service model, 29
Packet headers, 89, 285
Packet structure, 19–22
Packet switching, 11–14, 30
vs. circuit switching, 16–18
Packetization, 11–16
development of, 448
five essential tasks, 20
at IKEA, 16
and layering, 20–21
Packets
segmentation and reassembly of, 20–21

transferal of, 28
Padding, 181
PAN (personal-area networks), 470–472, 474
Passwords, best practices, 355
PAT (port address translation), 244. *See also* NAPT
Patches, 353
Patching, 353, 355, 359
Patents, 450, 456
Peering, 274
Perfmon, 407–411
Performance-focused design, 423
Personal finances
and compounding, 132
expenses comparison, 3–4
Personal-area networks (PAN), 470–472, 474
Phase, 60–61
Phase modulation (PM), 61–63
Phase shift keying, 63
Phone networks. *See* Cell phone networks; Telephone networks
Photo sharing, 220
Physical controls, 350–351
Physical layer
functions of, 41–43
overview, 41
and the smart grid, 78–79
special feature of, 43–44
Physical media/medium, 44
copper wire, 43–44, 45, 46–51
optical fiber, 45–46, 51–55
properties of, 44–46
Physical network design, 437, 438
Piconets, 472–473
Pigeon networks, 329
Piñera, Sebastian, 46
Ping, 154, 155–158
Pizza Tracker, 32–33
Plain text, 361
Platform as a Service (PaaS), 403
PM (phase modulation), 61–63
Point-to-point connections, 108, 328
POP (Post Office Protocol), 212, 214–215
Port addresses
TCP, 180–181
UDP, 183
Port conflicts, 172
Port numbers, 168, 170
Portals, 464–465
Ports, standard, 171
Post Office Protocol, 212, 214–215
Postal service delivery model, 29
Postel, Jon, 143
Predator UAV, 320–341
Presence, 222
Price-focused design, 423
Privacy, and small networks, 88
Private cloud, 402
Privileged mode, 383
Procedural controls, 350–351
Process and processes, 186, 385, 387, 389
Process isolation, 392
Process states, 385–386, 387
Protocol headers, 125
Protocols, 27–29, 186
ARP (Address Resolution Protocol), 247
application layer, 194, 224
BGP (Border Gateway Protocol), 280–281
CCMP (Counter-mode with cipher-block chaining message authentication code), 367
distance-vector, 281
e-mail, 211–218
FTP (File Transfer Protocol), 193
HTTP (Hypertext Transfer Protocol), 193
IM (instant messaging), 221–222
IMAP (Internet message access protocol), 211–212
in the IP header, 124
link-state, 281
MySQL, 207
NFS (network file system), 221
OSPF (open shortest path first), 280, 281
POP (Post Office Protocol), 211–212, 214–215
STMP (Simple Mail Transfer Protocol), 193
SMB (server message block), 221
SNMP (simple network management protocol), 441–443
SDP (sockets-direct protocol), 184
spanning tree, 107–108
SSH (Secure Shell), 366–369
TCP (Transmission Control Protocol), 162–167, 174–182, 186, 193
UDP (User Datagram Protocol), 162–163, 182–183, 186, 193
vendor driven, 221
PSTN (landline phone system), 499
Public cloud, 403
Push e-mail, 208

QAM (quadrature amplitude modulation), 62–63
QoS control, 467
Quadrature amplitude modulation (QAM), 62–63

Rackspace, 403
Radiation exposure, 460
Radios, AM/FM, 58
RAID storage, 372
Railroad standards, 446
RAND licensing, 447

RAND (reasonable and non-discriminatory) rates, 447
Random back-off, 93–94
RBOCs (Regional Bell Operating Companies), 492–494
Reading to learn, 43
Reasonable and non-discriminatory (RAND) rates, 447
Redundancy, 371, 372, 423
Regional Bell Operating Companies (RBOCs), 492–494
Regional registries, 143, 237
Registry, 142, 145, 154, 261, 298
Regulation, balancing with innovation, 450
Replication, 423, 425
Requests for comments (RFCs), 121
Reserved, 181
Resource encapsulation, 392
Retailing, 148–151
Retain, 479–480
RFCs (requests for comments), 121
Risk management, 348
RJ (registered jack) 45 jacks, 49–50
Roberts, Ed, 405
Robust hardware, 421–422
Route advertisement, 280
Route aggregation, 282–284, 311
Route print command, 278
Router configuration error, 258
Routers, 13, 16, 18, 21, 35–37
    for home use, 283
    in networks, 272
    wireless 261
Routing, 117–119
    in autonomous systems, 273–274
    default routes, 277
    as metaphor for the Aerotropolis 289–290
    overview, 269–270
    vs. switching, 270–272, 294
    viewing routes, 277–282
Routing protocols, 278
    exterior, 280–282
    interior 280–282
Routing tables, 275–277
Run-out, 479–480

SaaS (Software as a Service), 403
Satellite television, 31
Satellites, 478
Scandinavia, data centers in, 401
SDP (sockets-direct protocol), 184
Search advertising, 200
Secure Shell (SSH), 366–369
Securities and Exchange Commission (SEC), 221
Security issues. *See* Network security
Segmentation, 161–165
Sequence numbers, 181
Server Message Block (SMB), 221
Service delivery
    availability management, 417
    background, 414–415
    capacity management, 416
    continuity management, 416–417
    IT financial management, 417
    service level management, 415–416
Service level agreements (SLAs), 415–416, 420
Session initiation protocol (SIP), 224
Shadowcrew, 373
Shannon, Claude, 56
Shannon's Theorem 66
Shannon-Hartley theorem, 66
SHARE (Society to Help Alleviate Redundant Effort), 382
Shutterfly, 220
Signal conflict, 65
Signal repeaters, 51
Signaling, 7, 20–21, 41, 42
    categories of methods, 58–59
    examples of improved, 58
Signals, 56
    analog, 59, 60–62
    binary 63–66
    broadband, 494
    vs. data, 56
    defined, 42
    demodulating, 70, 74–75
    digital, 59–60, 64
    and their properties, 56–57
    ternary, 64–65
    traffic, 57
Signal-to-noise ratio (SNR), 66, 66n18
Simonyi, Charles, 86n3
Simple Mail Transfer Protocol (SMTP), 193–195, 211–214
Simple network maintenance protocol (SNMP), 441–442
Sine waves, 60–61, 63
    generation of, 61
    properties of, 61
Single mode fiber cables, 67
SIP (session initiation protocol), 224
SkyGrabber software, 341
Skype, 194
Slammer worm, 353, 359
Sliding window flow control, 177–178
Smart grids, 76–78
SMB (server message block), 221
Smileys, 225

Smith, Chris, 401
SMTP (simple mail transfer protocol), 193–195
SNMP (Simple network management protocol), 441–442
SNR (signal-to-noise ratio), 66, 66n18
Social networking, 196–197, 208
Sockets, 168
Sockets-direct protocol (SDP), 184
Software
  antivirus, 359
  ConstantContact, 371
  use of layering in, 18–19
  maintaining, 443–444
  reliability of, 423
  SkyGrabber, 341
  virtualization, 395
  weaknesses in, 353
Software as a Service (SaaS), 403
SOPA (Stop Online Piracy Act), 233–234
Spam, 210
Spanning tree protocol, 107–108
Spectrum auctions, 59
Spencer, Percy, 461
Spread Networks, 185
SQL, 374
SQL injection attacks, 353, 373–374
SSH (Secure Shell), 366–369
Standard ports, 171
Standards, 444–446
  development of, 456
  for digital projection, 447
  IEEE, 78, 462, 464, 476–477
Standards essential patents (SEP), 447
Stateful workloads, 426
Stateless transactions, 426
States, 386
Statistical multiplexing. *See* Statistically multiplexed WANs
Statistically multiplexed WANs, 331–333
STMP (Simple Mail-transfer Protocol), 211–214
Stop Online Piracy Act (SOPA), 233–234
Strowger switch, 11
Submarine cables, 185
Subnet design, 323
Subnet masks, 310–314, 320, 322–323
Subnetting, 299–300
  addressing a subnet, 308–310
  business motivation for, 300–302
  example case, 318–320
  in IPv6, 318
  network address block, 305–308
  representative computations, 316–317
  and route aggregation, 311
  within subnets, 314–316
  technical motivation for, 302
  and three-part IP addresses, 303–305
Supply-chain management, 148
  at Kmart, 149, 150–151
  at Wal-Mart, 148–150, 151
Support services, 235
  ARP, 247–249
  DHCP, 235–243
  DNS, 249–259
  NAPT, 244–247
Switched Ethernet, 106–107
Switches, 106–107, 111
Switching, 10–11
  manual, 10
  vs. routing, 270–272
  See also Circuit switching; Packet switching; Switches
System calls, 383–384
System instance, 420
System mode, 383

T.J. Maxx, 351, 352, 353–354, 372–374, 458
Target, 149
T-carriers (telecom carriers), 329–330, 514
TCP, 100
  checksum and layering purity, 182
  multi-path, 180
TCP/IP stack, 1, 23, 30, 194
  compared to OSI model, 26–27
  data delivery at layers, 28
  development of, 448
  as Internet protocol, 27–29
  layers and technologies, 24
  network layer names and tasks, 23
  and wide-area networks (WANs), 338–339
TDM (time-division multiplexing) WANs, 71, 337, 504
Technical controls, 350–351
Technology, networking, 24
Technology milestones
  multiplexing, 9–11
  packetization and packet switching, 11–14
  switching, 10–11
  telegraph, 6–8
Telecom carriers (T-carriers), 329–330, 514
Telecom glossary, 41–42
Telecommunications Act of 1996, 449, 493
Telecommunications infrastructure, 4
Telecommuting, 451–452
Telegraph machine 6–8, 42–43, 450
  multiplexing, 9
Telemedicine, 452–453
Telephone exchange, automatic, 10n16
Telephone industry, 10n15

Telephone lines, channel capacities, 67
Telephone networks, 3, 28
  cell phones, 496
  components, 488–489
  internet service, 493
  introduction, 487–488
  legal developments, 491–492
  overview, 487
  phone signals, 490–491
  See also Cell phone networks
Telephone numbers, 135–136, 241
Telephones
  landline, 499, 514
  See also Cell phones
Television, 58
  satellite, 31
  See also Cable TV
Television channels, 67
Television networks, 28
Teleworking, 451
Telnet, 195
Temporal key integrity protocol (TKIP), 367
Ternary signals, 65
Text messaging, 208, 290
Thread pool, 388
Threads, 386–389
Threats, 348, 349–350, 377
3G/4G, 180
Three-way handshake, 179
Thunderbird, 211
Time-division multiplexing (TDM), 71, 504
Timeline, data communications, 12
Time-to-live (TTL), 255, 258
TKIP (temporal key integrity protocol), 367
TLDs (top-level domains, 251–252
TLS (transport layer security), 365–366
T-Mobile, 502
Top-level domains (TLDs), 251–252
Total internal reflection, 52–54
Traceroute, 35–37, 338–339
Tracert, 277–278
Traffic signals, 57, 147
Transmission control protocol (TCP), 62–167, 174–182, 186, 193
  connection establishment, 177–180
  flow control, 174–177
  header, 180–182
  multiplexing, 168
  reliability, 166–167
  segmentation, 163–165
Transport layer, 161, 185–186
  need for, 161–163
  transmission control protocol (TCP), 162–182, 186, 193
  user datagram protocol (UDP), 162–163, 182–183, 186, 193
Transport Layer Security (TLS), 365–366
TrendyWidgets (IT Infrastructure Design Exercises)
  and active replication, 432
  CIDR address block requirements, 159
  and the cloud, 411
  Ethernet diagram, 115
  Ethernet use, 115
  and failover, 298
  firm details, 38
  identifying market leaders, 234
  identifying uses, 37–38
  Infrastructure diagram, 268
  media selection, 82–83
  network administrator for, 456
  network security at, 378
  office locations and staffing, 39
  physical medium for building network, 82
  subnet design for, 324
  switch to VoiP, 512
  WAN design, 345
  adding Wi-Fi, 485
Triple DES, 362
TTL (time-to-live), 255, 258
Twitter, 371n16

UAV (unmanned aerial vehicles), 339–341
UBS Warburg, 210
UNICODE, 8, 68
Uniform resource locator (URL), 205
Universal exchange formats, 37
UNIX, 448–449
Unmanned aerial vehicles (UAV), 339–341
Unreliable commodity PCs, 422–423
Unshielded twisted pair (UTP) cables. *See* UTP cables
Uptime, 431
Urgent pointer, 181
URLs (Uniform resource locators), 205–207
US Postal Service (USPS), 194
User datagram protocol (UDP), 162–163, 182–183, 186, 193
User mode, 383
Utilities
  BGPlay, 274, 286, 296–297
  dig, 255
  ipconfig, 257, 322–323
  netstat, 172, 188
  netstat –b, 172–173, 188
  netstat –e, 173

netstat –f, 188
netstat –r, 188
netstat –s, 174, 188, 189–190
nslookup, 255, 266–268
pathping, 157
perfmon, 407–411
ping, 155–158
tracert, 157, 277–278
UTP (unshielded twisted pair) cables. *See* Cables
Uzzy, Brian, 209

Vail, Alfred, 7n9
VAIO computer logo, 63
Very small aperture terminals (VSAT), 480
Video files, 4, 220
Virtual circuits, 333–335
Virtual hosts, 261–264
Virtual machines (VMs), 394, 395–397, 405
Virtual memory, 390
Virtual private networks (VPN), 365–366
Virtualization, 395–397
and MAC addresses, 104
Virtualization software, 395
Viruses, 353, 358–359
VMs (virtual machines), 394, 395–397, 405
Voice-over IP (VoIP), 194, 290, 292
VoIP, 194, 290, 292
von Neumann architecture, 379–381, 405
VPN (virtual private networks), 365–366
VSAT (very small aperture terminals), 480
Vulnerabilities, 348, 349, 443–444

Wal-Mart, 148–150, 151, 479
Walton, James L. (Bud), 149
Walton, Sam, 149, 150
WANs. *See* Wide-area networks (WANs)
Warehouse scale computers (WSCs), 394, 397–401, 405
Web
evolution of, 197–199
vs. Internet/internet, 196
and web marketing, 200
See also Internet
Web browsers, 2, 198
graphical, 449
Web defaults, 206
Web hosting, 261–264
Web mail, 218. *See also* E-mail
Web pages, 199–201
debugging, 344
zipped format, 204
Weblogs (blogs), 199
Websites, 261
blocking, 281
WeChat, 506
WEP (wired equivalent privacy), 367
Western Union, 492
Wide-area networks (WANs), 514
ATM (Asynchronous Transfer Mode), 336–337
categories of, 328
compared to road network, 326–328
dial-up networking, 328–329
FDM (frequency-division multiplexing), 337–338
Frame Relay, 336
and home prices, 326
introduction, 325–326
in relation to IP and LANs, 339
overview, 3225
point-to-point, 328–330, 332
ports, 273
professional 345
statistically multiplexed, 331–337
T-carriers/DS-signals, 329–330
and the TCP/IP stack, 338–339
TDM (time-division multiplexing), 337
virtual circuits in, 333–335
X.25, 336
Wi-Fi, 180, 485
in Africa, 505
in stadiums, 458
See also Wireless LANs
Wi-Fi protected access (WPA), 367
Wikipedia, 196
Window size, 176, 181
Wired equivalent privacy (WEP), 367
Wireless communication medium, 45
Wireless LANs (WLANs), 449, 461–462, 514
802.11 series, 462–464, 466–467
802.11ac, 469–470
802.11n, 468–469
architecture, 464–466
coexistence with WPAN, 475–476
personal area networks (802.15 series), 470–471
popular 802.11 technologies, 468
See also Wi-Fi
Wireless metropolitan-area networks (WMANs), 476–477
Wireless networking, 457–458. *See also* Computer networking; Networking
Wireless networks
categories of, 461–462
health effects of, 458

limitations of, 458
security issues, 367, 458
See also Networks
Wireless personal area networks (WPANs), 475–476
Wireless routers, 259, 261, 466
Wireless security, 367
Wireless signals, safety of, 458, 460
Wireless spectrum, 449
Wireless spectrum auctions, 59
Wireshark, 214, 227–233, 376
download page, 228
Follow TCP Stream window, 232
packet capture, 229, 230, 483–485
viewing and analyzing a capture, 230–232
welcome interface, 228
WLANs. *See* Wireless LANs
World Wide Web. *See* Internet; Web
Worldwide Interoperability for Microwave Access (WiMAX), 476–477
Worms, 353, 358–359
WPA (Wi-Fi protected access), 367
WPA2, 367
WSCs (warehouse scale computers), 394, 397–401, 405
Wuchty, Stefan, 209

X.25, 336

Yahoo, 205, 262–264
Yang, Jerry, 205
YouTube, 220

Zen Cart, 266
Zip codes, 136–138
Zones, 255
dead, 401
demilitarized, 356–357
metro, 477
war, 240–241
See also Domains
Zubulake, Laura, 210
Zubulake vs. UBS Warburg, 210

Made in the USA
Lexington, KY
15 August 2018